16th International Conference on Microscopy of Semiconducting Materials 2009

Journal of Physics: Conference Series Volume 209

Oxford, United Kingdom
17-20 March 2009

ISBN: 978-1-61738-268-0

Printed from e-media with permission by:

Curran Associates, Inc.
57 Morehouse Lane
Red Hook, NY 12571

Some format issues inherent in the e-media version may also appear in this print version.

Printed by Curran Associates, Inc. (2010)

Institute of Physics
Dirac House, Temple Back
Bristol BS1 6BE UK

Phone: 44 1 17 929 7481
Fax: 44 1 17 920 0979

techtracking@iop.org

Additional copies of this publication are available from:

Curran Associates, Inc.
57 Morehouse Lane
Red Hook, NY 12571 USA
Phone: 845-758-0400
Fax: 845-758-2634
Email: curran@proceedings.com
Web: www.proceedings.com

TABLE OF CONTENTS

INVITED PAPERS

SESSION A & B

SESSION C & D

SESSION E TO H

PREFACE

This volume contains invited and contributed papers from the 16th international conference on 'Microscopy of Semiconducting Materials' held at the University of Oxford on 17–20 March 2009. The meeting was organised under the auspices of the Royal Microscopical Society with support by the Electron Microscopy and Analysis Group of the Institute of Physics and the Materials Research Society. This conference series focuses on the most recent advances in semiconductor studies carried out by all forms of microscopy, with an emphasis on electron microscopy and related techniques with high spatial resolution. It was attended by 139 delegates from 18 countries world-wide.

As semiconductor devices shrink further both new routes of device processing and device characterisation need to be developed, and for the latter methods that offer subnanometre spatial resolution are particularly valuable. Electron microscopy in its various forms of imaging, diffraction and spectroscopy provides indispensable information on both microstructure and chemistry. Recent advances in instrumentation, from lens aberration correction in both TEM and STEM instruments, to the development of a wide range of scanning probe techniques, as well as new methods of quantification of the various signals recorded have been presented at this conference. Two examples of topics that have attracted a number of interesting studies at this meeting are the contrast interpretation in secondary electron images of cleaved semiconductor surfaces and the measurement of chemical composition on the nano-scale of quantum domain structures and devices investigated in cross-section.

Each manuscript submitted for publication in this proceedings volume has been reviewed by at least two referees and was modified accordingly. Finally, 70 manuscripts were accepted for publication. The editors are very grateful to the following colleagues for their rapid and careful reviewing of the papers:

M al Jassim, J Barnard, R Beanland, H Bender, E Carlino, D Cherns, D Cooper, A J Craven, R E Dunin-Borkowski, D Gerthsen, V Grillo, E Grünbaum, M Hecker, K Honda, R Hull, CJ Humphreys, M Jepson, H Kirmse, S Molina, A Norman, R Oliver, B Pecz, P Pongratz, F M Ross , K Tillmann, S Tsukamoto and E Verleysen.

We would also like to thank the staff of the Royal Microscopical Society for expert assistance in planning and organising this conference and in particular Victoria Lee for her great help during the event.

December 2009
T Walther
P D Nellist
J L Hutchison
A G Cullis

All papers published in this volume of *Journal of Physics: Conference Series* have been peer reviewed through processes administered by the proceedings Editors. Reviews were conducted by expert referees to the professional and scientific standards expected of a proceedings journal published by IOP Publishing

GaN devices based on nanorods

D Cherns[1], L Meshi[1], I Griffiths[1], S Khongphetsak[1], SV Novikov[2], RP Campion[2], CT Foxon[2], C Liu[3], P Shields[3] and WN Wang[3]

[1] Department of Physics, University of Bristol, Tyndall Avenue, Bristol BS8 1TL, UK
[2] Department of Physics and Astronomy, University of Nottingham, Nottingham NG7 2RD, UK
[3] Department of Electronic and Electrical Engineering, University of Bath, BA2 7AY

E-mail: d.cherns@bris.ac.uk

Abstract. Transmission and scanning electron microscopy are used to examine the role of an intermediate nanorod layer in reducing threading defect densities in GaN/(0001)sapphire. Films grown by molecular beam epitaxy under N-rich conditions showed Ga-polar nanorods growing out of a more compact N-polar layer. The nanorods sometimes contained extended threading defects, which were faulted dipoles lying on {10-10} planes with a displacement vector of ±1/2[0001], which act as sources for spiral growth. By overgrowing the nanorods under Ga-rich conditions, continuous epilayers were formed with threading defect densities down to 10^8 cm^{-2}. In a second approach, nanorods produced by etching through a self-organised layer of Ni islands were overgrown by metal-organic chemical vapour deposition, to produce overlayers with defect densities down to $5*10^7$ cm^{-2}. In both cases, the mechanisms by which the nanorod layer reduces the threading defect density are identified.

1. Introduction

A major issue which currently affects the efficiency of GaN devices is the high density of threading defects (up to 10^9 cm^{-2}), which arise through growing epilayers on highly mismatched substrates. Epitaxial lateral overgrowth (ELOG) is one solution to this problem. In ELOG, "seed" columns of GaN, defined by masking and etching procedures, and usually several micrometers across, are induced to grow laterally until a continuous overlayer is formed. However, although threading defect densities in the laterally-grown "wings" can be reduced to around 10^6 cm^{-2}, threading defect densities remain high in the seeds, which might comprise 20% of the layer. The wings also often exhibit grain rotations, typically up to 0.5-1°, which lead to new grain boundary defects [1]. In addition, the mask deposition, patterning and etching steps add to overall costs. In order to circumvent some of these problems, we have investigated alternative "nano-ELOG" approaches where the seed columns are a "compliant" layer of nanorods, whose diameters and spacing are in the range 20-100nm, i.e. about two orders of magnitude less than in conventional ELOG.

In this paper, transmission and scanning electron microscopy (TEM and SEM) are used to examine two different nano-ELOG methods. In the first method, molecular beam epitaxy (MBE), at the University of Nottingham, has been used to nucleate and grow nanorods under strongly N-rich conditions on (0001)sapphire. Lateral overgrowth is then initiated under more Ga-rich growth conditions to form a continuous overlayer. The morphology of growth, and the nature and role of a new type of threading defect present in the nanorods will be explained. When the nanorods grow together further threading defects are generated. The origin of threading defects present in the final overlayer will be examined and the factors controlling their density will be discussed. In the second nano-ELOG method, carried out in the University of Bath, nanorods are produced by etching a continuous GaN epilayer through a self-organised mask. Following removal of the mask material, metal-organic chemical vapour deposition (MOCVD) is used to produce lateral growth of the GaN to form a continuous layer. Again, the role of the nanorod layer in reducing the threading defect density will be discussed.

2. The MBE method

The growth of GaN nanorods by MBE has been investigated quite extensively (see [2] for a review). It is clear that, in general, nanorod growth is favoured under strongly N-rich conditions, is sensitive to surface preparation (particularly the presence or absence of an AlN interlayer) and can be non-catalytic. In addition, growth under Ga-rich conditions favours a transition to 2-dimensional growth.

In our case, samples were grown by plasma-assisted MBE in a Varian ModGen II MBE system with active nitrogen provided by an HD25 RF activated plasma source. An AlN buffer layer 2-10nm thick was grown at a temperature of 600°C prior to GaN growth which was carried out at substrate temperatures in the range 700-800°C. Growth was carried out for up to 6hrs under strongly N-rich conditions, and subsequently for 0-6hrs under Ga-rich conditions.

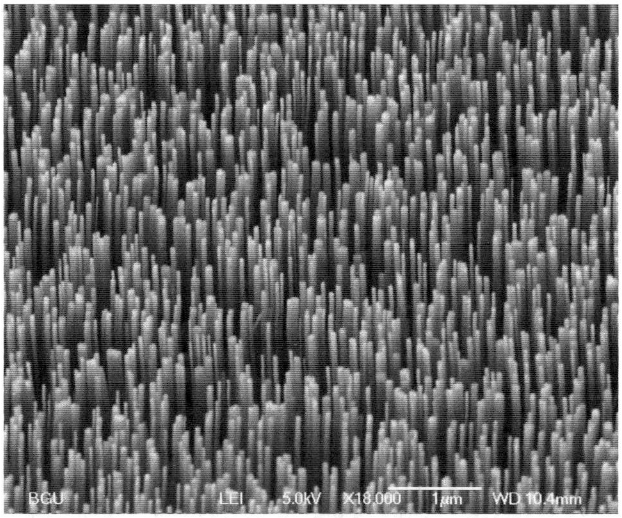

Figure 1. SEM image of nanorods grown by MBE under N-rich conditions

2.1 N-rich growth

Figure 1 shows a SEM image taken after N-rich growth, indicating an array of high aspect ratio nanorods. Figure 2 shows a more detailed view of the microstructure. The cross-sectional TEM image in Figure 2a reveals a bimodal microstructure in which parallel sided nanorods showing bend contours but little evidence of defects emerge from a more compact but rough layer. This compact, or intermediate, layer contains high densities of threading defects, some of which are in contrast in Figure 2a along with high densities of basal plane stacking faults which are out of contrast. From the SEM image in Figure 2b, it can be seen, at least qualitatively, that the nanorods mostly grow out of depressions or holes in the intermediate layer. The image in Figure 2c shows a plan-view sample obtained by coating the deposit in epoxy resin, and backthinning to reveal the tops of nanorods. It can be seen that the nanorods are facetted (with {10-10} surfaces), and that where nanorods have touched during growth, a liquid-like coalescence has taken place. The nanorods appear mostly perfect, but some have contrast indicative of threading defects, as discussed below.

Convergent beam electron diffraction (CBED) patterns, taken with the 0002 systematic row symmetrically excited, show strong asymmetry between the ±(0002) reflections, where [0001] is defined as the Ga-N bond direction. The asymmetry, which depends on double diffraction between systematic row reflections, can be used to distinguish the polarity of the GaN layer. CBED analysis showed that the nanorods in Figure 2 were Ga-polar (growing in the [0001] direction), while the intermediate layer was N-polar (see [3] for more details).

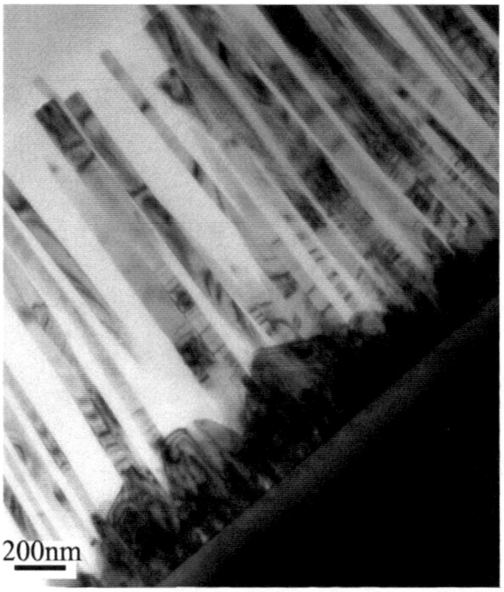

a.

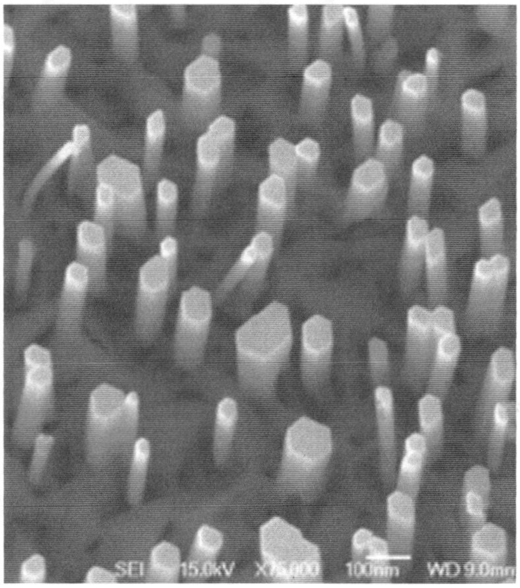

b.

c.

Figure 2. Samples grown under N-rich conditions (a) TEM cross-section revealing nanorods growing out of holes in a more compact layer, (b) SEM image suggesting that the nanorods emanate from holes in the compact layer, (c) TEM image of a plan-view sample intersecting the tops of nanorods. This shows good alignment of the nanorods, a liquid-like coalescence where nanorods are starting to merge. There is contrast in some of the nanorods suggesting the presence of threading defects.

2.2. Defects in nanorods

More in-depth studies of the structure of the nanorods suggested that, while many were perfect single crystals, others contained threading defects. Figure 3 compares perfect nanorods in Figure 3a with an imperfect nanorod in Figure 3b. The imperfect nanorod is seen to have an extended threading defect and terminates in a facetted growth surface, while the perfect nanorods appear to have relatively flat (0001) growth surfaces. By tilting the imperfect nanorods about their axes, it was found that the threading defects were planar. This is illustrated in Figure 4, which shows two nanorods which have been rotated by about 70° about [0001]. In both cases, 0002 bend contours cross the nanorods, and are displaced on crossing the threading defects. The left hand nanorod shows an extended defect which is obliquely inclined in one image and edge-on in the second (the projection of the defect is indicated by arrows), whereas, coincidentally, the defect in the second nanorod shows the opposite behaviour.

a. b.

Figure 3. (a) perfect nanorods, (b) a nanorod with a threading defect

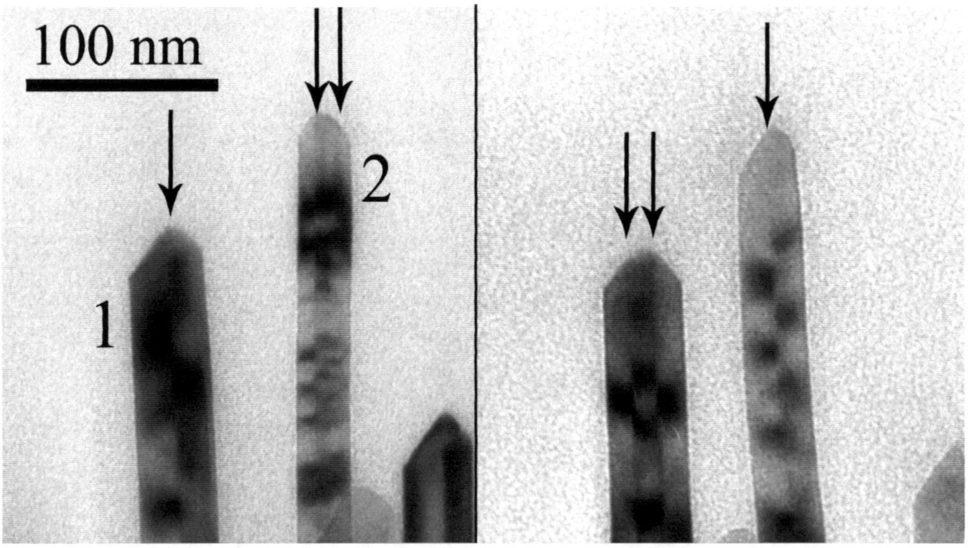

Figure 4. Nanorods with faulted planar defects, seen in images which are relatively rotated by about 70° about a vertical axis. The fault plane for nanorod 1 is approximately edge-on in the left hand image, while the fault in nanorod 2 is edge-on in the right hand image. The bend contours, which are of 0002 type, are displaced on crossing the bounding partials, which are arrowed.

Tilting experiments suggested that the defects in Figures 3b and 4 lie on {10-10} planes. The displacement of 0002 bend contours on crossing these defects suggests relaxation due to a displacement field parallel to [0001]. In contrast, the defects showed little contrast in a-type reflections suggesting that displacements in the basal plane were either zero or very small. In fact, an analysis of how the 0002 contours were displaced on crossing the defect suggested that the defects were bounded by a dislocation dipole with Burgers vector ±c/2 [4]. This implies that these threading defects were faulted, with a change in stacking sequence across the fault plane from ABABAB to BABABA.

Although this type of fault is unusual, it has been previously reported in GaN epilayers, for example by Tanaka et al [5] on the basis of high resolution images. We believe that the faults in nanorods provide a mechanism of nanorod growth as illustrated in Figure 5. This shows a top view in which the opposite c/2 partials associated with the fault provide a step on the (0001) growth surface one Ga-N layer in height, i.e. a single surface step, where two such steps represent a unit cell height (0.52nm). This step provides a spiral growth mechanism, with intermediate stages in a single repeat illustrated in the diagram. The facetted (and, in fact, vicinal) nature of the top surface can be readily explained by step flow from the source.

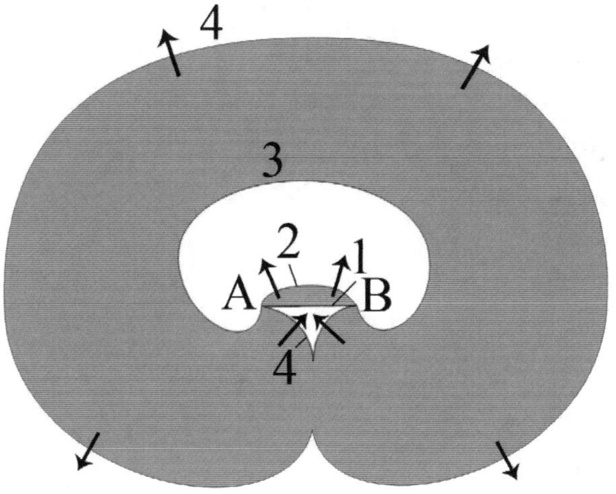

Figure 5. Schematic suggesting how the threading defects present in nanorods initiate a spiral growth mechanism. Viewed from the top, A and B are the emerging partials connected by a single c/2 step on the growth surface. Addition of adatoms to the step causes step flow in the sequence 1-4, whereupon the step is re-formed and the sequence repeats.

Close examination suggested that the faults emanated from points close to the sapphire surface. An example is illustrated in the two dark field images in Figure 6. The top image shows a general view. Threading defects are present along the length of the two arrowed nanorods. These nanorods are bent, such that the diffraction conditions change along their length. However, in places, the pairs of bounding partials associated with the threading defects can be seen quite clearly. These partials appear to converge to points close to the sapphire surface; the bottom image has been taken at a slight tilt to the top image to enhance the contrast of the partials in the boxed region as the sapphire surface is approached. However, the exact nucleation site remains unclear at present. One possibility is that nucleation takes place at steps on the sapphire surface, where a single step height of 0.22nm compares quite closely to the fault displacement of c/2 = 0.25nm. However, further work is needed to confirm this possibility.

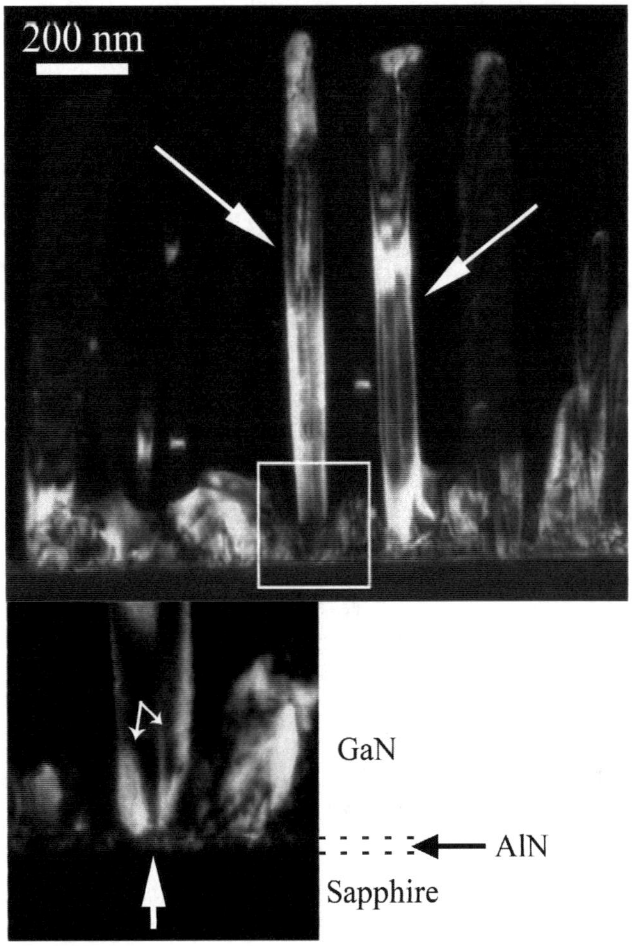

Figure 6. Dark field images in $\boldsymbol{g}$ = 0002. The top image shows two nanorods (arrowed) which contain threading defects. The defects extend right along the length of the nanorods, but are only partly visible due to bending of the nanorods (the bounding partials can be seen quite clearly in places). The image at the bottom, which is taken at a slight tilt to the top image, shows the boxed region at a higher magnification, and tracks the bounding partials (arrowed) to a point close to the sapphire interface.

2.3. Ga-rich growth

In samples subjected to N-rich followed by Ga-rich growth, it was found that the nanorods grew laterally to eventually form a continuous overlayer. Figure 7 shows an intermediate stage, where the nanorod diameter has increased, but nanorods remain mostly separate. Magnified images show the regions A and B more clearly. The region A shows two threading defects. One of these defects is seen to terminate. This could be due to the defect meeting the surface of the thin foil. However, termination of the threading defect is also consistent with recombination of the bounding partials, which have opposite Burgers vectors, to form a closed dislocation loop. In the region near B, it is apparent that the second defect has opened into a void. Again, this effectively terminates the threading defect, since there is no net dislocation, or fault, in the void. Thus, overgrowth of the void, should it eventually occur, would lead to perfect crystal with no net threading defect.

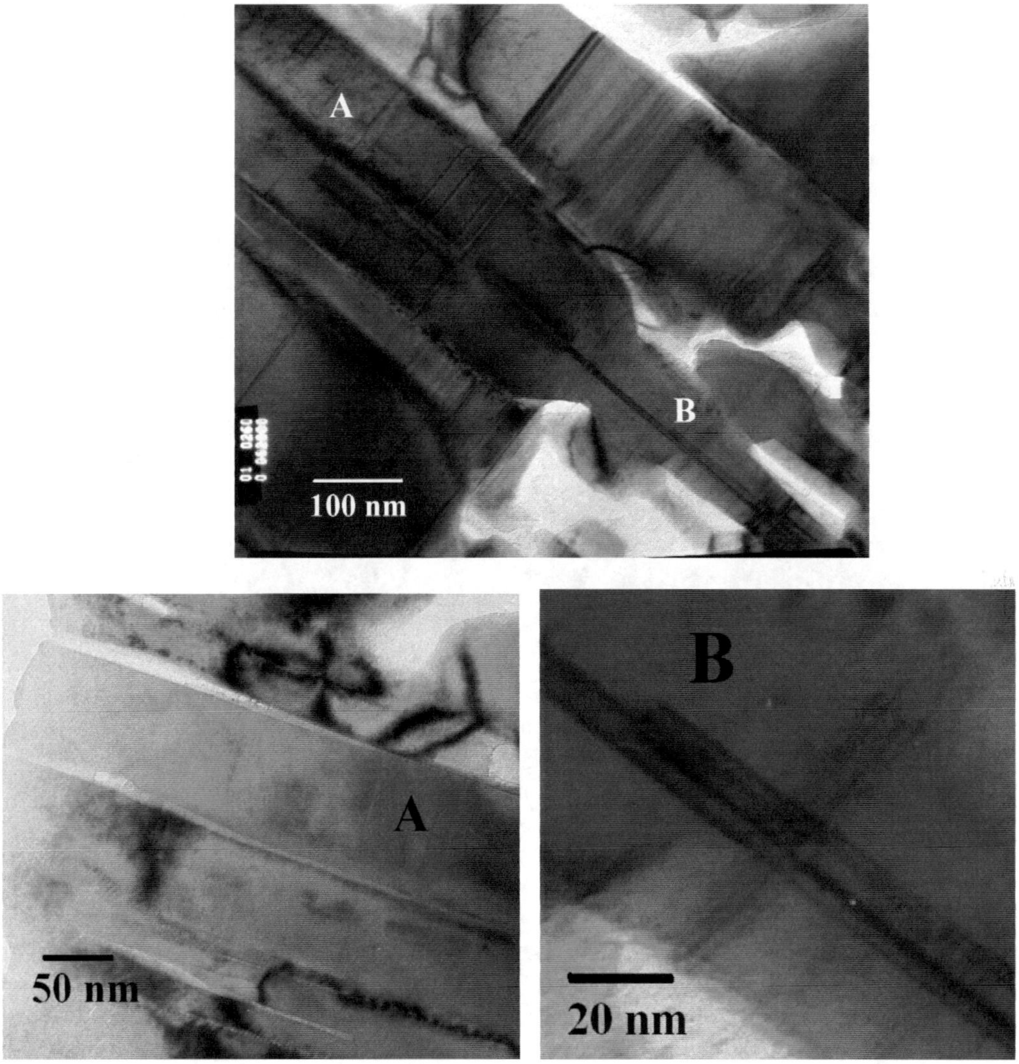

Figure 7 shows a sample grown by MBE under N-rich conditions followed by a period of growth under Ga-rich conditions. The top image is a general view showing that, under Ga-rich conditions, lateral growth is faster towards the tops of nanorods (i.e. towards the top left). The bottom images show expanded views of the regions near A and B.

Figure 8 shows an example where Ga-rich growth has led to a continuous overlayer. The image is taken with an *a*-type reflection strongly excited, to highlight misorientations due to relative rotations about the growth axis. The image shows clearly that many neighbouring nanorods coalesce to form regions of perfect crystal without threading defects. However, on a larger scale, there are grain boundaries present which contain threading defects. Overall, the density of threading dislocations was around 10^8 cm^{-2}, up to two orders of magnitude less than in continuous epilayers of an equivalent thickness.

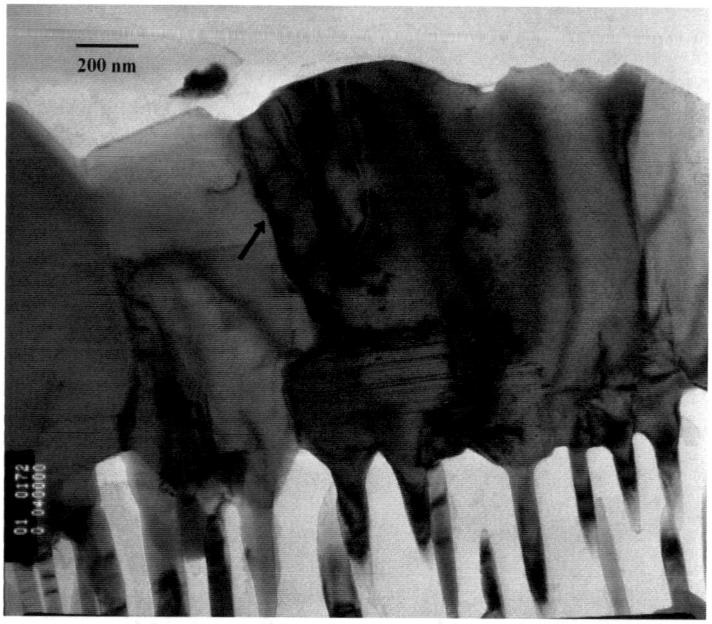

Figure 8. TEM cross-section showing a sample grown for 5hrs under N-rich conditions, followed by 5hrs under Ga-rich conditions to form a near-continuous overlayer. The image is taken under near 2-beam conditions with an *a*-type reflection operating, such that grain boundaries (arrowed) representing relative rotations about the growth directions are clearly visible. Basal plane stacking faults visible in the central region indicate that lateral growth extends about 1μm. Growth direction is upwards.

3. The MOVCD method

MOCVD has proved less effective than MBE for the growth of *c*-oriented nanorod arrays, but is well-established as a means of achieving lateral overgrowth. Thus, in this case, arrays of nanorods were fabricated by post-processing of continuous (0001)GaN epilayers grown on (0001)sapphire substrates by MOCVD. These epilayers, grown and subsequently processed in the University of Bath, generally had densities of threading defects in the range $1\text{-}3*10^8$ cm^{-2}. Nanorods were then produced in the epilayers by wet and dry-etching through a self-organised mask of Ni islands produced by annealing a thin vapour-deposited Ni layer [6]. Arrays of nanorods produced in this manner are illustrated in Figure 9, the density of nanorods being about 10^9 cm^{-2}. Since the density of nanorods was greater than that of the threading defects, most nanorods were free of threading defects, in contrast to the situation with optically defined masks.

Continuous overlayers produced by lateral overgrowth of the nanorod arrays by MOCVD showed a significant reduction in the density of threading defects compared with the original epilayers. In some cases lateral overgrowth preserved the airgaps between nanorods, while in other cases, complete closure of the airgaps resulted depending on the pre-processing treatment. Figure 10 shows an example of the latter, indicating extensive lateral migration of threading defects and, thereby, the region of lateral growth. Figure 11 shows an example where nanorods were etched partially through the GaN epilayer, and the airgaps have been partially closed by lateral overgrowth. Although the detailed structure in the region containing the nanorods is not clear, it can be seen that the overlayer has fewer defects than the original epilayer. Threading defects are seen to migrate laterally in the region just above the nanorods indicating again the region of lateral growth. In this region, it was found that dislocation reactions occurred, reducing the density of threading defects to 5×10^7 cm^{-2}, i.e. by nearly an order of magnitude compared with the original epilayer.

Journal of Physics: Conference Series **209** (2010) 012001 doi:10.1088/1742-6596/209/1/012001

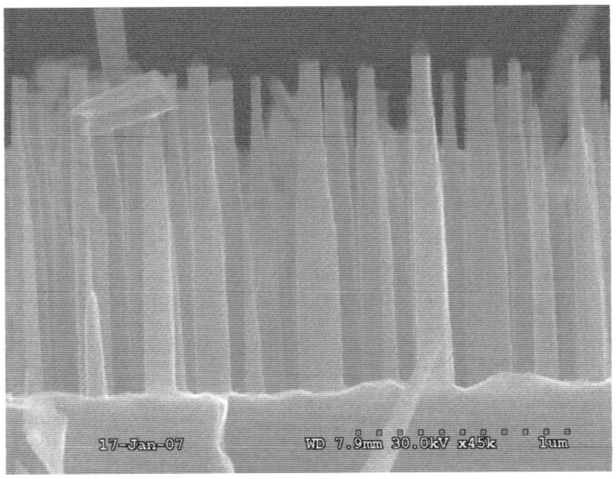

Figure 9. SEM image showing nanorods produced by etching through a mask of Ni islands

Figure 10. This shows nanorods produced by etching, which have been subsequently overgrown by MOCVD, generating extensive lateral migration of threading dislocations.

Figure 11. TEM cross-section, showing a sample where nanorods have been etched part-way through a GaN epilayer, and then overgrown by MOCVD. The overlayer shows some lateral migration of dislocations and a lower average density of threading defects than in the original epilayer.

4. Conclusions

The results show that both the MBE and MOCVD methods, although quite different, produce significant reductions in the density of threading defects. In the MBE approach, the nanorods provide seed columns, which are either perfect single crystals or contain threading defects. Some of the threading defects in nanorods terminate by recombination of the partials or in voids, such that relatively few of these defects are found in the final overlayers. There is a high density of threading defects in the imperfect intermediate layer, but these terminate at an airgap and do not propagate into the final overlayer. The majority of threading defects are formed when a continuous overlayer is formed, describing the misorientations and relative displacements between nanorods which must exist at nucleation. As groups of neighbouring nanorods coalesce to form defect-free regions, we must suppose that some of these displacements are accommodated elastically. However, elastic accommodation becomes less likely as the coalescence regions increase in size, leading eventually to the larger scale grain boundaries seen in Figure 7. The MOCVD approach shows some similar features. Although there are additional processing steps, the nanorods are again relatively defect-free. Lateral overgrowth in this case tends to close the airgaps, but lateral migration of the threading defects towards the tops of the nanorods again reduces the overall density of threading defects.

References

[1] Liliental-Weber Z and Cherns D 2001 *J. Appl. Phys.* **89**, 7833
[2] Calleja E et al. 2007 *Phys. Stat. Sol. b* **244,** 2816
[3] Cherns D et al. 2008 *Appl. Phys. Lett.* **92**, 121902
[4] Cherns D et al. 2008 *Appl. Phys. Lett.* **93**, 111911
[5] Tanaka S et al. 1995 *Appl. Phys. Lett.* **66**, 37
[6] Meshi L et al. 2008 *Phys. Stat. Sol. (c)* **5**, 1645

Journal of Physics: Conference Series **209** (2010) 012002 doi:10.1088/1742-6596/209/1/012002

Growth, structure and phase transitions of epitaxial nanowires of III-V semiconductors

F Glas, G Patriarche and JC Harmand

CNRS, Laboratoire de Photonique et de Nanostructures, Route de Nozay, 91460 Marcoussis, France

E-mail: frank.glas@lpn.cnrs.fr

Abstract. We review and illustrate the impact of TEM on the study of nanowires of non-nitride III-V semiconductors, with particular emphasis on the understanding of the thermodynamics and kinetics of their formation assisted by nano-sized catalyst particles. Besides providing basic information about the morphology of the nanowires and their growth rate as a function of diameter, TEM offers insights into the peculiar crystalline structure that they adopt. We discuss the formation of the unusual wurtzite hexagonal crystalline phase and that of planar stacking defects in these nanowires and show that they are kinetically controlled. We also demonstrate the transformation of wurtzite into cubic sphalerite upon epitaxial burying of the nanowires. Nanowires are particularly interesting in that they allow the fabrication of precisely positioned quantum dots with well-defined geometries. In this respect, we discuss the formation of strained quantum-size inclusions in nanowires, their critical dimensions and the kinetic and thermodynamic factors governing the changes of the crystalline structure that sometimes occur around a hetero-interface.

1. Introduction

Free-standing wires with diameters ranging from hundreds down to a few nanometers are nowadays commonly fabricated from a large range of semiconductor materials [1-6]. These 1D nanostructures have remarkable physical properties and many potential applications, yet the control of their growth and final characteristics is still far from optimal.

In this paper, we illustrate the impact of transmission electron microscopy (TEM) on the study of nanowires (NWs) of III-V semiconductors with particular emphasis on the understanding of the thermodynamics and kinetics of their formation. To be specific, we restrict the scope of our study in terms of materials and growth methods, by considering III-V compounds containing no nitrogen and grown by Molecular Beam Epitaxy (MBE) assisted by nano-sized gold "catalyst" particles. However, some of our conclusions remain valid for other material systems, growth methods and catalysts. We consider NWs growing on the $(\bar{1}\,\bar{1}\,\bar{1})$ surface (often called (111)B) of a III-V compound, which is by far the most common case. In all our images, (111)B points upwards. Growth on the (001) surface, which would be highly desirable for applications, has not been achieved reproducibly so far.

Let us first summarize the basics of our current understanding of metal-catalyst-aided NW growth by MBE. The catalyst nanoparticles deposited on the substrate before growth define the wire diameter. According to the vapour-liquid-solid (VLS) growth mechanism, the atoms are fed from the vapour phase to the solid wire through this particle (or droplet), which remains liquid during growth [7]. The growth of the NWs competes with that of the catalyst-free portion of the sample located between the NWs (the "non-activated" surface), which always occurs, albeit at a slower rate. This (111)B surface is typically fairly rough and grows by a combination of 2D nucleation and step flow whereas the NWs grow via the nucleation of new solid atomic layers at the top interface with the catalyst droplet. The droplet is a mixture of metal (here Au) with the wire elements. In most experiments, it is liquid during growth, as required for proper VLS. We checked this *in situ* for InAs NWs [8]. In some cases, the catalyst might however remain solid [3]. In MBE, the droplet is not a chemical catalyst which would facilitate the decomposition of precursor molecules, as in chemical vapour deposition methods, since the type-III elements are delivered as atomic beams and the type-V molecular beams are thermally decomposed. Instead, the droplet increases the growth rate at the top of the NW, probably by decreasing the nucleation barrier and by providing a dense feeding medium.

2. Basic investigations

2.1. Nanowire

The basic geometrical features of the NWs are not only important for the properties of the devices into which they may subsequently be incorporated. They also provide invaluable insights into the growth processes. To our knowledge, *in situ* TEM has only been performed for NWs of type-IV elements [9-10], whereas III-V NWs have only been examined *ex situ.* We shall see in section 3.2 that the way in which the growth is terminated before the NW is removed from the epitaxy chamber may marginally alter its geometry and structure, but in general the NWs remain largely unaffected by the transfer.

The first two dimensions to be determined are obviously the length and diameter of the NW. In the simplest case, each NW grows with a uniform diameter determined by the size of the interface between the top facet of the NW and the droplet, which itself depends on the droplet volume and on its contact angle with the top facet. Hence, when the Au nanodrops are formed by annealing, because of their size distribution, a range of NW diameters is obtained (ordered arrays of droplets of uniform sizes may however be deposited *ex situ*, for instance through a mask patterned by electron lithography). The length/diameter relationship is best determined by examining a large number of such NWs by scanning electron microscopy. It provides key information about the processes that limit the growth velocity of the NWs. For instance, the inverse relationship between length and diameter that is often observed indicates that growth is limited by the diffusion of the atoms to the NWs from the non-activated areas situated between the NWs [11].

Our NW volume measurements prove that the amount of material incorporated in the NW is larger than the amount impinging on the droplet. This confirms that part of the material reaches the top of the NW via diffusion from the substrate. In modern MBE machines the beams are inclined to the substrate normal (by about 30°), so that the NW material is also partly collected from the NW sidewalls. Initially overlooked in NW growth modelling, this fact is increasingly recognized as important [12].

In addition, individual NWs may have non-uniform diameters. The regular overall NW tapering observed at low growth temperature [6] indicates that the diffusing species tend to nucleate on the sidewalls of the NW before reaching the droplet. The localized tapering observed at higher growth temperature at the top of long NWs attests in turn that these species desorb if the NW length exceeds their diffusion length [4,12]. We sometimes observe double tapering at top and bottom; we attribute the former to a shadowing of the foot of the NW from the molecular beams by the neighbouring NWs.

2.2. Catalyst

In VLS growth, the catalyst droplet plays a critical part. Most often, the gold catalyst is first deposited as a film of roughly uniform thickness and then annealed to induce the formation of nanodroplets. Gold is usually deposited in a chamber separate from the epitaxy machine. On the contrary, we use an Au source located in the MBE chamber; this avoids any contamination of the coated substrate and significantly speeds up the fabrication cycle [4].

We have observed the catalyst particles not only after growth but also immediately after annealing, before exposing the sample to the growth fluxes. All chemical analyses performed in the TEM reveal a mixture of gold and type-III element. Type-V atoms are not detected, for two main reasons. First, their solubility in gold is very low (a few atomic percents at most). Second, since their equilibrium pressures are rather high, they probably escape from the droplet when the latter is cooled after annealing or after NW growth. Most often, the solid droplets are truncated spheres (spherical caps), another strong indication that they have been cooled from the liquid state.

Even after mere annealing, before being exposed to the molecular beams, the particles are not pure gold. The TEM lattice images show that around each particle a few atomic layers have been removed from the top of the substrate and the corresponding type-III atoms have been incorporated in the droplet (Fig. 1). In the case of a GaAs substrate, we have identified by transmission electron diffraction (TED) that the particles are Au_7Ga_2 crystals. This gives a lower limit to the Ga-content of the annealed droplets, since some of the dissolved Ga might have been lost during cooling. This content is consistent with the common view that VLS occurs at rather low temperature (430-620°C for GaAs in our MBE system) thanks to the formation of a eutectic between the catalyst and the substrate element(s). In the present case, gold liquefies by forming a eutectic with gallium, which may happen below 400°C for Ga concentrations of about 30 %.

We also analyzed the catalyst particles after growth of GaAs NWs [4]. Their composition depends much on the growth history and in particular on how growth was terminated. When the Ga and As fluxes are stopped simultaneously, we observe the Au_7Ga_2 or AuGa crystalline phases. This is consistent with composition measurements made by other authors [13]. Together with our calculation of a typical temperature drop of only a few kelvins between substrate and NW apex during growth [14], this strengthens the picture of a liquid catalyst. It is only when growth is terminated by stopping the Ga flux while maintaining the As flux during cooling that we observe almost pure gold. In this case, most of the Ga atoms of the droplet have been used to continue forming NW layers with the As atoms of the beam (see section 3.2).

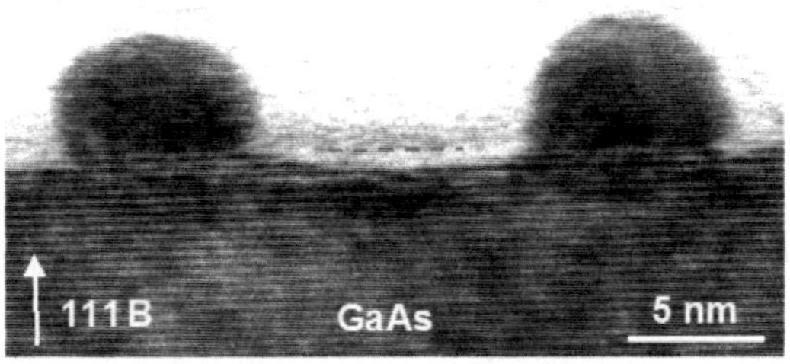

Figure 1. TEM lattice image of the catalyst droplets deposited on a GaAs substrate, after annealing and before nanowire growth. The substrate and the solidified Au_7Ga_2 catalyst are distinguished thanks to their differently spaced lattice planes, $(\bar{1}\bar{1}\bar{2}3)$ and (111) respectively. Dashes mark the top of the substrate before its local dissolution in the droplets.

3. Crystal structure of the nanowires

3.1. Prevalence of the wurtzite phase in nanowires of cubic III-V compounds

Under bulk form, all N-free III-V semiconductors crystallize in the cubic sphalerite structure, also known as zinc blende (ZB). A surprising feature of the NWs of these materials is that they very often adopt the hexagonal wurtzite (WZ) structure. This has been observed for most ZB III-V materials and growth techniques [1,3,4,15,16].

ZB and WZ are closely related structures. Both are stacks along a given direction (*e.g.* [111]B in ZB) of identical biatomic planes, or monolayers (MLs), which differ by their transverse positions. Three such positions (A, B and C) are allowed. ZB corresponds to a periodically repeated ABC sequence, whereas WZ consists of a periodic AB stack. ZB and WZ can be identified either by TED or by high resolution TEM (HREM), as shown in the inserts of Fig. 2. In HREM images, the 3-ML periodicity of ZB is clearly distinguished from the 2-ML periodicity of WZ (Fig. 2(b)).

More complicated periodic stacks are also encountered in NWs, for instance polytypes [16]. Indeed, any aperiodic sequence is allowed, provided it does not include twice the same letter (A, B or C) consecutively. A stacking defect is a local disruption of the ZB or WZ periodicity (stacking fault) or a boundary between differently oriented grains of the same phase. Although often dominantly of WZ structure, the NWs usually contain stacking faults and sequences of ZB structure (Fig. 3(a)). Since the coexistence of two phases impedes basic studies as well as applications, phase purity control has become one of the main challenges of III-V NW fabrication.

The prevalence of WZ (a high pressure phase observed in no other nanostructure) in NWs of III-V cubic materials is surprising. Until recently, its formation had not been explained satisfactorily. The difference in cohesive energy between ZB and WZ bulk III-V materials is typically of a few tens of meV per III-V pair (24 meV for GaAs). Several *ab initio* calculations have shown that the total energy of NWs can indeed be lower in the WZ structure than in the ZB structure because of a lower specific energy of some specific sidewall facets [17-19] or of their vertical edges [20]. However, for the contributions of these 2D or 1D features to compensate the difference in bulk cohesive energies, the NWs must be very thin: the calculated critical radii under which WZ might be favoured over ZB are only of a few nm, whereas WZ is observed in NWs up to radii of several 100 nm.

Moreover, the standard theories of NW formation are based on nucleation [21-23] and the occurrence of planar faults normal to the NW axis (see *e.g.* Figs. 2(c) and 3(a)) suggests that, once a nucleus of critical size forms at the solid/liquid interface, it rapidly spreads out laterally over the whole interface [5,16] unless the wire is very wide [23]. Hence, the reason for the formation of WZ must be searched for in the formation of WZ nuclei rather than in the total energies of fully constituted NWs. This is done in sections 3.2 and 3.3, where we consider the specific case of GaAs.

3.2. TEM observation of the systematic formation of zinc blende sections in wurtzite nanowires

Although WZ is largely prevalent in GaAs NWs, TEM demonstrates that ZB forms systematically during two particular stages of the growth. The first one corresponds to the very beginning of growth. By growing GaAs NWs for only a few seconds, we could observe the foot of each wire before its burying by the growth of the non-activated surface. The samples were thinned to obtain cross-sectional views, with the NWs still epitaxially attached to their substrate. The images identify unambiguously ZB as the sole phase formed in this initial stage (Fig. 2(a)). When the NW height reaches about 30 nm, growth switches abruptly to WZ stacking (Fig. 2(b)).

The second case where ZB forms systematically is when we terminate MBE growth by switching off the Ga flux while maintaining an As flux during cooling. Over a few tens of nm, the final section of the NW is of cubic structure (Fig. 2(c)). Persson *et al.* had already reported the formation of a short terminal ZB section in the case of chemical beam epitaxy, when growth was terminated under As only [3]. In both cases, growth continues in the absence of a fresh supply of Gallium by consuming the Ga atoms present in the drop, whereas As can only be provided by the beam (see section 2.2). Since the volume and the base area of the drop then decrease, the final section of the NW is tapered (Fig. 2(c)).

Our TEM experiments reveal that ZB forms systematically in two transient growth regimes during which the supersaturation in the liquid droplet (with respect to the solid) is less than during steady NW growth. Before growth, the deposited undersaturated Au droplets dissolve the substrate locally to achieve equilibrium with it (Fig. 1), and hence zero supersaturation. When vapour fluxes are turned on, the supersaturation increases until a permanent regime settles. The opposite happens during growth termination: then, the Ga concentration in the droplet, and hence supersaturation, decrease, since the atoms used to build the final NW MLs are not replaced. This shows that ZB systematically forms when the supersaturation is less than some critical value and suggests that, conversely, WZ formation requires a high supersaturation.

3.3. Nucleation at the triple phase line and wurtzite formation: a TEM-supported explanation

Following arguments developed by Ostwald at the end of the 19th century [24], the basic idea is that the phase that forms is not necessarily the stable bulk phase (ZB) but the one having the lowest nucleation barrier ΔG^{ϕ}, because the main dependence of the nucleation probability at temperature T is in $\exp(-\Delta G^{\phi}/k_B T)$. The formation of WZ may thus be searched for in its lower nucleation barrier as compared with ZB. Standard nucleation theory [25] gives the energy barrier for the formation of a 2D solid nucleus of radius r and height $h = 1$ ML of a given phase ϕ on the NW top facet:

$$\Delta G^{\phi} = -ar^2 h\Delta\mu^{\phi} + brh\gamma_{\ell}^{\phi} + ar^2\gamma_{SN}^{\phi} \tag{1}$$

where a and b are shape-dependent coefficients ($a=\pi$ and $b=2\pi$ for a disk), $\Delta\mu^{\phi} > 0$ is the difference of chemical potential between liquid and solid per unit volume of solid, γ_{SN}^{ϕ} is the interface energy between the top facet of the NW and the nucleus, and γ_{ℓ}^{ϕ} is the specific energy of the lateral interface between the nucleus and its environment.

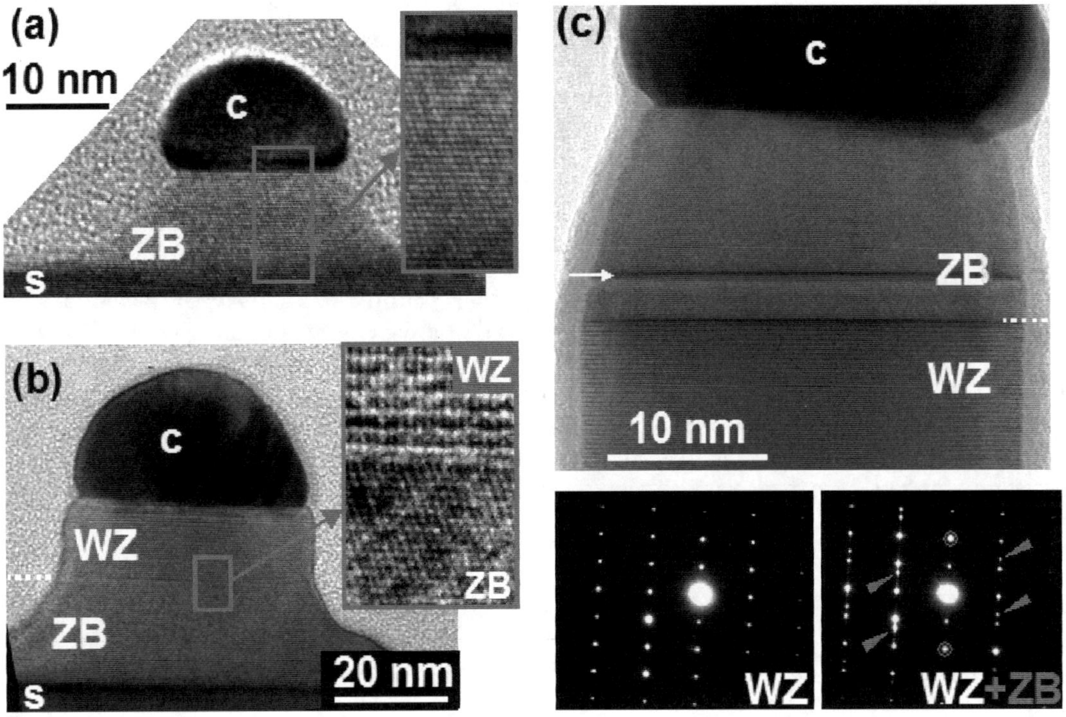

Figure 2. (a) TEM image of a very short GaAs NW with HREM close-up showing the pure ZB structure (s: substrate, c: catalyst). (b) A short GaAs NW soon after the ZB → WZ transition, with HREM image of the transition zone. (c) When growth is terminated under As only, the final tapered section of the GaAs NW (above dashed line) is pure ZB (with a stacking fault, pointed by arrow): the TED pattern from the main body of the NW (left) shows only WZ whereas the pattern from an area including also the terminal section (right) shows a superposition of WZ and ZB (arrowed spots).

We argued elsewhere that, whatever the crystalline phase, nucleation at the triple phase line (TPL), the border of the top facet of the NW (where solid, liquid and vapour coexist), is strongly favoured with respect to nucleation anywhere else on the top facet [26]. The reason is simply that, in the former case, part of the pre-existing liquid-vapour interface is destroyed and replaced by a nucleus-vapour interface. Then, $\gamma_{\ell}^{\phi} = (1-\alpha)\gamma_{\ell L}^{\phi} + \alpha\left(\gamma_{\ell V}^{\phi} - \tau\gamma_{LV}\right)$, where α is the fraction of the nucleus perimeter in contact with the vapour, $\gamma_{\ell L}^{\phi}$ and $\gamma_{\ell V}^{\phi}$ are the specific energies of the lateral nucleus-liquid and nucleus-vapour interfaces, γ_{LV} the liquid-vapour interface energy and $\tau \sim 1$ a factor accounting for the droplet geometry at the TPL [26]. Hence, nucleation at the TPL ($\alpha \neq 0$) may considerably reduce the nucleation barrier ΔG^{ϕ}. This is somewhat supported by recent *in situ* TEM experiments which show that Si crystals tend to nucleate at the edge of Au-Si droplets deposited on a membrane [27-28]

To compare the free energies of formation of ZB and WZ nuclei at the TPL, one may consider that, because of the difference in cohesive energies, $\Delta\mu^{\phi}$ is lower for WZ than for ZB. However, a 2D nucleus is not intrinsically of ZB or WZ structure. In accordance with the description of ZB and WZ as different stacking modes of otherwise identical MLs, it is only the lateral positioning of the nucleus that decides if the sequence formed with the underlying MLs is locally of WZ or of ZB type [26]. Thus, it is preferable to use the same $\Delta\mu^{\phi}$ in Eq. (1) for ZB and WZ and two different NW/nucleus interface energies $\gamma_{SN}^{ZB} = 0$ and $\gamma_{SN}^{WZ} > 0$ (the two approaches are equivalent since both contributions vary as r^2). Maximizing ΔG^{ϕ} with respect to r yields the radii of the critical nuclei of the two phases (or, rather, for nucleation in ZB and WZ positions) and the corresponding nucleation barriers ΔG_c^{ϕ} :

$$\Delta G_c^{ZB} = \frac{b^2}{4a}\frac{h\left(\gamma_\ell^{ZB}\right)^2}{\Delta\mu}, \qquad \Delta G_c^{WZ} = \frac{b^2}{4a}\frac{h\left(\gamma_\ell^{WZ}\right)^2}{\Delta\mu - \dfrac{\gamma_{SN}^{WZ}}{h}} \tag{2}$$

Eq. (2) shows that there are two conditions for the critical nucleus in WZ position to have a lower free energy of formation than the nucleus in ZB position. The first one is material-related: one must have $\gamma_\ell^{WZ} < \gamma_\ell^{ZB}$ Careful consideration of the atomic arrangements for the ZB and WZ nuclei positions indicates that this difference is not likely to arise from differences at the nucleus-liquid interface [26], although this may also play a part [29]. On the other hand, the nucleus-vapour interfaces that form during nucleation at the TPL are clearly different and the corresponding energy may be lower for the WZ nucleus position [26], as also suggested by calculations showing that the specific energies of some extended facets are lower for WZ than for ZB [17-19]. From Eq. (2), there is however a second condition for WZ nucleation to be favoured, namely that the supersaturation be high enough to overcome the larger NW-nucleus interface energy:

$$\Delta\mu \geq \Delta\mu_c = \frac{\left(\gamma_\ell^{ZB}\right)^2}{\left(\gamma_\ell^{ZB}\right)^2 - \left(\gamma_\ell^{WZ}\right)^2}\frac{\gamma_{SN}^{WZ}}{h} \tag{3}$$

To summarize, the frequently observed formation of the WZ phase in NW of cubic III-V materials is due to the preferential formation of nuclei in WZ lateral position at the triple phase line. This preferential formation in turns results from a lower energy of the edge of the nucleus when it is in WZ position as compared with ZB position. This condition recalls the arguments based on total energy calculations for fully formed NWs [18,20]. However, at variance with the latter purely thermodynamic explanation, ours is intrinsically kinetic, in that it is based on barrier-height-dependent nucleation probabilities. This kinetic character is also manifested by our second condition for WZ formation, namely the existence of a lower critical value of the chemical potential difference between liquid and solid. Even with differences $\gamma_\ell^{ZB} - \gamma_\ell^{WZ}$ as low as a few tens of $mJ\,m^{-2}$, we find that differences $\Delta\mu$ of the order of a few hundreds of meV per III-V pair are sufficient to promote WZ nucleation. This is indeed the order of the differences of chemical potential between vapour and solid that we estimate for our MBE growth conditions, although the actual $\Delta\mu$ between liquid and solid remains unknown.

The nucleation of each new ML has a probabilistic character. The formation of a stacking defect in WZ (Fig. 3(a)) may thus be interpreted as arising from the chance nucleation of a single ML in ZB position, occurring because $\Delta\mu$ is not sufficient to insure a near certain nucleation in WZ position. On the other hand, WZ formation requires nucleation of each ML in WZ position. This opens prospects of single-phase growth, by setting high supersaturation conditions for pure WZ or, alternatively, low supersaturation for pure ZB. However, our efforts in this direction have not been successful yet. In particular, supersaturation cannot be lowered under a limit below which the NWs cease to grow.

Nevertheless, several TEM-based results from other groups strengthen our arguments. Following the latter, Shtrikman *et al.* managed to obtain pure ZB GaAs NWs in MBE by deliberately attempting to achieve a low supersaturation [30]. Caroff *et al.* noticed that, in InAs NWs grown by MOVPE, the proportion of the WZ phase increases at low growth temperature [31]. We propose that this can be due to the increase of both the chemical potential difference $\Delta\mu$ and the relative difference of nucleation probabilities at low T. Conversely, if the material and supersaturation conditions are such that ZB formation prevails, lowering the temperature may again increase the difference of nucleation probabilities, this time in favour of ZB, and hence reduce the density of stacking defects, as observed by Johansson *et al.* [5] and Joyce *et al.* [32].

4. Phase transformation upon epitaxial burying

Despite some successes, it remains very difficult to fabricate reproducibly single-phase III-V NWs in a single growth operation. We have however devised a method to transform highly faulted hexagonal WZ GaAs NWs into perfect cubic ZB ones [33].

4.1. Experiments

The method is based on deliberately altering the relative growth rates of the NWs and of the non-activated substrate by changing the growth temperature T_G. Whereas in our standard growth conditions, for $T_G \leq 620\,°\mathrm{C}$, the NWs grow faster than the non-activated substrate (section 1), the reverse holds at $T_G \geq 620\,°\mathrm{C}$, so that any previously grown NW will get buried by the substrate.

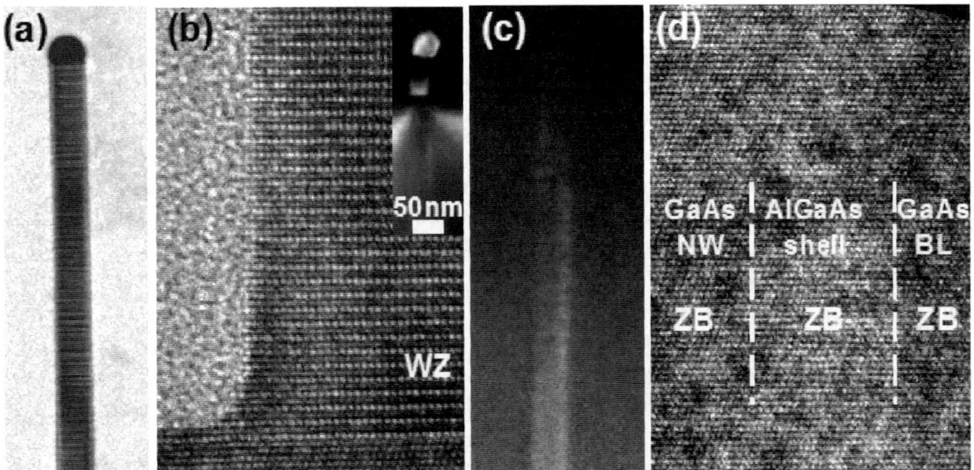

Figure 3. (a) TEM image of an unburied NW, with horizontal striations due to stacking defects. (b) HREM image of a partially buried NW (with low resolution inset). In the buried part, the dashed line marks the vertical interface between burying layer (BL) and NW (right). (c) Image of a fully buried NW. Note the absence of striation. (d) HREM image of the interface between NW, shell and burying layer after complete burying.

The standardly-grown NWs have the WZ structure with a high density of stacking defects (Fig.

3(a)). We first wrap them in a nm-thick protective $Al_xGa_{1-x}As$ shell to prevent their disappearance at elevated temperature. Because it results from epitaxial nucleation on the sidewalls of the NWs, the shell adopts the latter's WZ structure. By increasing the growth temperature to 640 °C, we then shift to a regime where the supporting non-activated cubic GaAs substrate grows faster than the wires, effectively burying them. The whole sequence is carried out in the same MBE chamber.

The samples were examined by TEM at various stages of the burying process. We first notice that the burying layer naturally adopts the cubic ZB structure of the GaAs substrate from which it grows. Examination of partially buried NWs (Fig. 3(b)) shows that the part emerging from the burying layer has not changed: it is still WZ with stacking defects. However, below the top level of the burying layer, the material is entirely of cubic structure. This means that the buried portion of any NW has changed to ZB. Hence, the growing non-activated substrate transforms the buried portion of the NW from WZ to ZB. This phase transformation is very fast, since the phase boundary is right at the level of the top ML of the burying layer (Fig. 3(b)).

Further growth leads to complete burying. Each NW can still be localized via the faint contrast of its AlGaAs shell (Fig. 3(c)). TED and HREM images (Fig. 3(d)) confirm that the whole sample is pure ZB. The burying is very efficient, in that it eliminates all the stacking faults (Fig. 3(c,d)). We thus end up with perfect cubic NWs embedded in a cubic matrix. Thanks to the AlGaAs shell, each NW retains its one-dimensional character.

4.2. The mechanism of the phase WZ → ZB phase transformation as revealed by HREM

Our TEM experiments on partially buried NWs show that, when the burying material reaches the periphery of a NW, it transforms the hexagonal structure of the latter into its own cubic one (Fig. 3(b)). To understand how TEM may also reveal the microscopic mechanism of the phase transformation, we must consider the burying process in more detail. First recall that, along the [111]B ZB or [0001] WZ direction, any type-III atom sits directly above the underlying type-V atom. The A, B, or C character of each III-V ML is therefore determined by the lateral positioning of its type-V atoms [26]. Let us assume that the burying layer has already transformed the lower part of the NW to its ZB stacking (Fig. 4(a), stage α) and consider how the next sequence of six yet unburied WZ MLs may transform into ZB.

When the advancing top ML of the burying GaAs reaches the NW sidewall (Fig. 4(a), α), the As atoms in the burying ML and in the corresponding ML of the NW are not in lateral registry. This fault in the NW (with respect to ZB) can be eliminated by a rigid shift of the emerging part of the NW (As plane included), which brings two more MLs in ZB position (Fig. 4(a), β). When the burying layer reaches the level of the next fault, two MLs above the previous one (Fig. 4(a), γ), the process repeats (Fig. 4(a), δ), and so on every other ML. Three shifts suffice to transform 6 MLs (Fig. 4(a), ε,φ).

The lateral contact between the top burying ML and the out-of-registry NW ML produces a Shockley partial dislocation (albeit only partly surrounded by material). Since our NWs have vertical sidewalls parallel to the six $\{10\bar{1}0\}$ WZ ($\{\bar{2}11\}$ ZB) planes, the edges of the NW ML are $\langle 1\bar{1}0\rangle$-oriented. The Burgers vector of the partial dislocation is one of three translations allowed at each step, namely $\frac{1}{6}[\bar{2}11]$, $\frac{1}{6}[1\bar{2}1]$ or $\frac{1}{6}[11\bar{2}]$ (or the complementary set, depending on the ZB variant). As the dislocation glides along the (111)B plane into the NW, the As atoms shift into ZB position. This translation of the whole As layer fully eliminates dislocation and stacking fault [33].

A type II transformation may also be imagined, whereby only one or two MLs of the NW are translated at each step, all other MLs retaining their lateral positions (Fig. 4(b)). The same partial dislocation is initially generated (Fig. 4(b), α) but the first step of the process is now the translation of the sole NW ML that it borders (Fig. 4(b), β). This eliminates the fault in this ML but produces a characteristic twin-like ABCBA sequence [34]. The next two NW MLs cannot be translated before the third one, which occurs when the burying material reaches its level (Fig. 4(b), γ), again by translation of a single ML, producing a "double-twin" BCBAC sequence (Fig. 4(b), δ). Finally, the two already buried intermediate MLs can transform (Fig. 4(a), ε,φ), albeit only if they are not yet entirely surrounded by the burying layer or if they can exchange atoms. The latter reservations make the type II process less likely than the type I.

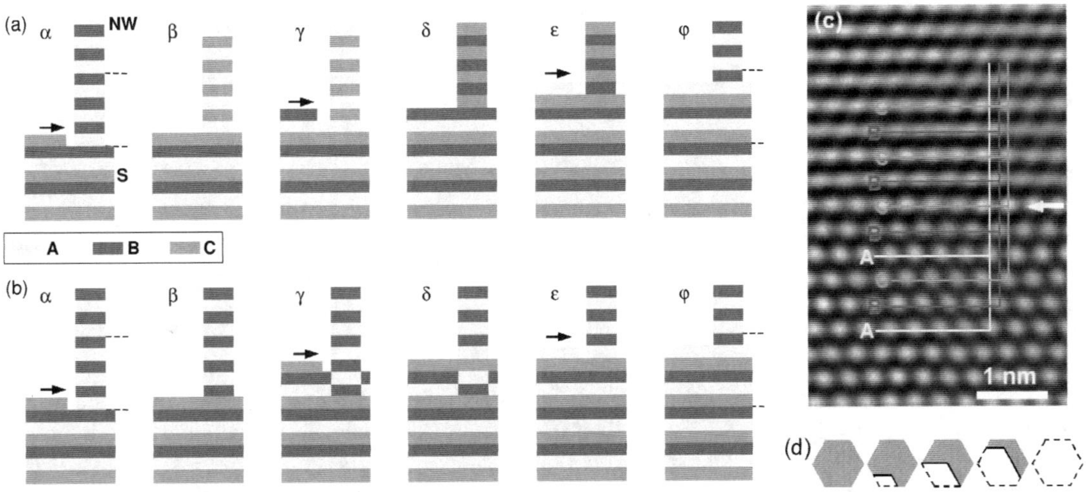

Figure 4. (a,b) Schematics of the transformation of a WZ nanowire into ZB. Type I (a) and II (b) processes for the transformation of the 6 MLs located between the dashed lines. Greek letters index the successive stages of each process. Each coloured rectangle represents a monolayer of the substrate (S) or NW. Its lateral position is colour-coded (see insert). Arrows give the step flow direction. (c) Fourier-filtered HREM image of the interface between the transformed and untransformed parts of a nanowire. The arrow marks the topmost ML in the ZB sequence. (d) Schematics of the transformation of a NW ML from WZ to ZB stacking positions (top view). The area not yet translated is shaded. Solid lines track the glide of the partial dislocation initially formed (dashed line) at the boundary between burying and NW MLs.

These two types of transformation of WZ into ZB can be distinguished by HRTEM via differences in stacking sequences at the interface between the untransformed and transformed parts of the NW. We first identify the topmost NW MLs that have already adopted the ZB stacking (Fig. 4(c)). At any stage of a type I process (Fig. 4(a)), the next NW MLs repeat the last two MLs of the ZB sequence (e.g. ABCA**BC/**BCBC... at stage δ) whereas during a type II process (Fig. 4(b)), the same sequence is found only at stage α or φ (ABC**AB/**ABAB...), other well-defined twin-like sequences occurring at stages β to ε. In the many HRTEM images examined, we found only the stacking sequence characteristic of the type I process (Figure 4(c)) and no evidence of the other sequences occurring in a type II process. This indicates that, at least for GaAs and in our burying conditions, type I is the dominant transformation mechanism, if not the only one.

4.3. Interpretation

The driving force for the type I phase transformation is the joint elimination of the partial dislocation that forms at the NW/burying material boundary and change of structure of the NW from WZ to ZB. The length of the partial dislocation segments formed between burying layer and NW and the area of the faulted interface between the transformed and untransformed parts of the NW determine the system energy. Fig. 4(d) describes schematically how these change during the transformation of each ML. As overgrowth proceeds, the NW sidewalls get surrounded by the top burying ML extending by step flow and the partial dislocation is generated at the periphery of the corresponding hexagonal-shape NW ML. As soon as the dislocation forms along at least two sides of the hexagon, any of its parts can glide inward without any length increase and therefore, without any energy increase. The area of the NW ML swept by the dislocation transforms to ZB stacking with respect to the underlying ZB portion of the NW. We calculate that eliminating the fault corresponds to a decrease of energy of about 28 $mJ \cdot m^{-2}$ [34]. Hence, the total energy of the defects decreases continuously during each elementary step of the transformation (Fig. 4). Of course, when a NW ML is already in the correct ZB position, no dislocation is created and hence no driving force for translating the upper part of the NW. This happens in particular when there is a stacking defect or a ZB segment in the yet unburied WZ NW and explains why all such pre-existing defects (Fig. 3(a)) get eliminated (Fig. 3(c)).

According to this simple analysis, there is no energy barrier to the transformation, in agreement with the fact that the translation of the whole ML occurs very soon after the burying layer reaches it, which leads to abrupt horizontal interfaces between the transformed and untransformed parts of the NW (Figs. 3(b) and 4(c)). In addition, there is no critical NW radius for the transformation, which explains why the structure of any NW shifts to ZB. The NW geometry makes the WZ $\rightarrow$ ZB transformation much easier than in bulk samples where, even for the type I transform, there is a barrier corresponding to the nucleation of a transformed domain of finite critical size [35]. Here, the dislocation is formed when the burying layer reaches the NW, and nothing hinders the translation of the top part of the NW. Since three translation (Burgers) vectors adding to zero are allowed, the net tilt of each NW remains nearly null (Fig. 3(c)). The former analysis however neglects the Peierls barrier impeding the glide of the partial dislocation, which might kinetically limit the transformation. This might be responsible for the incomplete transformation that we tend to observe in InP NWs.

5. Thermodynamics and kinetics of heterostructure formation in NWs

Nanowires are particularly interesting in that they allow the fabrication of precisely positioned quantum dots with well-defined geometries. In this section, we consider the formation of heterostructures in NWs as regards strain relaxation and crystalline structure.

5.1. Critical dimensions for mismatched axial heterostructures in nanowires

A major advantage of NWs appears when one compares the growth of a given misfitting material on a bulk substrate and on an already grown NW base. In the latter case, the heterostructure is termed axial to distinguish it from radial (core-shell) heterostructures. For a given substrate/layer misfit, the critical thickness for plastic relaxation is always larger in the NW case because the sidewalls allow the strains efficiently to relax [36], all the more if the NW diameter is small. Moreover, for each material couple, there exists a critical NW radius below which the critical thickness becomes infinite, which allows the growth of arbitrarily thick layers. We derived simple formulas for the radius-dependent critical thickness and for the critical radius [36]. The former are consistent with our own TEM experiments (*e.g.* the absence of dislocation in the long insertions of Fig. 5) and with the limited amount of experimental data available in the literature.

5.2. Fault suppression and structural phase changes at an axial hetero-interface

In VLS growth, sharp axial heterostructures are invariably fabricated by switching the type-V element (or modifying the fluxes of several type-V elements) because the type-III element, present in high concentration in the droplet (see section 2.2), tends to be incorporated even after its supply from the vapour phase is interrupted (see section 3.2). Changing the fluxes changes the thermodynamic growth conditions and may have profound effects not only on the composition of the grown material but also on its structure. We indeed frequently observe structural changes at hetero-interfaces.

A first example is given by a double $InP/InP_{1-x}As_x/InP$ axial heterostructure (Fig. 5(a)). A thick InP shell was later grown around the thin NW core. In the core, InP and $InP_{1-x}As_x$ ($x \sim 0.35$) both adopt the WZ structure. However, whereas InP is highly faulted, the InPAs insertion is free from any extended defect. Note that the stacking defects created during growth of the InP core propagate laterally in the WZ shell whereas the InP shell around the InPAs insertion is defect-free. The reason is that the shell material nucleates on the NW sidewalls, at variance with the burying material considered in section 4.

In light of the discussion of section 3.3, the high density of faults in the WZ InP core indicates that the difference of chemical potential $\Delta\mu$ is high enough to favour WZ but not sufficient to insure that each ML will nucleate in WZ position. On the contrary, the absence of any fault even in long InPAs insertions (Fig. 5(b)) indicates that, in this case, the probability for nucleation in ZB position is negligible. Since the insertion is grown by adding an As flux to an unchanged P flux, it is tempting to attribute this effect to an increased $\Delta\mu$. However, the deposited solid changes, and even though the liquid is still mainly Au-In, its composition might also change (as happens in chemical beam epitaxy [37]), so that the relation between the fluxes and $\Delta\mu$ is far from straightforward. Moreover, the contribution of the type-V elements to the chemical potential in the liquid is unknown. In addition, over and above the chemical potential condition, the balance between WZ and ZB nucleation also depends on the edge energy γ_ℓ^ϕ of the nucleus, as discussed in section 3.3. When nucleation occurs at the TPL, the edge is partly in contact with the vapour and the surface energies (and reconstructions) of III-V materials are known to depend on the vapour fluxes [38]. The changes of chemical potential and edge energy might thus both play a part in the absence of fault in the alloy segment.

The formation of the heterostructure may have an even more striking effect: when a $GaAs_{1-x}Sb_x$ alloy is grown axially above a GaAs WZ NW base, it adopts the ZB structure [39]. In this case, we observed by TEM that, in the GaAs cap above the insertion, the transition to (faulted) WZ is not abrupt but involves a section of the 4H polytype. Since in the latter, half the MLs nucleate in ZB position and the other half in WZ position, this phase can be considered as intermediate between ZB and WZ [34]. It seems more likely that such a delayed ZB $\rightarrow$ WZ transition is due to a gradual return of the chemical potential in the droplet to its pre-insertion value after the Sb flux is stopped, rather than to a surface effect which probably follows the changes of fluxes more closely in time.

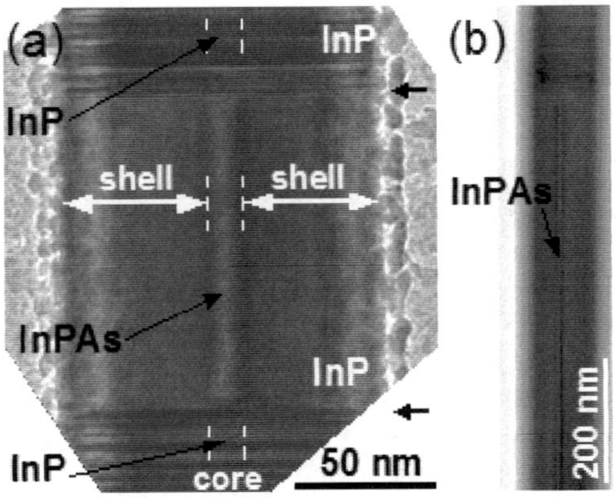

Figure 5. (a) TEM bright field image of an $InP_{1-x}As_x$ axial insertion about 100 nm high in an InP NW. The NW core diameter during axial growth (about 12 nm) is indicated by dashed lines and the hetero-interfaces by horizontal black arrows. The whole core was subsequently wrapped in a thick InP shell (double-ended white arrows). The contrast along the vertical edges of the insertion is due to strain relaxation. (b) A long fault-free $InP_{1-x}As_x$ insertion in a sample of similar structure.

Acknowledgements
The InP/InPAs/InP and the GaAs/GaAsSb nanowires discussed in section 5.2 were grown respectively by Linsheng Liu in our lab and by Dasa L. Dheeraj at NTNU (Trondheim, Norway). We thank them as well as Vladimir Dubrovskii and George Cirlin (Physics and Technology Centre for Research and Education of the RAS, St Petersburg, Russia) for enlightening discussions.

References

[1] Koguchi M, Kakibayashi H, Yasawa M, Hiruma K and Katsuyama T 1992 *Jpn. J. Appl. Phys.* **31**, 2061
[2] Duan X and Lieber C M 2000 *Adv. Mater.* **12**, 298
[3] Persson A I, Larsson M W, Stenstrom S, Ohlsson B J, Samuelson L and Wallenberg L R 2004 *Nature Mater.* **3**, 677
[4] Harmand J C, Patriarche G, Péré-Laperne N, Mérat-Combes M N, Travers L and Glas F 2005

Appl. Phys. Lett. **87**, 203101

[5] Johansson J, Karlsson L S, Svensson C P T, Mårtensson T, Wacaser B A, Deppert K, Samuelson L and Seifert W 2006 *Nature Mater.* **5**, 574

[6] Plante M C and LaPierre R R 2006 *J. Cryst. Growth* **286**, 394

[7] Wagner R S and Ellis W C 1964 *Appl. Phys. Lett.* **4**, 89

[8] Tchernycheva M, Harmand J C, Patriarche G, Travers T and Cirlin G E 2006 *Nanotechnol.* **17**, 4025

[9] Ross F M, Tersoff J and Reuter M C 2005 *Phys. Rev. Lett.* **95**, 146104

[10] Kodambaka S, Tersoff J, Reuter M C and Ross F M 2007 *Science* **316**, 729

[11] Dubrovskii V G, Sibirev N V, Cirlin G E, Harmand J C and Ustinov V M 2006 *Phys. Rev. E* **73**, 021603

[12] Tchernycheva M, Travers L, Patriarche G, Glas F, Harmand J C, Cirlin G E and Dubovskii V G 2007 *J. Appl. Phys.* **102**, 094313

[13] Johansson J, Karlsson L S, Dick K A, Bolinsson J, Wacaser B A, Deppert K and Samuelson L 2009 *Cryst. Growth Des.* **9**, 766

[14] Glas F and Harmand J C 2006 *Phys. Rev. B* **73**, 155320

[15] Mohan P, Motohisa J and Fukui T 2005 *Nanotechnology* **16**, 2903

[16] Soshnikov I P, Cirlin G E, Tonkikh A A, Samsonenko Y B, Dubovskii V G, Ustinov V M, Gorbenko O M, Litvinov D and Gerthsen D 2005 *Phys. Sol. State* **47**, 2121

[17] Leitsmann R and Bechstedt F 2007 *J. Appl. Phys.* **102**, 063528

[18] Galicka M, Bukala M, Buczko R and Kacman P 2008 *J. Phys.: Condens. Matter* **20**, 454226

[19] Hilner E, Håkanson U, Fröberg L E, Karlsson M, Kratzer P, Lundgren E, Samuelson L and Mikkelsen A 2008 *Nano Lett.* **8**, 3978

[20] Akiyama T, Sano K, Nakamura K and Ito T 2006 *Jpn. J. Appl. Phys.* **45**, L275

[21] Mutaftschiev B, Kern R and Georges C 1965 *Phys. Lett.* **16**, 32

[22] Givargizov E I 1975 *J. Cryst. Growth* **31**, 20

[23] Dubrovskii V G and Sibirev N V 2004 *Phys. Rev. E* **70**, 031604

[24] Ostwald W 1897 *Z. Phys. Chem.* **22**, 289

[25] Markov I V 2003 *Crystal Growth for Beginners* (Singapore: World Scientific)

[26] Glas F, Harmand J C and Patriarche G 2007 *Phys. Rev. Lett.* **99**, 146101

[27] Hoffmann S, Sharma R, Wirth C T, Cervantes-Sodi F, Ducati C, Kasama T, Dunin-Borkowski R E, Drucker J, Bennett P and Robertson J 2008 *Nature Mater.* **7**, 372

[28] Kim B J, Tersoff J, Kodambaka S, Reuter M C, Stach E A and Ross F M 2008 *Science* **322**, 1070

[29] Algra R E, Verheijen M A, Borgström M T, Feiner L F, Immink G, van Enckevort W J P, Vlieg E and Bakkers E P A M 2008 *Nature* **456**, 369

[30] Shtrikman H, Popovitz-Biro R, Kretinin A and Heiblum M 2009 *Nano Lett.* **9**, 215

[31] Caroff P, Dick K A, Johansson J, Messing M E, Deppert K and Samuelson L 2008 *Nature Nanotechnol.* **4**, 50

[32] Joyce H J, Gao Q, Tan H H, Jagadish C, Kim Y, Zhang X, Guo Y and Zou J 2007 *Nano Lett.* **4**, 921

[33] Patriarche G, Glas F, Tchernycheva M, Sartel C, Largeau L and Harmand JC and Cilin GE 2008 *Nano Lett.* **8**, 1638

[34] Glas F 2008 *J. Appl. Phys.* **104**, 093520

[35] Geilikman M B 1982 *Phys. Chem. Miner.* **8**, 2

[36] Glas F 2006 *Phys. Rev. B* **74**, 121302

[37] Fröberg L E, Wacaser B A, Wagner J B, Jeppesen S, Ohlsson B J, Deppert K and Samuelson L 2008 *Nano Lett.* **8**, 3815

[38] Moll N, Kley A, Pehlke E and Scheffler M 1996 *Phys. Rev.* B **54**, 8844

[39] Dheeraj D L, Patriarche G, Largeau L, Zhou H L, Hoang T B, Mosses A F, Grønsberg S, van Helvoort A T J, Fimland B O and Weman H 2008 *Nano Lett.* **8**, 4459

Bridging the length scales between lithographic patterning and self assembly mechanisms in fabrication of semiconductor nanostructure arrays

R Hull[1], JA Floro[2], M Gherasimova[3], JF Graham[2], JL Gray[4], A Portavoce[5], FM Ross[6] and J Thorp[2]

[1] Department of Materials Science and Engineering, Rensselaer Polytechnic Institute, Troy, NY 12180 USA
[2] Dept. of Materials Sci. and Eng., University of Virginia, Charlottesville, VA, USA
[3] Dept. of Physics, Southern Connecticut State University, New Haven, CT, USA
[4] Dept. of Mechanical and Materials Eng, University of Pittsburgh, PA, USA
[5] CNRS - IM2NP, Marseilles, France
[6] IBM Thomas J. Watson Research Center, Yorktown Heights, NY, USA

Email: hullr2@rpi.edu

Abstract. We employ focused ion beam patterning of single crystal Si(100) surfaces to template the assembly of Ge(Si) nanostructure arrays. The evolution and final structures of the templated arrays are determined by combinations of transmission electron, low energy electron microscope, focused ion beam and scanning probe microscopies. It is shown how the positions of individual nanostructures may be controlled to the order of 10 nm. However, to achieve controlled spacings between elements that are in the 10 nm range requires careful matching of the characteristic lengths scales of self assembly mechanisms to the length scales of the external lithographic "forcing functions".

1. Introduction

Several potential nanoelectronic architectures, such as quantum cellular automata [1,2] and coherent spin switches [3] require ultra high precision control of the position, dimension, spacing and electronic/ magnetic doping of individual semiconductor nanoscale elements in extended arrays. This manuscript focuses on the controlled assembly of such nanostructure arrays. Silicon (100) surfaces are patterned by an array of point-by-point focused ion beam pulses, and subsequent epitaxial growth of Ge(Si) nanostructures on the surface are templated to the original ion beam irradiation sites.

It is well established that growth of thin compressively strained epitaxial films leads to roughening of the surface and/or the formation of discrete nanoscale clusters of the epitaxial deposit. For the Ge_xSi_{1-x}/Si(100) system considered here, the set of transitions that occur during near-equilibrium growth have been established by the work of multiple groups as comprising formation of a thin planar wetting layer of the epitaxial Ge_xSi_{1-x}, followed by formation of coherent {510} faceted "hut clusters", transformation to predominantly {311} faceted coherent "dome clusters", and finally misfit dislocation formation within the clusters. For pure Ge/Si, the wetting layer is about 4 atomic monolayers thick and typical hut cluster dimensions are of order 30 nm. As strain is reduced through increasing Si concentrations in Ge_xSi_{1-x} epitaxial deposits, the relevant length scales (wetting layer thickness, cluster dimensions etc) increase. It also becomes necessary to increase the substrate temperature to enable adatom diffusion lengths to be sufficiently large to attain these quasi-equilibrium structures (for example, $Ge_{0.2}Si_{0.8}$ substrate temperatures of order 750° C enable this sequence of structural transitions to be maintained [4]), otherwise metastable morphological transitions may be observed, as will be discussed later in this manuscript.

The aim of the templating process is to transfer the symmetry of the focused ion beam (FIB) template pattern to the symmetry of the final nanostructure array as faithfully as possible. Thus for an initial ion distribution $i(x,y)$ on the surface of the Si(100) substrate, and a final nanostructure distribution $n(x,y)$, we want as simple as possible a relationship between these two distributions:

$$n(x,y) = i(x,y)*t(x,y) \qquad (1)$$

where $t(x,y)$ represents the "smearing function" between the processes and * is the convolution operator.

Analogous to treatments of the transfer of spatial frequencies through an objective lens in high resolution transmission electron microscopy, it is instructive to consider the arrays as a Fourier summation of periodic components, and to determine what range of spatial frequencies are faithfully transmitted from the "object" (i.e. the templating array, $i(x,y)$) to the "image" (the nanostructure array, $n(x,y)$) by a transfer function $T(u,v)$ in Fourier space.

$$n(x,y) = F[N(u,v)] = F[I(u,v)\bullet T(u,v)] = i(x,y)*t(x,y) \qquad (2)$$

Here $F[\ ]$ represents the inverse Fourier transform operator; u,v are the Fourier space coordinates corresponding to real space coordinates x,y; and the upper case function (e.g. $N(u,v)$) represents the Fourier transform of the corresponding lower case function (e.g. $n(x,y)$). For simplicity, we henceforth consider only real space and Fourier space variables, x and u, i.e. we will assume isotropic behavior of the templating process with respect to the surface azimuthal direction. Fundamental questions include the maximum spatial frequency u_{max} that can be transmitted, i.e. the closest spaced objects (this is critical with respect to electron tunneling and spin transmission between adjacent nanostructures for nanoelectronic applications), and the minimum transmissible frequency u_{min} (which speaks to the practical overall dimensions of the array, and the ability to maintain large areas of unoccupied "white space" as needed in parts of the array). Between these values, efficient transfer of all spatial frequencies is required to faithfully replicate complex aperiodic patterns. We point out, however, that one significant difference from contrast transfer function analysis in TEM phase contrast imaging is that for high spatial frequencies we are not defining the ideal transfer function $T(u)$ to be unity. Rather a major purpose of marrying lithographic templating with self assembly mechanisms is to access length scales *below* those that can be directly lithographically defined.

The major experimental limitations on the maximum spatial frequencies in the final nanostructure array, u^n_{max} are: (i) the minimum spacings accessible by the initial patterning function, $i(x,y)$, i.e. here by the spatial resolution of our focused ion beam pulses, corresponding to a maximum input spatial frequency, $u^i_{\max}$ (ii) the characteristic capture zone dimensions, R, for adatoms during epitaxial growth on the template sites, and (iii) the degree to which more complex nanostructures may be assigned to each lattice point defined by the template array. These last two factors speak to the balance of length scales between the initial lithographic pattern and the characteristic length scales over which nanostructures self assemble during the epitaxial growth process. The primary limitation on u^n_{min} will be the tendency for spontaneous nanostructure formation away from template sites. Suppression of this process requires either a large adatom diffusion length (although as described later, this can place lower limits on u^n_{max}), or a mechanism that inhibits nucleation away from templating sites.

2. Experimental

Structures are grown in a set of different UHV growth systems and examined by a range of microscopies; in several cases the growth and characterization systems are combined.

Ge quantum dot (QD) structures described in this manuscript are grown in one of two instruments: (i) A unique Hitachi UHV transmission electron microscope in the laboratory of Frances Ross at IBM

Thomas J. Watson Research Center [5]. This instrument enables digermane and disilane deposition in the objective pole-piece area, allowing direct TEM observation of the CVD growth process in real time. An FEI 611 FIB column is connected under UHV in an adjacent preparation chamber, and is typically operated at 25 keV, an ion current of 10 pA, and ion pulse times per template feature in the range of hundreds of μs (Figure 1(a)). The ion beam diameter is of order 50 nm. Atomically clean, electron transparent Si(100) samples are prepared using procedures previously described [6]. Growth of Ge QDs from digermane on unpatterned Si surfaces was performed at temperatures in the range 500-600° C and typical digermane partial pressures ~ 10^{-7} Torr. (ii) A SPECS P90 UHV Low Energy Electron microscope (LEEM) [7], equipped with multiple in-situ deposition and surface modification capabilities, with a base pressure of ~ 10^{-10} Torr. Samples are cleaned by in-situ oxide desorption at temperatures ~ 800-1100° C (the lower temperatures are employed to enable pre-existing FIB template patterns to be retained) and Ge deposition is from a Knudsen cell at a substrate surface temperature estimated to be ~ 500-550° C.

For growth and templating of Quantum Dot Molecule (QDM) structures, Ge_xSi_{1-x} films were grown using molecular-beam epitaxy (MBE) on (100) Si. 300 μm thick substrates were chemically cleaned and oxidized using a modified Shiraki procedure [8] before being loaded into the growth chamber. The resulting surface oxide was desorbed at a temperature of 750-820°C and a 1000Å Si buffer layer was then grown at 750°C. After buffer growth, the substrate temperature was lowered to the Ge_xSi_{1-x} growth temperature (usually 550° C). Following temperature equilibration, Si buffer (typically of order 7 nm thickness) was grown followed by Ge_xSi_{1-x} alloy growth at the same temperature.

Most of the FIB-templating of Ge(Si) QD and QDM sites on Si(100) surfaces described in this manuscript employs a standard Ga beam using an FEI 200 system, and an FEI 611 column integrated into a UHV-TEM system, as described above. We are also working with an Orsay Canion column [9] equipped with a mass selecting ExB filter, that allows separation of a far wider range of primary ion beams from liquid metal alloy sources (to date, we have generated Au, Si, B, As, Mn and Ge beams from AuSi, PdAsB and Mn(Au)Ge sources).

Samples are studied in-situ and ex-situ from growth by a set of different microscopies spanning lengths scales of interest in two and three dimensions, as summarized in Table I.

Technique / Instrument	Lateral resolution [a]	Vertical resolution [b]	2D projection, Surface, or 3D [c]	In situ or Ex-situ [d]
TEM / Hitachi UHV	~ 0.3 nm	Atomic at surfaces	2D Proj and Surface	In-situ
TEM / JEOL 2000 FX	~ 0.3 nm	N/A	2D Proj	Ex-Situ
LEEM / SPECS P90	~ 5 nm	Atomic	Surface	In-situ
FIB / FEI 200	~ 10 nm	~ 10 nm	Surface, 3D [e]	Ex-situ
AFM / Multiple	~ 1 nm	Atomic	Surface	Ex-Situ

Table 1. Imaging methods employed in this work. [a] Lateral resolution is in the plane of the sample surface. [b] Vertical resolution is perpendicular to the sample surface. [c] "2D Projection" is through a thin foil, "Surface" is for images sensitive to near surface monolayers only, "3D" is for a full tomographic reconstruction; [d] "In situ" means growth and imaging occur in parallel, "Ex situ" means imaging occurs in a different chamber after growth is completed. [e] For further details see reference [10].

3. Templating of Ge Quantum Dots

The ability to control the nucleation of individual Ge quantum dot clusters by pre-patterning the substrate by FIB features is demonstrated in Figure 1(b,c). Using Ga^+ ion pulses of order 25 keV energy, 10 pA current and 100 μs duration, a strong correspondence is observed between the initial template sites and the final positions of Ge clusters in the final array.

We have previously examined the fundamental mechanisms that govern this templating process [6,11-13] and concluded that the key process is the development of a transient morphology on the surface that evolves during post implant annealing. The initial implant dose in these particular experiments is insufficient to create substantial surface topography (the ion dose within the templating features is of order 10^{14} cm^{-2}, and the sputter yield is of order 2-3 for these conditions). In fact, under these ultra-low dose implant conditions, the surface actually swells due to damage and amorphization within the implanted volume. However, a fraction of a monolayer of the Si surface is sputtered away. On annealing, these "surface vacancies" aggregate to form initially a shallow annulus within each implant feature, which then evolves with further annealing into a surface pit, and then the surface eventually replanarizes [6,13]. If the surface nanostructure evolution is terminated at the pit stage (by stopping the anneal or initiating growth), then a strong correlation is observed between the position of the original template sites and the position of the final QDs.

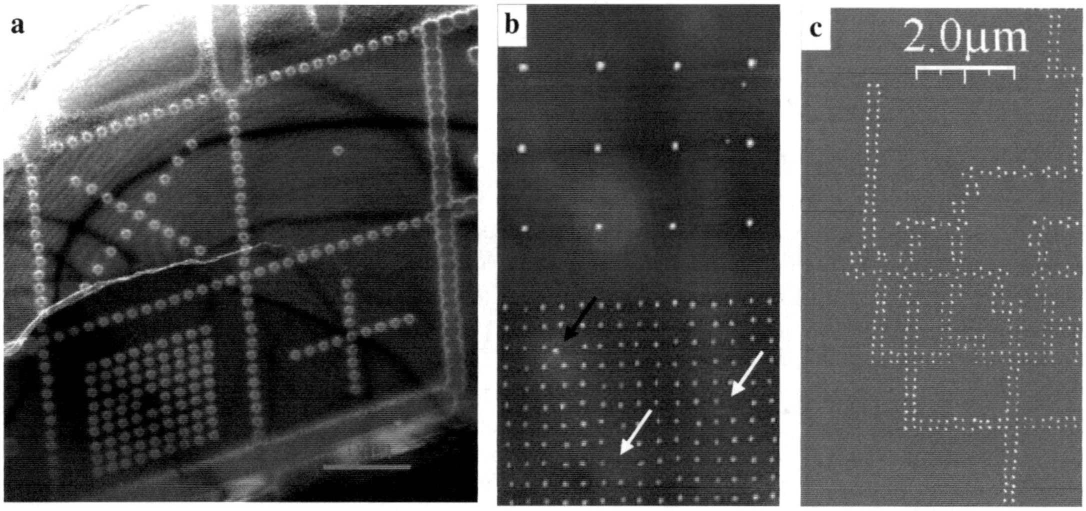

Figure 1. (a) As implanted features for 25 keV 10 pA 100 μs Ga^+ pulses imaged in weak beam in the UHV - TEM. (b) and (c) AFM images of examples of Ge QD arrays templated to the FIB sites. In (b) "vacancy" sites are indicated with white arrows and a "Frenkel defect" (interstitial QD adjacent to an unoccupied template site) by a black arrow. The pattern in (c) corresponds to part of the arrangement of a simple quantum cellular automata (QCA) adder circuit.

We quantify this correlation as follows. Let N_Q be the total number of quantum dots within a given area of substrate surface (we avoid the term "unit area" here, as the arrays we seek to fabricate may be of greatly varying local density, consistent with circuit architectures), and let N_t be the total number of template sites in the same area. Let F_n be the fraction of template sites that are occupied by exactly n quantum dots. Thus F_0 corresponds to the fraction of vacant sites, F_1 corresponds to the fraction of template sites that are occupied by exactly one quantum dot (generally the preferred state), and $F_{n>1}$ corresponds to the fraction of multiply occupied sites (n=2,3,4..etc.). Finally let N_i be the number of interstitial (i.e. not associated a template site) quantum dots within the given area. Then:

$$N_Q = N_t \, \Sigma \, nF_n + N_i \tag{3}$$

In general our goal is to suppress the terms F_n ($n \neq 1$) and N_i over the widest possible range of spatial frequencies, $F(u,v)$. The ability to achieve this goal will depend on both the energetic and kinetic landscapes encountered by surface adatoms during epitaxial growth. The energetic landscape will be determined largely by the strength of attachment of adatoms to the template sites, relative to the strength of attachment to other possible nucleation sites. The kinetic landscape will be dominated by adatom diffusion kinetics on the surface, which will be functions of substrate temperature and partial pressure of the depositing species.

Suppression of the N_i term requires suppression of other heterogeneous nucleation sites, and the presence of an energetic landscape that suppresses spontaneous nucleation on the surface. In the present system, the latter condition is achieved by controlling the wetting layer thickness. For unmodified Si(100) surfaces the wetting layer thickness is about 4 monolayers, depending on growth conditions. Local modification with the Ga^+ focused ion beam reduces the wetting layer thickness to about 2 monolayers [13]. Thus, if the net amount of Ge deposited is between 2 and 4 monolayers, nucleation occurs on template sites, but not spontaneously in between. This in principle eliminates the N_i term. Suppression of higher order (n>1) F_n terms requires tuning of the template dimensions to the QD size. In development of the template morphology by post-implantation annealing, we find that growth on features that have evolved to the annular stage can result in multiple nuclei per template site (n>1), while growth at the nano-pit stage corresponds to the n=1 condition [6,13]. The final factor in equation (3) is the n=0 or vacancy state. This is the dominant "defect" in our QD arrays (see for example Figure 1b), with populations that grow with increasing spatial frequency of the template sites, as will be subsequently discussed.

In an extensive series of experiments [14], we have examined the filling of templates sites as function of spatial frequencies in the pattern. For these experiments, we maintain a constant number of template sites per unit area, but vary the maximum spatial frequencies by adjusting the inter-dot spacings within templates of four quantum dots in a square lattice as illustrated in Figure 2. It is observed that as the inter-dot spacing d becomes smaller, the fidelity of pattern transmission decreases dramatically.

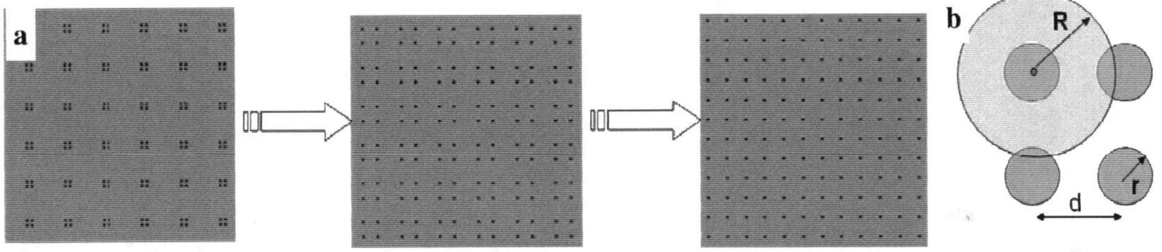

Figure 2. (a) Schematic of template array designed to investigate template filling as a function of template site separation. The spacing between QDs within a four-fold motif, d, varies but the total number of template sites per unit area of substrate remains constant. (b) Illustration of the simplified "capture length (R)" model. r is the radius of an individual template feature.

Figure 3(a) shows how as the spacing between features in the four fold cluster reduces, the fraction of "vacancy sites" substantially increases. This may be understood on the basis of the concept of a *capture zone*, that bridges the length scales between the lithographic pattern and the local propensity for self assembly of the Ge quantum dots. The capture zone, of radius R (capture length), is illustrated in Figure 2(b). The basic assumption of this model is that the nucleation of an island inhibits further nucleation within the capture length R (the diffusion length of Ge adatoms is expected to be large compared to the site separation under the growth conditions employed in our experiments [15]). Initially, we allow islands to nucleate anywhere with equal probability within circles of radius r centered on each nominal FIB implant site position, Figure 2(b). Once this first island has nucleated within its circle of radius r, it is then assumed that no other islands can nucleate within a radius R centered at the first island. Thus, the exclusion area can partially overlap with the neighboring nucleation sites, and the probability of another island nucleating on an adjacent site is assumed to be proportional to the remaining available area. As described in Reference 14 and illustrated in Figure 3(b), this model can reproduce the fundamental feature of our experimental observations.

Interestingly, as described in Reference [14], the transition from the low to high vacancy regions for a given capture length R occurs at template site spacings that are largely independent of the template feature size, r. Thus the challenge with increasing the fidelity of the templating process at high spatial frequencies is not predominantly a factor of improving r. This parameter (determined primarily by the quality of the focused ion beam optics in our experiments and the feature sizes that evolve at each template point during post implantation annealing [13]) controls the accuracy in positioning the nuclei (as it is assumed the initial nuclei can form anywhere within the template site) but not how closely they can be spaced. The minimum attainable spacing is primarily controlled by *R*. In comparing the results of our model to experimental observations, the best agreements between model and experiment occurs for values of *R* that are of order 100 nm. While the magnitude of *R* is expected to be governed by the same fundamental parameters, it does not appear to be equivalent to the adatom surface diffusion length, which from values in the literature (e.g. [15]) and from the fact that we can accurately populate much sparser template arrays (lower spatial frequencies), is considerably larger. Rather it appears to represent a critical length scale over which the local gradients in chemical potential are sufficient to drive Ge adatoms towards the clusters. However, given the finite time and temperature scales of these experiments, kinetic parameters clearly play a substantial role. For example, increasing the digermane partial pressure during growth reduces the "lifetime" of an adatom on a terrace before it encounters high densities of other atoms on the surface. This increases the probability that nucleation occurs in parallel on multiple closely spaced sites, and reduces the capture zone radius. As described in Reference [14], this indeed increases the fidelity of island attachment to template sites at higher spatial frequencies.

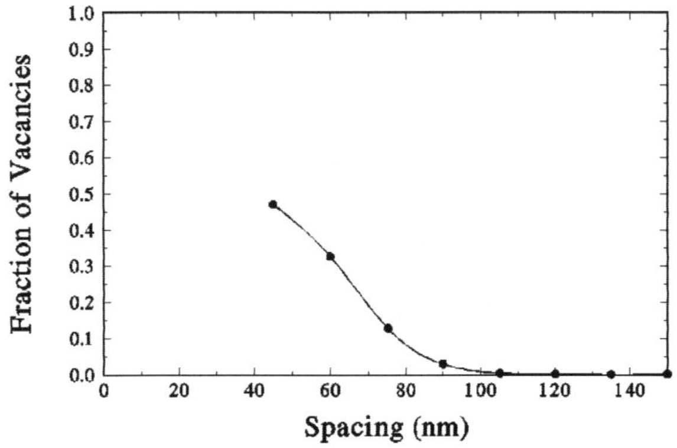

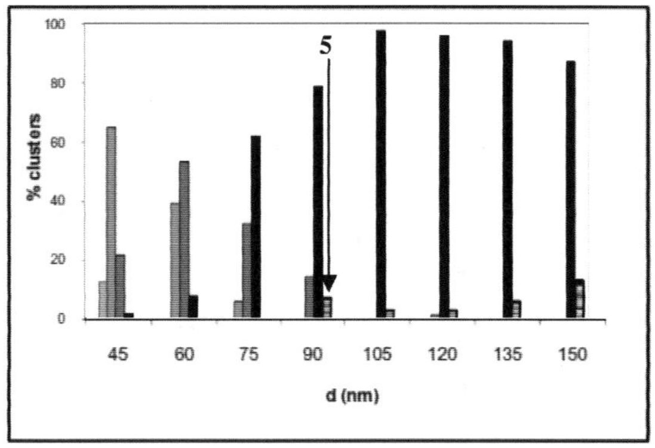

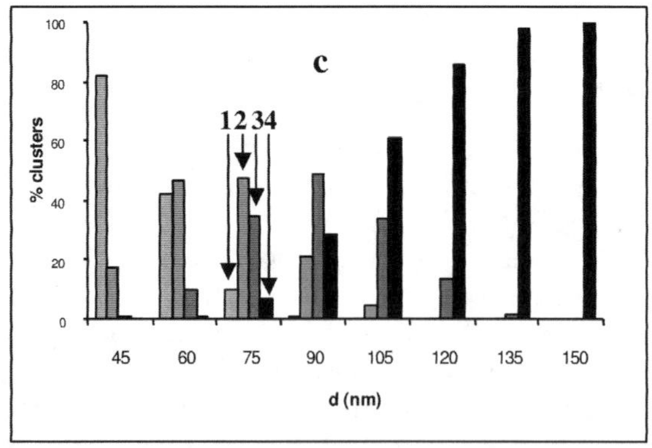

Figure 3. (a) Fraction of unoccupied template sites ("vacancies") by Ge QDs for 0.3 ms, 10 pA, 25 keV FIB template pulses, growth at 500° C, 1.6 x 10^{-7} Torr digermane . (b) Filling of four fold template patterns by 1,2,3,4 or 5 QDs. Upper plot corresponds to experimental data of (a). Lower plot corresponds to the model of Figure 2(b) with *r*= 30 nm, *R* = 90 nm. [14]

Another key parameter will be the diffusivity of adatoms between template sites. Of course, this is controlled partially by temperature, but it will also be controlled by surface structure. For example the Si(100) 2x1 surface reconstruction is known to have different adatom diffusivities parallel and perpendicular to the dimer rows [16]. The net adatom diffusivity will also depend critically upon the surface step density / terrace size between template sites, and recent LEEM results in our group suggest that this may be a potential route to controlling the size of the capture zone. Shown in Figure 4 is a LEEM image of the surface structure of a Si(100) surface that has been patterned with a Si

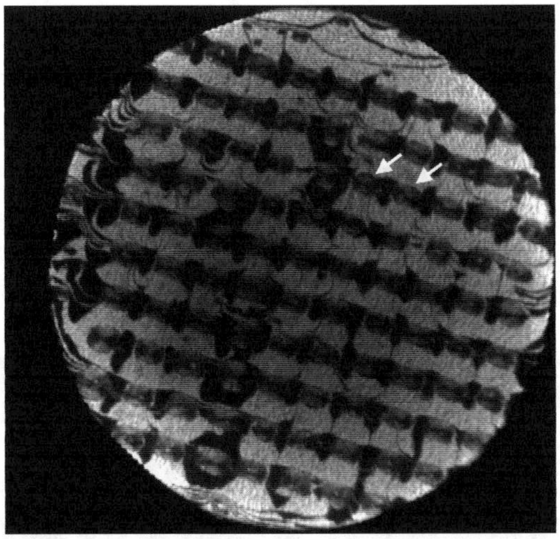

Figure 4. Dark field LEEM image (aperture around ½,0 reconstruction spot) from 2x1 reconstructed Si(100)surface which has been periodically patterned with pits (arrowed) by a Si focused ion beam. Field of view is 5 μm.

focused ion beam and then annealed to temperatures ~ 850° C in the LEEM. This temperature is determined from observation of the LEED pattern to be just sufficient to generate a reconstructed surface, without creating so much surface mobility that the FIB features planarize. Note also that the Si ion doses per FIB feature are much higher than we typically use for Ga FIB templating of Ge quantum dots. The images are formed by placing an aperture around one of the orthogonal 2x1 reconstruction reflections, such that terraces with the dimer rows aligned to diffract into the selected diffraction spot appear bright, and adjacent terraces with the dimer rows rotated 90 degrees appear dark. The boundaries between the bright and dark terraces thus represent monatomic steps. It is seen that large terraces extend between adjacent template sites, with very low step densities between the template sites. This is believed to be due to the pits acting as efficient sinks for surface steps, as previously demonstrated [17], and should provide the ability to "tune" adatom diffusivity by controlling step densities between template sites, through the length and temperature of annealing steps and through the initial pit dimensions (which will control its ability to absorb steps).

We can thus summarize our progress to this point by saying that the interplay between surface energetic, kinetics and lithographically induced topology currently limits the maximum attainable QD spatial frequencies in these patterns comprising a single QD element on each lattice basis point to $u^{n}_{max} \sim 10^7$ m^{-1}. The minimum attainable spatial frequency, u^{n}_{min} is controlled by the suppression of off-site "interstitial" nucleation between template sites that arises from the reduction of the Stranski Krastanov wetting layer thickness as a function of the implanted Ga density [13], as discussed earlier in this manuscript. For regularly spaced arrays of template sites, the wetting layer thickness changes from 1.75 ± 0.25 monolayers for a template site center-center spacing of 180 nm to 2.25 ± 0.25 monolayers for a template center-center spacing of 360 nm to 3.75 ± 0.25 monolayers for an unmodified Si (100) surface. The corresponding average Ga doses per unit area of substrate are 2x10^{13} cm^{-2}, 2x10^{12} cm^{-2} and 0. While it is somewhat surprising that such modest doses (only a fraction of which will, of course, reside at the surface) can modify the wetting layer thickness, this phenomenon is of great advantage in that by growing between about 2 and 4 monolayers (depending upon the exact template feature density), nucleation occurs on the template sites, but not on unpatterned regions. In principle, this extends the minimum transferrable spatial frequency down to zero, although the difference between wetting layer thicknesses for patterned and unpatterned features will become vanishingly small as the template feature density / dose is sufficiently reduced. However, in practice we find that nucleation is suppressed over length scales of at least microns to tens of microns, which in terms of the interaction distances for the mechanisms relevant to the nanoelectronic device architectures we are considering (spin exchange, tunneling etc) is essentially infinite.

In summary, lithography can be used to template Si(100) surfaces over spatial frequencies:

$$0 \leq u^{i} \leq 10^{8}\ \mathrm{m}^{-1} \tag{4}$$

The resulting transfer function limits the range of attainable spatial frequencies to:

$$0 \leq u^{n} \leq 10^{7}\ \mathrm{m}^{-1} \tag{5}$$

This attenuation of higher spatial frequencies is due to the characteristic length scales of self assembly, specifically the effective capture cross-section for adatoms of a nucleated cluster.

4. Templating of Complex Nanostructures: The Quantum Dot Molecule

For the nanoelectronic device architectures we are considering, the relevant interaction lengths (i.e. for spin exchange/interaction or electron/hole tunneling) are substantially less than the minimum QD spacings we can reliably transfer through the templating processes described in section 3. As a solution, we seek to assemble more complex nanostructures with the necessary length scales built into each component through self-assembly mechanisms. To develop the analogy of a crystallographic lattice, we would then be realizing the equivalent of a complex basis on each lattice point.

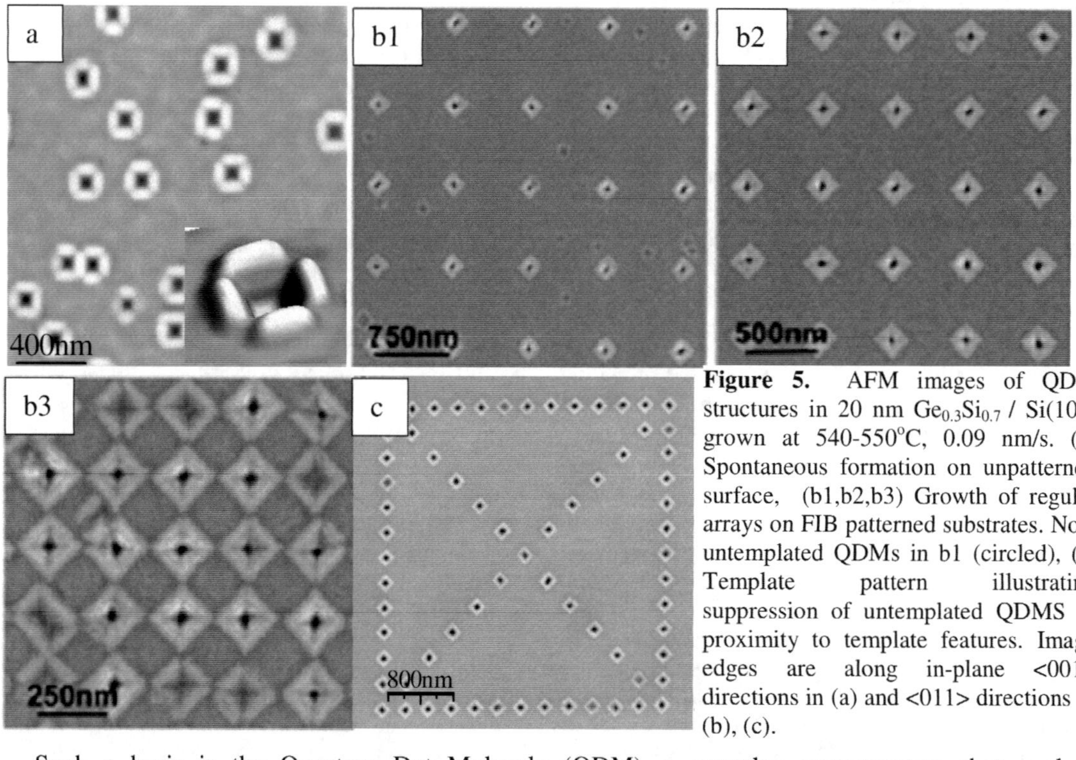

Figure 5. AFM images of QDM structures in 20 nm $Ge_{0.3}Si_{0.7}$ / Si(100) grown at 540-550°C, 0.09 nm/s. (a) Spontaneous formation on unpatterned surface, (b1,b2,b3) Growth of regular arrays on FIB patterned substrates. Note untemplated QDMs in b1 (circled), (c) Template pattern illustrating suppression of untemplated QDMS in proximity to template features. Image edges are along in-plane <001> directions in (a) and <011> directions in (b), (c).

Such a basis is the Quantum Dot Molecule (QDM), a complex nanostructure that evolves spontaneously during growth under conditions of limited adatom mobility in the Ge_xSi_{1-x} / Si(100) system [18-21]. It comprises a four fold set of quantum dots bound elastically to a central pit, Figure 5(a). The evolution of the QDM follows the following trajectory (illustrated with specific numbers for growth by Molecular Beam Epitaxy growth of $Ge_{0.3}Si_{0.7}$/Si(100) at 550° C and a growth rate of about 1 Angstrom s^{-1}). (i) First, the $Ge_{0.3}Si_{0.7}$ growth is planar up to a wetting layer thickness ~ 5 nm. (ii) At this point, strain relieving pits start to form, which are shallower than the film thickness and are thus assumed to not be directly associated with contamination at the original growth surface. (Although morphological relief of strain in compressively stressed films is more generally achieved through islanding, pits are also effective strain relief mechanism, e.g. [22]). (iii) With further film growth, the pit widens and deepens onto {510} facets and quantum dots start to grow at each pit edge. (iv) The quantum dots elongate along the edge of the pit and eventually form a continuous wall surrounding the pit. (v) At this point lateral growth of the QDM stops, and {510} facets define both the pit and QDs. On further growth, the structure maintains its shape and dimensions on the surface until the film thickness is large enough for formation of interfacial misfit dislocations. The QDM is then de-stabilized. For these specific film stoichiometries and growth conditions, the QDM size measured between outer edges of opposite QDs is 220 nm. This dimension scales with strain to produce increasingly smaller QDMs at higher Ge concentrations (e.g. 120 nm at $Ge_{0.5}Si_{0.5}$). X-Ray synchrotron experiments have demonstrated that the QDMs are Ge rich with respect to the surrounding matrix, and will thus confine carriers (holes) due to lower bandgap. [23]

While on an unpatterned surface, these nanostructures form quasi-randomly at densities ~ 10^8 cm^{-2}, we can template their formation by FIB patterning to "seed" the initial pits. Because we have no FIB column integrated with the MBE system in which these structures are grown (and because the growth rates and compositions are not accessible to the CVD processes in the UHV-TEM where the quantum dot arrays described in section 2 were grown), we FIB template ex-situ to the growth chamber, i.e. before substrate cleaning and growth. This process requires parallel optimization of many parameters [21]: the size of the ex-situ FIB seeding pits, the subsequent wafer cleaning conditions, the initial Si buffer layer growth conditions, and the final Ge_xSi_{1-x} growth conditions. We have established conditions where initial Ga^+ FIB template features of just a few nm deep can survive the substrate cleaning process, propagate through subsequent Si buffer layer and Ge_xSi_{1-x} film growth, and successfully template QDM structures.

Separation between individual QDs within the cluster can be controlled by arresting growth at the appropriate point of QD elongation along the pit edges, and we can thereby control the separation to be ≤ 10 nm, i.e. $u^n_{max} > 10^8$ m^{-1}. Equally, as the QDMs evolve to a narrow distribution of shapes and sizes [19], we can accurately control the spacing between QDMs, as illustrated in Figure 5(b).

The major limitation associated with QDM templating is that while we can successfully template QDMs with very high fidelity (i.e. the vast majority of templated sites result in growth of an associated QDM and there is thus a low "vacancy" concentration), there is no mechanism for suppressing off-site nucleation of QDMs ("interstitials") as they will continue to form randomly in unpatterned regions. Thus as illustrated in Figure 5(c) we can typically suppress QDM formation only within adatom diffusion lengths / capture lengths of templated sites. This limits the minimum accessible spatial frequency, u^n_{min}, in the FIB-templated QDM patterns. In practice this untemplated nucleation typically (but not always) occurs at length scales of order a μm or greater from template sites, and would then not interfere with operation of the intended nanoelectronic architecture. However, this lack of a mechanism inhibiting off-site nucleation is a limitation to using QDMs as a basis on the template lattices we construct.

In summary of this section, for lithographic templating over spatial frequencies:

$$0 \leq u^i \leq 10^7 \text{ m}^{-1} \tag{6}$$

We can assemble structures with a range of spatial frequencies

$$10^6 \leq u^n \leq 10^8 \text{ m}^{-1} \tag{7}$$

In this case control of self assembly enables extension of the maximum spatial frequencies that can be accessed in the final array, but the lack of a mechanism for suppressing nucleation off template sites limits the minimum spatial frequencies that can be transferred from the template pattern.

5. Summary and Continuing Work

We have demonstrated how focused ion beam templating of Si(100) surfaces enables controlled nucleation of Ge(Si) nanostructures in patterns of any desired complexity and with high fidelity transfer of a broad range of spatial frequencies. The maximum spatial frequency – which is the more critical limit for nanoelectronic device architectures - that can be transferred from template pattern to nanostructure array depends on subtle interplays between the length scales forced by the template pattern and the mechanisms driving cluster self assembly on the surface during growth. Depending on this interplay the maximum spatial frequency in the nanostructure array can be either enhanced or attenuated with respect to that contained within the template array. The minimum spatial frequency is largely dictated by the presence or absence of a mechanism that inhibits nucleation off the template sites. For the nanostructures discussed here, spatial frequencies from essentially zero to 10^8 m^{-1} can be designed into the final nanostructure pattern.

Our current work is focusing on the electronic and magnetic functionalization of these nanostructures. The work reported in this paper uses a Ga^+ focused ion beam to template the surfaces.

The use of Ga^+ is virtually ubiquitous in conventional liquid metal ion source (LMIS) FIBs because of its low melting point and low vapor pressure at the melting point. However Ga^+ is a shallow acceptor in Si and Ge, and will thus dope the substrate and QDs wherever it is present. For any practical fabrication of prototype nanoelectronic devices, we need to be able to template with electronically non-invasive species (such as Si or Ge) and to be able to intentionally dope with nanoscale precision to introduce extra carriers or spins into the system (e.g. with Mn, B, or As). Using a mass-selecting FIB column from Orsay Physics [9], we have succeeded in generating beams of Si, Ge, Mn, As and B from AuSi, Mn(Au)Ge and PdAsB sources respectively. To date beam resolution is of order 30-50 nm (limited primarily by vibrations in our home-built system) and we can deliver ion pulses from a few ions to numbers large enough for any templating or doping purposes. We are also exploring in detail structural, optical and electronic recovery of FIB implanted and annealed structures to understand recovery and activation in these structures.

Acknowledgements
We acknowledge invaluable collaborations and discussions with A. Berghaus (SPECS), M. Kammler (University of Essen), L. Bischoff (Insititute for Ion Beam Physics and Materials Reseach, Dresden), M. Copel and M. Reuter (IBM). We would like to acknowledge NSF-MRSEC, NRI SRC and IBM for supporting this work.

References

[1] Amlani I, Orlov AO, Toth G, Bernstein GH, Lent CS and Snider GL 1999 *Science* **284**, 289
[2] Orlov A O, Kummamuru R, Lent CS, Bernstein GH and Snider GL 2003 *Surface Science* **532**, 1193
[3] Hatano T, Stopa M, Yamaguchi T, Ota T, Yamada K and Tarucha S 2004 *Phys. Rev. Lett.* **93**, 066806:1-4
[4] Floro JA, Chason E, Freund LB, Twesten RD, Hwang RQ and Lucadamo GA 1999 *Phys. Rev. B* **59**, 1990
[5] Hammar M, LeGoues FK, Tersoff J, Reuter M C and Tromp RM (1995) *Surface Science* **349**, 129
[6] Portavoce A, Hull R, Reuter MC and Ross FM 2007 *Phys, Rev. B* **76**, 235301:1-12
[7] http://www.specs.de/cms/front_content.php?idcat=84
[8] Ishizaka A, Nakagawa K and Shiraki Y 1982 *Proc. 2nd International Symposium on MBE* (Japan. Soc. Apply. Phys., Tokyo, Japan) 183
[9] http://www.orsayphysics.com/fibexb.htm
[10] Kubis AJ, Vandervelde TE, Bean JC, Dunn DN and Hull R 2006 *Appl. Phys. Lett.* **88**, 263103:1-3
[11] Kammler M, Hull R, Reuter MC, and Ross FM 2003 *Appl. Phys. Lett.* **82**, 1093
[12] Portavoce A, Kammler M, Hull R, Reuter MC, Copel M and Ross FM 2004 *Phys. Rev. B* **70**, 195306:1-9
[13] Portavoce A, Kammler M, Hull R, Reuter MC and Ross FM 2005 *Nanotechnology* **17**, 4451
[14] Gherasimova M, Hull R, Ross FM and Reuter M 2008 *Appl. Phys. Lett.***93**, 023106:1-3
[15] Schwarz-Selinger T, Foo Y L, Cahill DG and Greene JE 2002 *Phys. Rev. B* **65**, 125317
[16] Mo YW, Kleiner J, Webb MB and Lagally MG 1992 *Surface Science* **268**, 275
[17] Blakely JM, Tanaka S and Tromp RM 1999 *J. Elec. Micr.* **48**, 747
[18] Gray JL, Hull R and Floro JA 2002 *Appl. Phys. Lett.* **81**, 2445
[19] Gray JL, Singh N, Elzey DM, Hull R and Floro JA 2004 *Phys. Rev. Lett.* **92**, 135504:1-4
[20] Gray JL, Atha S, Hull R and Floro JA *Nanoletters* **4**, 2447
[21] Gray JL, Hull R and Floro JA 2006 *J. Appl. Phys.* **100**, 084312:1-7
[22] Tersoff J and LeGoues FK 1994 *Phys. Rev. Lett.* **72**, 3570
[23] Leite MS, Gray JL, Hull R, Floro JA, Magalhaes-Paniago R and Medeiros-Ribeiro G 2006 *Phys. Rev. B***73**, 121308:1-4

Exploring semiconductor quantum dots and wires by high resolution electron microscopy

SI Molina[1], PL Galindo[2], L Gonzalez[3], JM Ripalda[3], M Varela[4] and SJ Pennycook[4]

[1] Departamento de Ciencia de los Materiales e Ing Metalúrgica y Q. Inorgánica, F. de Ciencias, Universidad de Cádiz, Campus Río San Pedro. 11510 Puerto Real (Cádiz), Spain
[2] Departamento de Lenguajes y Sistemas Informáticos, CASEM, Universidad de Cádiz, Campus Río San Pedro. 11510 Puerto Real (Cádiz), Spain
[3] Instituto de Microelectrónica de Madrid (CNM, CSIC), Isaac Newton 8, 28760 Tres Cantos, Madrid, Spain
[4] Materials Science and Technology Division, Oak Ridge National Laboratory, Oak Ridge TN 37831, USA

E-mail: sergio.molina@uca.es

Abstract. We review in this communication our contribution to the structural characterisation of semiconductor quantum dots and wires by high resolution electron microscopy, both in phase-contrast and Z-contrast modes. We show how these techniques contribute to predict the preferential sites of nucleation of these nanostructures, and also determine the compositional distribution in 1D and 0D nanostructures. The results presented here were produced in the framework of the European Network of Excellence entitled "Self-Assembled semiconductor Nanostructures for new Devices in photonics and Electronics (SANDiE)".

1. Introduction

The development and implementation of methods to extract structural and compositional information of semiconductor nanostructures at the nanoscale, and in some cases at the atomic scale, is a critical issue for improving their design and potential use in functional devices. These methods should help us to characterise the shape, size, composition and strain distribution of the nanostructures, in order to be able to know and relate these characteristics with their functional properties. Concerning our contribution to these methodologies, we present in this communication some methods based on high-resolution phase-contrast and Z-contrast electron microscopy techniques, and how they are applied to improve the measurement of the above mentioned structural characteristics of self-assembled semiconductor nanostructures (quantum dots and wires) and our capability to predict their preferential sites of nucleation.

With regards to strain measurements at the nanoscale, we have to mention that significant improvement has been reached in the last fifteen years, by geometric phase [1-2] and peak finding [3-9] methods, from the analysis of high resolution transmission electron microscopy (HRTEM) images. In the geometric phase method, local strain field components are calculated by derivation of the displacement obtained from two non-collinear components of the HRTEM image Fourier transform. In the case of peak finding methods, the displacement field is determined from a two-dimensional lattice built up from the set of intensity maxima in the HRTEM image, by measuring its distortion taking as a reference another two-dimensional lattice of an unstrained material of known lattice parameters. These methods can be applied to nanostructures and good results can be obtained, provided that finite element calculation of surface relaxation effects is taken into account.

On the other hand, the development of aberration correctors [10-11] has broadened the possibilities that both phase [12] and Z-contrast [13] high resolution electron microscopy offer for analysing materials with unprecedented high spatial resolution. Nowadays, electron probes with sub-angstrom size are becoming conventionally used. This characteristic, in combination with higher probe currents and better signal-to-noise ratios, larger probe forming and collector apertures, and improved depth resolution, converts the aberration-corrected electron microscopes into powerful analytical tools [14].

The use of these methodologies has demonstrated to be an excellent approach, with the support of other analytical and theoretical tools, to improve the understanding and design of self-assembled semiconductor nanostructures. We have developed both phenomenological [15] and simulation [16] approaches to determine the composition with atomic column resolution from aberration-corrected high-angle annular dark-field (HAADF) scanning transmission electron microscopy (STEM) images, that can be applied to a wide range of materials problems, ranging from compositionally homogeneous phases to heterogeneous nanostructures including millions of atoms. The strain distribution of nanostructures can be determined from the obtained compositional maps, and compared with strain maps obtained by analysing phase-contrast HRTEM images. With regard to this aspect, we have created the Peak Pairs algorithm [17], a new procedure to extract quantitative strain maps from HRTEM images working in real space.

The application of these approaches in combination with other analytical methods has allowed us to successfully predict the nucleation sites of stacked quantum wires for different growth conditions [18-19], learning about the nucleation mechanism of these nanostructures [20] and the defect nucleation mechanism in quantum dots [21]. Good correlation has been found in these nano-wires between the strain maps obtained directly from the strain analysis of HRTEM images and the strain calculated by finite element methods using the compositional maps determined by HAADF-STEM and electron energy-loss spectroscopy (EELS) [22]. In relation to the EELS technique, we have to mention that the analysis of the low-loss signal of the spectra has demonstrated to be particularly powerful for the analysis of quantum dots, as a technique complementary to HAADF-STEM. This combination of techniques helped us to discover the occurrence of a GaSb-InAs intermixing process in GaSb capped InAs quantum dots [23].

Exciting new methodologies for 3D imaging of nanostructures are now emerging due to the improved depth resolution that aberration-corrected electron microscopes offer. Individual atoms have been detected inside the bulk of materials by using aberration-corrected STEM [24-25] and optical sectioning is more extensively used during the last few years [26-27]. This approach is usually applied to heavy atoms or clusters embedded in a light scattering matrix, as for example a light amorphous material. However, with the assistance of image simulation, we have been able to detect heavy atoms located at particular interstitial configurations in crystalline nanowires [28]. We also will show in this communication some results and prospects for 3D information extraction from semiconductor nanostructures. The possibility of working in true confocal conditions in a double aberration-corrected electron microscope [29-30] is another strategy that is giving its first results and appears to be very promising to fulfil this aim.

2. Peak Pairs method

High-resolution transmission electron microscopy (HRTEM) combined with image analysis can be used to analyze materials at the atomic-column level, measuring local displacements and strain, and revealing lattice translations, dislocations and/or rotations. Strain maps can be used to identify defects and/or quantify distortions in the image with respect to a reference area, offering the researcher a valuable tool to locate different phases or analyze defects and even to rectify distorted images. Due to some effects, such as thin foil relaxation, local crystal tilts and thickness and/or composition variation across the material, the reliability of strain mapping measurements is somewhat problematical. Only when all these factors are taken into account does the method offer a reliable way of extracting quantitative strain information at an atomic scale.

This restriction arises because strain determination is based on the assumption of a direct relationship between the intensity maxima in the image and the position of atomic columns in the sample, and so great care should be taken to ensure that a proper analysis is applied. This assumption may be too restrictive, but in some cases, the use of finite element modeling for predicting sample behaviour when it is subject to deformation has been shown to produce good results.

Several methodologies have been described in the literature [9] to determine strain field mapping from HRTEM images, in either real or Fourier space. The Fourier transform fringe analysis was introduced by Takeda et al. [31] and adapted to 2D images by Bone et al. [32], along with a set of very interesting observations regarding the inherent errors of the technique and some suggestions about how these could be overcome. This technique was applied to strain mapping under the name of Geometric Phase Analysis (GPA) [2]. While GPA has become a powerful tool to study the displacements in HRTEM images, it cannot be applied to any image. The basic assumption of GPA assumes that the phase varies slowly and this happens only for particular images, like dot-like HRTEM images. This means that in the case of strain gradients, unexpected oscillations appear in strain maps [33-34], and at crystal discontinuities such as interfaces, GPA averages on both sides of the discontinuity and its measurements are not correct [35]. What is more, when more than one atom at each lattice position is present, i.e. cation-anion pairs in [110] oriented wurtzite samples, the method cannot be directly applied, because the use of only two reflections for the determination of strain in GPA is not enough to represent the complexity of the HRTEM image of a strained crystal structure.

In order to overcome these constraints, a recent methodology called Peak Pairs Analysis (PPA) has been developed [17]. As a preliminary step the image is filtered, not only to reduce noise in the image but also to smooth the image around the peaks, allowing us to improve the precision by a factor of 2–4, as it was noted in a previous work [5]. The usual approach is to apply a Wiener-Bragg filter sequence, but other filters, such as low-pass and/or adaptive Wiener filters are also often very useful. A Wiener filter [36] is designed so as to minimize the mean square error, and if applied to an image adaptively, it tailors itself to the local image variance, preserving edges and other high frequency parts of the image. After Wiener filtering, Bragg filtering is usually applied, consisting of the application of a mask around one or more Bragg peaks in the Fourier transform of the image followed by an inverse Fourier transform. The size of the kernel mask determines the range of frequencies that are kept by the filter. The number of kernels is chosen by the user, and depends on the complexity of the analysed image, two being enough for dot-like patterns.

Once the image has been filtered, all the intensity peaks are automatically located by examining the local neighbourhood of each pixel to see if the pixel has a maximum value in the local neighbourhood (usually an 8-connected neighbourhood is enough). Peak position is refined by fitting a 2D quadratic function to grey levels in the neighbourhood of each peak, and solving the resulting equations for the maxima, thus achieving sub-pixel resolution.

The next step is to define a reference area in an unstrained area of the micrograph and determine two non-collinear basis vectors that will define the position of the peaks of maximum intensity in the reference area. Therefore, the image should contain non-deformed areas far away from defects so that they can be taken as a reference.

Then an affine transformation defined using these two basis vectors is applied to all peaks. In the affine transformed space, peaks will arrange forming a square grid of points regularly spaced at a distance equal to unity, both in x and y direction. Then, for each intensity maximum (x,y), we look for its associated peaks, thus forming "pairs of peaks". The application of this transformation allows us to use the Euclidean distance in order to determine the most probable ''partner'' of a given peak, assuming Gaussian probability distributions centred at the peaks and considering equal covariance matrices, thus greatly reducing the appearance of errors in "peak pairs" identification.

The identification of pairs of peaks along two non-collinear directions enables us to determine precisely the displacement and strain at each peak, and by cubic interpolation in 2D, displacement and strain at any desired point in the image is calculated.

The accuracy of the strain map strongly depends on the accuracy of the basis vectors of the undistorted structure. The original basis vectors can be refined for a specified reference region with high accuracy by adjusting the initially estimated displacement maps. Thus, the displacement field, and hence the strain field, can be evaluated in terms of the selected reference structure in the image. At this point, by defining a region corresponding to a reference structure, we can update the base vectors and get refined displacement maps. The reference area should be taken on the same image, but far away from deformed regions. Using the refined displacement maps, the strain field is calculated with high accuracy after some mathematical treatments as described in [17].

Fig. 1 shows an example of application of the PPA method to an InAs/GaAs (001) interface. The green colour in the strain map (Fig. 1b) corresponds to the unstrained GaAs area, which is also the material defined as the reference and set to 0% strain. The orange area corresponds to the InAs material showing about 7% strain in reference to GaAs along the *y* direction, which is perpendicular to the InAs/GaAs interface.

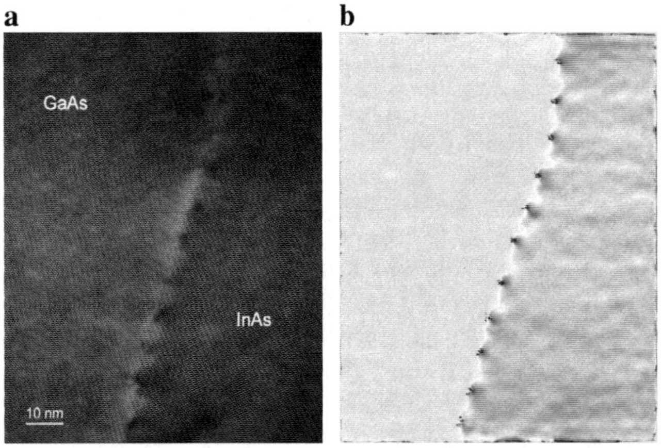

Figure 1. (a) HRTEM image of an InAs/GaAs (001) interface. (b) Component ε_{yy} of the strain tensor showing the presence of two well distinguished materials (GaAs and InAs) and a misfit dislocation network just at the InAs/GaAs interface (reproduced from Microscopy and Analysis, **23** (2009) 23-25, with permission of the copyright owner).

The strain field obtained by applying strain mapping has been compared to that calculated by solving the anisotropic elastic theory equations by Finite Element Analysis (FEA), giving good agreement. In [21], the correct application of both methods has allowed the localization of critical strain zones in semiconductor In(Ga)As/AlGaAs quantum dot (QD) samples, predicting the nucleation of defects. In [22], experimental strain field maps of stacked nanostructures show a good correspondence with the strain values calculated by FEA. The agreement existing between determined and simulated strain values reveals that these techniques are consistent enough to qualitatively characterize the strain distribution of nanostructured materials. Larsson et al. [37] has also used Peak Pairs to measure the strain distribution in heterostructured wurtzite InAs/InP nanowires, and results are in good agreement with FE calculations and were also validated by comparing pure FE data with strain mapping data from a simulated image. Cushley compared strain in Si/SiGe samples and BST/SRO/MgO capacitor structures using convergent beam electron diffraction (CBED) and Strain Mapping using GPA and Peak Pairs [38].

The strain mapping technique has opened up a number of exciting new possibilities in the field of HRTEM characterization of nanostructured materials. For example, in [39] this technique has been applied to the characterisation of the misfit dislocation network in highly mismatched interfaces of InN/GaN quantum dots, demonstrating the tendency of edge misfit dislocations in high misfit systems to bend towards the free surface, producing a network of threading edge dislocations. Tomioka et al. have analyzed the growth directions of InAs nanowires on Si substrates, showing that misfit dislocations with local strains were accommodated in the interface [40].

Strain mapping is often used not only to identify defects and/or distortions in the image, but also to rectify distorted images. CTEM projector lens systems, STEM scanning systems, CCD cameras or scanners often distort the images they form. When the distortions are systematic, it is possible to eliminate these distortions by measuring them using the image of an unstrained specimen, and using the information contained in the resulting displacement maps to define a warp function, that when applied to a distorted image restores the undistorted one [41-43]. Peak Pairs Analysis has been implemented as a plug-in for DigitalMicrograph, the industry leading software for electron microscopy, and is distributed by HREM Research [44]. Its integration provides the user with complete access to all the other powerful functions and useful features present in DigitalMicrograph.

3. Quantitative Z-contrast imaging

The compositional quantification with high spatial resolution from Z-contrast scanning transmission electron microscopy images can be analysed by phenomenological and theoretical approaches. We have contributed to the advancement of this task by both approaches. In relation to the first approach, we have developed a phenomenological method to determine the composition of materials, with atomic column resolution. This method is based on the analysis of integrated intensities of aberration-corrected Z-contrast images [15].

In order to develop this method, we have studied the dependence of integrated intensities on the composition and specimen thickness in samples with calibrated compositions. For this purpose, a number of thick $InP/InAs_xP_{1-x}$ epilayers were grown by molecular beam epitaxy. The alloy composition and the strain of these epilayers were obtained from the analysis of X-ray diffractograms. The integrated intensities are measured around the anionic columns both from the InP substrate and $InAs_xP_{1-x}$ epilayers, from high resolution Z-contrast images taken from both materials. We have found that working with normalized integrated intensities, defined as the quotient (R) between the integrated intensities in $InAs_xP_{1-x}$ and InP, is very useful because of their low dependence on the specimen thickness and their linear dependence with the alloy composition.

Similar advantages have been obtained when we work with normalized intensities integrated in whole projected unit cells, though this entails a decrease of the spatial resolution because of the stronger influence of intensities coming from neighbouring atomic columns. The measurement of the R quotients column-by-column can be done automatically by software, localising In atomic columns by applying intensity maxima identification methods [17]. Hence, the composition can be determined column-by-column using the experimental dependence of the quotient R on the alloy composition, which has been obtained by analysing intensities measured in several hundreds of atomic columns of Z-contrast images taken from composition-calibrated materials. This method has been applied to determine the composition with atomic column resolution in semiconductor heterostructures and nanostructures. In the case of nanostructures, it is important to notice that the strain influence is an important variable to take into account to estimate composition following this approach. For 2D heterostructures, such as thin quantum wells, this effect is not so important, since the strain responsible for compositional errors, in the case of the $InAs_xP_{1-x}$, is as small as 2%. We have performed Z-contrast image simulations of the studied semiconductor materials and we have been able to explain quantitatively the dependence of the experimental R quotients versus the alloy composition for the used range of experimental thicknesses. This fact opens the possibility to quantify column-by-column the composition of a wider range of alloys following a similar procedure based on Z-contrast image simulations, which can be validated with experimental Z-contrast images taken from adequate calibrated samples.

In order to work with semiconductor nanostructures, the simulation of Z-contrast images gets more complicated because a supercell representing the atomic positions of a real nanostructure and its surroundings has to be built up, and this implies working with a high number of atoms, typically in the order of hundreds of thousands or millions. It is well-known that simulation of Z-contrast images can take a lot of time even for simple structures using a personal computer or a small cluster. Thereby, simulation of Z-contrast images of nanostructures formed by millions of atoms demands calculation time of the order of years.

We have developed parallel software (SICSTEM) that affords Z-contrast image simulations of nanostructures composed of several millions of atoms in a manageable time [16]. Supercells representing nanostructures are usually built using two approaches: (i) refining atomic positions by calculating local atomic displacements using finite element models, and (ii) determining atom positions by minimizing potential energy using molecular dynamics models. Images are calculated taking into account elastic scattering [45] and incoherent thermal diffuse scattering (TDS) [46]. TDS intensity is calculated based on Ishizuka´s [47] and STEM-SIM [48] codes, using the multislice algorithm and a local TDS absorptive potential approach. The calculation of each pixel of the image requires a complete multislice simulation. However, as the intensity of each pixel can be independently calculated, division of tasks in parallel results in a very efficient process. The code has been optimised to run in the supercomputer installed at the University of Cadiz, operating with 320 cores in parallel and achieving a peak performance of 3.75 TFLOPS. Fig. 2 shows an example of a high resolution Z-contrast image of a nano-wire represented by a supercell of 422184 atoms, which has been simulated using the SICSTEM software.

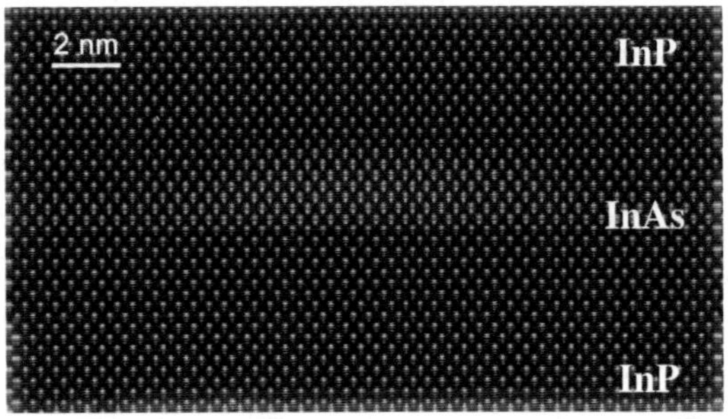

Figure 2. High resolution Z-contrast image (560 x 1024 pixels) simulated with the SICSTEM software for an $InAs_xP_{1-x}$ nanowire (adapted from [16] with permission of the copyright owner).

4. Nucleation of quantum wires

Strain energy at the growth surface is the driving force for nucleation of self-assembled epitaxial $InAs_xP_{1-x}$/InP (001) quantum wires, both for single and stacked layers of these nanostructures. During the epitaxial growth by molecular beam epitaxy of these wires, the existence of intrinsic strain asymmetry built in at the InAs/InP interface under group V element stabilized growth conditions explains the spontaneous formation of quantum wells instead of quantum dots [49]. This intrinsic strain asymmetry explains the formation of wires instead of islands, but it does not explain the location of the nucleation sites where these nanostructures nucleate.

The analysis of atomic force microscopy images of uncapped InAs/InP nanowires during the initial stages of their formation shows that the majority of the nanowires are located on the upper terraces of the growth surface steps close to the step edge. These nanowires are formed by a 2D-3D Stranski-Krastanov growth mode. We have calculated, using anisotropic elasticity theory, the strain at the growth surface on the upper and lower terraces of diatomic steps during the very initial stage (2D) of InAs deposition. After depositing one, two or three InAs monolayers, we found a larger positive strain at the upper terrace surface. The existence of larger lattice parameters at the upper terrace explains why the chemical potential for the further growth of InAs is lower on this terrace, and so why nanowires preferentially nucleate on the upper terrace of the diatomic step just in the 2D-3D transition. This explanation is supported by the analysis of high-resolution Z-contrast images of nanowires located just on $InAs_xP_{1-x}$/InP (001) interfacial steps. This analysis shows that InAs-rich regions are situated on the upper terrace of interfacial diatomic steps, in good agreement with AFM observations and strain calculations [20].

In order to produce efficient materials from $InAs_xP_{1-x}$/InP nanowires, it is important to control their size, shape, composition, distribution and density. The density of these nanostructures can be increased by stacking, along the [001] growth direction, several layers of quantum wires. The above mentioned parameters of the nanowires and their nucleation sites on each stacked layer will depend on the strain field at the growth surface. The InP capped $InAs_xP_{1-x}$ nanowires of each layer underneath each new stacked layer of nanowires will control the distribution of the strain at the InP growth surface. From the analysis of high-resolution Z-contrast images of stacked nanowires, complemented in some cases by the analysis of spatially resolved electron energy loss spectrum images, we have determined the compositional distribution for a number of samples grown at different conditions. These conditions are the thickness and the growth temperature of the InP layers separating each successive layer of stacked nanowires.

Models representing the $InAs_xP_{1-x}$/InP nanowires were built from the obtained compositional maps. From these results, the 3D strain and stress distributions were obtained for each studied sample, by solving the equations of anisotropic elasticity theory by finite element methods. The determined strain and stress distribution allowed us to determine the distribution of these properties at the growth surface for each stacked layer of nanowires, helping us to locate the sites where the chemical potential has minimum values, that is, where new stacked nanowires must nucleate. We have found a very good correlation between the preferential sites of nucleation of stacked nanowires determined by this procedure and the positions of stacked nanowires measured from compositionally sensitive images taken from these stacked nanostructures [18-19]. Fig. 3 presents a graphical scheme of the process followed to find the preferential nucleation sites of stacked epitaxial nanowires. Strain has also been mapped in the same stacked layers of nanowires, by PPA and GPA methods, and we found quantitative correlation with the strain values based on experimental compositional maps [22].

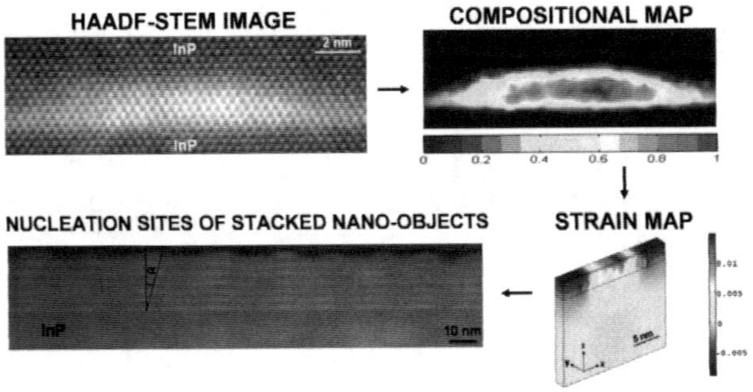

Figure 3. Scheme of the sequential process to predict preferential nucleation sites of stacked InAs/InP nanowires by the analysis of high-resolution HAADF-STEM images (reproduced from [19] with the permission of the copyright owner).

5. Semiconductor quantum dots

We review in this section two examples where analysis of high-resolution Z-contrast images has been a key methodology to understand the formation process and behaviour of quantum dots composed of III-V semiconductors. The first example has to do with GaSb/GaAs quantum dots. This material has a type-II band alignment, which is very attractive for the fabrication of new devices such as quantum information devices and memories. However, incorporation of Sb in GaAs is complicated because it tends to float on the growth surface during the epitaxial growth process. Previous works carried out on epitaxial GaSb/GaAs quantum dots show that the GaAs capping process substantially alters the quantum dots. We have studied, by sub-angstrom Z-contrast imaging, GaAs capped GaSb/GaAs quantum dots grown by molecular beam epitaxy. The use of very high spatial resolution, as in our case, allows resolution of the cationic (Ga) and anionic (As-Sb) atomic columns. Through the statistical analysis of the integrated intensities associated with both atomic columns we have been able to map the Sb-rich regions close to the interface. Fig. 4 shows a column-to-column intensity map obtained from a high-resolution Z-contrast image of a GaAs capped GaSb/GaAs nanostructure. Our analysis provides evidence for a strong reduction of the size of the GaSb nanostructures as a consequence of the GaAs capping process, by comparison of uncapped (by

AFM) and capped (by Z-contrast) nanostructures [50]. This process is understood because Sb segregation and floating happens during epitaxial growth of the capping layer. The strong nanostructural changes occurring in these nanostructures during the capping process have important consequences in their optical behaviour. Optical characterisation of these quantum dots has shown holes to be strongly confined but electrons to be weakly confined [51].

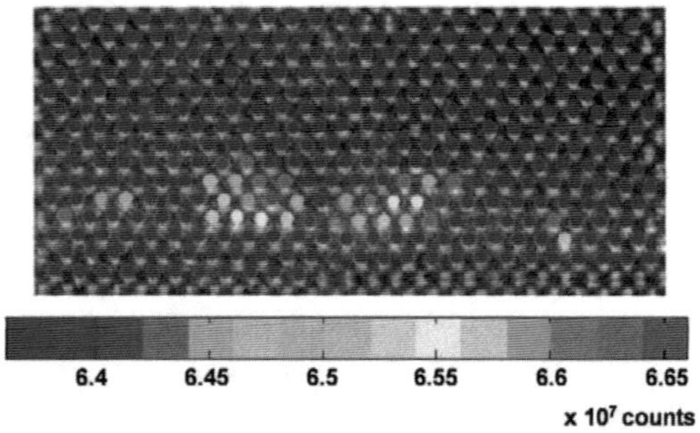

Figure 4. Column-to-column intensity map obtained from a high-resolution Z-contrast image of a GaAs capped GaSb/GaAs nanostructure. Blue circles correspond to atomic-column contrasts that statistically do not differ from GaAs (reproduced from [49] with the permission of the copyright owner).

Compositional analysis with atomic-column resolution has also been demonstrated to be a useful approach to study the effect of Sb irradiation and GaSb capping on InAs/GaAs quantum dots grown by molecular beam epitaxy. Sb incorporation by exposure to an elemental Sb flux or Ga(As)Sb capping on top of InAs/GaAs quantum dots is an interesting approach to tune the electronic structure of quantum dots for optoelectronic devices emitting at telecommunication or even longer wavelengths. Though incorporation of Sb in InGaAs quantum dots appears to be a difficult task, because intermixing in the quaternary InGaAsSb alloy is limited by the existence of a thermodynamical miscibility gap, the existence of strain changes the compositional range in which a significant incorporation of Sb is possible. We have analyzed the ratio of integrated intensities in anionic (As-Sb) and cationic (In-Ga) positions in high resolution Z-contrast images of InAs quantum dots irradiated by Sb or capped with GaSb layers, and we have demonstrated, with the help of low loss EEL spectra, that Sb is incorporated in the QDs [23]. These studies prove the existence of an InGaAsSb quaternary alloy inside the dots, which is related to an intense photoluminescent signal at 1.3 μm, highly interesting for telecommunication applications.

6. Summary

The application of methodologies based on high-resolution Z-contrast imaging, complemented with other analytical techniques, such as spatially resolved electron energy loss spectroscopy, and other methods based on image analysis (e. g. strain determination at the sub-nanometre scale from high resolution transmission electron microscopy images), constitutes a solid approach to improve the understanding and design of semiconductor nanostructures with functional applications.

The advances produced in the last decade in equipment based on aberration-corrected electron microscopes and the improved methodologies for quantitative analysis of high resolution electron microscopy images, let us predict important contributions in the next few years to quantify in 3D the strain and composition in semiconductor nanostructures. These advancements combined with the necessary theoretical inputs (electron microscopy image simulations, ab-initio atomic configurations, electronic and optical properties calculations) will pave the way for the creation of fundamental knowledge and concepts for the new generation of functional nanostructures.

Acknowledgements

This work was supported by the Spanish MCI (TEC2008-06756-C03-02/TEC, Consolider-Ingenio 2010 CSD2006-00019), the CAM (S 0505ESP 0200), the Junta de Andalucia (PAI research groups TEP-120 and TIC-145; projects P08-TEP-03516 and PAI05-TEP-00383) and the Office of Basic Energy Sciences, Division of Materials Sciences and Engineering, US DOE (MV and SJP). The authors would like to thank their collaborators and other co-authors of the articles reviewed here: D. L. Sales, T. Ben, A. M. Beltran, A. M. Sanchez, J. Pizarro, E. Guerrero, M. P. Guerrero, A. Yañez, A. G. Taboada, Y. Gonzalez, D. Fuster, A. Rosenauer, M. F. Chisholm and S. Kret.

References

[1] Hytch M J and Plamann T 2001 *Ultramicroscopy* **87**, 199
[2] Hytch M J, Snoeck E and Kilaas R 1998 *Ultramicroscopy* **74**, 131
[3] Bierwolf R, Hohenstein M, Phillip F, Brandt O, Crook G E and Ploog K 1993 *Ultramicroscopy* **49**, 273
[4] Bayle P, Deutsch T, Pilles B, Lancon F, Marty A and Thibault J 1994 *Ultramicroscopy* **56**, 94
[5] Jouneau P H, Tardot A, Feuillet B, Mariette H and Cibert J 1994 *J. Appl. Phys.* **75**, 7310
[6] Kilaas R, Paciornik S, Schwartz A J and Tanner L E 1994 J. Comput. Assist. Microscopy **6**, 129
[7] Robertson M D, Corbett J M, Webb J B, Jagger J and Currie J E 1995 *Micron* **26**, 521
[8] Seitz H, Ahlborn K, Seibt M and Schroter W 1998 *J. Microscopy* **190**, 184
[9] Kret S, Ruterana P, Rosenauer A and Gerthsen D 2001 *Phys. Status Solidi* (b) **227**, 247
[10] Haider M, Uhlemann S, Schwan E, Rose H, Kabius B and Urban K 1998 *Nature* **392** 768 [11] Batson P E, Dellby N and Krivanek O L 2002 *Nature* **418**, 617
[12] Kisielowski C, Freitag B, Bischoff M, van Lin H, Lazar S, Knippels G, Tiemeijer P, van der Stam M, von Harrach S, Stekelenburg M, Haider M, Uhlemann S, Muller H, Hartel P, Kabius B, Miller D, Petrov I, Olson E A, Donchev T, Kenik E A, Lupini A R, Bentley J, Pennycook S J, Anderson I M, Minor A M, Schmid A K, Duden T, Radmilovic V, Ramasse Q M, Watanabe M, Erni R, Stach E A, Denes P and Dahmen U 2008 *Microsc. Microanal.* **14**, 469
[13] Nellist P D, Chisholm M F, Dellby N, Krivanek O L, Murfitt M F, Szilagyi Z S, Lupini A R, Borisevich A, Sides W H and Pennycook S J 2004 *Science* **305**, 1741
[14] Pennycook S J, Varela M, Chisholm M F, Borisevich A Y, Lupini A R, van Benthem K, Oxley M P, Luo W, McBride J R, Rosenthal S J, Oh S H, Sales D L, Molina S I, Sohlberg K and Pantelides S T 2008 *Scanning transmission electron microscopy of nanostructures. In "Frontiers in Nanoscience and Nanotechnology"* (Narlikar, A. V., ed, pp. 273-286. Oxford University Press, Oxford, U. K.)
[15] Molina S I, Sales D L, Galindo P L, Fuster D, González Y, Alén B, González L, Varela M and Pennycook S J 2009 *Ultramicroscopy* **109**, 172
[16] Pizarro J, Galindo P L, Guerrero E, Yáñez A, Guerrero M P, Rosenauer A, Sales D L and Molina S I 2008 *Appl. Phys. Lett.* **93**, 153107
[17] Galindo P L, Kret S, Sanchez A M, Laval J-Y, Yañez A, Pizarro J, Guerrero E, Ben T and Molina S I 2007 *Ultramicroscopy* **107**, 1186
[18] Molina S I, Ben T, Sales D L, Pizarro J, Galindo P L, Varela M, Pennycook S J, Fuster D, González Y and González L 2006 *Nanotechnology* **17**, 5652
[19] Molina S I, Varela M, Ben T, Sales D L, Pizarro J, Galindo P L, Fuster D, González Y, González L and Pennycook S J 2008 *J. NanoSci. Nanotechnol.* **8**, 3422
[20] Molina S I, Varela M, Sales D L, Ben T, Pizarro J, Galindo P L, Fuster D, González Y, González L and Pennycook S J 2007 *Appl. Phys. Lett.* **91**, 143112
[21] Sales D L, Pizarro J, Galindo P L, Garcia R, Trevisi G, Frigeri P, Nasi L, Franchi S and Molina S I 2007 Nanotechnology **18**, 475503
[22] Ben T, Sales D L, Pizarro J, Galindo P L, Fuster D, González Y, González L, Varela M, Pennycook S J and Molina S I 2008 *Microsc. Microanal.* **14**, (Suppl 2) 344
[23] Molina S I, Sánchez A M, Beltrán A M, Sales D L, Ben T, Chisholm M F, Varela M, Pennycook S J, Galindo P L, Papworth A J and Goodhew P J 2007 *Appl. Phys. Lett.* **91**, 263105
[24] Voyles P M, Muller D A, Grazul J L, Citrin P H and Gossmann H J L 2002 *Nature* **416**, 826
[25] Lupini A R and Pennycook S J 2003 *Ultramicroscopy* **96**, 313
[26] van Benthem K, Lupini A R, Kim M, Baik H S, Doh S, Lee J H, Oxley M P, Findlay S D, Allen L J, Luck J T and Pennycook S J 2005 *Appl. Phys. Lett.* **87**, 034104

[27] Marinopoulos A G, van Benthem K, Rashkeev S N, Pennycook S J and Pantelides S T 2008 *Phys. Rev.* B **77**, 195317
[28] Oh S H, van Benthem K, Molina S I, Borisevich A Y, Luo W D, Werner P, Zakharov N D, Kumar D, Pantelides S T and Pennycook S J 2008 *Nano Lett.* **8**, 1016
[29] Nellist P D, Behan G, Kirkland A I and Hetherington C J D 2006 *Appl. Phys. Lett.* **89**, 124105
[30] Nellist P D, Cosgriff E C, Behan G and Kirkland A I 2008 *Microsc. Microanal.* **14**, 82
[31] Takeda M, Ina H and Kobayashi S 1982 *J. Opt. Soc. Am.***72**, 156
[32] Bone D J, Bachor H A and Sandeman R J 1986 Appl. Opt. **25**, 1653
[33] Guerrero E, Galindo P L, Yañez A, Ben T and Molina S I 2007 *Microsc. Microanal.* **13**, 320
[34] Chung J and Rabenberg L 2008 *Ultramicroscopy* **108**, 1595
[35] Rouviere J-L and Sarigiannidou E 2005 *Ultramicroscopy* **106**, 1
[36] Wiener N 1949 *Extrapolation, Interpolation and Smoothing of Stationary Time Series* (Wiley, New York, USA)
[37] Larsson M W, Wagner J B, Wallin M, Hakansson P, Froberg L E, Samuelson L and Wallenberg L R 2007 *Nanotechnology* **18**, 015504
[38] Cushley M 2008 *J. Phys.: Conf. Ser.* **126**, 012086
[39] Lozano J G, Sanchez A M, Garcia R, Gonzalez D, Herrera M, Browning N D, Ruffenach S and Briot O 2007 *Appl. Phys. Lett.* **91**, 071915
[40] Tomioka K, Motohisa J, Hara S and Fukui T 2008 *Nano Lett.* **8**, 3475
[41] Sanchez A M, Galindo P L, Kret S, Falke M, Beanland R and Goodhew P J 2006 *Microsc. Microanal.* **12**, 285
[42] Sánchez A M, Galindo P L, Kret S, Falke M, Beanland R and Goodhew P J 2006 *J. Microsc.* **221**, 1
[43] Recnik A, Mobus G and Sturm S 2005 *Ultramicroscopy* **103**, 285
[44] HREM PPA website: http://www.hremresearch.com/Eng/plugin/PPAEng.html
[45] Nellist P D and Pennycook S J 1999 *Ultramicroscopy* **78**, 111
[46] Xu P, Loane R F and Silcox J 1991 *Ultramicroscopy* **38**, 127
[47] Ishizuka K 2002 *Ultramicroscopy* **90**, 71
[48] Rosenauer A and Schowalter M 2007 *Springer Proc. Phys.: Microsc. Semicond. Mater.* (Springer, New York, USA) **120**, 169
[49] Garcia J M, Gonzalez L, Gonzalez M U, Silveira J P, Gonzalez Y and Briones F 2001 *J. Cryst. Growth* **227**, 975
[50] Molina S I, Beltrán A M, Ben T, Galindo P L, Guerrero E, Taboada A G, Ripalda J M and Chisholm M F 2009 *Appl. Phys. Lett.* **94**, 043114
[51] Alonso-Álvarez D, Alén B, García J M and Ripalda J M 2007 *Appl. Phys. Lett.* **91**, 263103

Quantitative Z-contrast atomic resolution studies of semiconductor nanostructured materials

E Carlino
Centro di Microscopia Elettronica (CME) TASC-INFM-CNR National Laboratory, Area Science Park Basovizza Ed. MM, SS 14 Km 163.5, 34012 Trieste, Italy

E-mail: carlino@tasc.infm.it

Abstract. High angle annular dark field (HAADF) scanning transmission electron microscopy has demonstrated the capability to achieve sub-Angstrom resolution in the study of the structure of materials. Furthermore, the sensitivity of HAADF imaging to fine variations of the chemistry of the specimen allows one to derive the relevant chemical map from the intensity distribution in an image. Here, a general approach to calculate the HAADF image intensity for an alloy is derived and applied to experimental images to measure quantitatively the distribution of the chemical species. The calculations of HAADF image contrast have been performed by *multi-slice* methods in the framework of the *frozen-phonon* approximation by developing and using a parallel code to strongly reduce the computing time necessary to obtain a reliable simulation of realistic specimens. The parameters that influence the HAADF image contrast have been studied and their role has been quantified. Experimental examples of quantification of the chemistry of semiconducting heterostructures will be shown. Attention will be focused on the different parameters that influence the HAADF image contrast depending on the material system and on the specimen composition.

1. Introduction

Transmission Electron Microscopy (TEM) has been a main methodology for the study of the solid-state matter for more than half a century. The success of TEM is related to its capability to provide reliable information on the properties of the materials, at the highest spatial resolution, and to its flexibility to tackle a wide range of problems providing, over the years, different methods to solve old and new issues in the knowledge of solid state matter. Certainly, the capability of electron microscopy to provide images capable to *show* and measure important specimen properties is a big part of its success and popularity. The goal is to achieve knowledge of the crystal structure and perfection, chemistry, electronic structure and magnetic properties, at the highest spatial resolution. In the last years, high angle annular dark field (HAADF) imaging in scanning transmission electron microscopy (STEM) has played a big role in this direction. HAADF provides images of crystals, oriented in a given zone axis, achieving sub-Angstrom resolution [1]; the position of the atomic columns, projected on the observation plane, is visible as white dots for a relatively wide range of objective lens defoci and specimen thicknesses [2]. These features often make the interpretation of high resolution HAADF images much more straightforward compared to phase contrast high resolution TEM (HRTEM), where the distribution of the intensity in the image is a complex function of the interference of the Bloch waves, generated in the specimen by the primary electron beam, convoluted with the phase contrast transfer function of the objective lens. This latter image formation process gives results that are highly sensitive even to small variation of lens defocus and specimen thickness and a simulation of the experimental results is needed to have reliable data [3]. On the contrary, in many cases basic information on specimen structure and perfection can be achieved by atomic resolution HAADF without image simulation. Nevertheless, more subtle information, as for example the kind of bond between the atoms in the specimen, can also be addressed by HAADF imaging but requires the modelling of the structure and the relevant image simulation [4]. One of the most desirable results for the nano-science, in particular in the field of semiconducting materials, would be to have an atomic resolution picture of both structure and chemistry of the specimen, especially at the interface between two materials, as the properties of materials and devices are strongly dependent on the structure and chemistry of their interfaces [5,6]. The HAADF image contrast is highly sensitive to the atomic number of the scattering species [7] and the proper choice of the experimental conditions allows one to obtain a direct high-resolution picture of the chemical distribution within the specimen [8], or to detect the presence of dopants within an atomic column [9] even without the computer simulation of the experimental results. Nevertheless, to quantify the chemistry it is necessary to simulate the Z-contrast image intensity [10,11]. Here we show how atomic resolution HAADF imaging coupled with multi-slice computer simulation allows one to obtain atomic resolution quantitative measurement of the chemistry of a specimen also in the case of an interface between two different materials. The studies performed on different semiconducting material systems emphasize how the parameters influencing the chemistry quantification are dependent on the compositions and on the kind of system considered.

2. Experimental

Two materials systems were considered in detail in this study: a Si monolayer buried in [001] GaAs substrate and InGaAs quantum wells, with different known In concentrations of 5%, 12% and 24% respectively, grown on [001] GaAs substrate. The samples were grown by molecular beam epitaxy following a procedure described in details in previous papers [8,12]. The choice of these case studies will make evident how the weight of the parameters involved in the HAADF image contrast formations are strongly dependent on the materials system and on the alloy concentration.

Specimen preparation for quantitative atomic resolution profiling of chemical species by STEM HAADF needs some special care as the surfaces of the thin sample play a basic role. Three points are of particular importance: (i) the specimen surfaces have to be smooth with no clustering, (ii) the amorphous layer due to the ion milling process has to be negligible and (iii) the eventual hydrocarbon contaminants should not have any influence during the image acquisition time with the highly focused electron probe necessary for atomic resolution HAADF imaging. STEM specimens for our studies were prepared in <110> cross section geometry. Milling conditions and geometry were chosen in order to minimize differential sputtering between substrate and epilayer [5]. In particular two ion-guns were used - one for each side of the STEM specimen - in order to avoid redeposition of sputtered material on the opposite surface. Furthermore, the use of a low milling angle and high Z-number ions reduces the dependence of the sputtering process on the specimen atomic number. In this case, the milling geometry was chosen in order to have the ion beam direction forming an angle of 4° relative to the {110} plane. Such precautions were employed together with a discontinuous milling process during rotation of the specimen holder. In particular, the guns were switched on only when the ion beams where within ± 30° of the direction normal to the substrate/epilayer interface. Hence, no sputtering was allowed when the ion beams were parallel to heterostructure interface, as such geometry is known to enhance preferential sputtering of the interface region. The milling procedure was performed by using Xe^+ ions, with energy of 4.5 keV for the initial part of the milling process, and progressively reduced down to 2keV up to the formation of a small hole. Then, to remove the layers damaged by the high-energy ions, the specimens were ion milled by Xe^+ ions by using hot cathode ion guns, specially designed to operate down to 200 V, with an incidence milling angle of 2°. Finally, the specimens and specimen holders were treated in low-energy, high frequency plasma of argon and oxygen to remove hydrocarbon contamination, prior to being inserted in the electron microscope vacuum.

HAADF experiments were performed at room temperature using a JEOL 2010F UHR TEM/STEM field-emission gun electron microscope, operating at 200 kV with a measured objective lens spherical aberration coefficient C_s of 0.47±0.01 mm and interpretable resolution limit in HRTEM of 1.9 Å [3].

For Z-contrast imaging, the STEM attachment was equipped with a YAP HAADF detector. The resolution achievable in Z-contrast mode was 0.126 nm [13,14]. All Z-contrast micrographs were recorded digitally with a collection angle $84 \leq 2\theta \leq 224$ mrad. The convergence angle, and hence the optimum aperture of the condenser lens, is related to the spherical aberration coefficient of the pre-field of the objective lens [15,16]. In our case a convergence angle of 14 mrad was chosen as the best compromise between the aperture cut-off of the detectable frequencies and the signal-to-noise ratio in the image. A much larger condenser aperture would produce a distorted image due to the spherical aberration, whereas a much smaller condenser aperture would reduce the highest frequencies in the images and therefore the visibility of the smallest lattice spacing. In between these two limits the choice of the aperture diameter was made taking into account the signal-to-noise ratio in the image. The use of an annular STEM detector with a collection angle larger than about 80 mrad allows acquisition of mainly incoherent electrons, since thermal diffuse scattering dominates at high scattering angles [17,18]. This choice significantly reduces the contribution to the image of the scattering due to a possible strain at the interfaces [15]. The simulations of the HAADF image contrast were performed according to these parameters of the microscope.

3. Simulations and Measurement of an alloy by HAADF imaging

The experimental measurement of the spatially resolved chemical composition of the specimen is based on the simulations of the HAADF image intensity for a model structure as a function of the concentration of the relevant chemical species [19]. To achieve a reliable calculation all the parameters that could influence the image contrast, such as specimen thickness, distortion of the lattice cell, microscope objective lens defocus or the atomic size effects in the alloys [20], should be considered. As will be shown in the examples below the physical values influencing the image contrast can vary depending on the materials' system or on the concentration of the species to be measured. The images were simulated in [110] zone axis according to the experimental conditions used and the image intensities were measured on both experimental and simulated images with the same criteria, in particular the image intensity was averaged on a half projected unit cell [19]. Atomic resolution HAADF image contrast can be simulated by Bloch-waves [21] or multi-slice approaches [22]. The Bloch-waves based methods are relatively fast, and represent a good tool to understand basic and subtle details of the physics of the interactions involved in the HAADF image formation process. The influence of thermal diffuse scattering (TDS), which plays a basic role in the HAADF image formation [23], is not naturally included in the calculation based on Bloch-waves approaches and requires ad hoc solutions [24-26]. Furthermore, Bloch-wave methods are not convenient whenever a non-periodic object is considered, as for example in the case of the description of a defect, or an interface in the material. In this case the use of multi-slice methods has been recognized as more effective also for the HRTEM image simulation. TDS effect has been introduced in the multi-slice calculations by several authors [27-31]. Nevertheless, STEM HAADF image contrast simulation by multi-slice with the "frozen-phonon" model has proven to be the most accurate method reaching an accuracy better than 2%, in the simulations of the HAADF image contrast, by considering 20 atomic configurations [32,19]. This approach has been used for the calculations presented here. Unfortunately, the approach is extremely time-consuming and image simulation considering models with several atoms, as in the case of large supercells, or realistic STEM specimen thickness, becomes impractical as it could require hundreds of hours of computing time. Indeed, the structure of the calculation of the image contrast by multislice in the frozen-phonon approximation is naturally parallel and can be modified and rewritten for parallel computing reducing the computing time of orders of magnitude [33]. We caution the reader that the use of virtual crystal approximation [34] in the description of the alloy being simulated could produce the formation of artefacts in the image intensity

[19]. In fact, the description of the alloy within the virtual crystal approximation considers a certain position inside the crystal cell as occupied by a combination of potentials weighted according to the concentration to be modelled. The crystal potential V is given by $V = \sum_{i=specie} x_i V_i$, where x_i is the partial occupation coefficient for the i-th species with potential V_i. In the frozen phonon approximation the effect of thermal diffuse scattering is simulated by offsetting the position of each atom by a small random amount and then performing the standard multislice calculations [30]. This procedure together with the virtual crystal approximation could produce an artificial splitting of the crystal potential and hence the formation of artefacts in the simulated HAADF image.

3.1. Measurement of the Ga(Si)As alloy composition

The intensity of the HAADF image is sensitively dependent on the order of the atoms in the atomic column, as shown in figure 1. These data were obtained by calculating the HAADF image intensity of a Ga(Si)As alloy, averaged over half a unit cell projected on the (110) plane, as a function of the position of the Si atoms within the atomic colum. The calculations were performed by taking into account the tetragonal distortion of the unit cell due to the presence of the Si in the GaAs matrix [19], and for a STEM specimen thickness of 10 nm. In particular, the dots in figure 1 were obtained by considering Si atoms randomly distributed on Ga sites, the squares were obtained by considering the atoms on both Ga and As sites and the triangles were obtained by considering the Si atoms of the alloy all located at the beginning or at the end of the atomic column. The latest configuration produces a strong difference in the calculated intensity showing, from one side, the high sensitivity of HAADF to the presence of clusters within the crystalline matrix but, from the other side, raises concerns on the accuracy of the chemistry measurement by HAADF. In fact, the same concentration of Si could produce different intensities in the image.

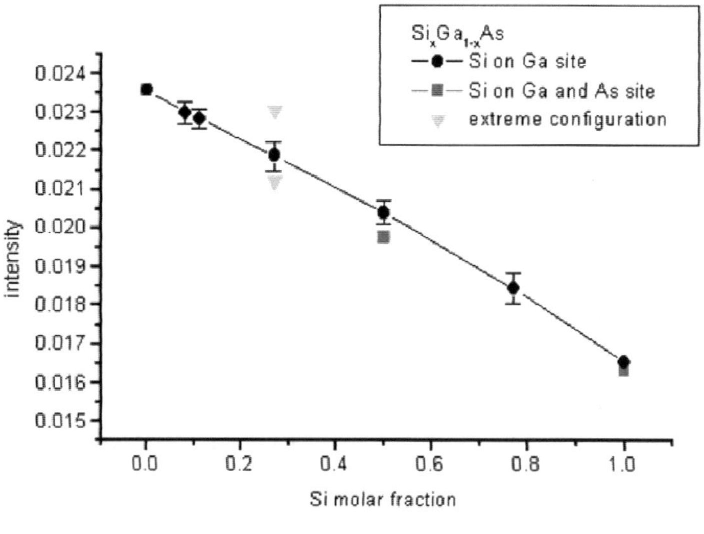

Figure 1. Calculated HAADF image intensity for Ga(Si)As alloy as a function of the atomic species distribution within the atomic column.

The more straightforward way to tackle this problem is to average the intensity I within the atomic column on all the possible atomic configurations, for each Si content, and to calculate $\tilde{I} = \frac{\sum_{i=config.} I_i}{N_{config}}$. This averaging introduces an intrinsic uncertainty, of about 2%, in the measurement of the concentration of Si in Ga(Si)As from HAADF imaging, calculated as the standard deviation of each intensity value from the mean value.

The HAADF experiments indicate that, in most of the cases, the images are acquired for a specimen thickness between 10 nm and 40 nm. Figure 2 shows the calculation of the HAADF image intensity for Ga(Si)As alloy, as a function of Si concentration, for four STEM specimen thicknesses.

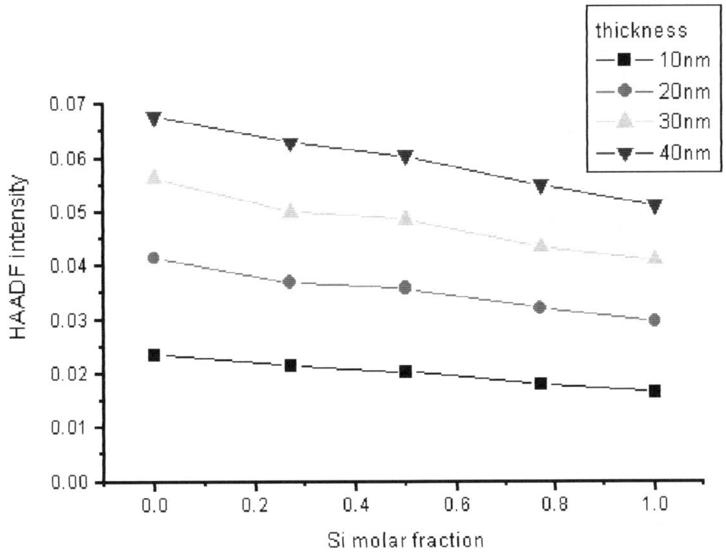

Figure 2. Calculated HAADF image intensity for Ga(Si)As alloy as a function of Si concentration and STEM specimen thickness.

It should be noted that for each thickness the image intensity varies nearly linearly with the Si concentration. This is not a general behaviour but it is peculiar of this material system. A similar behaviour was also reported for GaAlAs system [10]. If we define the maximum contrast as: $C_{100\%} = \frac{I_{GaAs} - I_{SiAs}}{I_{GaAs}}$, this value varies from 0.30 to 0.26 going from a specimen thickness of 10 nm to 40 nm, resulting in a relatively small role of the specimen thickness on the quantification of the composition of the alloy by HAADF imaging.

It is now worthwhile to remark that, experimentally, the atomic resolution HAADF images in scanning transmission electron microscope, not equipped with correctors for spherical aberration, are obtained with a defocus value relatively close to the optimum defocus [35]. Experiments and simulations of atomic resolution HAADF images have shown that the role of the variation of the objective lens focus value, in an interval of some tens of nm around the optimum defocus, has a negligible influence on the quantification of the chemistry by HAADF imaging [35]. Another aspect that should be considered in the simulation of the HAADF image contrast of an alloy is related to the influence of the static atomic displacements (SDs) due to atomic size effects in alloys with atoms having different covalent or ionic radii. The calculation performed on Ga(Si)As system in a range of concentrations of Si below 15% has shown an influence smaller than 1% [20]. This aspect will be discussed in more detail below.

Figure 3 shows a) an experimental HAADF image of a nominal one monolayer of Si atoms buried in a GaAs matrix with b) the relevant quantification of the Ga(Si)As alloy as derived from the measure of the image intensity. The HAADF image gives an immediate and intuitive picture of the formation of a Ga(Si)As region within the GaAs matrix. In fact, the formation of the Ga(Si)As alloy is evidenced by the dark region, in the middle of figure 3a, due to the smaller scattering factor of Si atoms compared to Ga and As atoms. A close inspection to the high-resolution image shows the well known cation-anion dumbbells of the sphalerite structure seen in [110] projection and spaced by 0.14 nm. The inset in figure 3a is the experimental intensity profile, averaged over the opaque marked region to increase the signal to noise ratio. The average along the direction of the interface reduces the resolution to 0.28 nm.

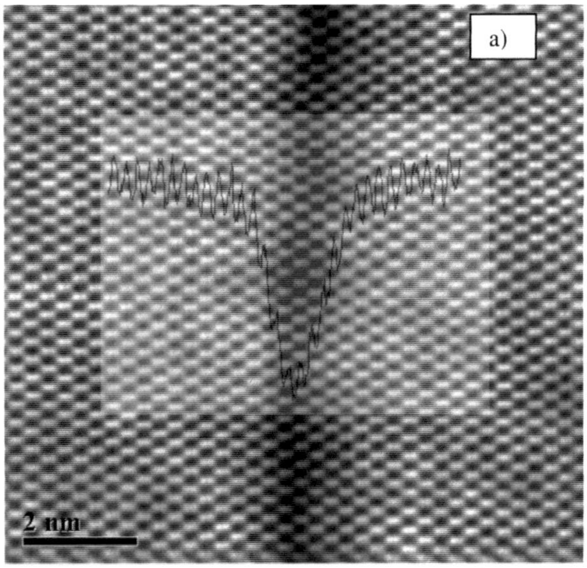

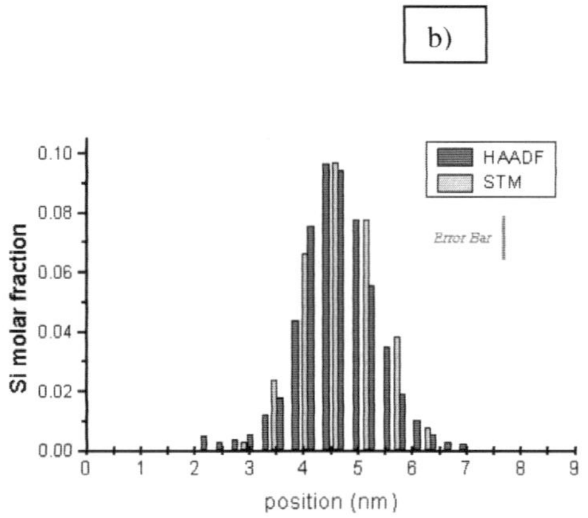

Figure 3. a) HAADF image of a nominal 1ML of Si buried in GaAs matrix. The inset of a) is the image intensity line profile averaged on the 5 nm region parallel to the interface sketched in the figure; b) Si concentration profile as measured by HAADF and X-STM

It is worthwhile to remark that in the calculated images the intensity it is also averaged over half a unit cell, removing the need of the intensity subtraction, reported by other authors [36] and due to the delocalised scattering in the specimen and to the electron and mechanical instabilities of the instrumentation. To evaluate the Ga(Si)As experimental image contrast, the high angle annular dark field detector background has been subtracted and the Ga(Si)As intensity value has been normalized to the GaAs regions on both sides of the dark Ga(Si)As region. Indeed, the experimental images should be acquired close to a hole in the specimen to adjust the HAADF detector gain to have zero counts, or the lowest value of counts, in the hole as a high value of the dark counts could produce the formation of artefacts in the image intensity [37]. Figure 3b shows the Si concentration profile as measured from figure 3a, together with the Si concentration measurement performed by cross-sectional scanning tunnelling microscopy (XSTM) on the same specimen [8]. The total amount of Si in the alloy corresponds to about two monolayers, instead of the nominal one monolayer, and results in the formation of an alloy distributed over about 3 nm. The error bar reported in the figure 3b could be a bit overestimated and it was calculated considering the sum of the uncertainties in the calculations (2%), the intensity averaging over all the possible atomic configurations (2%) and the uncertainty on the STEM specimen thickness (2%).

3.2. Measurement of InGaAs alloy composition

As will be seen in the following, the features of HAADF images for InGaAs are rather different with respect to what has been reported so far for Ga(Si)As. To study the InGaAs system, specimens with three different concentrations of In were considered: (5±1)%, (12±1)% and (24±1)% respectively, as measured by photoluminescence (PL) and, in-situ during the molecular beam epitaxy growth, by X-ray photoelectron spectroscopy (XPS). Figure 4 shows the simulated HAADF image intensity ratio for an alloy of InGaAs, calculated as a function of the In concentration, taking into account the effect of the static atomic displacements, I_{SD}, or without the effect of static atomic displacements, I_0, for two specimen thicknesses of 10 nm and 40 nm respectively. The supercells with SDs were calculated by Valence Force Field (VFF) model [38-41] which reproduces accurately the SD field in semiconductor alloys [38]. The supercells were then used to calculate the atomic resolution HAADF image contrast following a procedure described in a previous paper [20]. The $\frac{I_{SD}}{I_0}$ gives an evaluation of the weight of the SD effect on the simulation of the HAADF image intensity of InGaAs. From figure 4 it can be seen that the effect of static displacements is highest for In concentrations of 50%, as expected, and has a parabolic behaviour, as can be seen from the solid line obtained by fitting the simulated data with a parabola whose fitting parameters are also shown in figure 4. It should be noted that the SDs and the corrections to the structure factors due to the SDs also follow a parabolic law as a function of the In concentration [41-44]. Hence, SDs in the case of InGaAs play a role that cannot be neglected if the chemistry of the specimen is derived from HAADF image.

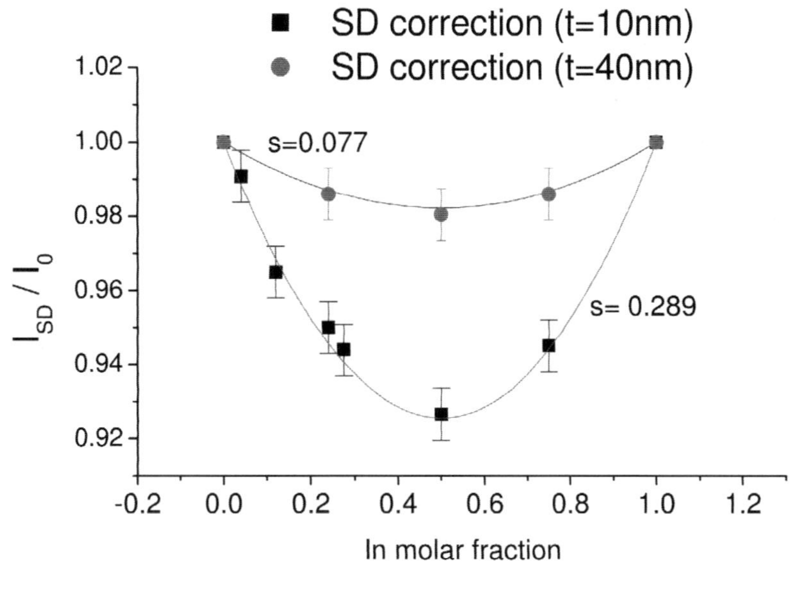

Figure 4. Ratio of the HAADF intensities for InGaAs, calculated with and without taking into consideration the effect of the atomic static displacements, as a function of the In content and for two specimen thicknesses of 10nm and 40nm.

Figure 5a is a low magnification high resolution image, in [110] zone axis of InGaAs QWs/GaAs (001), and has been acquired by achieving the condition for atomic resolution, with objective lens focus close to Scherzer's value, and then decreasing the magnification to show all the QWs of InGaAs, and adjusting the detector gain to optimize the visibility of all the features of interest: the three QWs, the GaAs barriers and the black level in the hole of the specimen to evaluate the dark count of the detector to be subtracted for the image intensity quantification. The intensity profile in figure 5a immediately reveals the kind of In concentration in the three QWs as the high angle scattering increases with increasing the In content in the alloy. The ratios $\frac{I_{InGaAs}}{I_{GaAs}}$ between the HAADF image intensity of InGaAs QWs and the relevant intensity of the GaAs barriers have been measured for different thickness of the STEM specimen and are reported, as squares, in figure 5b, together with the relevant experimental errors, for the different In concentrations. The calculated HAADF image intensities for the three wells, taking into account the effect of SDs (solid lines) and without the effect of SDs (dotted lines), as a function of the STEM specimen thickness are also reported in figure 5b. The figure 5b demonstrates that the image intensity is strongly dependent on the specimen thickness, showing a pronounced dip in its value around a thickness of 10 nm, and hence a reliable quantification of the chemistry in the InGaAs alloy requires an accurate measurement of the specimen thickness.

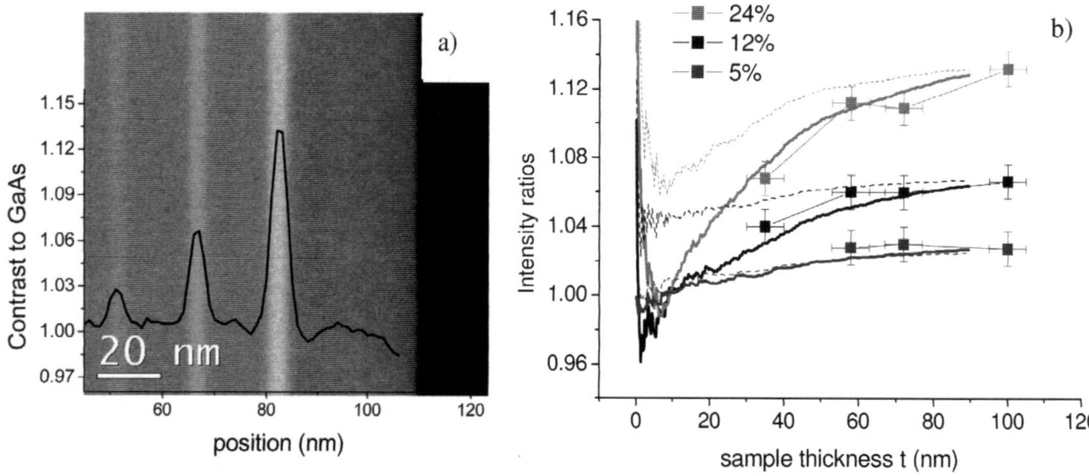

Figure 5. a) low magnification high resolution [110] zone axis HAADF image of three quantum wells of InGaAs, with In content of (5±1)%, (12±1)% and (24±1%) respectively, as measured by photoluminescence and in-situ photoemission spectroscopy; b) intensity ratio I_{InGaAs}/I_{GaAs} measured on the experimental images (square dots) for each QW, and calculated with the effect of SDs (solid lines) and without SDs (dotted lines)

Furthermore, the role of the SDs cannot be neglected as, for example in the case of the well with 24% of In, it could produce an error of 10% in the absolute measurement of the In concentration. The figure 5b demonstrates that taking into account the SDs produces a good agreement between simulated and experimental curves. The results of the experiments and simulations also indicate that, for reliable quantification of the InGaAs alloy composition, STEM specimen thicknesses larger than 20 nm should be preferred.

The examples shown so far demonstrate how the specimen thickness plays a subtle role in the quantification of the specimen chemistry by HAADF. In fact, from one side the precise knowledge of the specimen thickness is not very important for the quantification of Si in Ga(Si)As, for the Si concentrations here considered, and hence thin specimen could be preferred to maximize the incoherent component of the HAADF image [24]; on the other side, chemistry quantification for other materials system, like InGaAs or SiGe [45,46], needs larger specimen thickness to avoid non-monotonic behaviour of image contrast as a function of thickness, observed for small sample thicknesses.

4. Conclusions

HAADF images contain detailed chemical information that can be quantified thus providing, at the same time, structural and chemical information at atomic resolution. The chemistry of the specimen can be measured within an accuracy of few percent if the parameters influencing the HAADF image contrast have been properly addressed and simulated by reliable calculations. In particular, in the modelling of the alloy, the use of virtual crystal approximation can produce the formation of artefacts in the images. Furthermore, it has been observed that different atom positions within atomic columns, with the same overall composition, can produce different image intensity and hence limit the intrinsic accuracy of the chemistry measurement by HAADF imaging. The parameters influencing the measurement are: the STEM specimen thickness, the crystal cell distortion induced, for example, by the presence of a guest species in a host matrix, or by the mismatch at the interface between the substrate and the epilayer, and the influence of the static atomic displacements due to atomic size effects in alloy with atoms having different covalent or ionic radii. These parameters have different weight in the accuracy of the measurement depending on the material system considered and, within the same system, also as a function of the chemical species concentration. The results reported here show that for Ga(Si)As alloy, with a maximum Si content of 15%, an accuracy of few percent in the chemistry quantification can be achieved with an uncertainty on the specimen thickness between 10 nm and 40 nm and neglecting the role of SDs. On the other side the example of InGaAs demonstrates that completely unreliable chemistry quantification can be done neglecting the role of SDs and without an accurate STEM specimen thickness measure. The amount of uncertainty depends on the In concentration: for example for a specimen thickness of 30 nm, neglecting the SDs produces an absolute error on the In concentration of 10% in an alloy containing 24% of In whereas, for the same specimen thickness, it produces an absolute error of 1% in the alloy containing 5% of In. In all the cases, an uncertainty on the specimen thickness in the range between 10nm and 40nm produces completely unreliable determination of the chemistry of the InGaAs alloy. The simulations of the HAADF image intensity suggest the strategy to be used for STEM experiments as they allow one to derive the specimen thickness necessary to obtain quantitative analytical information from HAADF images. Finally, atomic resolution HAADF provides in an image a reliable and well reproducible quantitative description of the structure and chemistry of the specimen and represents a further powerful tool, now available also on standard field emission gun TEM/STEM microscopes, to study in a direct way the properties of the solid-state matter at the highest spatial resolution.

Acknowledgements

I would like to thank E. Cociancich, A. Franciosi, F. Glas, V. Grillo, V. Rosato and S. Rubini for their contributions to the ideas and results reported in this paper. B. Davidson and M. Catalano are acknowledged for the accurate reading of the manuscript. Finally, I would like to thank P. G. Merli for his suggestions and examples.

References

[1] Nellist P D, Chisholm M F, Dellby N, Krivanek O L, Murfitt M F, Szilagyi Z S, Lupini A R, Borisevich A, Sides W H jr. and Pennycook S J 2004 *Science* **305**, 1741
[2] Pennycook S J in *Advances in imaging and Electron Physics* 2002 **123** Academic Press New York, 140
[3] Spence J C H *Experimental High-resolution electron microscopy* 2nd ed. 1988 Oxford University Press
[4] Yan Y F, Al-Jassim M M, Chisholm M F, Boatner L A, Pennycook S J and Oxley M 2005 *Phys. Rev. B* **71**, 041309

[5] Colli A, Carlino E, Pelucchi E, Grillo V and Franciosi A 2004 *J. Appl. Phys.* **96**, 2592
[6] Rafferty B and Pennycook S J 1999 *Ultramic.* **78**, 141
[7] Pennycook S J and Jesson D E 1990 *Phys. Rev. Lett.* **64**, 938
[8] Carlino E, Modesti S, Furlanetto D, Piccin M, Rubini S and Franciosi A 2003 *Appl. Phys. Lett.* **83**, 662
[9] Voyles P M, Muller D A, Grazul J L, Citrin P H and Gossmann H-J L 2002 Nature **416**, 826
[10] Anderson S C, Birkeland C R, Anstis G R and Cockayne D J H 1997 *Ultramic.* **69**, 83
[11] Liu C P, Preston A R, Boothroyd B and Humphreys C J 1999 *J. Microsc.* **194**, 171
[12] Rubini S *et al.* 2006 *Appl. Phys. Lett.* **88**, 141923.
[13] James E M and Browning N D 1999 *Ultramic.* **78**, 125
[14] Carlino E, Grillo V 2005 *Proceedings MCEM VII* ISBN 961-6303-69-4 159
[15] Pennycook S J and Nellist P D 1999 *Impact of Electron Microscopy on Materials Research* Kluwer Academic Publisher 161
[16] Treacy M M J, Howie A and Pennycook S J 1980 *Inst. Phys. Conf. Ser.* **52**, 261
[17] Kittel C 1986 *Introduction to Solid State Physics* 6th edition J. Wiley & Sons, Inc, New York 602
[18] Rafferty B, Nellist P D and Pennycook S J 2001 *J. Electr. Mic.* **50**, 227
[19] Carlino E, Grillo V 2005 *Phys. Rev. B.* **71**, 235303
[20] Grillo V, Carlino E and Glas F 2008 *Phys. Rev. B.* **77**, 054103
[21] Metherell A J F 1976 *Diffraction of Electrons by Perfect Crystals* in *Electron Microscopy in Materials Science* Ed. U. Valdre' and E. Ruedl – EUR 5515e Brussels and Luxemburg
[22] Goodman P, Moodie A F 1974 *Acta Cryst. A* **30**, 280
[23] Muller D A, Edwards B, Kirkland E J and Silcox J 2001 *Ultramic.* **86**, 371
[24] Pennycook S J and Jesson D E 1991 *Ultramic.* **37**, 14
[25] Allen L J, Rossouw C J 1993 *Phys. Rev. B* **47**, 2446.
[26] Watanabe K, Yamazaki T, Hashimoto I, Shiojiri M 2001 *Phys. Rev. B* **64**, 115432
[27] Wang Z L 1999 *Phil. Mag. B* **79**, 37
[28] Ishizuka K 2002 *Ultramic.* **90**, 71
[29] Dinges C, Berger A, Rose H 1995 *Ultramic.* **60**, 49
[30] Kirkland E J 1998 *Advanced Computing in electron microscope* Plenum Press, New York
[31] Croitoru M D, Van Dyck D, Van Aert S, Bals S, Verbeeck J 2006 *Ultramic.* **106**, 933
[32] Xu P, Loane R F, Silcox J 1991 *Ultramic.* **38**, 127.
[33] Carlino E, Grillo V and Palazzari P 2007 *Microscopy of Semiconducting Materials* Cambridge 2007 Springer Proc. in Phys. **120** Eds. Cullis A and Midgley P Springer Netherlands, 177
[34] Yu P Y and Cardona M 1996 *Fundamentals of Semiconductors* Springer-Verlag Berlin Heidelberg New York 182
[35] Grillo V and Carlino E 2006 *Ultramic.* **106**, 603
[36] Klenov D O and Stammer S 2006 *Ultramic.* **106**, 889
[37] Yu Z, Batson P E and Silcox J 2003 *Ultramic.* **96**, 275
[38] Glas F 1995 *Phys. Rev. B* **51**, 825
[39] Keating P N 1966 *Phys. Rev. B* **145**, 637
[40] Martin R M 1970 *Phys. Rev. B* **1**, 4005
[41] Podgorny M, Czyzyk P M, Balzarotti A, Letardi P, Kisiel A and Zimnal-Starnawska M 1985 *Solid State Commun.* **55**, 413
[42] Rosenauer A, Schowalter M, Glas F and Lamoen D 2005 *Phys. Rev. B* **72**, 085326
[43] Glas F 2004 *Phil. Mag.* **84**, 2055
[44] Petkov V and Billinge S J L 2001 *Physica B* **305**, 83
[45] Walther T and Humphreys C J 1997 Inst. Phys. Conf. Ser. **153**, 303
[46] Grillo V and Carlino E *Microscopy of Semiconducting Materials* 2007 Springer Proc. in Phys. **120** Eds. Cullis A and Midgley P Springer Netherlands, 173

Journal of Physics: Conference Series **209** (2010) 012006 doi:10.1088/1742-6596/209/1/012006

On the incorporation of indium in InAs-based quantum structures

D Gerthsen[1], H Blank[1], D Litvinov[1], R Schneider[1], A Rosenauer[2], T Passow[3], A Grau[3], P Feinäugle[3], H Kalt[3], C Klingshirn[3], M Hetterich[3]

[1] Laboratorium für Elektronenmikroskopie and Center for Functional Nanostructures (CFN), Universität Karlsruhe, D-76128 Karlsruhe, Germany
[2] Institut für Festkörperphysik, Universität Bremen, D-28359 Bremen, Germany
[3] Institut für Angewandte Physik and CFN, Universität Karlsruhe, D-76128 Karlsruhe, Germany

E-mail: dagmar.gerthsen@kit.edu

Abstract. Transmission electron microscopy (TEM) was applied to study In segregation during molecular beam epitaxy (MBE) growth of InAs-based quantum-dot (QD) structures with a focus on achieving QD luminescence at the technologically relevant wavelength of 1.3 μm. Quantitative composition analyses were carried out with the composition evaluation by lattice fringe analysis (CELFA) technique and In-segregation efficiencies were determined on the basis of atomic-scale composition profiles. The effect of the InAs-deposition rate on the In distribution was studied in detail which needs to be low, in the 0.005 ML/s regime to achieve long-wavelength emission. Further minimization of In-segregation induced intermixing can be achieved by lowering the substrate temperature during GaAs cap layer growth. Finally, a procedure is presented to extract the real composition of QDs taking into account their three-dimensional morphology and embedding in a GaAs cap layer.

1. Introduction

The properties of InAs/GaAs quantum-dot (QD) structures continue to be of interest due to numerous applications that have emerged during the past decade. In addition to established (opto)electronic devices like laser diodes and single-photon emitters [1,2] new applications in semiconductor spintronics are envisaged where InAs QDs could be promising for future devices in quantum information processing [3,4]. Selforganized QD formation is based on the Stranski-Krastanow (SK) growth mode which is characterized by a transition between two-dimensional (2D) layer and three-dimensional (3D) island growth. The critical InGaAs thickness for the 2D/3D growth-mode transition has been studied excessively over about two decades. While the SK growth was first solely attributed to strain relaxation in lattice-mismatched heterostructures, more recently the amount of segregated indium on the growth surface was discussed as a decisive factor for the growth-mode transition [5-7].

The resulting structures always consist of a continuous wetting layer (WL) and islands. However, the chemical composition of the WL and islands and QD properties such as size, shape and density are strongly dependent on the epitaxy conditions and deposited amount of InAs as demonstrated in comprehensive photoluminescence (PL) and transmission electron microscopy (TEM) studies, e.g. [8,9]. Segregation was recognized as an important mechanism in epitaxial growth of semiconductor nanostructures which may lead to a significant difference between the real and intended composition as already demonstrated in the pioneering work of Moisson et al. [10] for AlGaAs and Muraki et al. [11] for InGaAs. Muraki et al. proposed a phenomenological model which was shown to be well suited for the description of experimentally derived composition profiles of InGaAs quantum wells and wetting layers in InAs QD structures [7,12]. In this model, the segregation efficiency *R* characterizes the fraction of In atoms which segregate from to uppermost crystalline InGaAs layer into the layer deposited on top which is considered as a weakly bonded floating layer. The segregation efficiency depends on the temperature but also on other parameters like - in the case of molecular beam epitaxy (MBE) growth - the V/III-flux ratio, surface reconstruction and the type of As-source molecule (As_4 or As_2). High *R* values between 0.65 and 0.93 are observed in the temperature interval between 370 and 560 °C according to [13] and references therein.

Strong In segregation in InGaAs has consequences with respect to achieving luminescence of QD-based laser diodes in the technologically interesting 1.3 and 1.5 μm range which requires large QDs (small confinement energies) with the highest possible In concentration. Optimum MBE conditions for PL with a peak at 1.3 and 1.5 μm are obtained at very low growth rates around 0.005 monolayer (ML)/s, temperatures around 500 °C and a deposition of 2 to 4 ML InAs which favours the formation of large islands [14,9]. However, the high substrate temperature T_s is not well suited to minimize segregation. Since lowering of T_s leads to a reduction of the QD size [9], different strategies are used to optimize the QD structures with respect to long-wavelength emission as e.g. capping by InGaAs instead of GaAs [15,16] or embedding of InAs QDs in InGaAs quantum wells [17].

Although TEM is frequently applied for the characterization of InAs-based quantum structures, quantification of the composition is scarcely found in literature even though of the group-III element distribution is essential for the understanding of the growth mechanisms and device properties. Techniques like electron energy loss spectroscopy [5,6,18] and quantitative scanning tunnelling microscopy [19] have been used to analyze the composition of InGaAs QD structures but quantification on the basis of the chemically sensitive {200} reflections is particularly suited in the case of InGaAs. Quantitative composition analyses on an atomic-scale were carried out in this study by applying the composition evaluation by lattice fringe analysis (CELFA) technique [20] which works well for planar InGaAs quantum-well type structures. However, quantitative analyses of QDs have been hampered up to now by the 3D morphology of the QDs which are embedded in a GaAs matrix. In a TEM cross-section sample with finite thickness, the measured In concentration will be inevitably too low for QDs due to an averaging effect along the electron beam.

In this study we compare in §3.1 InAs QD samples which were grown under identical MBE conditions apart from the InAs deposition rate. The reduction of the InAs deposition rate to a low value of 0.0056 ML/s is essential for achieving a PL peak at 1.3 μm. In §3.2 we will present a procedure that allows the correction of averaged composition data of InGaAs QDs embedded in GaAs which is accomplished by the determination of the sample thickness and QD shape. Finally, results on the effect of lowering the substrate temperature T_{cap} during GaAs cap-layer growth are presented in §3.3. The latter experiments are motivated by the aim to reduce undesired In-segregation induced intermixing during overgrowth to eventually shift the PL towards even longer wavelengths.

2. Experimental techniques

All investigated samples were grown in a Riber Compact 21 MBE system equipped with effusion cells for Ga and In, and a valved cracker cell for As on GaAs(001) substrates covered by a GaAs buffer. The GaAs buffer growth was interrupted to reduce the substrate temperature from 580 °C to the value chosen for InAs deposition. Several sample series were investigated as outlined in detail in [9]. With respect to achieving 1.3 μm luminescence we focus here on the effect of the InAs deposition rate which was varied between 0.0056 ML/s (sample denoted as S_{low}) and 0.08 ML/s (sample S_{high}) with an As:In beam-equivalent pressure ratio (BEPR) of 80:1, As_4 as an As precursor and an optimized value for T_s=500 °C. The amount of deposited InAs was 2.4 ML and a growth interruption of 10 s was introduced before the GaAs cap-layer growth.

In §3.3 we present results of a sample series where the substrate temperature was lowered after the InAs deposition (2.4 ML InAs, deposition rate 0.09 ML/s, As:In BEPR = 10:1, T_s=485 °C). Three samples were fabricated with identical InAs QD layers but different temperatures T_{cap} during the deposition of the 50 nm GaAs cap layer. The capping temperature was 485 °C for the first sample. The initial 20 ML of the GaAs cap layer of the second sample were deposited at 410 °C before the temperature was raised to 485 °C. In the third sample, T_{cap} was lowered to 350 °C for the first 20 ML and the rest of the GaAs cap-layer growth took place at 570 °C.

TEM cross-section samples were prepared along the [010]-zone axis using the procedure outlined by Bravman et al. [21]. Plan-view samples were obtained by chemical etching from the substrate side using a solution of NaOH (1mol/l) and H_2O_2 (30%) with a ratio of 5:1 to prevent defect formation during sample preparation.

TEM micrographs were recorded using a 200 keV Philips CM 200 FEG/ST transmission electron microscope. The determination of the In concentration with CELFA is based on HRTEM lattice-fringe images taken under {200} two-beam conditions close to the [010]-zone axis. Choosing an <100>-type zone-axis instead of a <110>-zone axis is essential to avoid contributions of the {111} reflections due to dynamical electron diffraction and nonlinear image formation. The selected {200} reflection is centered on the optical axis of the electron microscope to minimize delocalization. The local In concentration is determined by evaluating the local amplitude of the {200} Fourier component of the image intensity which is normalized with respect to the respective Fourier component in the adjacent GaAs. The local normalized {200} amplitudes are then compared with values calculated by the Bloch-wave method using structure factors according to [22] which comprise the effect of static atomic displacements in InGaAs. Due to the strong bending of the (002) planes in the QD area we use the (200) planes for the lattice-fringe imaging in §3.2. These planes are oriented perpendicular to the buffer/QD-interface and less affected by distortions. More details on the accuracy of CELFA and sources of error can be found in [23]. Scanning transmission electron microscopy (STEM) with a high-angle annular dark-field (HAADF) detector was carried out using a ZEISS 922 Omega with a LaB_6 emitter to obtain composition sensitive (Z contrast) images.

Low temperature (5 K) PL spectra were acquired using an InGaAs detector and a spectrometer equipped with a 600 mm^{-1} grating. The excitation was carried out by the 442 nm and 325 nm lines of a HeCd laser. The signal was amplified by a standard lock-in technique.

3. Results and discussion

3.1. Influence of the InAs deposition rate on the morphology of InGaAs QD structures and distribution of group-III elements

The optimization of the parameters substrate temperature, deposited amount of InAs and growth rate is essential with respect to achieving 1.3 and 1.5 μm luminescence from InAs QD structures. We focus here on the effect of the InAs deposition rate for otherwise already optimized conditions (see §2 and [9]) which needs to be reduced to extremely low values. A strong effect is observed on the PL peak position which shifts from 1.21 eV for sample S_{high} (0.08 ML/s) to 1.04 eV for S_{low} (0.0056 ML/s) at 6 K. Bright-field (BF) TEM micrographs of plan-view samples are presented in figure 1(a) for S_{low} and in 1(b) for S_{high}. The growth-rate reduction is correlated with a significant decrease of the QD density from 1.1×10^{11} cm^{-2} to 2.0×10^{10} cm^{-2} and the strain field associated with the QDs changes from a small dot-like shape for S_{high} to a square-shaped one with a larger extension for S_{low}.

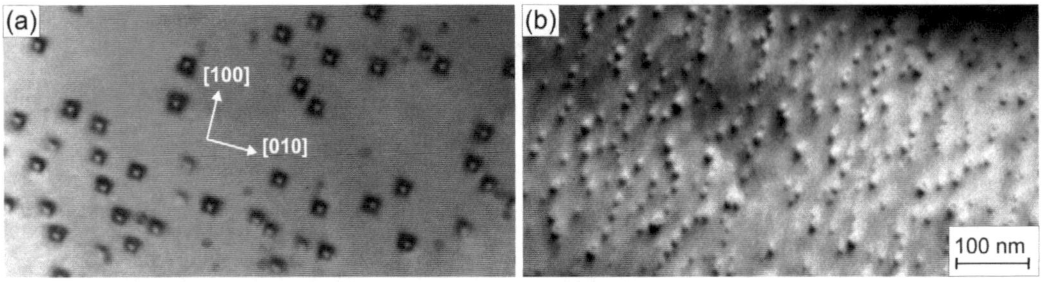

Figure 1. Plan-view BF images taken along the [001]-zone axis of samples with an InAs deposition rate of (a) 0.0056 ML/s and (b) 0.08 ML/s. The magnification in (a) and (b) is identical.

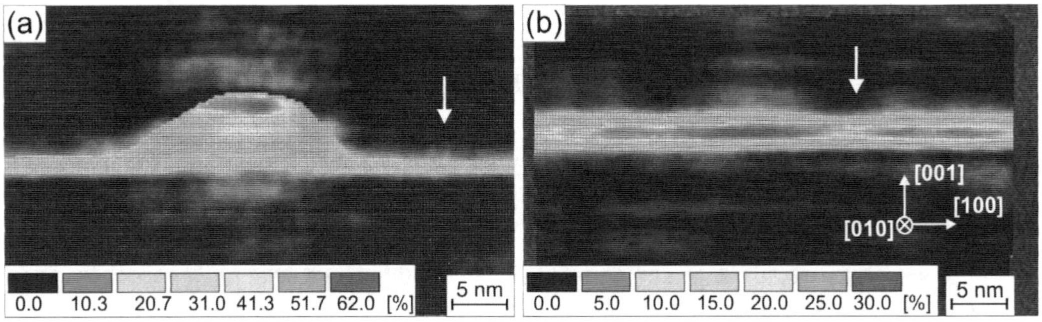

Figure 2. Colour-coded cross-section maps of the In distribution in the samples (a) S_{low} and (b) S_{high}. Note the different scaling of the colour coding for the In concentration in (a) and (b). The arrows mark WL regions where the composition profiles in figures 3(a,b) are extracted.

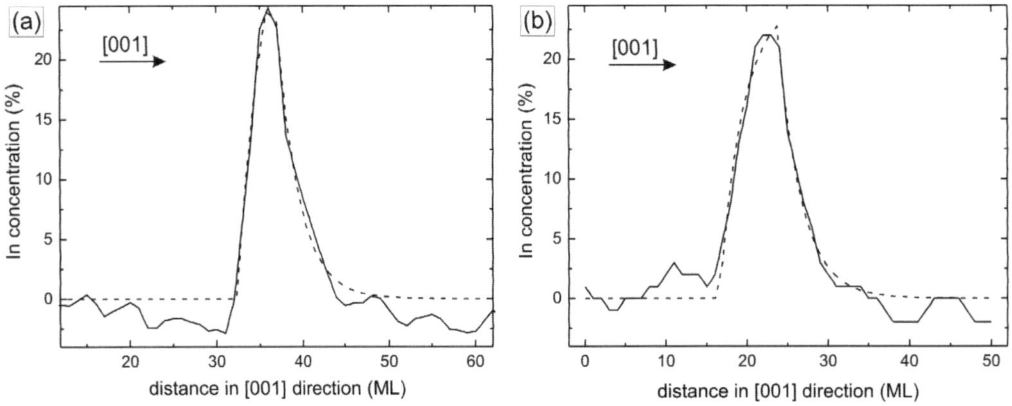

Figure 3. In-concentration profiles as a function of distance in growth direction in the WL regions indicated by arrows in figures 2(a,b) for samples (a) S_{low} and (b) S_{high}. Dashed lines are fit curves according to Eqs. (1) and [11].

Colour-coded cross-section views of the In distribution obtained by CELFA are presented in figure 2(a) for S_{low} and in 2(b) for S_{high}. Note that a different scaling of the colour coding was used. The QDs with a height of about 10 nm clearly emerge from the WL in the low-growth-rate sample. The In concentration in the QD in S_{low} increases gradually to about 60 % at the top of the QD. However, the real In concentration will be higher because we measure average values given by the composition of the QD and the embedding GaAs cap layer. In contrast, Figure 2(b) shows a layer with fluctuating In concentration up to only 30 % with a slightly varying thickness around 6 nm for S_{high}. The regions with high In concentration are assigned to the QDs but the interpretation of figure 2(b) is not straightforward because the high QD density has to be considered with respect to the finite TEM sample thickness. QDs may therefore overlap along the electron-beam direction. However, it is clear that the height of the QDs is rather small in S_{high} and the QDs are almost embedded completely in the WL. The apparent high In concentration (light blue colour) in the buffer and cap layers above and below the QDs is an artefact induced by the strong bending of the (002) planes in the vicinity of the QDs.

Composition profiles along the [001]-growth direction are presented for S_{low} and S_{high} in figure 3. The profiles are measured in the planar WL regions indicated by arrows in figures 2(a,b). The asymmetry of the profiles is a characteristic feature of In segregation. Maximum In concentrations between 20 and 25 % are found independently of the deposition rate. The total amount of indium in the WL corresponds to 1.7 ML in both samples which is obtained by the integration of the composition profiles.

The measured In concentrations x in figure 3 were fitted by applying the segregation model of Muraki [11] using the equations

$$x(n) = \begin{cases} 0 & : \quad n < 1 \\ x_0(1-R^n) & : \quad 1 \le n \le N \\ x_0(1-R^N)R^{n-N} & : \quad n > N \end{cases} \tag{1}$$

where n is the ML number in growth direction, x_0 the nominal In concentration, R the segregation efficiency and N the total amount of deposited indium expressed in MLs of InGaAs. The parameters x_0, N and R are considered as fit parameters. As these parameters define different characteristics of the profile (x_0 and R the slope of the increasing part and the maximum value, R the shape of the decreasing part, and N the position of the maximum), unique values of the fit parameters are obtained. From several fitted profiles in both samples, In-segregation efficiencies of 0.69 ± 0.03 for S_{low} and 0.7 ± 0.03 for S_{high} are derived which agree within the error margins.

The evaluation of the concentration profiles demonstrates that the properties of the WL and the segregation efficiency do not depend on the InAs deposition rate. The change of the island density and size results solely from the redistribution of deposited InAs exceeding 1.7 ML. The slow growth promotes In redistribution which leads to an accumulation of indium in large QDs. Fast surface diffusion is necessary for the redistribution process which limits T_{sub} to relatively high values around 500 °C. However, to model quantitatively the PL energy, the real In concentration in the QDs must be known. It is in particular interesting to see if binary InAs is contained in the top region of the QDs. A procedure is presented in the next subsection which takes into account the embedding GaAs cap layer. Considering the substrate temperature of 500 °C during InAs deposition, the measured segregation efficiency of 0.7 is low compared to values compiled in [13] which can be attributed to the high BEPR of 80 applied during the growth of these samples. Using a lower substrate temperature of 475 °C and a BEPR of only 4, Muraki et al. [11] found a high segregation efficiency of 0.83 which underlines the strong effect of the BEPR on In segregation.

3.2. Group-III element distribution in InGaAs quantum dots

The determination of the real composition of embedded QDs requires – in addition to the CELFA data – information on the local sample thickness as well as shape and size of the QDs. The following analysis was carried out for QDs in sample S_{low} because the QDs in S_{high} are too small. The local sample thickness was determined by recording a tilt series of (002) dark-field images. The sample was tilted around an axis parallel to the [100] direction in steps of 5° starting close to the [010]-zone axis. With increasing tilt angle the projection of the thin WL broadens. By measuring the width of the projected WL at a given tilt angle the local sample thickness can be calculated using simple trigonometry in analogy to the procedure described in [16]. Images taken at large tilt angles also show whether a QD is contained completely in the TEM sample or whether it is sectioned by the sample surface. The latter would be useless for further study because the measured In concentration would always be too low.

The square symmetry of the QD strain field in figure 1(a) suggests a pyramidal island structure with edges aligned along the [100] and [010] directions. This shape is verified by HAADF STEM cross-section images as shown in figure 4(a) where a pyramidal-type island with a truncated top is seen. Angles of about 45° between the side faces and base suggest {101} side faces.

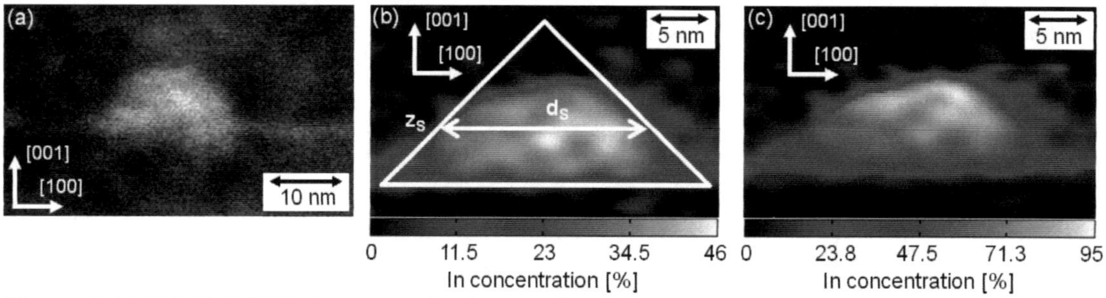

Figure 4. (a) HAADF STEM cross-section image of an InGaAs island and colour-coded cross-section maps of the In-distribution obtained by CELFA (b) before correction and (c) after correction. Note the different scaling of the colour coding in (b) and (c).

A colour-coded map of the In distribution in a QD before processing is presented in figure 4(b) which shows a maximum In concentration of 46 % in the center of the QD. To minimize artefacts due to lattice-plane bending (200) lattice-fringe images were evaluated. To correct the CELFA data for the thickness-dependent averaging effect, a projection of an {101}-facetted pyramid was fitted on the QD in figure 4(b). The relevant parameters for the correction are the sample thickness d = 32 nm (in this case) and the distance d_S between the {101} facets at the vertical position z_s. The ratio V between the sample thickness d and the local thickness of the QD d_s (see figure 4(b)) is given as a function of the vertical coordinate z_s, $V(z_s) = d / (d_B + 2\cdot(z_B - z_s))$ with the basis width of the pyramid d_B at the position z_B. Multiplication of V with the In concentration obtained by CELFA yields the corrected In concentration displayed in figure 4(c) (note the different scale of the colour coding in figures 4(b,c)). A maximum In concentration of 95 % is reached at the top of the island. The thickness correction also leads to a rearrangement of the In distribution with the highest In concentration in the top region of the QD as opposed to the uncorrected data. However, a large fraction of the QD still contains significantly lower In concentrations starting at about 50 % near the base. This is in rather good agreement with earlier investigations on similar QDs by different methods like electron energy loss spectroscopy by Wang et al. [18], who used a similar procedure for the correction of the measured In concentration, and cross-sectional scanning tunnelling microscopy [19].

There are several factors which affect the accuracy of the data. First, the QD shape may deviate from the ideal shape of an {101}-facetted pyramid which applies especially to the top of the QD which is not completely flat. Second, a non-quadratic QD base cannot be detected in cross-section samples and would lead to an error for the corrected In concentrations. Another contribution to the error is caused by the thickness measurement because the boundaries of the projected WL are blurred, limiting the accuracy of the width measurement and the derived sample thickness. Apart from the intrinsic errors of CELFA [23], the In concentration should be measurable with a relative error of about 10 % using the suggested procedure with a more precisely known sample thickness and neglecting the fact that shape information along the electron-beam direction for a specific QD is not accessible.

3.3. Influence of the substrate temperature during GaAs cap-layer growth

In the following we present results which illustrate the effect of the reduction of the substrate temperature during GaAs cap layer deposition. The investigation of the plan-view samples (not presented here) shows that the sample capped at 485 °C contains only defect-free islands while a small density of relaxed islands with defects is observed for the sample capped at 410 °C. The formation of relaxed islands can be attributed to In redistribution during the growth interruption after InAs deposition that was necessary to reduce the substrate temperature. Figures 5(a-c) show colour-coded maps of the local In concentration for the samples overgrown at T_{cap} = 485 °C (sample S_{485}), T_{cap} = 410 °C (sample S_{410}) and T_{cap} = 350 °C (sample S_{350}). Again, different colour scales were chosen in figures 5(a-c) for a more differentiated visualization of the In distribution. In-concentration profiles along the [001]-growth direction in the WL regions (at the position of white arrows in figures 5(a-c)) of all samples are displayed in figures 5(d-f).

According to figure 5 the reduction of T_{cap} has a strong effect on the In distribution. Lowering T_{cap} from 485 °C to 410 °C leads to a decrease of the thickness of the WL with a full width at half maximum (FWHM) for sample S_{485} of 10 ML to 5 ML for S_{410}. The reduction of the WL width is compensated by a corresponding increase of In concentrations with maximum values in the WL of 13% in S_{485} and 27% in S_{410} (figures 5(d,e)). These effects are a clear indication of the reduction of In segregation.

Sample S_{350} does not show the expected behaviour (figures 5(c,f)) because a strong broadening of the In distribution (WL with a FWHM of 13 ML) and a general reduction of the In concentration is observed with an overall maximum In concentration of only 14 %. We assume that interdiffusion occurred in this sample for the following reasons. A high concentration of point defects was probably introduced due to the low T_{cap} which is supported by the weak PL intensity of this sample. A potentially high vacancy concentration may lead to strong interdiffusion of In and Ga atoms after increasing T_{cap} from 350 °C after the deposition of 20 ML GaAs to 570 °C. The dominant influence of interdiffusion on the In distribution is also supported by the symmetrical composition profile in figure 5(f) in contrast to the asymmetrical ones for S_{485} and S_{410}.

Segregation efficiencies were determined on the basis of the measured profiles using Eqs. (1). The fit curves (dashed lines) are included in figures 5(d,e). We obtain $R = 0.76 \pm 0.02$ for sample S_{485} which exceeds $R = 0.7$ obtained for the samples in §3.1 deposited at 500 °C. However, here a low V:III BEPR of only 10 was used. For sample S_{410} $R = 0.59 \pm 0.09$ is derived. This value with a large error has to be considered with caution because the temperature was changed during the growth. It was 485°C for the ascending part of the profile during InAs deposition and 410 °C for the descending part associated with the cap-layer growth. The amount of InAs in the WL is 1.7 – 1.8 ML in all samples in agreement with the values obtained for the samples in §3.1.

Lowering T_{cap} to 410 °C leads to an obvious reduction of segregation-induced intermixing which could indeed be utilized as a technique to maximize the indium content in QDs which are grown under optimized conditions for long-wavelength luminescence.

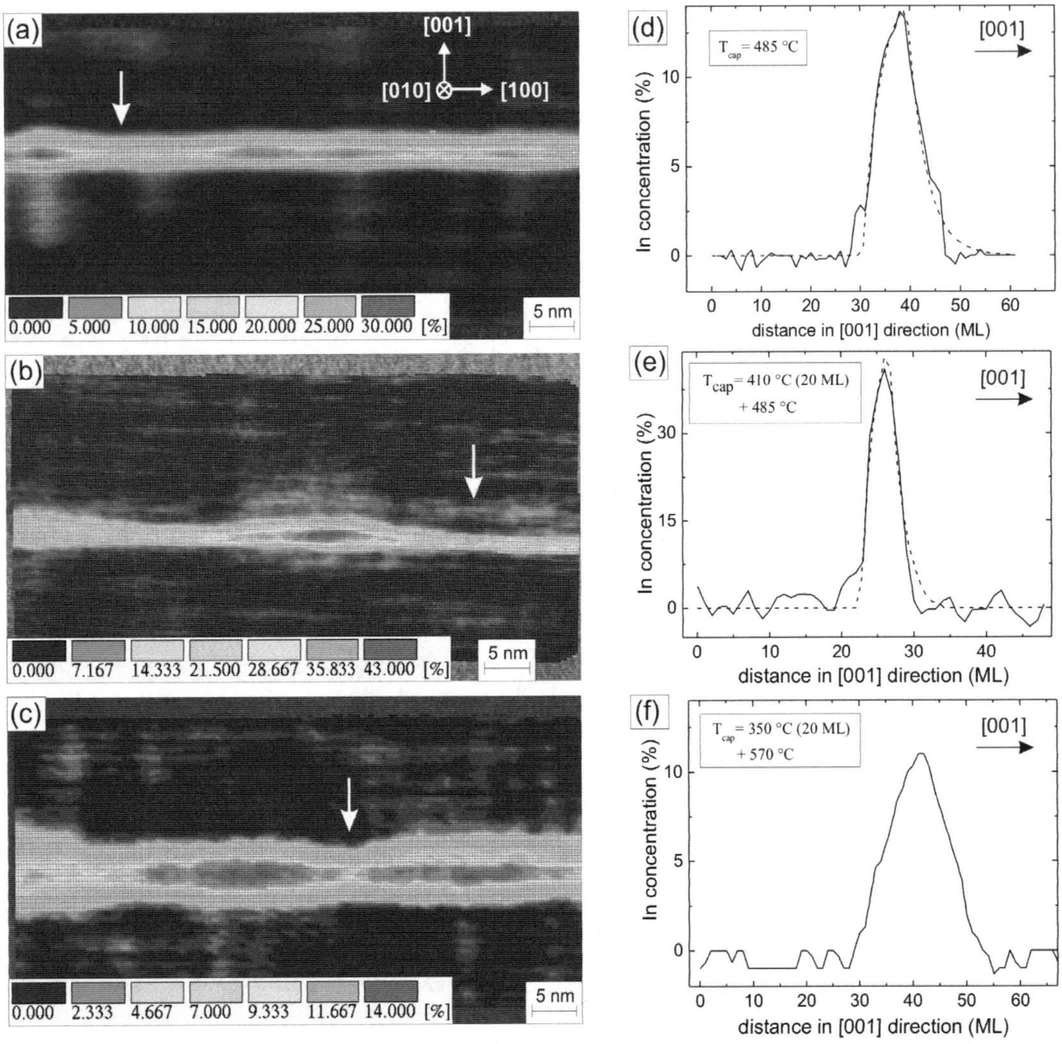

Figure 5. Colour-coded cross-section maps of the In distribution obtained by CELFA for samples (a) S_{485}, (b) S_{410} and (c) S_{350} and corresponding In-concentration profiles extracted from the WL regions marked by the arrows in the composition maps for samples (d) S_{485}, (e) S_{410} and (f) S_{350}.

4. Summary

A TEM and PL study of MBE-grown InAs QD structures was performed with respect to obtaining large QDs with a high In concentration which are required for luminescence in the 1.3 μm regime. Special attention was paid to In segregation which strongly affects the distribution of group-III elements in these structures. In-segregation coefficients were derived from atomic-scale composition profiles obtained on the basis of {200} lattice-fringe HRTEM images by applying the CELFA evaluation technique.

The effect of the InAs deposition rate was studied under otherwise optimized conditions for large QDs. A high substrate temperature of 500 °C and a low growth rate of 0.0056 ML/s are essential to allow In redistribution during the deposition process. The In-segregation efficiency of 0.7 is low compared to literature data which is attributed to the high V:III BEPR of 80 which is – in addition to the temperature – an important parameter for In segregation. The same In-segregation efficiency was observed for a deposition rate of 0.08 ML/s. A procedure was developed for the determination of the real In concentration in InAs QDs which takes into account the finite TEM sample thickness and the embedding of the QDs in a GaAs cap layer. The analyses were carried out for QDs grown under optimized conditions for long-wavelength luminescence. It is found after correction of the measured In distribution that the In concentration increases from about 50 % at the bottom to almost 100 % at the top of the QD.

Finally, the effect of the temperature during cap-layer growth was studied. The InAs QD layers were deposited under identical conditions but the GaAs capping was performed at 485, 410 and 350 °C. The In concentration in the QD structure increases strongly due to the reduction of In segregation during cap-layer growth at 410 °C. A further decrease of T_{cap} to 350 °C leads to strong broadening and intermixture between InAs and GaAs. This effect is attributed to interdiffusion due to a high vacancy concentration introduced at 350 °C and the increase of the temperature to 570 °C after the deposition of 20 ML GaAs. Nevertheless, lowering of the substrate temperature during cap-layer growth can be a reasonable procedure to maximize the In concentration in InAs QDs.

Acknowledgements
This work has been performed within project A2 of the DFG Research Center for Functional Nanostructures (CFN). It has been further supported by a grant from the Ministry of Science, Research and the Arts of Baden-Württemberg (Az: 7713.14-300)

References
[1] Arakawa Y and Sakaki H 1982 *Appl. Phys. Lett.* **40**, 939
[2] Michler P, Kiraz A, Becher C, Schoenfeld W V, Petroff P M, Zhang, Hu E, and Imamoğlu A 2000 *Science* **290**, 2282
[3] Fathpour S, Holub M, Chakrabarti S, and Bhattacharya P 2004 *Electron. Lett.* **40**, 694
[4] Löffler W, Tröndle D, Fallert J, Kalt H, Litvinov D, Gerthsen D, Lupaca-Schomber J, Passow T, Daniel B, Kvietkova J, Grün, M, Klingshirn C, and Hetterich M 2006 *Appl. Phys. Lett.* **88**, 062105
[5] Walther T, Cullis A G, Norris D J, and Hopkinson M 2001 *Phys. Rev. Lett.* **86**, 2381
[6] Cullis A G, Norris D J, Walther T, Migliorato M A, and Hopkinson M 2002 *Phys. Rev. B* **66**, 081305R
[7] Litvinov D, Gerthsen D, Rosenauer A, Schowalter M, Passow T, Feinäugle P, and Hetterich M 2006 *Phys. Rev. B* **74**, 165306
[8] Chu L, Arzberger M, Böhm G, and Abstreiter G 1999 *J. Appl. Phys.* **85**, 2355
[9] Passow T, Li S, Feinäugle P, Vallaitis T, Leuthold J, Litvinov D Gerthsen D, and Hetterich M 2007 *J. Appl. Phys.* **102**, 073511
[10] Moison J, Guille C, Houzay F, Barthe F, and Rompay V M1989 *Phys. Rev. B* **40**, 6149
[11] Muraki K, Fukatsu S, Shirakia Y, and Ito R 1992 *Appl. Phys. Lett.* **61**, 557
[12] Rosenauer A, Gerthsen D, Van Dyck D, Arzberger M, Böhm G, and Abstreiter G 2001 *Phys. Rev. B* **64**, 245334
[13] Piscopiello E, Rosenauer A, Passaseo A, Montoya Rossi E H, and van Tendeloo G 2005 *Phil. Mag.* **85v** 3857
[14] da Silva M J, Quivy A A, Martini S, Lamas T E, da Silva E C F, and Leite J R 2003 *Appl. Phys. Lett.* **82**, 2646
[15] Nishi K, Saito H, Sugou S, and Lee J 1999 *Appl. Phys. Lett.* **74**, 1111
[16] Litvinov D, Blank H, Schneider R, Gerthsen D, Vallaitis T, Leuthold J, Passow T, Grau A,Kalt H, Klingshirn C, and Hetterich M 2008 *J. Appl. Phys.* **103**, 083532
[17] Ustinov V M, Maleev N A, Zhukov A E, Kovsh A R, Egorov A Yu, Lunev A V, Volovik B V, Krestnikov I L, Musikhin Yu G, Bert N A, Kop'ev N A, Alferov Z I, Ledentsov N N, and Bimberg D 1999 *Appl. Phys. Lett.* **74**, 2815

[18] Wang P, Bleloch A L, Falke M, Goodhew P J, Ng J, and Missous M 2006 *Appl. Phys. Lett.* **89**, 072111

[19] Bruls D M, Vugs J W M, Koenraad P M, Salemink H W M, Wolter J H, Hopkinson M, Skolnick M S, Long F, and Gill S P A 2002 Appl. Phys. Lett. **81**, 1708

[20] Rosenauer A and Gerthsen D 1999 *Ultramicroscopy* **76**, 49 and Rosenauer A 2003 *Transmission Electron Microscopy of Semiconductor Nanostructures - An Analysis of Composition and Strain* (Heidelberg, Springer), Springer Tracts in Modern Physics **182**

[21] Bravman J and Sinclair R 1984 *J. Electron Microscopy Technique* **1**, 53

[22] Rosenauer A, Schowalter M, Glas F, and Lamoen D 2005 *Phys. Rev. B* **72**, 085326

[23] Rossi E H M, Rosenauer A, and van Tendeloo G 2007 *Phil. Mag.* **87**, 4461

Journal of Physics: Conference Series **209** (2010) 012007 doi:10.1088/1742-6596/209/1/012007

STEM-EELS investigations of high-k dielectrics

AJ Craven, M MacKenzie[1], P Longo and S McFadzean

Department of Physics and Astronomy, University of Glasgow, Glasgow, G12 8QQ, UK

E-mail: a.craven@physics.gla.ac.uk

Abstract. Metal (e.g. TiN, Ta_2N) inserted hafnia-based high-k gate stacks on Si as well as gallium oxide/gallium gadolinium oxide gate stacks on III-V semiconductors (GaAs, InGaAs, AlInAs) are being investigated for current and future MOSFET applications. The combination of light, medium and heavy elements in multilayers with widths down to a nanometre is a major challenge for high spatial resolution EELS spectrum imaging, especially as the interfaces have significant roughness on this scale. However, it is possible to identify layers significantly thinner than the widths of the profiles that result from beam broadening and/or roughness. To deal with the range of edges present, it is essential to collect edges out to losses of ~3keV. Edges from some elements are difficult to extract but, in suitable case, the ELNES on other edges can be used to give information on such elements. Recording the low loss region of the spectrum as well as the core loss regions at each pixel allows much fuller processing of the data including determination of the absolute numbers of atoms per unit volume. In this way, the distributions of the different phases can be estimated and the overall accuracy is excellent. The results are also in good agreement with earlier work using ELNES.

1. Introduction

In the continued drive to reduce the size of digital semiconductor devices, the traditional SiO_2 gate dielectric has become so thin that leakage due to tunnelling has imposed a limit. This was mitigated to some extent by moving to Si(O,N) but the International Technology Roadmap for Semiconductors (ITRS) has shown the requirement for high-k dielectrics over a number of editions [1]. Much work has gone into their investigation throwing up a wide range of problems. These have led to a steady reduction in the targets in each subsequent ITRS edition. Despite the many problems, INTEL has launched commercial devices which include HfO_2-based dielectrics [2]. There is still intense research activity on HfO_2-based dielectrics but a range of other materials is also being investigated, often involving rare earths [e.g. 3, 4]. The thermal processing required in making a device, including a spike anneal to activate the dopants, causes both physical and chemical changes in the system including re-growth of SiO_2 at the substrate-high-k dielectric interface. While decreasing the effective dielectric constant, if done in a controlled manner, such re-growth can be beneficial as it separates the carriers in the channel from the trapped charge and defects in the high-k dielectric and so stops degradation of the channel mobility [e.g. 5].

[1] Now at Institut für Mineralogie, Westfälische Wilhelms-Universität Münster, Germany

The situation is complicated further by the desire to remove the depletion capacitance associated with the traditional highly-doped poly-Si gate electrode and also to prevent diffusion of the dopant through the thin dielectric to the channel. Both are achieved by replacing the poly-Si electrode with a metallic one. A metal with an effective work function near the conduction band edge is required for n-type devices while one with an effective work function near the valence band edge is required for p-type devices. Again changes of effective work function occur during thermal processing and it seems that one dielectric/metal pair may need to be used for n-MOS devices and a second pair for p-MOS devices [6]. A wide range of metal electrodes has been investigated with TiN, TiAlN, Ta_2N, $Ta_2(C,N)$ and Ru_2O being amongst them.

Looking further into the future, semiconductors with higher channel mobility are under investigation including III-V materials for n-type devices and Ge for p-type devices with co-integration of both with Si envisaged to get ultimate CMOS. Passlack *et al* showed that deposition of Ga_2O_3 under the correct conditions unpins the Fermi level in GaAs allowing MOSFET action but that it gives an unacceptably high leakage current [7]. Incorporating Gd into the oxide after the first ~1nm, gives gallium gadolinium oxide (GGO) on top of a thin Ga_2O_3 template layer and this both unpins the Fermi level and gives acceptably low leakage. Using this, world leading MOSFETS have been made on GaAs-based heterostructures incorporating a buried InGaAs channel [8]. However, to scale down such devices to the size needed beyond the 22nm technology node, two issues must be addressed. The first is to increase the mobility in the channel. The second is to achieve sufficiently low access resistance. Only if both are achieved will the necessary drive current be achieved. To increase the mobility, an increase in the In content in the channel seems the most promising approach but raises issues with the dielectric. While the Ga_2O_3/GGO combination is still a promising candidate, a range of other dielectrics is under investigation [9-12].

The wide range of atomic numbers present and the fine scale on which they are grown produces a challenge for any characterisation technique. Here the information available from nanoanalytical electron microscopy is considered and, in particular, from electron energy loss spectroscopy (EELS) carried out in the scanning transmission electron microscopy (STEM) mode using spectrum imaging (SI) [13]. Not only does this provide information on elemental distributions and composition from the intensities of the ionisation edges but the electron loss near edge structure (ELNES) also gives information on the local chemistry. The challenge is increased significantly by the fact that the interfaces are frequently rough on the nanometre scale rather than being atomically flat.

2. Experimental Methods

The Ga_2O_3/GGO on the III-V heterostructure discussed below was grown by MBE on a 400 semi-insulating GaAs substrate, using a dual chamber system and the method is described in [14, 15]. The III–V heterostructure was grown in one chamber and consists of a 100 nm GaAs/AlGaAs superlattice layer, a 65 nm AlGaAs buffer layer, a 10 nm InGaAs layer, a 5 nm AlGaAs barrier layer and a 1 nm GaAs spacer layer. The heterostructure was transferred under ultra-high vacuum conditions to a second chamber, where oxide deposition took place. The Ga_2O_3/GGO gate stack consists of a template layer of ~1nm of Ga_2O_3 with a carefully controlled interface which unpins the GaAs surface. This is followed by ~10nm of GGO. After removal from the MBE chamber, Pt was deposited on top of the GGO by electron beam evaporation [16].

The metal inserted "hafnium silicate" (HfSiO) dielectric gate stack on silicon sample discussed was grown in the following way. The native oxide was removed from a Si (100) wafer by etching in $HF:H_2O$ (1:100) solution. The wafer was then put through an O_3/DI water cleaning sequence resulting in the growth of ~1 nm chemical oxide. Metal organic chemical vapour deposition (MOCVD) was used to deposit approximately 4nm of HfSiO (a co-deposition of 70% HfO_2 and 30% SiO_2). The wafer was then given a standard degas treatment at 330 °C for 40 s and ~10 nm of TiN was deposited by physical vapour deposition (PVD). This was followed by 100 nm of amorphous Si and the resulting stack annealed at 1000 °C for 10 s to simulate dopant activation. During this step, the amorphous Si crystallised into poly-Si and it was likely that the HfSiO phase separated into HfO_2 and

SiO_2 rich phases. The wafer surface was exposed to the clean room atmosphere as it was moved between process tools. In particular, this occurred after the HfSiO and metal deposition steps. However, there was no vacuum break between the degas stage and the metal deposition.

Cross-sectional TEM specimens were prepared by cutting, grinding, polishing, dimpling and ion milling. A Gatan PIPS ion mill was used for the thinning. The III-V sample was finished by using 200eV energy ions in a Technoorg-Linda GentleMill. The specimens were examined in an FEI Tecnai TF20 operated at 200 keV. Both conventional transmission (CTEM) and scanning transmission (STEM) modes were used. In the latter mode, a probe half angle of ~8 mrad was used with a spot size in the range 0.2 to 0.7 nm depending on the particular experiment. Electron energy loss spectroscopy (EELS) was performed using a Gatan ENFINA spectrometer. The data was collected in the form of spectrum images using the spectrum imaging plug-in of DigitalMicrograph and the scan was provided by DigiScan via DigiScanII hardware. The dispersion, collection angle and the integration time were chosen in the light of the edges that were to be collected.

Where possible, the edge intensities were extracted after removal of the background using a power law model fitted in front of the edge. However, the edge overlaps and edge shapes often require more detailed processing to extract the intensities and the processing parameters required may change as the composition changes. More details are given with the results below.

3. Results and discussion

3.1. Ga_2O_3/GGO on InGaAs/AlGaAs/GaAs

Analysis of the GGO itself has been reported previously [16, 17]. Study of the dielectric on a range of heterostructures is now underway. Here an InGaAs/AlGaAs/GaAs heterostructure is used as an example [16]. Figure 1a shows a STEM annular dark field (ADF) image of region around the interface. The sensitivity of the ADF signal to the atomic number of the elements clearly delineates the InGaAs, the AlGaAs, the GaAs, the Ga_2O_3 and the GGO. However, it is also clear that the interfaces are not atomically flat. Figure 1b shows a lower magnification image with the area selected for the spectrum image acquisition. A 0.7 nm probe was scanned over 12x22 pixels with a step size of 1.4 nm. The half angle of collection was 22 mrad and the dispersion was 1 eV/ch. The spectra were recorded over the energy range 480 eV to 1820 eV. The contribution to the spectrum from the stray scattering in the Schottky field emission gun [18] was minimised by using a small non-beam defining C1 aperture and the extraction voltage and the gun lens settings were chosen so that the stray signal stopped ~200 eV before the Ga $L_{2,3}$ edge threshold. GGO undergoes radiation induced crystallisation which is accompanied by some phase separation. To obtain better signal to noise ratio while keeping within an acceptable dose, sub-pixel scanning was used so that the dose was spread over the 1.4 nm x 1.4 nm pixel rather than being concentrated within the area covered by the 0.7 nm probe. To further improve the signal to noise ratio, the spectra from pixels in lines parallel to the interface were summed. This gives a loss of spatial resolution parallel to the interface but maintains it perpendicular to the interface.

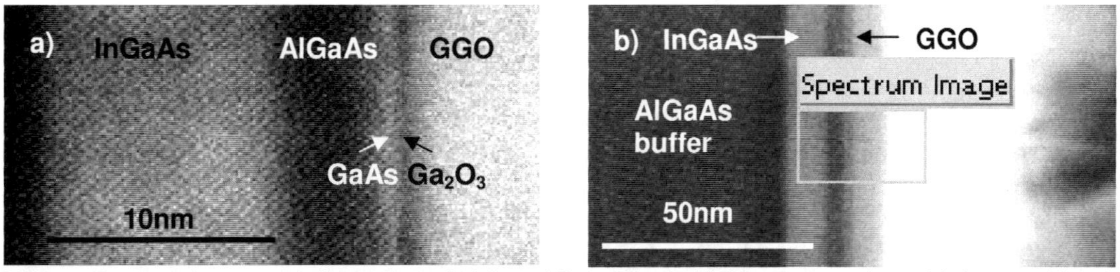

Figure 1. ADF images of the InGaAs/AlGaAs/GaAs/Ga_2O_3/GGO gate stack: a) high magnification showing details of the layers; b) lower magnification showing the region of the spectrum image.

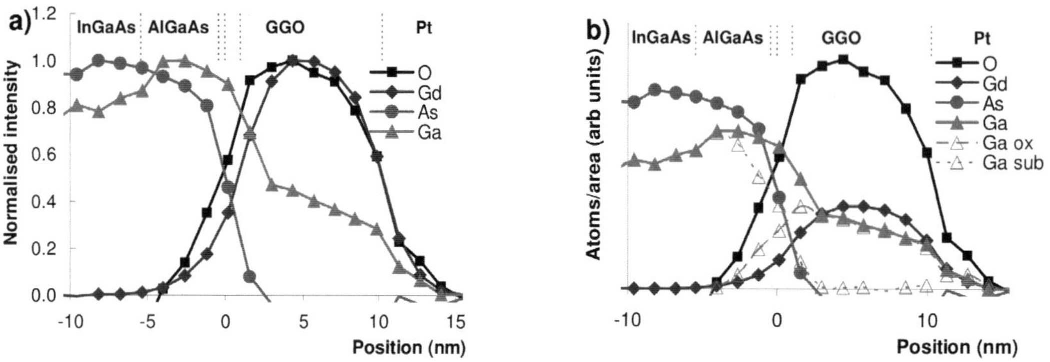

Figure 2. Elemental profiles across the InGaAs/AlGaAs/GaAs/Ga_2O_3/GGO gate stack: a) profiles normalized to a maximum intensity of unity; b) profiles normalized by cross-sections.

Figure 2a shows a section of the intensity profiles for the As $L_{2,3}$, Ga $L_{2,3}$, Gd $M_{4,5}$, and O K edges. The profiles have been normalised so that the maximum intensity of each is unity and the origin is chosen at the position where the O profile is ~0.5. This is the position where the As profile is also ~0.5 and so is a good indicator of position of the interface between the substrate and the dielectric. Power law backgrounds were removed from under the As, Ga and O edges. The background fitting windows were chosen in regions of the spectrum image where perturbations are not significant. However, their use to process the whole spectrum image results in negative intensities for O in pixels containing little or no O but significant In (from -10 nm to -5 nm in figure 2a) and also for As in regions containing little or no As but significant Ga and Gd (from 3 nm to 11 nm in figure 2a) These negative counts are not shown. Since the Ga $L_{2,3}$ and Gd $M_{4,5}$ are close together, their intensities were separated by scaling the Ga $L_{2,3}$ shape from GaAs to the region of the Ga edge between the Ga and Gd thresholds [17]. More recent work has allowed the extraction of the In $M_{4,5}$ and the Al K edge intensities and this will be reported elsewhere.

It is clear from figure 2a that the Ga extends to the right of the As at the interface while the Gd also rises to the right of the O, showing the presence of the Ga_2O_3 template layer. Because the Ga is in both the substrate and the GGO layer, it is hard to estimate the width of the template layer from this profile. However, assuming that no Gd penetrates into it, the separation of the O and Gd profiles gives a direct measure of the template layer thickness of 1.2 nm with an uncertainty of less than 0.2 nm. How is this possible given that that the probe size is ~0.7 nm and the step size in the spectrum image is 1.4 nm? The answer is that the interface is actually sampled on a much finer scale because of a number of effects. Interface roughness, beam spreading and slight misorientation of the spectrum image with respect to the interface may all play a role in this. Since the displacement of the O and Gd profiles is essentially constant, the relative positions of the two profiles is a measure of the template layer width and the uncertainty involved is controlled by the uncertainty with which the displacement can be measured rather than by the widths.

Using GaAs, $Gd_3Ga_5O_{12}$ and Ga_2O_3 as standards, the intensities at each pixel can be made proportional to the number of atoms per unit area. Figure 2b shows the number of atoms per unit area obtained by this method in arbitrary units. Because the low loss intensity was not recorded, it is not possible to correct for the effects of image contrast and so the constant of proportionality will vary with position. Nonetheless, the atomic ratios at each pixel can still be evaluated. Assuming the oxide has the form $Gd_xGa_{0.4-x}O_{0.6}$, it is straightforward to determine the amount of Ga in the oxide (Ga_{ox}) and hence the amount in the substrate (Ga_{sub}) and these are also shown on figure 2b. Ga_{ox} extends to the left of the Gd at the interface, as expected, because it is obtained from the original O and Gd profiles where the O extends to the left of the Gd. However, the Ga_{sub} profile follows the As profile closely at the interface, which confirms that the interface between the oxide and the substrate is sharp even if rough. There is a clear difference in the Ga:As ratio in the InGaAs and the AlGaAs and the convergence of the Ga_{sub} and As profiles at the interface hints at the presence of the two monolayers of GaAs at the interface.

3.2. TiN on HfSiO on Si

After the final thermal processing stage, this system is much less radiation sensitive than the III-V system above. HfO_2 and HfSiO dielectrics and poly-Si, TiN and Ta_2N gates have all been studied successfully and ELNES has been used to separate out the different phases and identify reactions at the metal/poly-Si interface [19-24]. Here data from the Si/SiO_2/HfSiO/TiN/poly-Si gate stack are considered. The data were recorded as a line spectrum image using the Glasgow dual EELS system [25]. This allows both the core loss and low loss regions of the spectrum to be recorded at each pixel before the probe is moved to the next pixel. The ADF signal is recorded at each pixel as part of the dataset. The data were acquired using a step size of 0.51 nm, an energy dispersion of 0.5 eV/channel and energy offsets of 56.5 eV and -543.5 eV for the core and low loss parts of the spectrum, respectively. The core loss and low loss data were each recorded using 15 integrations of 0.5 sec. For the low loss part of the spectrum, the fast beam switch was used to limit the time electrons entered the spectrometer to 3 msec per integration.

Recording both the core loss and low loss regions of the spectrum at the same pixel offers a number of advantages [25]. As a first step, the position of the zero-loss peak can be determined to sub-channel accuracy by fitting a peak to it. Combining this with the value of the voltage applied to the drift tube of the spectrometer when recording the core loss region, all of the spectral data can be aligned in energy to this accuracy giving an accurate energy calibration. If the energy ranges overlap, the two sections of the spectrum can be spliced together. Figure 3a shows the elemental profiles extracted from this spliced spectrum using power law backgrounds and the Si $L_{2,3}$, N K, Ti $L_{2,3}$ and O K edges. Each intensity was divided by the appropriate cross-section. Hartree-Slater cross-sections were used for the Si, N and O edges while a SIGMAL cross-section with white line correction was used for the Ti [26]. They were calculated using the routines in DigitalMicrograph. Thus the values are in units of atoms/nm^2 multiplied by the number of electron in the spectrum.

There are a number of issues with the values in figure 3a. The negative values for the Si and O in the TiN layer are due to background perturbations caused by the Ti in combination with multiple scattering. The number of atoms/nm^2 of Ti obtained using the SIGMAL cross-section is greater than the that for N despite the fact that comparison to a TiN standard shows that they should be the same provided there is no surface oxidation of the specimen [23]. An arbitrary scaling can be applied to force them to be equal, as in the lower Ti profile in figure 3a. However, there is an issue with surface oxidation and this is considered further below. Finally, it was not possible to extract a reliable Hf edge intensity from the Hf edges in the dataset.

It is possible to deal with these problems. Having the spliced spectrum, it can be Fourier logarithmically deconvoluted [26]. This removes the background problem in front of the Si $L_{2,3}$ edges. It also makes it relatively simple to use the Ti $L_{2,3}$ edges from a TiN standard to extract the O K edge when it is sitting on the Ti $L_{2,3}$ edges. However, in this particular case, there was some surface oxidation and so the underlying Ti contribution was subtracted using a Ti edge shape from the TiN layer in the stack. Thus any O present in the TiN layer does not appear in the resulting profiles. The TiN standard also allows scaling of the Ti to the N. The shape of the O K edge can be used to identify the relative amounts of SiO_2 and HfO_2 using MLLS fitting of edge shapes from standards. Finally, the profiles can be normalised by the low loss intensity to give the absolute values of the numbers of atoms per nm^2, as shown in figure 3b.

Now that the composition is known at each point, it is possible to calculate the local inelastic mean free path, λ, using the parameterised sum rule method of Malis et al [27]. Combining this with the t/λ profile from the low loss data gives the absolute thickness (t) profile. In fact there is significant thickness variation due to differential thinning between the TiN and the other phases. Dividing the profiles in figure 3b by the absolute thickness profile gives the absolute numbers of atoms per nm^3 and

Journal of Physics: Conference Series **209** (2010) 012007 doi:10.1088/1742-6596/209/1/012007

these are shown in figure 3c. The ADF image intensity recorded as part of the data set is also shown with a vertical offset. The position of the peak in the ADF signal clearly matches that of the Hf peak.

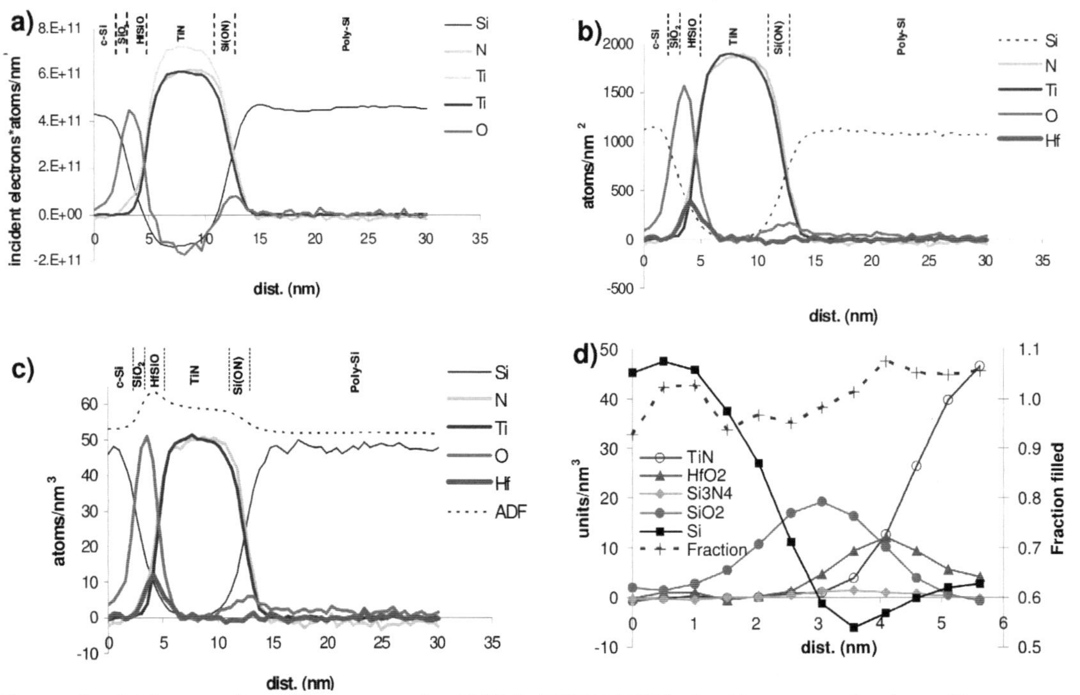

Figure 3. a) Elemental profiles across the Si/SiO_2/HfSiO/TiN/poly-Si gate stack; a) profiles extracted from the un-deconvoluted spectra and normalized by cross-sections; b) profiles extracted from the deconvoluted spectrum normalized by both cross-sections and the low loss intensity to give the absolute numbers of atoms per nm^2 along with the Hf profile extracted using ELNES; c) profiles after further normalizing by the local thickness to give the absolute numbers of atoms per nm^3 with the ADF signal shown offset; d) the distribution of phases in the dielectric region based on a mixture of stoichiometric compounds plus the calculated fraction of space filled by this distribution.

From figure 3c, the average number of Si atoms/nm^3 in the poly-Si region is 47.8 with a standard deviation of 0.9 compared to the bulk value of 49.9 for crystalline Si (c-Si). However, there is also an average of 1.6 O atoms/nm^3 in this region with a standard deviation of 0.8 atoms/nm^3. This is likely to be the result of oxidation of the TEM sample surface. The presence of this O will reduce the number of Si atoms/nm^3 below that for c-Si. If it is assumed that the O is in the form of SiO_2, then there are 0.8 units of SiO_2 per nm^3 and 47.0 atoms/nm^3 in the form of c-Si in the poly-Si. From the bulk density of quartz, the volume occupied by 0.8 SiO_2 units is 0.031 nm^3 while, from the bulk density of c-Si, 47.0 Si atoms occupy 0.941 nm^3. Thus both phases together occupy 0.97 nm^3 with a standard deviation of 0.02 nm^3 compared to the ideal value of unity i.e. on these assumptions, the accuracy is excellent in the poly-Si region.

However, there is clearly a fortuitous cancellation of systematic errors. The accuracies quoted for the mean free paths and the cross-sections are in the range 10% to 20%. The edge intensities are divided by the product of a cross-section and a mean free path, which is essentially a ratio of cross-sections. While this may itself help to cancel the systematic error, the core loss cross-sections and the inelastic mean free paths are evaluated by two very different approaches with a range of possibilities for the introduction of other systematic effects. In addition, the densities and stoichiometries of thin amorphous films can differ significantly from those in the bulk crystalline forms e.g. the density of quartz is higher than that of amorphous SiO_2 and it is not clear that the Si on the surface of the TEM specimen is fully oxidised. However, in the poly-Si region, these effects will only give a small correction to a small correction and so are not very significant.

Turning to the TiN layer, in the centre there are 50 TiN units per nm^3 while there are 53 in stoichiometric TiN. Here, there are some unresolved issues with oxidation of the TiN and how it should be treated and these are likely to be the reason for the low value. Until these issues are resolved, it is not possible to know whether the level of agreement will match that in the poly-Si region or whether the degree of cancellation of the systematic effects depends on the local composition.

The interfaces between the TiN and poly-Si in Hf-based gate stacks have been investigated previously [20-24]. As in the case of the III-V sample above, there is a separation of the elemental profiles showing an additional layer where the N extends to the right of the Ti in the region from 10-15 nm in figure 3c. In the same region, the O profile also goes through a peak. MLLS fitting using edge shapes from TiN, SiN_x and SiO_2 showed the presence of a Si(O,N) reaction layer in such a case [20,22].

In the dielectric region between the c-Si substrate and the TiN, the elemental distribution is quite complex. The O peaks 0.5 nm to the left of the Hf. The N also extends significantly to the left of the Ti in the region of the Hf peak. This was not observed when the dielectric was HfO_2 [20-23] and is therefore associated with nitriding of the HfSiO and, in particular, the phase separated SiO_2. The c-Si itself shows some degree of oxidation. If it is assumed that the phases present are c-Si, SiO_2, HfO_2, Si_3N_4 and TiN, it is possible to obtain the distribution of these phases, as shown in figure 3d. The numbers of molecular units per nm^3 are given on the left hand axis. Note that the Si_3N_4 profile is associated with three times more Si than the SiO_2 profile and so its height is "artificially" suppressed in such a plot. As a cross-check, it is possible to calculate the fraction of space filled assuming the bulk densities of these phases, as was done above for the poly-Si. This is plotted as the dashed line and its scale is given on the right hand axis. Its average value is 1.00 with maximum excursions of ± 0.07 but it is clear that systematic effects dominate random ones. Given the complexity of the elemental distributions and the uncertainties involved, this result shows excellent self-consistency.

Looking at the distributions of the phases themselves, the c-Si goes negative in the region of the O peak in figure 3c and is most negative at the position of the maximum of the O peak, suggesting a systematic error in the scaling. The HfO_2 profile peaks closer to the TiN while the SiO_2 profile peaks closer to the c-Si. This is to be expected since there is always some re-growth of SiO_2 by oxidation of the substrate during thermal processing. The Si_3N_4 profile peaks between those of the HfO_2 and the SiO_2. This is consistent with N diffusing in from the TiN side. The original HfSiO was deposited to give 70mol%HfO_2:30mol%SiO_2. However, since there is phase separation of the HfSiO in three dimensions and additional growth of SiO_2, the original HfSiO composition is not reflected directly in a line profile and it is not clear that it would be reflected even in a two dimensional map. What can be said is that the HfO_2 concentration is much higher than the SiO_2 concentration adjacent to the TiN. The HfO_2 and TiN distributions overlap significantly indicating that they overlap along the beam path. This is due to the roughness of the interface, which has been seen clearly in images e.g. [21]. At the peak of the HfO_2 profile, figure 3c shows that the Si:Hf ratio is close to unity. Overall, these observations are consistent with the HfSiO phase separating by SiO_2 moving towards the existing SiO_2 on the substrate and leaving the HfO_2 against the TiN. Such behaviour has been observed in images of HfSiO after annealing [28]. As the substrate is approached, the composition of the dielectric tends towards SiO_2.

4. Conclusions

The results presented show that it is possible to analyse nanometre-scale multilayer systems containing elements with a wide range of atomic number using EELS spectrum imaging with great success. Relative shifts of the elemental profiles that are smaller than the scale of the interface roughness or the spectrum image step size can often be detected. This can be particularly useful where radiation sensitivity is a limiting factor. Use of suitable standards can convert edge intensities into relative elemental concentrations. However, unless the low loss region of the spectrum is collected as well as the core loss region, corrections cannot be made for the effects of image contrast and local thickness. When using sub-nanometre step sizes, it is essential that the core loss and low loss regions of the spectrum are recorded at the same pixel before moving to the next pixel. If this is not done, the spatial registration is unlikely to be accurate enough to cope with the rapid spatial variations of the signals that are frequently encountered. Access to the complete spectrum allows accurate alignment and calibration of the energy scale and deconvolution of the multiple scattering. This often enables the use of a power law background subtraction where it had not previously been possible and allows use of similarly deconvoluted edge shapes from standards as references for separating overlying edges in the data. Once a procedure has been established for extracting the intensities of the edges, they can be converted to the number of atoms per unit area of specimen by dividing by the intensity of the low loss region and by an appropriate absolute cross-section. This procedure also corrects for the image contrast. In some cases, ELNES on an extractable edge can be used to provide information on an element whose edge intensities cannot be measured directly. If the areal concentration of all the elements is obtained, the composition at each pixel can be determined. This in turn allows the inelastic mean free path to be evaluated and used with the low loss spectrum to calculate the absolute thickness at each pixel. Hence the absolute number of atoms per unit volume can be evaluated. The accuracy here is limited not only by potential systematic errors in the processing but, more importantly, by the systematic errors in the cross-sections and inelastic mean free paths. For the former, reasonable values are available for K cross-sections and moderate ones for L cross-sections. Cross-sections for other edges need to be determined from standards containing the edge of interest and a suitable K (or possibly L) edge. Fortuitously, there seems to be reasonable cancellation of the systematic errors involved in calculating the number of atoms per unit volume and there is good agreement for bulk silicon and reasonable consistency in regions with complex elemental distributions. However, this is an area where improved accuracy is required. Whether this can be achieved directly using theoretical approaches or whether a case by case approach using standards is the only way forward is not yet clear.

The ultimate limit is set by the signal to noise ratio that can be obtained in the data without introducing significant electron beam induced change in the samples. To this end, it is essential that signal acquisition is orchestrated in such a way that the maximum information is recorded per incident electron. This means that the coupling of the specimen to the spectrometer is crucial with the collection half angle being big enough to accept a large fraction of the EELS signal. This problem is exacerbated when going to the very small probe size made possible by aberration correction because the larger probe half angle involved requires a larger acceptance angle. In fact, using aberration correction to obtain a higher probe current rather than a smaller probe is likely to be a better approach provided radiation damage is not the factor limiting the spatial resolution.

If it is elemental analysis that is the prime concern, a low dispersion (>1eV/channel) and the use of losses above 1kev is often useful [24]. For the study of ELNES, high dispersion is needed. Current CCD cameras used for EELS can only record a dynamic range of about 100 if the signal to noise ratio at the low intensity end is to be adequate for signal processing. Thus it is necessary to record two or three sections of the spectrum with different energy ranges and integration times in order to acquire data over the energy range 0 to 3keV, which includes useful edges for all elements up to Bi. To do this without introducing excessive deadtime needs a different approach to that used by the Glasgow dual EELS system. Better approaches are possible e.g. that of Tencé *et al* [29] and a commercial implementation of one of them is urgently need.

Instrumental artifacts also play a significant role in perturbing the data obtained. Stray scattering within a typical Schottky gun leads to perturbations of the background shape [18] as does scattering within the spectrometer itself. Ideally, a spectrum recorded when the probe passes through a hole in the specimen should show only intensity from the zero-loss peak and elsewhere the intensity should be negligible relative to that in an EELS spectrum from the specimen itself. Commercial systems rarely achieve this and there is always significant intensity present over the whole energy range, arising from scattering within both the gun and the spectrometer. This is an area where manufacturers must continue to strive for improvement.

Acknowledgments
The authors would like to thank: Prof I G Thayne and Dr M Holland of Glasgow University for the III-V materials; Prof S de Gendt of IMEC for the Si based materials; Mr B Miller for specimen preparation; EPSRC for funding under grants GR/S44280/01 and EP/F002610/1.

References

[1] ITRS web site: http://www.itrs.net/reports.html
[2] Intel web site: http://www.intel.com/pressroom/archive/releases/20070128comp.htm.
[3] Adelmann C, Van Elshocht S, Franquet A, Conard T, Richard O, Bender H, Lehnen P and De Gendt S 2008 *Appl. Phys. Lett.* **92**, 112902
[4] Tsoutsou D, Lamagna L, Volkos S N, Molle A, Baldovino S, Schamm S, Coulon PE and Fanciulli M 2009 *Appl. Phys. Lett.* **94**, 053504
[5] Fischetti MV, Neumayer DA,and Cartier EA 2001 *J. Appl. Phys.* **90**, 4587
[6] Zhang Z, Song SC, Huffman C, Hussain M M, Barnett J, Moumen N, Alshareef HN, Majhi P, Sim JH, Bae SH and Lee BH 2005 *Electrochem. Solid-State Lett.* **8**, G271
[7] Passlack M, Medendorp N, Gregory R and Braddock D 2003 *Appl. Phys. Lett.* **83**, 5262
[8] Hill RJW, Moran DAJ, Li X, Zhou H, Macintyre D, Thoms S, Asenov A, Zurcher P, Rajagopalan K, Abrokwah J, Droopad R, Passlack M and Thayne IG 2007 *IEEE Electron. Device Lett.* **28**, 12
[9] Xuan Y, Wu YQ and Ye PD 2008 *IEEE Electron. Device Lett.* **29**, 4
[10] Zhao H, Shahrjerdi D, Zhu F, Kim H-S, Ok I, Zhang M, Yum J H, Banerjee SK and Lee JC 2008 *Appl. Phys. Lett.* **92**, 253506
[11] Goel N, Heh D, Koveshnikov S, Ok I, Oktyabrsky S, Tokranov V, Kambhampati R, Yakimov M, Sun Y, Pianetta P, Gaspe CK, Santos MB, Lee J, Datta S, Pand M and Tsai W *Proc IEDM* 2008, 363
[12] Sun Y, Kiewra EW, de Souza JP, Bucchignano JJ, Fogel KE, Sadana DK and Shahidi S 2009 *IEEE Electron. Device Lett.* **30**, 5
[13] Jeanguillaume C and Colliex C 1989 *Ultramicroscopy* **28**, 252
[14] Droopad R, Abrokwah J, Rajagopalan K, Adams L, England N, Uebelhoer D and Passlack M 2007 *J. Cryst. Growth* **301–302**, 139
[15] Holland M, Stanley CR, Reid W, Patterson GW, Long AR, Longo P, Scott J, Craven AJ and Thayne IG 2007 *J. Vac. Sci. Technol.* B **25**, 1024
[16] Longo P, Craven AJ, Holland MC, Moran DAJ and Thayne IG 2009 *Microelectronic Engineering* **86**, 214
[17] Longo P, Craven AJ, Scott J, Holland M C and Thayne IG 2008 *Springer Proceedings in Physics* **120**, 317
[18] McComb DW and Weatherly GC 1997 *Ultramicroscopy* **68**, 61
[19] MacKenzie M, Craven AJ, McComb DW and Hamilton DA 2006 *Appl. Phys. Lett.* **88**, 022108
[20] MacKenzie M, Craven AJ, McComb DW and De Gendt S 2006 *Appl. Phys. Lett.* **88**, 192112
[21] MacKenzie M, Craven AJ, McComb DW and De Gendt S 2006 *J. Electrochem. Soc.* **153**k F215
[22] MacKenzie M, Craven AJ, McComb DW, De Gendt S, Docherty FT, McGilvery C and McFadzean S 2007 *Electrochemical and Solid-State Letters* **10**, G33
[23] Docherty FT, MacKenzie M, Craven AJ, McComb DW, De Gendt S, McFadzean S and McGilvery CM 2008 *Microelectronic Engineering* **85**, 61
[24] Craven AJ, MacKenzie M and McComb DW IPFA 2008: *Proceedings of the 15th International Symposium on the Physical and Failure Analysis of Integrated Circuits (Singapore)* 269
[25] Scott J, Thomas PJ, MacKenzie M, McFadzean S, Wilbrink J, Craven AJ and Nicholson WAP 2008 *Ultramicroscopy* **108**, 1586

[26] Egerton RF 1996 *Electron Energy Loss Spectroscopy in the Electron Microscope 2nd edition* (Plenum Press, New York)
[27] Malis T, Cheng SC, Egerton RF 1988 *J Elctron Microsc Tech* **8**, 193
[28] McGilvery CM, McFadzean S, MacKenzie M, Docherty FT, Craven AJ, McComb DW and De Gendt S 2008 *Springer Proceedings in Physics* **120**, 325
[29] Tencé M, Pinna H, Birou T, Guiraud L, Mayet A, Pertel C, Serin V and Colliex C 2006 *Proc of the 16th Int Microscopy Congress (Sapporo, Japan)* 824

Journal of Physics: Conference Series **209** (2010) 012008 doi:10.1088/1742-6596/209/1/012008

Strain distribution analysis in Si/SiGe line structures for CMOS technology using Raman spectroscopy

M Hecker[1], M Roelke[1], P Hermann[1,2], E Zschech[1] and V Vartanian[3]

[1] GLOBALFOUNDRIES Dresden Module One LLC & Co. KG, Wilschdorfer Landstrasse 101, D-01109 Dresden, Germany
[2] Fraunhofer Center Nanoelectronic Technologies CNT, Königsbrücker Strasse 180, D-01099 Dresden, Germany
[3] International Sematech Manufacturing Initiative (ISMI), 257 Fuller Road Suite 2200, Albany, NY 12203, USA

E-mail: michael.hecker@globalfoundries.com

Abstract. Strained silicon underneath the field-effect transistor gate increases significantly the charge carrier mobility and thus improves the performance of leading-edge Complementary Metal Oxide Semiconductor (CMOS) devices. For better understanding of the structure-strain relationship on the nanoscale and for optimization of device structures, the measurement of the local strain state has become essential. Raman spectroscopy is used in the present investigation to analyze the strain distribution in and close to silicon/embedded silicon-germanium (SiGe) line structures in conjunction with strain modeling applying finite element analysis. Both experimental results and modeling indicate the impact of geometry on the stress state. An increase of compressive stress within the Si lines is obtained for increasing SiGe line widths and decreasing Si line widths. The stress state within the Si lines is shown to be a mixed one deviating from a pure uniaxial state. Underneath the SiGe cavities, the presence of a tensile stress was observed. To investigate a procedure to scale down the spatial resolution of the Raman measurements, tip-enhanced Raman scattering experiments have been performed on free-standing SiGe lines with 100nm line width and line distance. The results show superior resolution and strain information not attainable in conventional Raman scans.

1. Introduction

Strained silicon thin film structures have excited an increasing interest during the previous years in basic research as well as for applications in high-performance microprocessor devices. This interest is related to the significant performance gain attainable by utilizing strained films in active transistor regions [1], which will become more important as in future the scaling of device dimensions cannot be continued without changing materials, technology and device architecture. Thus strain has become a key approach for extending CMOS to the next technology nodes.

One potential technique for introducing strain into the channel silicon regions of CMOS devices is employing a lattice mismatch of epitaxially grown film structures. In particular, silicon-germanium (SiGe) solid solution allows the fabrication of films with tunable lattice constants, and thus, it has been used to generate dedicated strain in Si films for improved CMOS performance [2]. To understand the impact of strain on device structures and to be able to optimize it, local measurements of the mechanical strain in such structures are mandatory. A useful technique for strain determination in silicon is Raman spectroscopy, enabling fast and nondestructive measurements with sub-µm resolution. This technique has the potential of sub-100 nm resolution in combination with near-field approaches [3]. Though still not downscaled to the lateral resolution of single MOSFET channel regions, Raman spectroscopy can be successfully applied to test structures with typical dimensions in the 100 nm range. For data evaluation, Raman measurements on patterned structures require combination with modeling, since usually the Raman signal does not yield the full tensor information related to the strain distribution. Such investigations taking into account the tensorial character of the strain state have been performed e.g. regarding the average strain state in series of parallel SiGe lines, where the asymmetric relaxation, i.e. the presence of larger strains parallel to lines and smaller strains perpendicular to lines was analyzed by ion scattering [4], X-ray scattering [5] and also Raman spectroscopy [6]. However, for understanding of the strain distribution in and close to structures relevant for strained devices, local strain profiles have to be measured, as achieved e.g. by Raman spectroscopy in silicon substrates underneath silicon oxide lines and silicon nitride lines with a µm-scale pitch [7,8]. The strain state in freestanding SiGe lines on a Si substrate was investigated by Jain et al. utilizing analytical modeling and Raman spectroscopy, showing compressive stresses in the lines increasing with reducing line heights [9,10]. This tendency is different for embedded SiGe line structures, as applied for strained Si CMOS technology. Here, an increasing height of the cavities supports the compressive strain in between the SiGe regions. However, above a critical thickness [11] defect formation leading to plastic relaxation can limit the strain level. Though less is known regarding the experimentally determined details of the strain distribution in such structures, valuable information has been obtained by TEM based approaches such as nano-beam diffraction [12] or electron holography [13], which require substantial sample preparation efforts.

In this study, we applied a combination of strain distribution determination by Raman spectroscopy and finite-element modeling (cf. e.g. [14]) on embedded SiGe / Si line structures with varying pitch and line widths ranging from the µm range down to 100 nm in order to study the impact of the line geometry onto the strain state. The combination of both techniques allows deriving predictions from the modeling also for smaller dimensions. This approach can thus provide helpful information to support technology progress and to optimize processes for CMOS manufacturing.

2. Experimental details

For sample preparation, a series of parallel SiO_2 lines was deposited onto (001) Si wafers, followed by etching of 45 nm deep cavities into the silicon between the lines and subsequent deposition of SiGe with a germanium content of 30 at% using a chemical vapour deposition (CVD) process. Thus, patterns of embedded SiGe/Si line structures were obtained with different SiGe line widths w_{SiGe} varying between 1 µm and 3 µm and Si line widths w_{Si} varying between 100 nm and 420 nm (cf. figure 1a). The lines with a length of 100 µm were arranged on pads of respective size (100 µm × 100 µm).

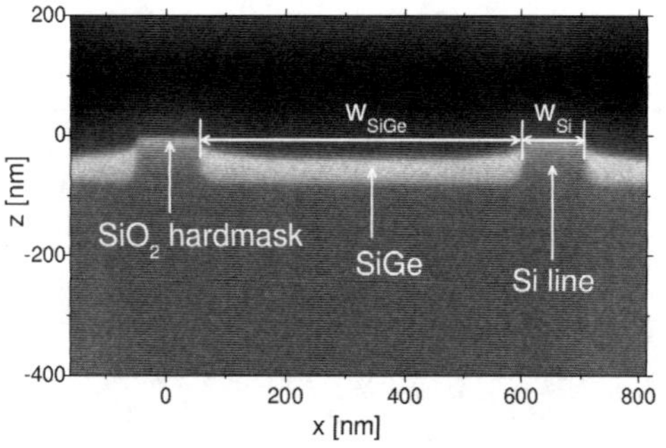

Figure 1. SEM cross section of the Si/SiGe line structures (image source: ISMI).

For the Raman measurements on these structures a Jobin Yvon HR800 spectrometer was used, equipped with 458 nm Ar excitation laser and a 100x objective (NA = 0.9) focusing the laser beam to a spot size of about 1 μm. The penetration depth of the laser radiation is approximately 300 nm in the Si and about 200 nm in the regions capped with SiGe. Near-field Raman measurements were performed also in a backscattering setup on a MV 2000 system (Renishaw / Nanonics) using a 488 nm Ar-ion laser excitation and Au/Ag coated quartz tips for near-field enhancement.

Additional high resolution X-ray diffraction (XRD) reference measurements were performed utilizing a Bruker AXS D8 diffractometer equipped with copper rotating anode and video laser alignment system, enabling measurements with a beam size of 100 μm.

3. Measurement results

Figure 2 shows both the XRD measurement results and the Raman spectrum obtained on an unpatterned SiGe region. The XRD measurement of the Si and SiGe *004* reflections enables an accurate determination of the SiGe film thickness t and of the Germanium concentration x_{Ge}. It indicates a fully strained state of the SiGe due to the pronounced film wiggles. This strain state is quantitatively confirmed by the relaxation scan performed on the SiGe *224* reflection (inset). For evaluation of the Raman data (figure 2b) with respect to composition and strain, the positions of the SiGe film related peaks denoted as "Si-Ge" and "Si-Si", and of the reference Si peak "Si-Si bulk" were used. Whereas the positions of the film peaks are shifted with respect to those of an unstrained (fully relaxed) SiGe film, no shift of the Si bulk peak position with respect to that of an unstrained pure Si wafer was detected on this unpatterned SiGe region. Thus for scans on patterned structures, the shifts of the peaks compared to the unstrained Si bulk peak position given in figure 2b were determined.

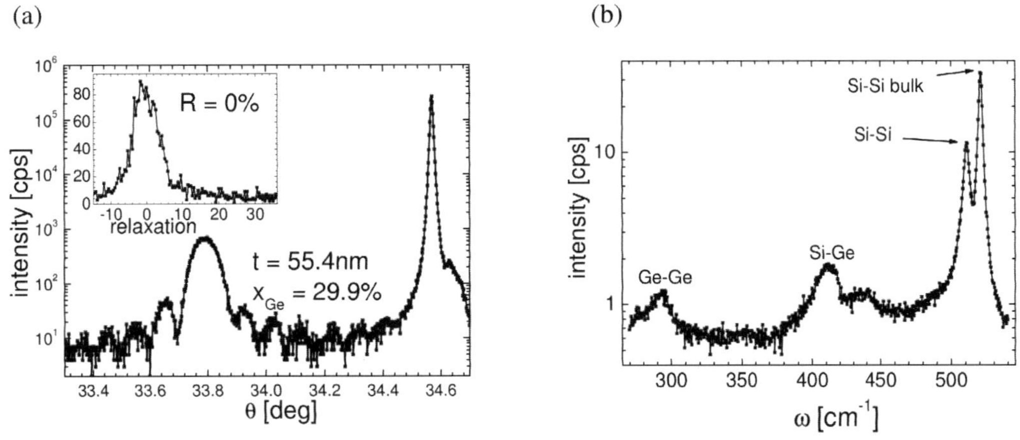

Figure 2. XRD measurement of the SiGe 004 reflection and the 224 reflection (inset) as measured on an unpatterned SiGe pad without Si lines (a, left), and Raman spectrum measured as reference in the center of this pad (b, right) containing the Si bulk peak and three SiGe related main peaks. Apart from these first-order phonon modes, a weak SiGe-related peak at 440 cm^{-1} is observed (cf. [15]).

To investigate the strain profile of the Si and SiGe line structures, scans oriented perpendicular to the lines across structures with different line widths w_{Si} and w_{SiGe} were performed. The Raman peak intensities show maxima of the Si signal and minima of the SiGe signal on the Si line positions, matching with the SEM measured geometry. Accordingly, also a characteristic dependence of the Raman peak positions on the scan coordinates was observed as shown for the example of two patterns with different pitch and line width in figure 3a,b. The dashed and dotted horizontal lines represent the positions of the reference peaks measured on the unpatterned regions, corresponding to the positions of the unstrained silicon and fully strained SiGe, respectively. The wavenumber of the Si bulk peak (black squares, upper curves) is maximum on the Si line positions and minimum in between them, with its average close to the reference bulk Si wavenumber. In contrast, the SiGe film peak position (grey circles, lower curves) is minimum at the Si lines and maximum on the SiGe cavity centers with a remaining offset to its reference position for all points on the pattern. The resulting question whether this offset is due to an onset of plastic relaxation or due to elastic strain in these structures has been analyzed making use of FEM modeling as shown below. Comparing figures 3a and b, it is clearly visible that the variation of the peak position depends sensitively on the line geometry. For smaller line widths (figure 3b), reduced variation of the peak positions is observed. Furthermore, also the average position of the film peak shifts towards the (fully-strained) reference value for smaller Si line widths.

(a) (b)

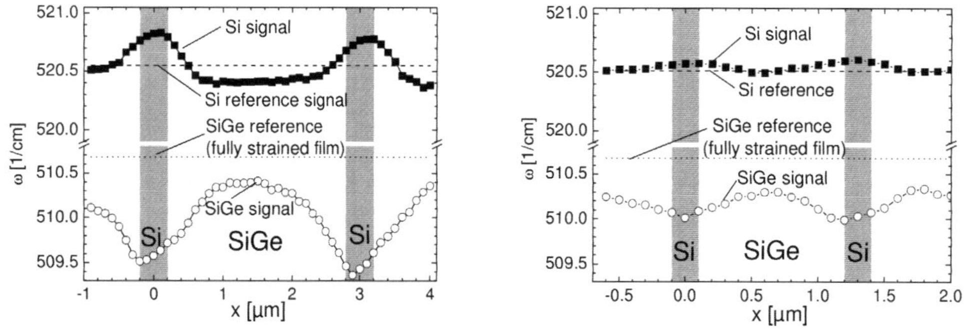

Figure 3. Variation of the Raman wave number ω for scans across two Si lines in between embedded SiGe regions for the geometry parameters w_{Si} = 420 nm, w_{SiGe} = 2550 nm (a, left) and w_{Si} = 100 nm, w_{SiGe} = 1000 nm (b, right). The upper curves show the Si bulk peak shift, and the lower ones the film peak shift. The coordinates x and y are aligned perpendicular and along the lines, respectively (cf. figure 1).

4. Evaluation of Raman measurements and finite-element modeling

To analyze the mechanical strain state related to the measured Raman shifts $\Delta\omega$ in a silicon crystal, generally three eigenvalues of a matrix [16], leading to a cubic equation have to be determined. For the given geometry of backscattering from a Si (001) surface, only one mode (denoted as longitudinal phonon mode) ω_3 is observable [17]. Whereas for the complex strain state within line structures the general secular equation has to be solved [16], for the special case of the unpatterned SiGe pad significant simplifications can be utilized. Assuming the silicon to be bisotropically strained, i.e. equal-axis biaxially strained with $\sigma_{xx} = \sigma_{yy}$ (cf. [18]) and vanishing remaining stress components, a factor coupling the measured peak shift and the Si stress can be derived [3]:

$$\sigma_{xx} = \sigma_{yy} = \text{-235 MPa cm } \Delta\omega_3. \tag{1}$$

For the SiGe film tentatively also assumed to be bisotropically strained, the approach [3] based on calibration measurements by high-resolution X-ray diffraction can be used for determining the parameters germanium content x_{Ge} and the degree of relaxation R:

$$x_{Ge} = \frac{\beta_2(\omega_{Si-Si} - \omega_{Si-Si,0}) - \beta_1(\omega_{Si-Ge} - \omega_{Si-Ge,0})}{\alpha_1\beta_2 - \alpha_2\beta_1}, \tag{2a}$$

$$R = 1 - \frac{\omega_{Si-Si} - \omega_{Si-Si,0} - \alpha_1 x_{Ge}}{\beta_1 x_{Ge}}, \tag{2b}$$

where $\alpha_1 = -62\text{cm}^{-1}$, $\beta_1 = 30.5\ \text{cm}^{-1}$, $\alpha_2 = 19.5\ \text{cm}^{-1}$ and $\beta_2 = 23.9\ \text{cm}^{-1}$ are fixed parameters, $\omega_{\text{Si-Si}}$ and $\omega_{\text{Si-Ge}}$ denote the SiGe film peak positions of the Si-Si and Si-Ge mode, respectively, and $\omega_{\text{Si-Si},0}$ and $\omega_{\text{Si-Ge},0}$ are the corresponding reference peak positions for vanishing Ge content. Thus, for the stress in the bisotropically strained SiGe results

$$\sigma_{xx} = \sigma_{yy} = (1-R)\ \sigma_0\ , \tag{3}$$

where σ_0 denotes the stress for a fully strained SiGe film.

The application of the equations (2) to the measurement on the unpatterned position (figure 2b) yields a Ge content of 30.0 % and a relaxation parameter $R = 0$. Correspondingly, for the biaxial compressive stress within the SiGe we obtain a value of $\sigma_{xx} = \sigma_{yy} = -1910$ MPa. To analyze also the Raman scans across the line structures, this information can be used to reduce the noise in the scans. If according to equations (2) both SiGe peaks are evaluated, due to the low intensity of the SiGe mode and the limited measurement time per spectrum during a line scan, a significant scattering of the obtained data for x_{Ge} and R results. The scattering of R can be reduced when assuming the Ge content within the pattern to be constant and thus all variations of the Raman peak positions to be strain related. With this assumption used in the following, only the Si-Si modes are considered for strain evaluation on the pattern, and a significant noise reduction is achieved (cf. figure 4). Note that on the patterned structure the occurrence of R-values does not necessarily indicate the onset of plastic relaxation, which will be discussed in detail below (figure 7). Due to the averaging of the Raman signal across the laser spot, there also appears a remnant SiGe signal on the positions of the Si lines.

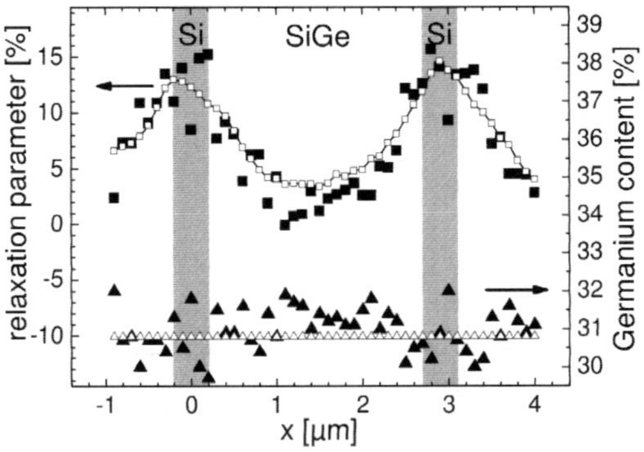

Figure 4. Ge concentration x_{Ge} and relaxation parameter R for the Raman scan shown in figure 3a. Solid symbols are results as derived from evaluating both the Si-Si and Si-Ge modes according to equation (4), and open symbols correspond to the evaluation of only the Si-Si Raman mode and additionally utilizing as input the Ge concentration derived from the unstructured SiGe film (figure 2b).

The shift of the SiGe curves on the positions of the Si lines towards lower wavenumbers (figure 3a,b) corresponds to a reduction of compressive stress within the SiGe, whereas the upshift of the Si curves at these positions indicates compressive stress within the Si lines (cf. equation (2b)). To analyze these features in more detail, FEM calculations were performed using ANSYS 9.0 with a geometry according to figure 1, i.e. a free surface on top and periodic boundary conditions on the right and left. The lattice mismatch between Si and SiGe was modeled by a thermal strain, i.e. by introducing different coefficients of thermal expansion and rising the temperature accordingly as initial condition. As a result of the modeling, nearly fully compressively strained SiGe regions with compressively stressed Si lines in between and regions of tensile strain underneath the layer were obtained (figure 5).

Within the Si line, the amount of stress σ_{xx} reduces with increasing depth. At a depth of about 500 nm beneath the surface of the Si line, a change from compressive to tensile stress occurs. In contrast, the tensile stress in the silicon beneath the SiGe regions is concentrated only in a thin stripe. Thus, according to the penetration depth of the laser, only a small shift of the bulk silicon peak is observed in the Raman scans. At the edges of the SiGe regions, a stress concentration occurs, in agreement with the observed stress concentrations close to edges of patterned strained SiN film structures [8].

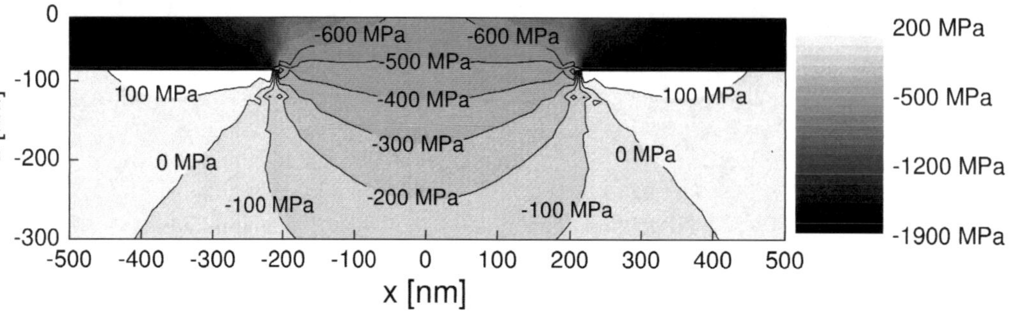

Figure 5. Calculated distribution of the stress component σ_{xx} for a geometry according to figure 1 (w_{Si} = 420 nm, w_{SiGe} = 2550 nm). The Si line is on the top center surrounded by the (dark) SiGe regions.

For all measured line geometries, the experimentally derived stress data utilizing the bisotropic assumption (equations (2)) are matching the FEM calculations for the SiGe regions. The geometry of the SiGe cavities yields a stress state which is at their center close to that of an infinite biaxially strained film. However, a clear mismatch between the stress values derived from equation (1) and the FEM calculations results for the Si regions. This gives rise to reconsider the assumption of a simple bisotropic strain state within the Si for the Raman scan evaluation. For more detailed analysis of the stress tensor, we calculate a parameter S with

$$S(x) = \frac{|\sigma_{xx}| - |\sigma_{yy}|}{|\sigma_{xx}| + |\sigma_{yy}|}, \tag{4}$$

where S = -1 or +1 corresponds to a uniaxial stress state in y or x direction, respectively, and S = 0 results for the bisotropic stress state. The stress component σ_{zz} is small compared to σ_{xx} and σ_{yy} (cf. figure 8b) and thus neglected in the following.

The contour plot of the parameter S shows that the assumption of biaxial stress in SiGe is reasonable ($S \approx 0$), but not further valid for the silicon regions (figure 6). As the SiGe region gets narrower, the assumption of biaxial stress is not valid furthermore. However, in our experiments the width of the SiGe regions was always larger than the Si line width, yielding a deviation of less than 3% at the SiGe center and less than 30% at the edges from the bisotropic stress state ($|S|$ = 1).

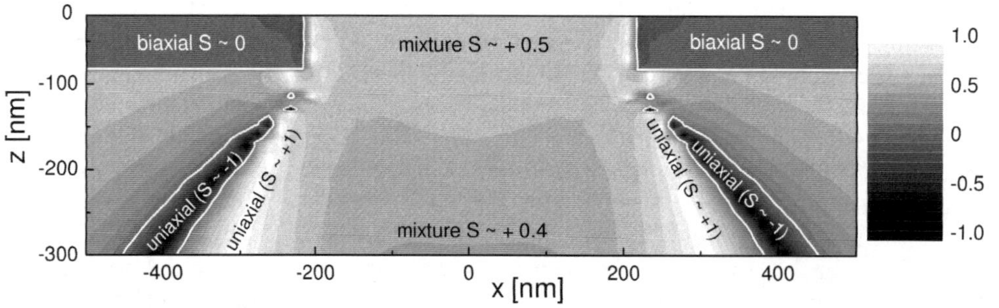

Figure 6. Contour plot of the stress distribution parameter S for the geometry according to Figure 1.

The stress state in the silicon regions shows a different behaviour, i.e. S varies between -1 and +1 there. According to the performed line scans, along horizontal paths the parameter S(x) can be calculated and utilized for a generalization of equation (3), yielding for the component σ_{xx}

$$\sigma_{xx} = (1 + S)\,(-235 \text{ MPa cm})\,\Delta\omega_3\,. \tag{5}$$

Using this approach for the experimentally derived stress distribution, both for the Si and for the SiGe regions, curve matching with the FEM modeling was obtained (figure 7). Consequently, the elastic strain state in the structures is concluded to be fully responsible for the measured Raman peak shifts; i.e. no onset of plastic deformation is detected (neither on the unstructured pad nor on the line structures). Therefore, the R parameter calculated in equation (2b) is a formal quantity reflecting here the elastic strain state, in contrast to its usual interpretation on unpatterned films as parameter for the plastic deformation related strain relaxation.

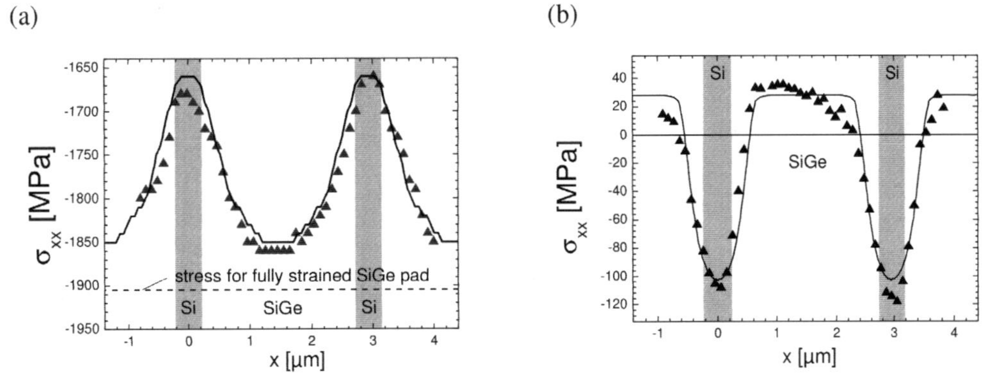

Figure 7. Comparison of experimentally derived stress data and FEM results for a scan across the line structure according to figure 1. Evaluation of the Raman SiGe peak (a, left) and the Si peak (b, right) was performed. The dashed line represents the SiGe stress value as measured on the unstructured pad (fully strained SiGe film). Due to the averaging of the Raman signal across the laser spot, there appears also a remnant SiGe signal on the positions of the Si lines, representative for the edges of the SiGe regions.

According to the size of the laser spot of ~ 1µm, for the comparison with experimentally determined curves an intensity weighted average of all calculated local peak shifts was considered (cf. [19, 20]). The variation of the geometry parameters of the investigated line structures enabled to investigate directly the impact of geometry onto the mean stress within the Si and SiGe lines. For the SiGe lines, the absolute mean values of the compressive stress $|\langle\sigma_{xx}\rangle|$ are increasing with increasing width of the SiGe lines (figure 8a). Accordingly, a similar tendency is obtained for the stress components σ_{xx} and σ_{yy} within the Si lines, whereas σ_{zz} is negligibly small (figure 8b). In real device structures the lateral extension of the SiGe regions is limited due to the positions of adjacent MOSFETs. However, the present experiments show also a gain of compressive Si line stress for reduced Si line widths. An upper limit of Si line stress is defined for a given Ge content in the SiGe regions by the strain of a correspondingly fully strained SiGe film. For small Si line widths below 50 nm, it has to be considered that the approximation of vanishing vertical stress σ_{zz} is not longer valid. A significant amount of vertical stress in the tensile range is obtained, which has (beneficial) influence onto the hole mobility in the PMOS channel [14]. The performed Raman line scans give direct evidence for this impact of downscaling onto the silicon stress and prove experimentally that local stress introduction will be an important approach also for future technology nodes with further reduced channel dimensions.

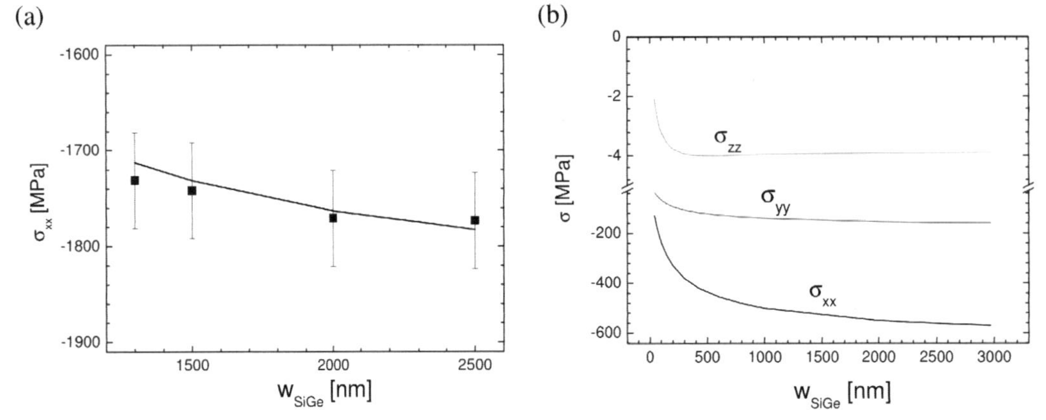

Figure 8. Mean SiGe stress (a, left) and the stress at the center of the Si line in 20 nm depth (b, right) as a function of the SiGe line width w_{SiGe} (w_{Si}=420nm). Experimental values are marked by squares and error bars.

5. Raman investigation with enhanced spatial resolution

For strain measurements with enhanced spatial resolution, near-field approaches are increasingly considered, since they can maintain such advantages of optical techniques as low sample preparation effort and high strain sensitivity while having the potential for nanoscale spatial resolution. In particular, tip-enhanced Raman scattering (TERS) was utilized successfully on silicon structures, as shown e.g. by Sun and Shen on patterns with a relatively large periodicity length (>500 nm, cf. [21]). Improved resolution on patterns with a pitch down to ~250 nm pitch was obtained with Ag coated tips on strained silicon line structures [22]. Downscaling the strain resolution in silicon is difficult compared to other material systems not only due to technical issues (regarding e.g. an accurate control of the special tip geometry needed for backscattering and thus control of the near-field), but also due to the limited knowledge of the real strain distribution in the used silicon nanostructures. Thus, the above presented FEM approach was adapted to the samples used in the near-field experiments and carefully compared with the measurement results. For these experiments, on a fully strained SiGe film with 23at% Ge content, oxide lines were drawn by local anodic oxidation with an atomic force microscope (AFM) and used as mask for subsequent wet-chemical etching. As a result, a group of five free-standing SiGe lines with a height of 40 nm, a width of 100 nm and a line distance of 100 nm were obtained. Since the etching of the SiGe was stopped 10 nm above the interface to the silicon underneath, a thin fully strained SiGe film was retained on the sample surface as reference. Figure 9a shows the intensity distribution of Raman scans performed across these lines both without a tip in the laser beam and with the Au coated tip in feedback. Interestingly, not only the SiGe signal but also the Si signal is enhanced on the structures for both measurement modes. The enhancement of the Si signal can be explained by a scattering effect on the edges of the SiGe lines, resulting in an increased Raman scattering from the Si regions close to these edges. A striking feature is the clear resolution of all the lines in the measurements with the tip in feedback, which is not achievable in the conventional Raman scans. In the measurement performed with the tip in feedback, also in the Raman peak shift the signature of the single lines is visible. The increased spatial resolution of the scans with the tip in feedback can be explained by the excitation of plasmons in the apex of the Au tip, yielding an enhancement of the field in a region close to the tip. In the experiments, an enhanced SiGe film signal of about 25% compared to the conventional Raman measurement was obtained. By comparison of the measurement results with the FEM calculations (figure 10), a range for the field enhancement of about 200 nm has to be assumed to obtain reasonable agreement between the measurements and the strain calculations.

(a) (b)

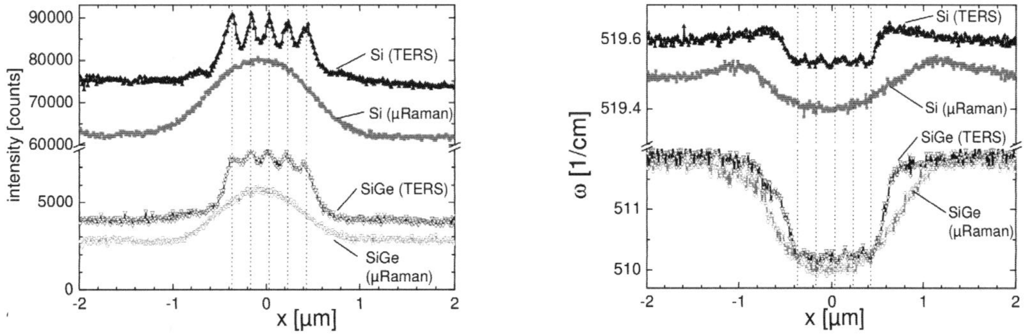

Figure 9. Conventional μRaman and TERS scans across free standing SiGe lines: Intensity (a) and peak shift (b). For details of the TERS scan setup cf. [22].

(a) (b)

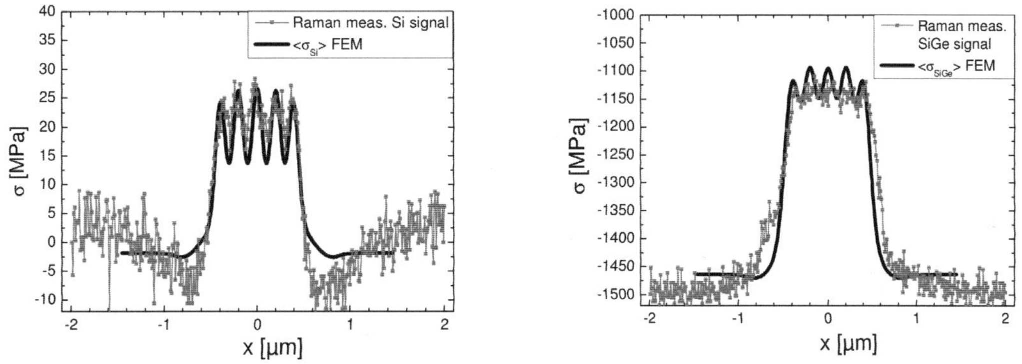

Figure 10. Average profiles, derived from the TERS-scan, for the Si stress (a) and the SiGe stress (b).

The strain distribution of free-standing lines differs essentially from that of embedded SiGe lines (figure 7). The downshift in the Raman signals for the freestanding lines indicates a relaxation of the compressive SiGe stress according to an average relaxation value of $R = 20\%$ within the lines, and an average tensile stress in the silicon underneath. This tensile stress is enhanced at the positions just underneath the lines (figure 10). The results indicate that the description of the strain state with the used FEM approach reflects the measurement results appropriately and can thus be used for the analysis of the local strain distribution in strained silicon structures.

6. Summary and conclusions

Raman line scans across Si / SiGe line structures revealed local strain profiles in conjunction with finite element modeling, enabling a systematic investigation of stress relaxation effects compared to the stress in an unpatterned SiGe film. In the Si lines between the SiGe, a compressive stress is obtained, whereas underneath the SiGe a tensile Si stress was observed. As shown by careful comparison with modeling, the stress relaxation for the here investigated embedded SiGe structures is of elastic nature, i.e. obtained without onset of additional plastic relaxation. The stress state within the SiGe filled cavities can be assumed roughly to be a bisotropic one with $\sigma_{xx} = \sigma_{yy}$, whereas in the silicon a mixed stress state is present which can be described by a parameter S. The measured compressive stress in the silicon lines is sensitive to the line geometry. It can be increased e.g. by increasing the SiGe cavity widths and by reducing the Si line widths, supporting an increase of charge carrier mobility and thus the performance of CMOS devices. Significantly increased spatial resolution can be obtained in Raman measurements utilizing tip-enhanced Raman scattering, as shown on free-standing SiGe line structures. A strain resolution for SiGe lines separated by a distance of 100 nm was obtained, which can be further improved by optimizing the tip geometry and still more accurately controlling the tip position within the laser beam.

Acknowledgements
The authors are grateful to Yvonne Ritz, Inka Zienert, David Lehninger, and Martin Weisheit, all with Globalfoundries, for performing additional measurements and helpful discussions. We also acknowledge funding of the investigations in line with the technology funding for regional development (ERDF) of the European Union, and by funds of the Free State of Saxony.

References

[1] Thompson SE, Armstrong M, Auth C et al. 2004 *IEEE Elect. Device Lett.* **25**, 191
[2] Horstmann M, Wei A, Kammler T et al. 2005 *IEEE Techn. Dig.* 233
[3] Hecker M, Zhu L, Georgi C, Zienert I, Rinderknecht J et al. 2007, *AIP Proc.* **931** *(Frontiers of Characterization and Metrology for Nanoelectronics)* 435
[4] Buca D, Hollaender B, Feste S, Lenk S et al. 2007 *Appl. Phys. Lett.* **90**, 032108
[5] Hashemi P, Gomez L, Hoyt JL et al. 2007 *Appl. Phys. Lett.* **91**, 083109
[6] Wormington M, Lafford T, Godny S, Ryan P et al. 2007 *AIP Proc.* **931** (*Frontiers of Characterization and Metrology for Nanoelectronics*) 220
[7] Liu R and Canonico M 2004 *Microelectr. Engn.* **75**, 243
[8] de Wolf I 2003 *Raman spectroscopy* **15**, 6
[9] Jain SC, Dietrich B, Richter H, Atkinson A and Harker AH 1995 *Phys. Rev. B* **52**, 6247
[10] Jain SC, Maes H and Picardi K 1997 *Thin Solid Films* **292**, 218
[11] People R and Bean JC 1985 *Appl. Phys. Lett.* **47**, 322
[12] Usuda K, Numata T, Irisawa T et al. 2005 *Mat. Sci. Engn. B* **124-125**, 143
[13] Hue F, Hytch M, Houdellier F et al. 2008 *Mat. Sci. Engn. B* **154-155**, 221-224
[14] Yeo Y and Sun J 2005 *Appl. Phys. Lett.* **86**, 023103-3
[15] Alonso MI and Winer K 1989 *Phys. Rev. B* **39**, 10056
[16] Ganesan S, Maradudin AA and Oitmaa J 1970 *Annals of Physics* **56**, 556
[17] Loudon R 1964, The Raman effect in crystals *Advances in Physics* **13**, 423
[18] Borblik VL 2007 *Journal of Physics: Condensed Matter* **19**, 456219
[19] Narayanan S, Kalidindi SR and Schadler LS 1997 *J. Appl. Phys.* **28**, 2595
[20] Atkinson A and Jain SC 1999 *J. Raman Spectrosc.* **30**, 885–891
[21] Sun WX and Shen ZX 2003 *Ultramicroscopy* **94**, 237
[22] Zhu L, Georgi C, Hecker M, Rinderknecht J et al. 2007 *J. Appl. Phys.* **101**, 104305

Measurement of composition profiles in III-nitrides by quantitative scanning transmission electron microscopy

A Rosenauer[1], K Gries[1], K Müller[1], M Schowalter[1], A Pretorius[1], A Avramescu[2], K Engl[2] and S Lutgen[2]

[1] University of Bremen, Institute for Solid State Physics, Bremen, Germany
[2] OSRAM Opto Semiconductors GmbH, Regensburg, Germany

E-mail: rosenauer@ifp.uni-bremen.de

Abstract. In this paper we demonstrate a quantitative method for composition evaluation based on comparison of normalized image intensity with simulations carried out with the frozen lattice approximation. The method is applied to evaluate composition profiles of $Al_xGa_{1-x}N$/GaN layers. We measure ratios of image intensities obtained in regions with unknown and with known Al-concentration x, respectively. We show that estimation of specimen thickness combined with evaluation of intensity ratios allows quantitative measurement of composition profiles. Delocalization effects at interfaces due to instrumental resolution and dynamic electron diffraction are simulated. These effects can well be described by convolution with a Lorentzian. Measured intensity profiles can be corrected for delocalization effects using statistical parameter estimation so that deconvolution is avoided.

1. Introduction

Optical and electronic properties of light emitting diodes and laser diodes are governed by morphology and composition of semiconductor heterostructures. Therefore, research as well as quality control in industrial production require analysis of elemental concentration at high spatial resolution. Scanning transmission electron microscopy (STEM) using a high-angle annular dark field (HAADF) detector allows chemically sensitive imaging. We suggest quantitative analysis of composition based on comparison of normalized [1] measured detector intensity with image simulation [2] carried out with the frozen lattice or frozen phonon approximations [3].

2. Experimental procedures

STEM experiments were performed on a FEI TITAN 80/300 TEM/STEM equipped with an HAADF detector (Fischione Model 3000). Detector intensities are normalized with respect to the intensity of the incident electron beam similar to the method suggested by Le Beau et al. [1] but using the internal signal amplifiers of the TEM. Using different spot sizes as well as different settings of brightness and contrast of the amplifiers we found that good linearity is ensured for spot sizes between 6 and 11 if the detector dark signal is larger than zero and the signal corresponding to the incident electron beam is kept below $5 \cdot 10^4$. These values were measured by scanning the incident electron beam over the HAADF detector. The test structure investigated was grown by OSRAM with MOVPE and TEM specimens were prepared by focused ion beam (FIB) thinning.

3. Simulation of detector intensities

As measurement of specimen thickness and composition is based on a comparison with theoretical calculations, we simulated area scans for specimen thicknesses up to 200 nm and Al concentrations between 0 and 1 with the STEMSIM program [4]. The acceleration voltage was 300 kV, the spherical aberration of the probe forming lens was 1.2 mm, the convergence angle of the probe was 8.8 mrad and the detector covered an angular range between 33 mrad and 200 mrad. The non-homogeneous sensitivity of the detector was measured and taken into account. The simulations were carried out in the frozen phonon approximation for the $[2\bar{1}\bar{1}0]$ electron beam direction using a supercell with 7×7 unit cells. We obtained a converged calculation if the k-space ranged up to 400 mrad, corresponding to 1512×1596 pixels. The values for the mean square displacements were 3.24×10^{-3} Å^2 for Al and 3.59×10^{-3} Å^2 for N in AlN, as well as 3.47×10^{-3} Å^2 for Ga and 3.92×10^{-3} Å^2 for N in GaN, obtained by density functional theory. Linear combinations of mean square displacements were used for the ternary semiconductors. Each area scan covered one crystal unit cell and consisted of 20×20 calculation points. The detector intensity was then averaged over the image and plotted vs. the specimen thickness. The results are shown in figure 1. The accuracy of the GaN curve was checked with a FIB-lamella by comparison of specimen thickness evaluated from STEM Z-contrast images with the thickness directly measured in the FIB. We obtained an accuracy of the measurement of specimen thickness in GaN of ±15 nm.

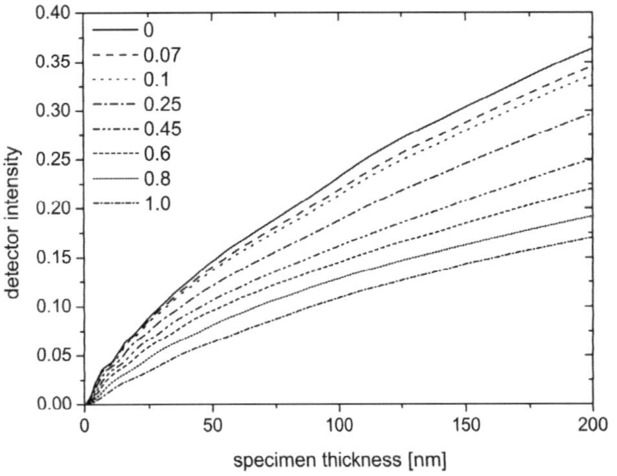

Figure 1. Simulated intensity averaged over one unit cell plotted vs. the specimen thickness. The curves have been computed with the frozen lattice approximation using 5 displacement configurations. The different curves correspond to the Al concentrations listed in the legend.

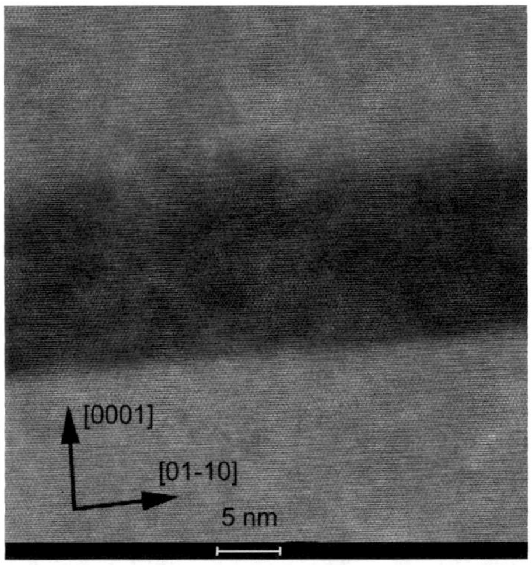

Figure 2. Experimental high-resolution STEM HAADF Z-contrast image of an $Al_{0.2}Ga_{0.4}N$ layer buried in GaN. The zone axis is $[2\bar{1}\bar{1}0]$ and the growth direction is [0001].

4. Experimental measurement and result

Figure 2 shows the experimental high-resolution STEM HAADF Z-contrast image whose evaluation is described in the following. First, we extracted the line scan shown in figure 3a, where we averaged along the $[01\bar{1}0]$ direction over a 20 nm wide region. The intensity oscillations mark the (0002) lattice planes. The intensity profile was Fourier transformed and the 0002 and 0001 Fourier components were deleted. The result was inverse Fourier transformed which led to the profile of the averaged image intensity shown as thick grey curve in figure 3a. Then we fitted a third-order polynomial to the GaN-regions. The resulting reference intensity is drawn as the dashed curve, corresponding to a hypothetical GaN crystal with the actual specimen thickness. Comparison with the solid curve in figure 1 yields the specimen thickness depicted in figure 3b. The intensity ratio shown in figure 3c was calculated by dividing the normalized raw intensity by the normalized reference intensity. The Al-concentrations were computed for the local specimen thickness as given in figure 3b by comparing the profile of the experimental intensity ratio with simulated ratios obtained from figure 1. Positions of (0002) lattice planes were derived from the maxima positions of the intensity oscillations (black solid curve in figure 3a). Finally, the evaluated Al-concentration was averaged over half a unit cell for each (0002) lattice plane. The result is shown in figure 3d, where each box corresponds to one (0002) lattice plane. The composition profile indicates that the lower interface is sharper than the upper interface.

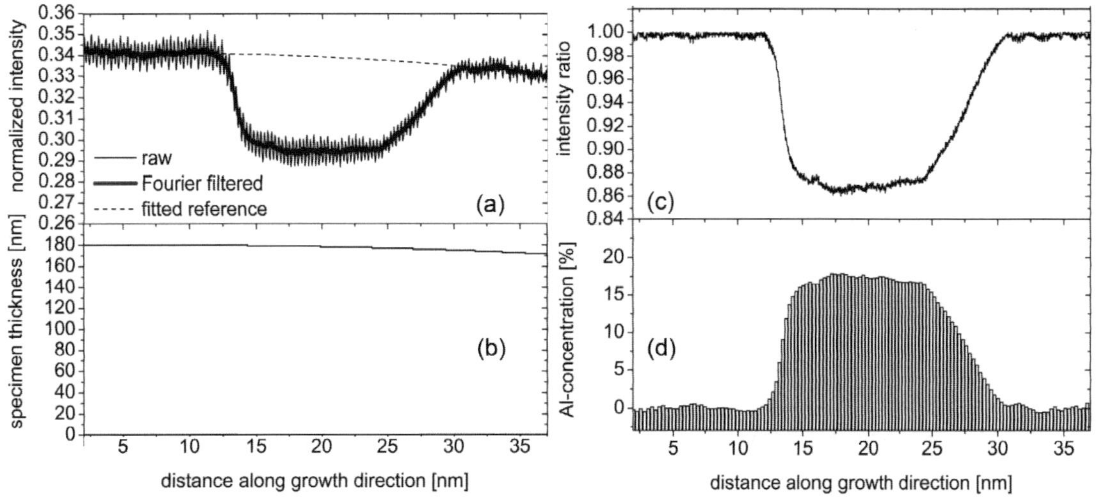

Figure 3. a) Normalized intensity obtained from averaging inside a 20 nm wide region in figure 2. The dashed curve is the reference intensity used for calculating intensity ratios as well as for evaluation of specimen thickness. b) Measured specimen thickness, c) intensity ratio and d) evaluated Al-concentration profile, where each bar corresponds to an (0002) lattice plane.

5. Effect of instrumental resolution and dynamic electron diffraction

5.1. Simulation

To check the influences of instrumental resolution and of a broadening of the electron beam by dynamic electron diffraction, we simulated an atomically sharp interface between GaN and $Al_{0.2}Ga_{0.8}N$ using the absorptive potential method. Figure 4 shows the obtained profile of normalized intensity. The maxima of the oscillation correspond to the positions of (0002) lattice planes. Then, the maxima positions and minima positions, respectively, were fitted by appropriate functions. The sum of maxima curve and minima curve divided by 2 gives the profile of the mean intensity indicated by the thick grey curve. This curve was used to calculate the delocalization function as demonstrated in figure 5. First, we defined an ideal intensity profile that contains a linear transition at the interface. This ideal profile was then convoluted with a Lorentzian $f(x) = 1/\left(\pi\delta^2\left[1+(x/\delta)^2\right]\right)$, where δ describes its width. The parameter δ was optimized so that the ideal profile convoluted with $f(x)$ fitted the simulated intensity profile.

5.2. Correction

The measured composition profile was corrected for delocalization effects without using deconvolution. For that, a model composition profile $m(x)$ was optimized so that $m(x)\otimes f(x)\otimes g(x)$ shows minimum deviation from the measured composition profile depicted in figure 3d. The function $g(x)$ is a Gaussian that describes the source size dependent broadening. Its half-width at half maximum value of 0.07 nm has been measured as described in [5]. Figure 6 shows

Journal of Physics: Conference Series **209** (2010) 012009 doi:10.1088/1742-6596/209/1/012009

the corrected composition profile in comparison with the uncorrected profile for a small region containing the lower interface visible in figure 2. As the width δ is only 0.13 nm at 180 nm specimen thickness, the correction leads to small changes of the concentration profile only.

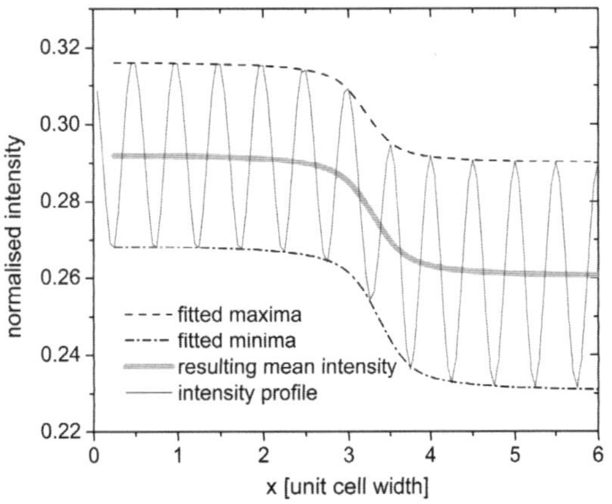

Figure 4. Profile of normalized intensity simulated for an atomically sharp interface between GaN (left) and $Al_{0.2}Ga_{0.8}N$ (right). The mean intensity profile was obtained as the mean of the fitted maxima and minima curves. The specimen thickness was 180 nm.

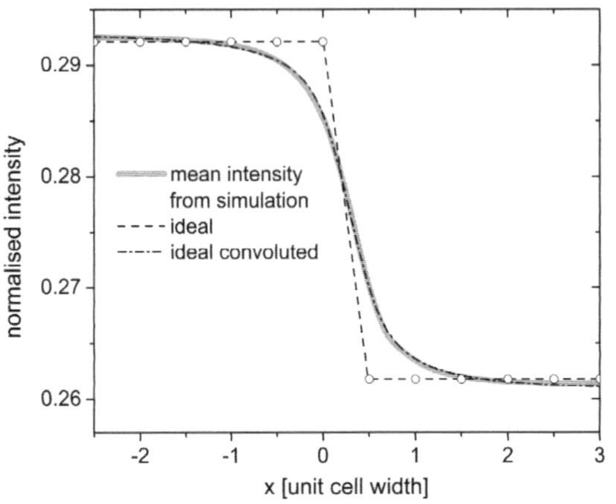

Figure 5. The dashed line shows the ideal intensity profile. The thick grey curve is the profile of the mean intensity shown in figure 4. The dash-dotted curve was obtained by convoluting the ideal profile with a fitted Lorentzian delocalization function.

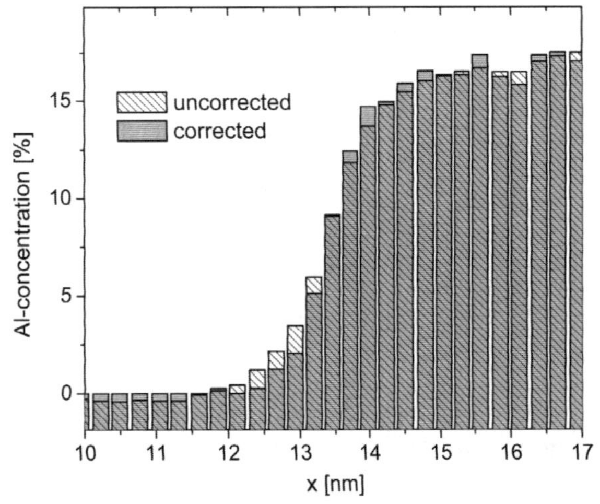

Figure 6. Corrected and uncorrected Al-concentration profiles plotted vs. the distance along the growth direction. Each bar corresponds to one (0002) lattice plane. The hatched bars show the original profile as depicted in figure 3d. The grey bars correspond to the model concentration profile $m(x)$, whose convolution with $g(x)$ and $f(x)$ yields the uncorrected Al-concentration profile. The function $g(x)$ describes the source-size dependent broadening and $f(x)$ is the delocalization due to instrumental resolution and dynamic diffraction of the electron beam in the specimen.

Acknowledgements
This work was supported by the Deutsche Forschungsgesellschaft under project II-2 of the research group FOR 506 and by project RO2057/4-1. A.R. thanks S. Stemmer for valuable discussion.

References

[1] LeBeau JM, Stemmer S 2008 Ultramicroscopy **108** 1653
[2] Grillo V, Carlino E, and Glas F 2008 Phys. Rev. B **77** 054103
[3] Rosenauer A, Schowalter M, Titantah JT, Lamoen D 2008 Ultramicroscopy **108** 1504
[4] Rosenauer A, Schowalter M 2007 Springer Proceedings in Physics **120** 169
[5] Rosenauer A, Gries K, Müller K, Schowalter M, Avramescu A, Engl K, and Lutgen S, submitted to Ultramicroscopy

Structural Properties of GaN Nanowires and GaN/AlN Insertions Grown by Molecular Beam Epitaxy

C. Bougerol[1], R. Songmuang[1], D. Camacho[2], Y.M. Niquet[2], B. Daudin[1]

[1]CEA-CNRS group « Nanophysique et Semiconducteurs », Institut Néel/CNRS-UJF and CEA Grenoble, INAC, SP2M, 17 rue des Martyrs, 38 054 Grenoble, France
[2] CEA Grenoble, INAC, SP2M/L-Sim, 17 rue des Martyrs, 38 054 Grenoble, France

catherine.bougerol@cea.fr

Abstract. The microstructure of GaN nanowires (NWs) at the Si(111) substrate interface and the strain state of GaN insertions in GaN/AlN nanowire heterostructures have been studied by high resolution transmission electron microscopy. We have found that GaN NWs are relaxed from the beginning of the growth, due to a plastic relaxation mechanism. Concerning the GaN insertions in AlN, the presence of dislocations in the AlN capping layer can explain their partially relaxed state and the apparent discrepancies between experimental results and theoretical modelling.

1. Introduction

It is well known that the efficiency of nitride-based optoelectronic devices is limited by the high densities of threading dislocations (TDs), typically up to 10^9-10^{10} cm^{-2}, generated by the use of highly mismatched substrates. Therefore, the growth of nanowires (NWs), which can be considered as a juxtaposition of non-coalesced grains, appears very promising since TDs formed by the coalescence of adjacent grains in the usual 2D layer growth should be suppressed. Furthermore, in case of axial heterostructures, such as AlN/GaN, the presence of free surfaces at the side should allow lateral relaxation and consequently prevent the formation of misfit dislocations at the interfaces. These two characteristics make nitride wire-like heterostructures potentially very interesting for the realization of efficient light emitting devices (LEDs) in the visible and UV range [1,2]. However, the details of the NWs' growth and especially their nucleation are still far from being clear. Moreover, due to the reduced diffusion length of Al compared to Ga and to its enhanced incorporation probability on the NWs' sides, the formation of AlN/GaN wire-like heterostructures is accompanied by the growth of a thin AlN shell around the GaN insertions [3-5]. This experimental feature raises the issue of strain relaxation in GaN/AlN core-shell heterostructures through dislocation formation beyond a given critical thickness, which has recently been addressed from a theoretical point of view [6,7].

This has motivated us to investigate by TEM the microstructure of nitride NWs, and especially the strain state, directly related to the optical properties. We have focused on two main points which are reported in the following, namely the interface with the substrate and the AlN/GaN insertions.

2. Experimental

Samples were grown catalyst-free by plasma-assisted molecular beam epitaxy (PA-MBE) on a thin (0001) AlN buffer layer deposited on (111) Si. After a base of standard GaN NWs, between 500 and

Journal of Physics: Conference Series **209** (2010) 012010 doi:10.1088/1742-6596/209/1/012010

700 nm thick [8], several tens of nm of AlN containing insertions of GaN were grown according to the procedure described in [5]. The nanowires were either scrapped off the substrate and dispersed on a Cu microscope grid covered with a lacey carbon film, or covered with glue and then prepared in cross-sectional orientation by mechanical polishing followed by Ar-ion milling. During this last process, the top part of the NWs was often removed. The TEM investigations were performed on a Jeol 4000EX microscope operated at 400kV (Cs=1mm). For strain maps determination by the Geometrical Phase Method (GPA) [9], off-axis high resolution images, i.e. tilted from the main [11-20] zone axis, were recorded. This configuration was chosen in order to enhance the signal to noise ratio [10,11] and therefore increase the reliability of the analysis.

3. Nanowire/substrate interface

A typical HREM image of a NW base, taken along the [11-20] zone axis, shows the Si/AlN and AlN/GaN interfaces (Figure 1a). No amorphous layer is present at these interfaces, contrary to the case of direct growth of GaN NWs on Si [12]. We observe a coincidence relationship between Si and AlN (4 to 5 interplanar distances (inset)) as already reported in [13]. An almost complete strain relaxation in AlN grown on Si is therefore expected.

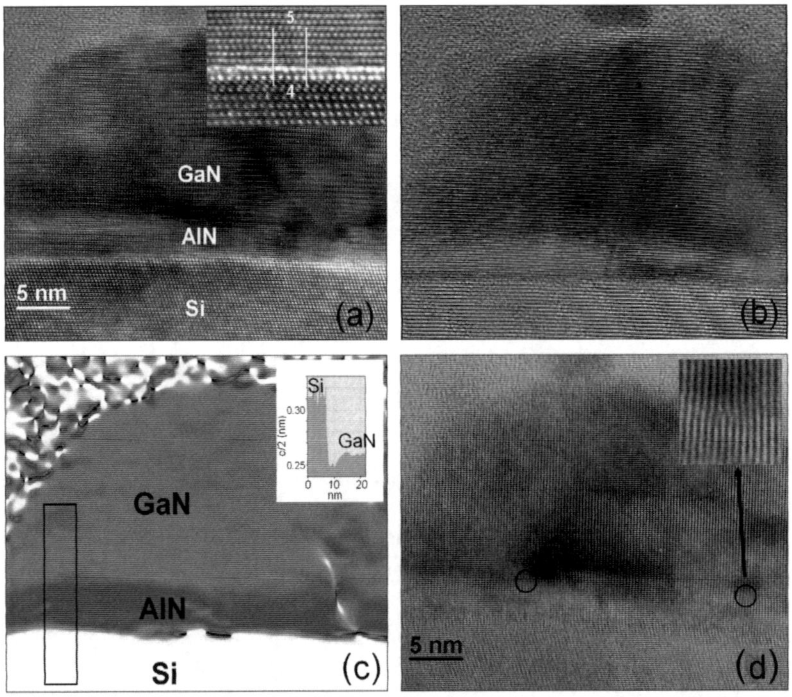

Figure 1. (a) HREM image of a NW base taken along the [11-20] zone axis. The inset shows the coincidence between Si and AlN
(b) c-tilt off-axis HREM image of the same NW
(c) corresponding map of the c lattice parameter and profile from Si to GaN (inset)
(d) a-tilt off-axis image showing the presence of dislocations at the AlN/GaN interface

The NW was then tilted around the c and a axes successively in order to get off-axis images. The c-tilted image (Figure 1b) has been analyzed by the GPA method to obtain a map of the c/2 lattice parameter (Figure 1c). Si was chosen as the reference region. The inset represents a profile from Si to GaN integrated over the black rectangle area. Surprisingly, the c/2 values of AlN and GaN correspond to almost fully relaxed material (0.249 and 0.259 nm respectively). On the other hand, the a-tilted image, figure 1d, clearly reveals the presence of two misfit dislocations at the AlN/GaN interface.

This indicates that a plastic relaxation mechanism occurs in the present case, which is markedly different from what is observed in the case of Stranski-Krastanow (SK) QDs where elastic relaxation takes place. We can therefore conclude that the thin AlN buffer layer behaves as a compliant substrate. These results confirm that GaN grown on AlN is fully relaxed from the beginning and are in total agreement with those recently obtained from in-situ grazing incidence synchrotron X-ray diffraction experiments [14].

4. AlN/GaN insertions

The investigated sample consisted of successive GaN inclusions, 30 nm in diameter and 2.5 nm in height, separated by 12 nm high AlN regions, as shown in figure 2a. The presence of a thin AlN shell around the wire should be noticed. The strain component along the nanowire axis, ε_{zz}, was obtained by applying the GPA method, the reference region being chosen as the AlN side part and taken as fully relaxed. ε_{zz} was calculated as $\varepsilon_{zz} = (d_{002} - d_{002ref}) / d_{002ref}$ with $d_{002ref}=0.5c_{AlNbulk}$, and is given in figure3b. Concerning AlN, we note an in-plane tensile strain in the central part with respect to the reference side, leading to a decrease of the AlN c- lattice spacing (≈-1%) with respect to the reference region. Concerning the GaN inclusions, we notice that ε_{zz} is maximal in the central part where it reaches about 6% and that it remains almost constant along the growth direction (Figure 2b). On the other hand, a lateral gradient is observed, with a decrease from the central part to the side where it reaches about 4%, a value close to the GaN relaxed one. This behaviour is different from the case of SK QDs, for which the highest strain is localized at the bottom of the QD and in the wetting layer as well [11].

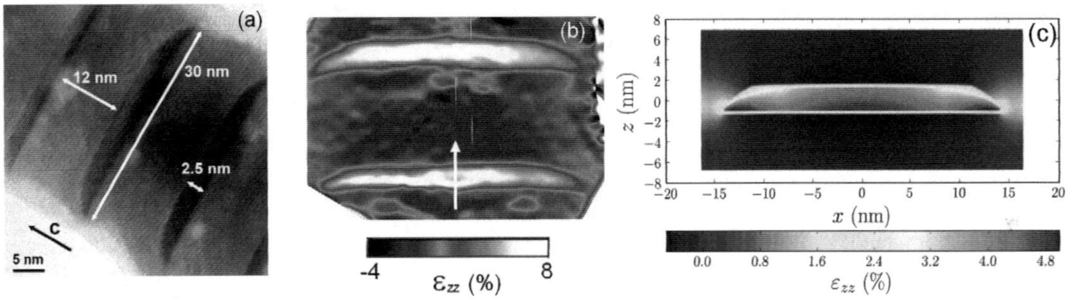

Figure 2. (a) HREM showing the GaN insertions in AlN
(b) Mapping of strain component ε_{zz} obtained from GPA. The reference region was taken in the AlN side part
(c) Mapping of ε_{zz} computed with a valence force field model

These experimental results were compared with computed ones obtained by using a nearest neighbour valence force field based on Keating's model [15]. We considered {0001} oriented AlN nanowires with hexagonal cross section (radius R ~15 nm) and {1-100} facets. The GaN insertion was modelled as a slice of a sphere. Periodic boundary conditions were applied along the c axis on a supercell of height L ~ 13 nm. The elastic energy of the nanowire was minimized assuming coherent growth (no dislocation). In order to make the calculation easily comparable to HRTEM results, the atomic columns were then projected in a (1-100) plane, and the deformation ε_{zz} was computed from the average column positions using a finite difference scheme, taking AlN as the reference material. The ε_{zz} map in a 2.5 nm thick GaN insertion with radius R_{GaN} = 30 nm is plotted in figure 2c. As expected, the GaN layer mostly undergoes biaxial compressive strain from the AlN nanowire and therefore tends to be dilated along thc c axis (ε_{zz} > 4%) despite the AlN shell [16]. The strain appears to be maximal at the basis of the GaN layer where it peaks in the corners. The decrease of ε_{zz} toward the upper edge results from strain relaxation and from the increase of the AlN content in each projected atomic column. As a whole, this feature deviates markedly from the GPA results, especially in the vicinity of the corners of GaN insertions.

These discrepancies could be explained by the presence of a relaxation mechanism not taken into account in the calculations, such as the presence of dislocations. In fact, the appearance of dislocations in nanowires has been theoretically predicted in either axial [6] or coaxial heterostructures [7], depending on the lattice mismatch and wire radius, or core radius and shell thickness, respectively. We indeed observed the insertion of an extra (0002) plane at the GaN/AlN interface, as illustrated in figure 3 for three successive inclusions. Furthermore, from this image, one can point out that the dislocations occur at the same position along the oblique side of the three inclusions, corresponding to the same AlN shell thickness, namely about 3 nm. This is in qualitative agreement with what has been predicted [7] for coaxial nanowire heterostructures.

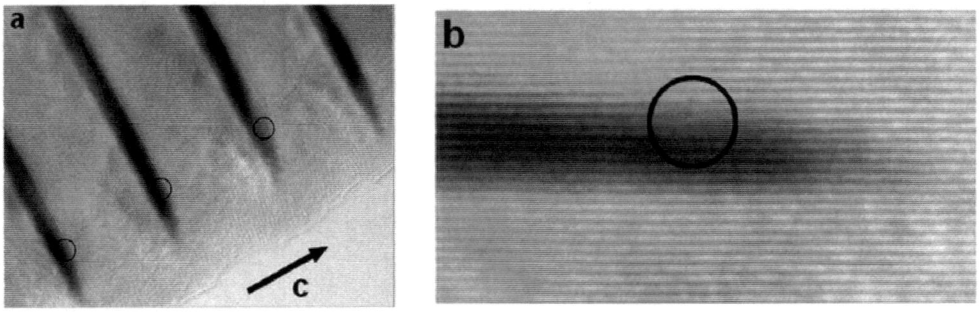

Figure 3. (a) HREM image showing a dislocation at the AlN/GaN interface for three successive inclusions. (b) enlargement showing the insertion of an extra (0002) plane in AlN

5. Conclusion

Our TEM studies have shown that the GaN NWs are relaxed from the beginning of the growth, which has been assigned to the compliant role of the thin AlN buffer layer grown on the Si substrate and the presence of dislocations at the AlN/GaN interface. This is markedly different from the classical case of QDs. We have also found that GaN insertions in AlN are partially relaxed, the strain release occurring through the formation of misfit dislocations during the capping process by AlN. This is probably due to the rounded shape of the GaN insertions which leads to a local radial thickening of the AlN capping layer beyond the critical thickness.

References

[1] Kikuchi A, Kawai M, Tada M, Kishino K 2004 *Jpn. J. Appl. Phys.* **43**, L1524
[2] Sekiguchi H, Kishino K, Kikuchi A 2008 *Electronics Lett.* **44**, 151
[3] Ristic J, Calleja E, Sanchez-Garcıa M A, Ulloa J M, Sanchez-Paramo J, Calleja J M, Jahn U, Trampert A, Ploog K H 2003 *Phys. Rev. B* **68**, 125305
[4] Tchernycheva M, Sartel C, Cirlin G, Travers L, Patriarche G, Harmand J C, Le Si Dang, Renard J, Gayral B, Nevou L, Julien F 2007 *Nanotechnology* **18**, 385306
[5] Renard J, Songmuang R, Bougerol C, Daudin B, Gayral B2008 *Nanoletters* **8**, 2092
[6] Ertekin E, Greaney P A, Chrzan D C, Sands T D 2005 *J. Appl. Phys.* **97**, 114325
[7] Raychaudhuri S, Yu E T 2006 *J. Appl. Phys.* **99**, 114308
[8] Songmuang R, Landré O, Daudin B 2007 *Appl. Phys. Lett.* **91**, 251902
[9] Hÿtch M, Snoeck E, Kilaas R 1998 *Ultramicroscopy* **74**, 131
[10] Rouvière JL, Bayle-Guillemaud P, Radtke G, Groh S, Briot O, 2001 Inst. Phys. Conf. Ser. **169**,17
[11] Sarigiannidou E, Andreev AD, Monroy E, Daudin B, Rouvière JL 2006 *Phys. Stat. Sol.* **C 3** 1667
[12] Ristic J, Calleja E, Fernandez-Garrido S, Cerutti L, Trampert A, Jahn U, Ploog K H *J. Cryst. Growth* **310**, 4035
[13] Bourret A, Barski A, Rouvière J L, Renaud G, Barbier A 1998 *J. Appl. Phys.* **83**, 2003
[14] Landré O, Bougerol C, Renevier H, Daudin B 2009 *Nanotechnology* **20**, 415602
[15] Keating P N 1966 *Phys. Rev.* **145**, 637
[16] Niquet Y M 2007 *Nano Letters* **7**, 1105

GaN, AlGaN, HfO_2 based radial heterostructure nanowires

L Lari[1,2], T Walther[1], K Black[2], RT Murray[2], TJ Bullough[2], PR Chalker[2], C Chèze[3], L Geelhaar[3] and H Riechert[3]

[1] Dept. Electronic & Electrical Eng., University of Sheffield, Sheffield, S1 3JD, UK
[2] Department of Engineering, University of Liverpool, Liverpool, L69 3GH, UK
[3] Paul-Drude-Institut für Festkörperelektronik, 10117 Berlin, Germany

E-mail: l.lari@sheffield.ac.uk

Abstract. Gallium nitride based core/shell nanowire heterostructures are promising components for UV optoelectronic nanodevices. A better understanding of their growth mechanisms is required to tailor the growth and the properties of these structures. We report an investigation of GaN-based radial nanowire heterostructures grown by Molecular Beam Epitaxy combined with Atomic Layer Deposition of HfO_2 dielectric cladding. The structural quality of GaN nanowire samples coated with different materials, the elemental distribution along the nanowires' length, as well as the elemental incorporation mechanism are discussed.

1. Introduction

In the last few decades research in the field of nitride materials has been focused towards controlling the growth and morphology of structures at the nano-scale. The combination of GaN, $Al_xGa_{1-x}N$, and AlN in various heterostructures has potential applications in optoelectronic devices in the far ultra-violet and in high frequency devices [1]. In addition, the performance of AlGaN / GaN high electron mobility transistors has been found to improve using HfO_2 to passivate surface states [2]. It has been demonstrated previously that Plasma Assisted - Molecular Beam Epitaxy (PA-MBE) is a promising technique for the growth of free-standing GaN nanowire (NW) structures [3] that can be used as components of nano-scale devices. One of the next challenges is to control the growth of NW radial heterostructures incorporating cores of GaN and shells of different materials whose device structure and physical properties can be appropriately tailored. A critical issue is the control of the uniformity of the coating material. In this contribution, we present a study of nanowire samples with GaN cores coated with a single shell of either AlGaN or HfO_2 or a double shell of AlGaN-GaN.

2. Experimental

Two samples which consist of core/single-shell (sample A) and core/double-shell (sample B) nanowire structures were grown on α-sapphire substrates using plasma assisted MBE. GaN cores of the NWs of both samples were obtained using Ni catalyst seeds to promote NW columnar growth under group V-rich conditions (N:Ga ratios of 9:2 for 1h 15min at 730 °C). For sample A, an $Al_{0.2}Ga_{0.8}N$ shell was deposited for 3.5 minutes at 730 °C after switching to group III-rich conditions (N : Ga flux ratio 7:10 and Al : Ga flux ratio of 1:4). Details and mechanisms of NW growth by PA-MBE under different conditions are described in [2]. For sample B, an additional deposition of GaN followed the preparation of the GaN/AlGaN structure (as for sample A) to form an outer shell under Ga-rich conditions (N:Ga ratio of 7:10 for 3.5 minutes). A third sample, C, was obtained through a combination of PA-MBE growth followed by Atomic Layer Deposition (ALD). For sample C the first step involved GaN NWs growth on silicon (001) under the group-V rich conditions used for sample A. The second step consisted of depositing a coating of HfO_2 using [$(MeCp)_2HfMe(OMe)$] (MeCp =

$CH_3C_5H_4$) precursor in toluene (0.05M) using water vapour as the oxidant. 150 cycles of atomic layer deposition were used to obtain a 6 nm thick coating.

Samples A and B were analysed using a HB601VG cold-FEG STEM microscope operating at 100kV (minimum probe size ~1nm), which is equipped with a windowless Si(Li) EDX spectrometer. Sample C was analysed using a 2010F FEG-TEM operating at 197kV, equipped with a scanning unit and an EDX detector. TEM samples A and B were prepared by mechanically scraping the NWs from the substrate surface into an ultrasonic bath of iso-propanol before depositing the NWs in suspension on a lacey carbon film supported by a TEM Cu grid. Samples B and C were prepared with the classic sandwich technique followed by mechanical polishing and Ar^+ ion milling to electron transparency.

3. Results and discussion

Energy-dispersive X-ray (EDX) point analyses and line scans were performed on sample A, to spatially quantify the Al concentration and distribution along the width and length of the NWs. Figure 1a) shows a typical EDX point analysis. The Al concentration was quantified using *K*-factors calculated with the Zaluzec and Mott-Massey atomic models assuming the thin film approximation. A general decrease of the Al : Ga ratio is observed from the NW growth tip (~6 at% Al) towards the root end (~1 at% Al) in the NWs analysed.

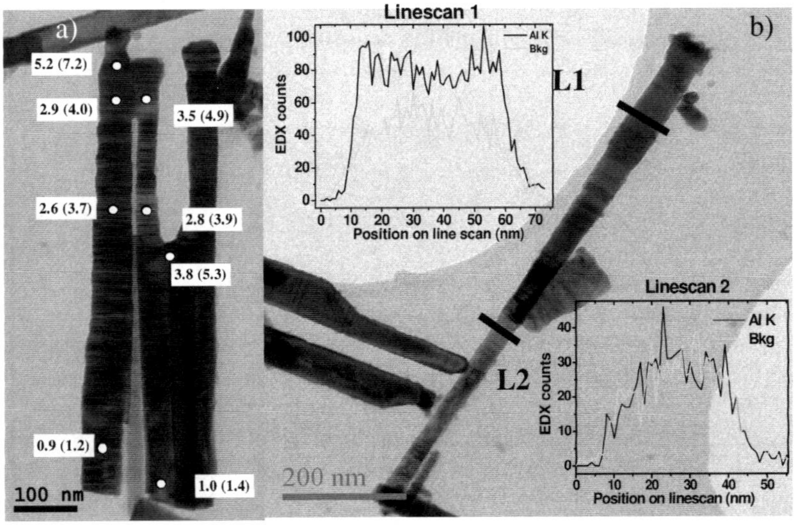

Figure 1. (a) STEM BF image of a GaN/AlGaN core/shell NWs. White spots indicate the positions of EDX point analyses. Numbers represent the values of at% Al, as % of group-III elements quantified using *K*-factors calculated by the EDAX software using the atomic models of either Zaluzec or Mott-Massey (in brackets). (b) STEM BF image with Al and background profile inset.

Transverse EDX Al line-profiles were obtained from several NWs by integrating the X-ray counts within a 100 eV wide energy window centred on the Al-K peak at 1.56 keV. It was not feasible to reliably eliminate the background contribution using the EDAX software automatic background subtraction routine because of the very low count rates. The small excitation volumes and the short acquisition times, required to avoid beam damage to the sample, generate particularly low count rates for the Al signal. To assess the background component of the Al profile a second acquisition, with a 100eV wide window, was set just above the Al peak and acquired simultaneously. As shown in Figure 1(b) the resultant Al profile is above the background only near the NW tip (linescan L1) with lateral maxima compatible with a projected volume of a shell containing Al. Near the substrate end of the NW (linescan L2) the Al and background signals are similar, indicating a decrease in shell thickness along the NWs length. A general enlargement in diameter of the NW growth tip is observed similar to pure GaN NWs grown with the same growth sequence (group-V rich followed by group-III rich conditions) [3]. The Al distribution (decreasing from the tip downwards) follows a reverse trend with respect to the group-V rich growth of pure AlGaN [4] NWs. This observation is attributed to the different growth mechanism of the NW under group-III growth conditions [3] combined with the lower mobility of Al under group-III rich growth conditions with respect to Ga [5]. A shadowing effect of neighbouring wires of the incoming atomic fluxes might also play a role.

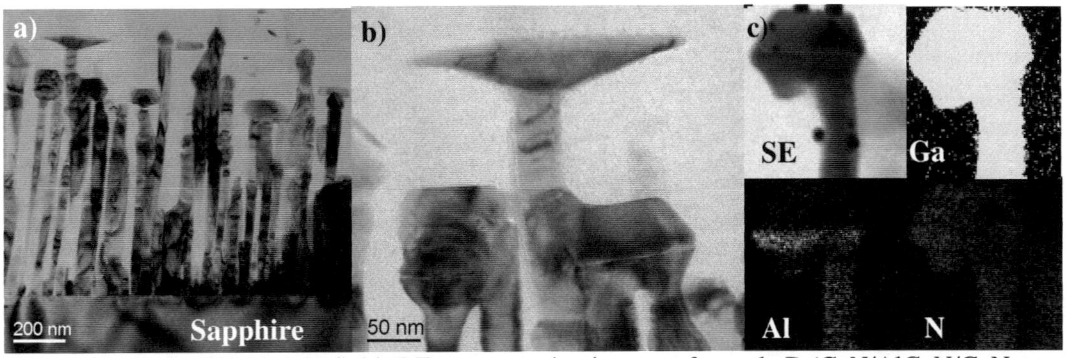

Figure 2. (a), (b) TEM bright field (BF) cross section images of sample B (GaN/AlGaN/GaN core/ inner shell/ outer shell structure) showing NW coalescence and 'hat' structures. (c) STEM secondary electron image of a NW tip and EDX maps corresponding to Al-K, Ga-L, N-K X-rays. Al is at background level in the upper part of the 'hat' structure.

The NW tip broadening effect is more drastic for sample B. Bright field images of a TEM cross-section (Figures 2a, b) show a non-uniformity of the NW diameters that often coalesce especially near the top of the wires where large "hat" structures tend to form. NW coalescence can be attributed to a progressive and uneven deposition of material particularly during the last two stages of radial growth (shell formation under group-III rich conditions). EDX maps of harvested nanowires show that the Al distribution stops abruptly at the 'hat' structure (Figure 2c).

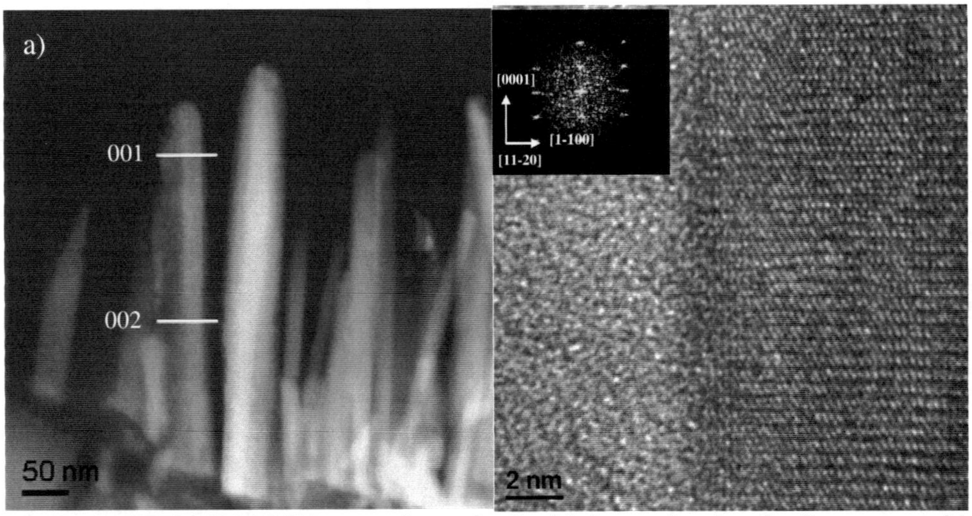

Figure 3. (a) STEM HAADF image of sample C with insets showing the line scan positions of Figure 4. (b) HREM image of the NW tip imaged along [11-20] zone axis with diffractogram. The lattice planes visible are d_{0002}=0.26 nm (horizontal fringes) and $d_{1\text{-}100}$=0.27 nm (vertical fringes).

Al appears here at background level suggesting that the 'hat' formation mainly takes place during the growth of the outer GaN shell. Point analysis along the NWs (not shown here) indicates an increase of the Al : Ga ratio from the substrate towards the NW growth tip confirming an Al distribution similar to that in sample A. To overcome this limitation of the MBE growth mechanism,

ALD has been used as an alternative technique to deposit HfO_2 on GaN NWs. HREM images (Figure 4b), taken along [11-20] wurtzite direction, show a single crystalline NW with sharp lateral walls.

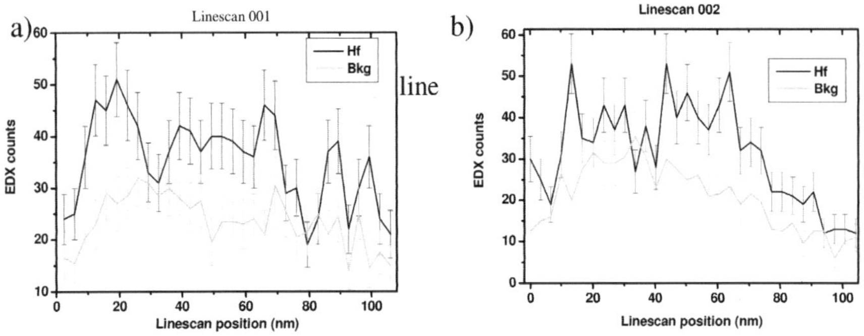

Figure 4. (a) EDX scan 001 across the upper part of the NW of Fig. 3a). (b) EDX line scan 002 across the lower part of the same NW.

The core/shell structure is not visible from the high-angle annular dark field (HAADF) image (Figure 3b) but a small EDX Hf signal was observed above the background along the whole length of the wires, which is shown in the line scans of Figure 4. For these EDX analyses the Hf signal was obtained integrating over a 300 eV wide window centred on the Hf-Lα1 peak, and the background was evaluated averaging a pre-peak and a post-peak window of roughly the same size. The thickness of HfO_2 deposited (~1nm) was thinner than a thin film (~ 6nm) grown under the same conditions. Although ALD is a 'saturative' cyclic growth process, the difference in deposition rates is probably due to limited adatom surface diffusion and the higher surface-to-volume ratio of NWs compared with planar 2D layers. Further investigation and tuning of the ALD growth process is needed to obtain a thicker and uniform radial coverage of NWs but, as a preliminary result, it suggests a possible way of producing free standing dielectric-semiconductor heterostructures.

4. Conclusions

Core/single shell GaN/AlGaN NWs showed a partial coverage of AlGaN with a thickness of the Al-rich shell decreasing toward the growth substrate. A similar Al distribution was found for core/double shell GaN/AlGaN/GaN radial NW structures and attributed to an Al incorporation mechanism dominated by a decreased diffusion along the NW walls, with a shadowing effect from neighbouring NWs. During the growth of the outer GaN shell, the deposition of material at the NW tip formed "hat"-like structures. HfO_2 deposited onto GaN NWs was found to be uniformly distributed along the NW length but with a lower thickness than expected. This preliminary observation suggests that ALD is potentially a complementary deposition method to provide uniform dielectric isolation of MBE grown NWs.

Acknowledgements

Financial support by EPSRC grant EP/F02374X/1, EU PARSEM contract MRTN-CT-2004-005583 and IST project NODE 015783 is gratefully acknowledged.

References

[1] Nitride semiconductors-Handbook on Materials and Devices eds. Ruterana P, Albrecht M and Neugebauer J (2003. Wiley-VCH, Berlin)
[2] Liu C, Chor EF and Tan LS *Thin Solid Films* **515** (2007) 4369
[3] Geelhaar L et al. *Appl. Phys. Lett.* **91** (2007) 093113
[4] Lari L et al. *Mater. Res. Soc. Symp. Proc.* **1026** (2008) C01
[5] Bertness KA et al. *J. Cryst. Growth* **287** (2006) 522
[6] Kehagias T et al. *Proceedings of the Conference on "Microscopy of Semiconducting Materials XV"* (2007, Cambridge, UK, Springer Proc. in Physics 120) p. 225

Basal-plane Stacking Faults in Non-polar GaN Studied by Off-axis Electron Holography

Lewis Z-Y Liu[1], DV Sridhara Rao[1], MJ Kappers[1], CJ Humphreys[1] and D Geiger[2]

[1] Department of Materials Science and Metallurgy, University of Cambridge, Pembroke Street, Cambridge, CB2 3QZ, United Kingdom
[2] Triebenberg Laboratory, Institute for Structure Physics, Technische Universitaet Dresden, D-01062 Dresden, Germany

E-mail: ZL249@cam.ac.uk

Abstract. We have studied basal-plane stacking faults in a non-polar (11-20) GaN epilayer using high-resolution electron microscopy and off-axis electron holography. The microstructure of the basal-plane stacking faults (BSFs) has been determined to be I_1 type from high-resolution TEM images. High-resolution holograms along the [11-20] zone axis were obtained by off-axis electron holography on a Cs-corrected TEM, providing ~2 Å spatial resolution in the reconstructed amplitude and phase images. Phase fluctuations across the stacking faults were detected, suggesting the presence of a built-in electric field. The uncertainties in the experiments and their interpretation are discussed.

1. Introduction

Gallium nitride based materials and structures are conventionally grown along the polar crystallographic *c*-axis. The spontaneous and piezoelectric polarization in this direction generates internal electric fields, which lower the radiative recombination efficiency in the quantum well (QW) structures. The growth of non-polar structures on *a*-plane (11-20) GaN eliminates the electric fields across the QWs, but the thin films generally have more defects such as basal-plane stacking faults (BSFs), prismatic stacking faults (PSFs) and partial dislocations (PDs).

BSFs have been proposed theoretically to act as zinc-blende quantum wells in wurtzite GaN [1], and their luminescence has been observed [2]. A built-in electric field across the BSF region has been calculated [3] and is supported by recent studies on THz radiation from non-polar GaN with high BSF density [4]. Relevant studies of stacking faults in other wurtzite materials, such as SiC, have also shown evidence of built-in electric fields [5]. However, the nature and scale of the internal electric field has not been directly determined. Here we carried out studies using high-resolution electron microscopy and electron holography to determine the microstructure of the BSFs in non-polar GaN and in particular to detect the electrostatic potential distribution across the BSF regions.

2. Experimental

The non-polar (11-20) GaN sample was grown by metal-organic vapour phase epitaxy (MOVPE) on (1-102) *r*-plane sapphire using trimethyl gallium (TMG), silane and ammonia (NH_3) as precursors with hydrogen as the carrier gas. A more detailed description of the growth is given elsewhere [6].

The plan-view TEM specimens were prepared by mechanical polishing and Ar^+ ion milling. The ion milling voltage was lowered from 5 kV to 2 kV for the last 10 minutes to minimize surface damage. TEM imaging was carried out on a Philips CM30 at 200 kV and a JEOL 4000EX-II at 400 kV. A FEI Tecnai F20/Cs-corrected TEM with a Möllenstedt biprism was used to acquire high-resolution holograms at 200 kV. The fringe contrast of the reference hologram was 21-24% with a carrier frequency of 14.5 nm^{-1}.

3. Results and discussion

3.1. Microstructure study

The plan-view specimens of the non-polar (11-20) GaN layer were examined using bright field imaging with the ***g*** vector [1-100] to show the BSFs in strong contrast (figure 1). BSFs are either terminated by PDs, or jump to an adjacent plane through the generation of PSFs. The average BSF density in this region is of the order of magnitude of 0.1 nm^{-1}, which is later shown to be appropriate for high-resolution electron holography study.

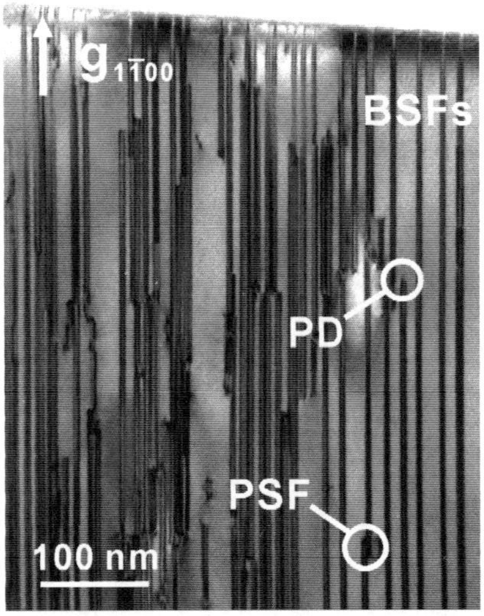

Figure 1. Bright field TEM image showing BSFs bounded by PDs and PSFs

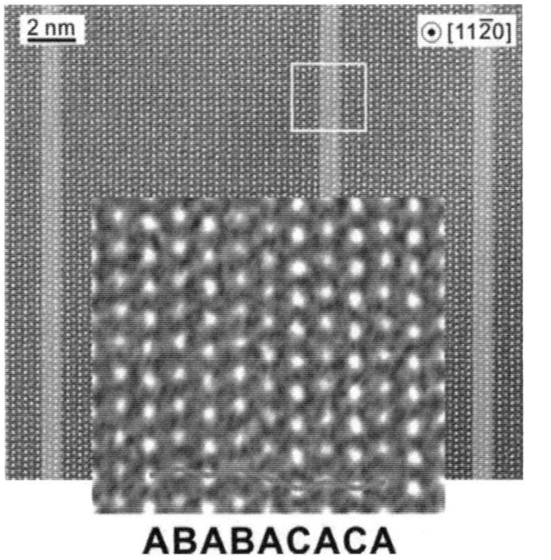

Figure 2. High-resolution TEM image showing the structure of the BSFs in the non-polar GaN

High-resolution [11-20] plan-view TEM studies show the structure of the BSFs (figure 2). Although it is not straightforward to confirm the true positions of the atom columns simply from the image, the stacking sequence of the close packed planes is shown to be ...ABA*BAC*ACA..., corresponding to the so-called I_1 type stacking fault with the displacement vector ***R***=1/6 [20-23]. This has the lowest energy among all types of stacking faults in the wurtzite structure. Over 40 BSFs have been examined with high-resolution imaging on different TEM samples of this material, all of which were found to be I_1 type.

3.2. Electron holography investigation

High-resolution electron holography with a fringe spacing under 0.7 Å provides the spatial resolution of ~2 Å in the reconstructed amplitude and phase image, sufficient to resolve the phase distribution to atomic level. Figure 3a is a high-resolution electron hologram along the [11-20] zone axis. It shows two BSFs (indicated in regions b and c, respectively) with the stacking sequence changing from BA to AC (figure 3b), and then to CB (figure 3c). The spacing between the two BSFs is 4 nm, and no other BSFs were observed in the adjacent areas of 10 nm.

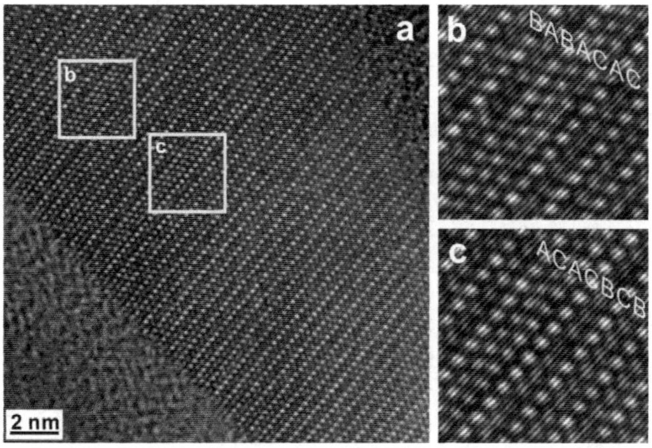

Figure 3.
(a) High-resolution electron hologram taken along the [11-20] zone axis showing two BSFs; the areas in the squares are shown magnified in:
(b) Stacking sequence changes from BA to AC
(c) Stacking sequence changes from AC to CB

The reconstruction of the hologram is shown in figure 4a and b. The phase image has been linearly tilted -0.16 rad/nm and 0.38 rad/nm in the x and y directions respectively to compensate the phase variation generated by the uniform thickness increase away from the edge. Both the amplitude and phase images show uniform and well-defined lattice planes of the specimen. Figure 4c shows the line profile perpendicular to the BSFs of the phase image. The periodic sharp peaks correspond to shielded nuclei, while the background shows the average distribution of bonding electrons as well as free carriers. Fluctuations in the background are observed at the two fault areas, which indicate that locally there may be free carrier segregation. It is possible that in the 4 nm narrow region between the two BSFs free carriers cannot fully screen the spontaneous polarization and thus the background still shows fluctuations, while in the region further away from them, the spontaneous polarization is fully screened, which leads to a flat background in the phase profile. These provide evidence for the presence of built-in electric fields across the BSF regions.

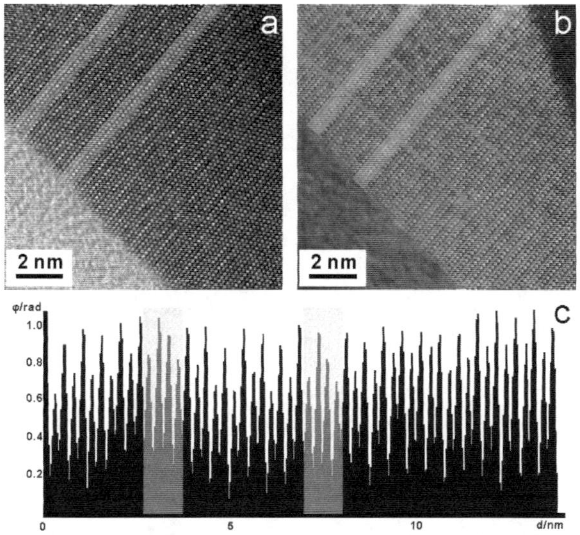

Figure 4.
(a) Amplitude and
(b) Phase image reconstructed from the electron hologram shown in figure 3a
(c) Integrated line profile of the quoted area in b. The BSFs are highlighted in all three figures.

There are, however, many uncertainties in this speculative interpretation. Contrast given by diffraction is always a concern and has not been properly removed here. Apart from this effect, however, given the spontaneous polarization of the wurtzite structure, a pair of charge "sheets" are expected at the faces of a BSF to show some phase contrast. Moreover, the peaks of the background fluctuation at the top faces of the BSFs indicate hole segregation, and therefore "valleys" in the background are expected at the bottom faces. However, these features have not been observed in our results. On the other hand, such an experiment aiming at measurement of phase with a precision of 0.1 rad is highly sensitive to the quality of the specimen. Any artifact on or inside the sample, local thickness variation as well as charging effects under the electron beam could significantly affect the results. More experiments are necessary in this respect.

The origin of the internal electric field across a BSF is normally referred to as the result of cubic layers terminating the wurzite structure. However, there must be a change in the inter-planar spacings along the *c*-axis from the fault layers to the bulk material, otherwise there would simply be no internal electric field. We plan to fit the atomic positions in the high-resolution TEM image to determine the change in *c*-plane spacings across the BSFs, which may be as large as 5 to 10 pm [7].

4. Summary and outlook

We have studied basal-plane stacking faults in *a*-plane (11-20) epitaxial lateral overgrown GaN using high-resolution electron microscopy and electron holography. The structures of over 40 BSFs have been observed, all determined to be I_1 type. High-resolution amplitude and phase images were obtained using off-axis electron holography. Phase fluctuations were detected at the stacking fault areas, indicating the presence of built-in electric fields. No quantitative results have been derived, however, due to uncertainties in the interpretation of the phase images.

Electron wave simulations under various tilt and defocus conditions in order to subtract the diffraction contrast are planned. DFT simulations could provide more detail about the potential distribution. For both of the simulations, precise *c*-plane spacing values are required, which could be obtained from atomic position fitting of the HR-TEM images.

Acknowledgement

The research was funded by the EPSRC (EP/E035167/1) and the European Union Framework 6 program under a contract for an Integrated Infrastructure Initiative (Reference 026019 ESTEEM). LZYL thanks Homerton College for financial support.

References

[1] Rebane YT, Shreter YG and Albrecht M 1997 *Phys. Status Solidi* (a) **164**, 141
[2] Liu R, Bell A, Ponce FA, Chen CQ, Yang JW and Khan MA 2005 *Appl. Phys. Lett.* **86**, 021908
[3] Majewski JA, Zandler G and Vogl P 2000 *Phys. Status Solidi* (a) **179**, 285
[4] Metcalfe GD, Shen H, Wraback M, Hirai A, Wu F and Speck JS 2008 *Appl. Phys. Lett.* **92**, 241106
[5] Juillaguet S, Camassel J, Albrecht M and Chassagne T 2007 *Appl. Phys. Lett.* **90**, 111902
[6] Hollander JL, Kappers MJ, McAleese C and Humphreys CJ 2008 *Appl. Phys. Lett.* **92**, 101104
[7] Leszczynski M, Teisseyre H, Suski T, Grzegory I, Bockowski M, Jun J, Porowski S, Pakula K, Baranowski JM, Foxon CT and Cheng TS 1996 *Appl. Phys. Lett.* **69**, 73

Atomic resolution imaging of *in situ* InAs nanowire dissolution at elevated temperature

RS Pennington[1], JR Jinschek[2], JB Wagner[1], CB Boothroyd[1] and RE Dunin-Borkowski[1]

[1] Center for Electron Nanoscopy, Technical University of Denmark, DK-2800 Kongens Lyngby, Denmark
[2] FEI Company, Achtseweg Noord 5, 5600 KA Eindhoven, Netherlands

E-mail: robert.pennington@cen.dtu.dk

Abstract. We present a preliminary study of the *in situ* heating of InAs nanowires in a gaseous environment in an environmental transmission electron microscope. Nanowire dissolution, accompanied by dynamic reshaping of crystalline Au-containing catalyst particles at the ends of the wires, is observed, accompanied by indium oxide crystallite formation nearby.

1. Introduction
Catalytic growth of semiconductor nanowires is of considerable interest [1], but is poorly understood at the atomic level. Some nanowires can be grown below the eutectic temperature of the catalyst-semiconductor alloy, leading to the proposal of a vapor-solid-solid growth mechanism [2]. Here, we examine InAs nanowires, whose growth by chemical vapor deposition is known to drop abruptly above ~500 °C [3]. We image nanowires just above this temperature in the transmission electron microscope (TEM) in a gaseous environment, as a preliminary experiment aimed at understanding the behavior and imaging properties of semiconductor nanowires in an environmental TEM, in advance of attempting true *in situ* III-V nanowire growth. We observe a complex dynamic nanowire dissolution process accompanied by subsidiary crystallite formation.

2. Experimental Details
InAs nanowires containing InP barriers were grown by chemical beam epitaxy at 390°C. A bright-field TEM image showing a selection of nanowires, which are each capped by a Au catalyst particle and supported on a C film for TEM imaging, is shown in Fig. 1. The nanowires were examined at 300 kV in an image-aberration-corrected FEI Titan 80-300 TEM equipped with a differentially-pumped environmental cell. Motivated by recent reports that the surfaces of similar nanowires can be cleaned using atomic hydrogen [4], the nanowires were examined at elevated temperature in a H_2 atmosphere.

3. Results
The InAs nanowires were heated to 520 °C in 1.3 mbar H_2 *in situ* in the TEM. Unexpectedly, under these conditions, the Au-containing particles changed shape dynamically while remaining crystalline, as confirmed by the presence of lattice fringes in the image shown in Fig. 2. Dissolution of the wires proceeded, apparently at the catalyst-nanowire interface, with the wires shrinking in length. New crystallites were observed to form on the support film adjacent to the nanowires.

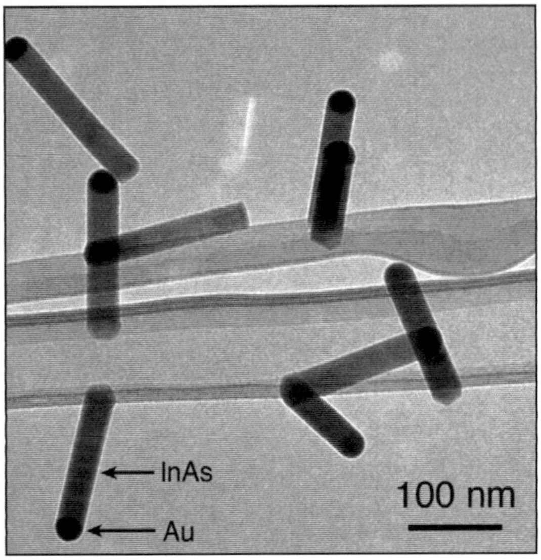

Fig. 1. Bright-field image of nanowires imaged at room temperature.

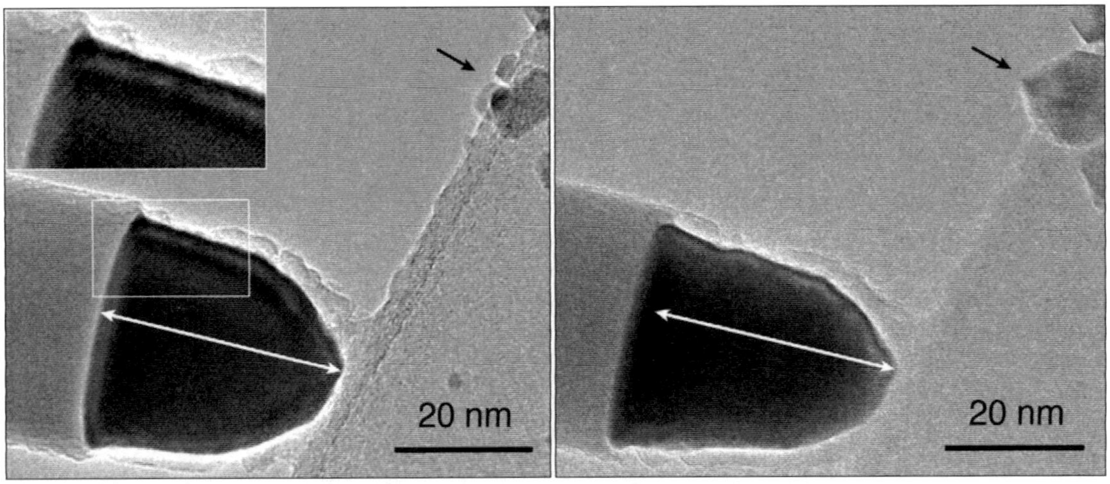

Fig. 2. Images acquired 3 minutes apart at 520 ºC in 1.3 mbar H_2, showing lattice fringes in a Au-containing particle at the end of an InAs nanowire. The white arrows, which have the same length, illustrate the elongation of the catalyst particle. New crystallites form close to the wire at the position indicated by the black arrow.

Fig. 3 shows three snapshots taken from a movie illustrating dynamic changes in the shape of a crystalline Au-containing particle during dissolution of an InAs nanowire. The brighter region in the wire is a segment of InP. Dissolution of the InP segment is observed to proceed faster than that of the InAs. Crystallites are observed to form in the upper-right corner of each frame. Lattice fringe contrast in both the particle and the nanowire is visible in Fig. 4, which shows the presence of a ledge at the interface during nanowire dissolution. Fig. 5 shows high-resolution lattice images acquired both during and after crystallite formation, as well as an electron energy-loss spectrum, confirming that the crystallites are In_2O_3. Neither As nor P were detected in the crystallites. The original thin amorphous coating on the nanowire surface, which is visible in Figs 2 and 3, remained as a product next to the Au particle. This coating was subsequently identified using electron energy-loss spectroscopy as being composed primarily of C.

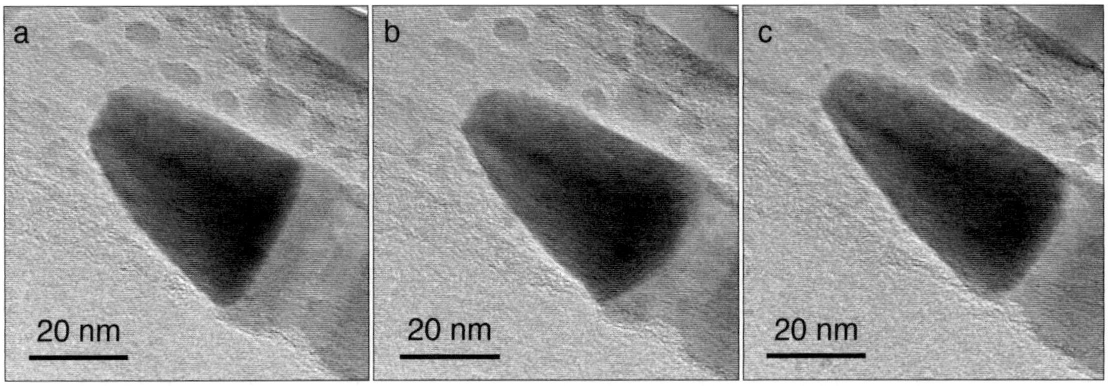

Fig. 3. Three frames, separated by 7 seconds, taken from a movie of InAs nanowire dissolution. A short segment of InP is visible as a lighter-contrast segment of the wire.

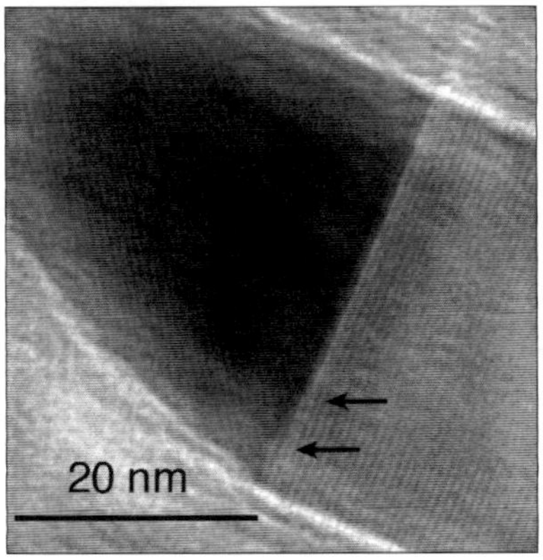

Fig. 4. One Fourier-filtered frame taken from a movie acquired during InAs nanowire dissolution. Lattice fringe contrast is visible both in the Au-containing particle and in the InAs nanowire, indicating that they are both crystalline at this temperature (520 °C). The image also confirms that lattice fringe contrast can be observed in the presence of 1.3 mbar of H_2 in the differentially pumped section of the electron microscope column. Ledges are visible at the particle-nanowire interface (arrowed).

4. Discussion
The crystalline nature of the Au-containing particle at the temperature used in the present experiments, which is above the growth temperature of the InAs nanowires, suggests that the particle may also have been crystalline during nanowire growth at lower temperature. The interaction between the Au-containing particle and the support film may play a role in the observed shape changes. The formation of In_2O_3 crystallites is surprising as there is no intentional source of oxygen in the microscope. Some native oxide on the nanowire is likely, but residual gas in the environmental cell is a more likely source given the amount required for particle formation. The present observations illustrate the complexity of even a nominally straightforward heating experiment carried out in an environmental TEM. Parameters that may affect the observed dissolution process include specimen temperature, gas environment, the interaction between the Au particle and the carbon support film and the effect of the incident electron beam both on the specimen and on the surrounding gas.

5. Conclusions
Preliminary results from *in situ* heating of InAs nanowires in H_2 in an environmental TEM reveal, at atomic resolution, a complicated nanowire dissolution process, which appears to proceed at the nanowire-catalyst interface. During this process, the Au-containing catalyst particle remains crystalline but changes shape dynamically. At the same time, In_2O_3 crystallites form adjacent to the wire. All aspects of such a simplified experiment should ideally be understood before proceeding to the additional complication of introducing reactive gases to grow III-V nanowires *in situ* in the TEM.

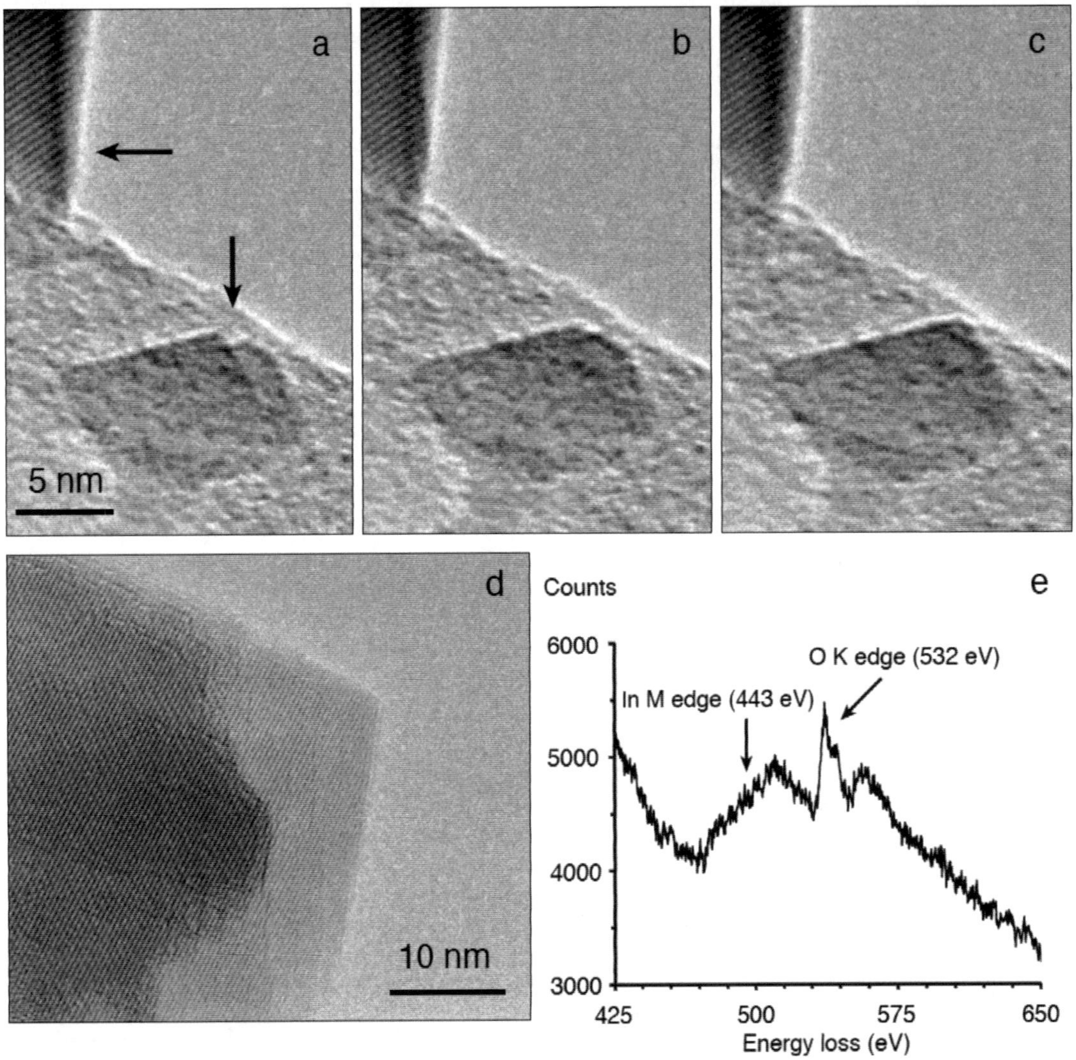

Fig. 5. a) - c) High-resolution lattice images showing crystallite formation adjacent to a nanowire. d) High-resolution lattice image and e) electron energy-loss spectrum acquired from crystallites after nanowire dissolution, with In and O peak edges indicated [5], confirming the formation of In_2O_3.

Magnus Borgström and Linus Fröberg from Lund University are acknowledged for useful discussions and for growing and providing the nanowires used for this study, respectively.

References

[1] K A Dick 2008 *Progress in Crystal Growth and Characterization of Materials* **54**, 138
[2] K A Dick, K Deppert, T Mårtensson, B Mandl, L Samuelson and W Seifert 2005 *Nano Letters* **5**, 761
[3] A I Persson, M W Larsson, S Stenström, B J Ohlsson, L Samuelson and L R Wallenberg 2004 *Nature Materials* **3**, 677
[4] E Hilner, U Håkanson, L E Fröberg, M Karlsson, P Kratzer, E Lundgren, L Samuelson and A Mikkelsen 2008 *Nano Letters* **8**, 3978
[5] C C Ahn and O L Krivanek 1983 *EELS Atlas* (Gatan)

Mg dopant distribution in an AlGaN/GaN p-type superlattice assessed using atom probe tomography, TEM and SIMS

SE Bennett[1], PH Clifton[2], RM Ulfig[2], MJ Kappers[1], JS Barnard[1], CJ Humphreys[1] and RA Oliver[1]

[1] Department of Materials Science and Metallurgy, University of Cambridge, Pembroke St., Cambridge, CB2 3QZ, UK

[2] Imago Scientific Instruments Corporation, 5500 Nobel Drive, Madison, WI, 53711, USA

Email: sb534@cam.ac.uk

Abstract. P-type conducting layers are critical in GaN-based devices such as LEDs and laser diodes. Such layers are often produced by doping GaN with Mg, but the hole concentration can be enhanced using AlGaN/GaN p-type superlattices by exploiting the built-in polarisation fields. A Mg-doped AlGaN/GaN superlattice was studied using SIMS. Although the AlGaN and GaN were nominally doped to the same level, the SIMS data suggested a difference in doping density between the two materials. Atom probe tomography was then used to investigate the Mg distribution. The superlattice repeats were clearly visible, as expected and, in addition, significant Mg clustering was observed in both the GaN and AlGaN layers. There were many more Mg clusters in the AlGaN layers than the GaN layers, accounting for the difference in doping density suggested by SIMS. To evaluate the structural accuracy of the atom probe reconstruction, layer thicknesses from the atom probe were compared with STEM images. Finally, future work is proposed to investigate the Mg clusters in the TEM.

1. Introduction

GaN-based light emitting diodes (LEDs) are becoming ubiquitous as their use expands from niche applications towards everyday home and office lighting. These devices and others, such as laser diodes, require p-type and n-type layers for injection of the charge carriers that then recombine to produce light. Although n-type layers can be produced easily, often using Si as the shallow donor, p-type conductivity is much harder to achieve. Mg was the first dopant added to realise p-layers in GaN [1] and continues to be used in modern devices. The efficiency of such doping, however, is intrinsically low due to the deep acceptor state that Mg forms. By using an AlGaN/GaN p-type superlattice, the innate polarisation fields result in a periodic oscillation of the valence band edge. This oscillation allows acceptors to be ionised wherever the acceptor level is below the Fermi energy, resulting in a hole concentration that is enhanced by up to 10 times.

A variety of characterisation techniques have been used to study a Mg-doped p-type superlattice. Although grown with nominally the same doping level in both the AlGaN and GaN layers, initial secondary ion mass spectrometry (SIMS) evaluation suggested that the AlGaN layers have a greater Mg content. Here, we present three dimensional (3D) atom probe tomography (APT) data to show the Mg distribution within the superlattice. The APT data are then compared with transmission electron microscopy (TEM) data to enhance the understanding of the superlattice structure.

2. Methods

The AlGaN/GaN sample was grown on a c-plane sapphire substrate using a 6x2" Thomas Swan close-coupled showerhead metal-organic vapour phase epitaxy (MOVPE) reactor. Trimethylgallium (TMG), trimethylaluminium (TMA), bis(cyclopentadienyl)magnesium (Cp_2Mg) and ammonia (NH_3) were used as precursors, with hydrogen (H_2) as the carrier gas. First, a ca. 30 nm GaN nucleation layer was deposited at 540 °C, followed by a 2 μm GaN buffer layer grown at 1005 °C. Next, the superlattice was grown at 1000 °C with 24 repeat periods of ca. 10.5 nm $Al_{0.11}Ga_{0.89}N$ and ca. 10.5 nm GaN. The Cp_2Mg flux during superlattice growth was 0.364 μmol/minute. The sample was then annealed at 780 °C in N_2 for 1200 s. For the purpose of atom probe sample preparation, a non-intentionally doped 500 nm capping layer of GaN was added at 1005 °C.

To study the Mg content in the layers of the superlattice, SIMS was conducted at Loughborough Surface Analysis Ltd. using a Cameca IMS 3f. Standard samples for Mg in AlGaN were not available at the time of testing, therefore only relative changes in the level of Mg can be reliably obtained from these data.

For the APT study of 3D Mg distribution, needle-shaped samples were prepared in a dual focused ion beam (FIB)/scanning electron microscope (SEM) by the standard lift-out and annular milling method described in Miller *et al.* [2]. An FEI Company™ Nova NanoLab™ with an Omniprobe Autoprobe™ 200 micromanipulator was employed. To remove areas of the sample damaged by ion beam milling at 30 kV, a low energy clean-up mill at 5 kV was used as the final preparation step. It has been shown that this minimises the ion beam induced damage to the APT sample [3].

The superlattice was then studied by APT on an Imago Scientific Instruments Corp. LEAP 3000X Si™. Under high vacuum and at low temperature (25 K), the sample was held at a standing voltage close to the threshold needed for field evaporation of the ions at the sample tip. A laser was focused onto the tip and pulsed at 200 kHz to induce field evaporation. Ion position and time-of-flight data were analysed using the Imago IVAS™ software. Once compositional identities were assigned to mass spectrum ranges and a number of other reconstruction parameters were defined, a 3D atomic map was produced.

Further structural characterisation was carried out using scanning TEM with a high-angle annular dark field detector (STEM-HAADF) on an FEI Tecnai F20-G2 with a 200 kV field emission gun. Sample preparation for the TEM foil was via standard mechanical polishing, followed by ion milling to electron transparency with a Gatan PIPS™.

3. Results and Discussion

Figure 1 shows the SIMS data for Mg and Al in the superlattice. The thick line is the Al concentration and the thin line is the Mg concentration. Although nominally doped to the same level in both the AlGaN and GaN layers, the Mg content appears to vary layer-to-layer. The dotted line in figure 1 illustrates that the AlGaN layers seem to have greater Mg contents. Although the Mg fluctuations appear to decrease with time, this is partially due to intermixing between layers as the primary ion beam penetrates further into the sample.

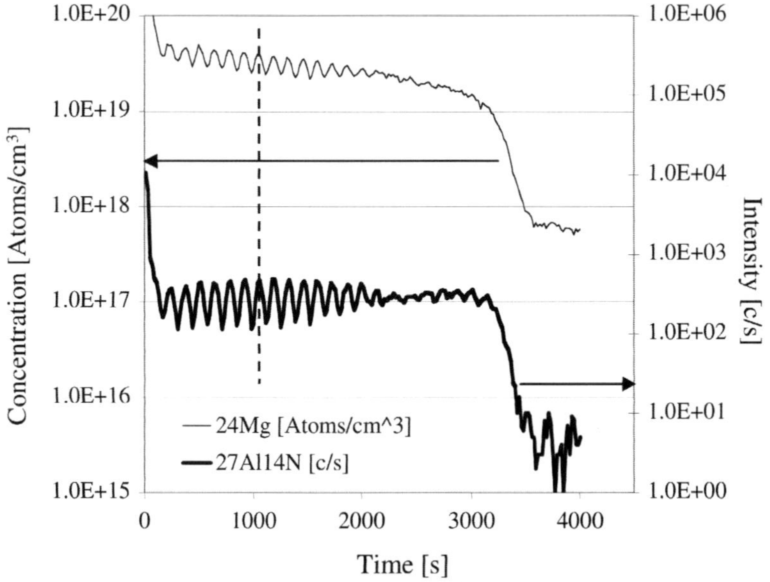

Figure 1. SIMS data for the superlattice with Mg doping to nominally the same level throughout. The thick line shows the Al and the thin line shows the Mg. There appears to be greater Mg content where there is more Al, corresponding to the AlGaN layers.

One of the strengths of APT is its ability to detect low concentrations of light elements such as Mg. Here, this ability is particularly relevant, since the Mg content of the superlattice is approximately 0.01 atomic % (at%); however, as shown in figure 2, it is not uniform. Ten percent of the Al ions collected are shown (dots) to indicate the AlGaN layers and a 4 at% isoconcentration surface is shown around the Mg ions. The areas of higher Mg concentration can be clearly seen and these will be referred to as 'clusters'. To investigate the difference in Mg content between the AlGaN and GaN layers, the cluster density in each type of layer was analysed. There are many more clusters in the AlGaN layers, with 14.7 clusters on average in the AlGaN layers as compared to 1.7 clusters on average in the GaN layers (per layer volume of ca. 124,000 nm^3). Outside these clusters, the Mg content did not vary significantly between the two types of layers, therefore we conclude that the difference in Mg content suggested by SIMS is due to the greater density of clusters in the AlGaN.

To evaluate the structural accuracy of our reconstruction, the layer thicknesses measured from APT were compared with TEM. STEM-HAADF was used to image the superlattice, as shown in figure 3. The lighter contrast is GaN and the darker contrast is AlGaN. The layers were very regular, with an average thickness of 9.5 ± 0.5 nm for AlGaN and 10.7 ± 0.2 nm for GaN. The APT data for AlGaN layer thickness agreed well with the STEM,

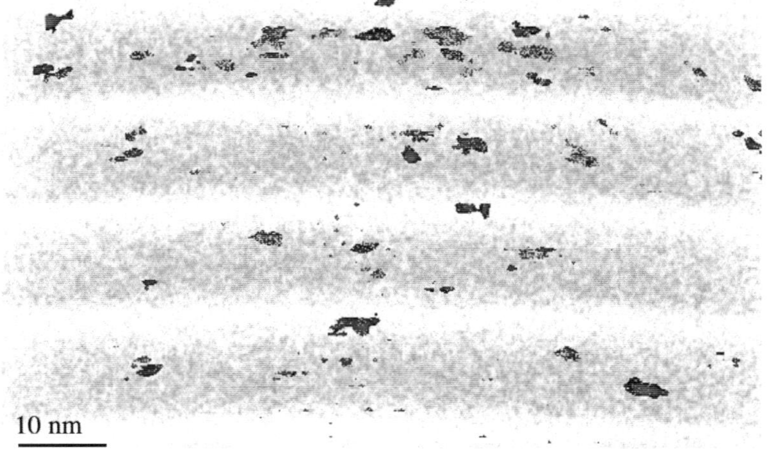

Figure 2. APT data showing 10% of all Al ions collected (dots) and a 4 at% isoconcentration surface around the Mg ions (clusters). There are more Mg clusters in the AlGaN layers, accounting for the difference in Mg content suggested by the SIMS data.

giving an average thickness of 10.9 ± 0.2 nm, indicating the APT reconstruction parameters were accurate for this material. The GaN layer thicknesses were more inaccurately reconstructed in the APT, with an average thickness of 6.0 ± 0.2 nm, as currently released software allows only one global reconstruction field. It is important to note that if the reconstruction parameters were changed to alter the GaN thickness, this would not change the placement of the Mg clusters, as their positions would be similarly altered.

Figure 3. STEM-HAADF image showing the entire superlattice. The lighter contrast is GaN and the darker contrast is AlGaN.

Further TEM-based characterisation is planned to investigate the nature of the Mg clusters. Mg-rich, pyramidal defects in p-type GaN have been reported in the literature [4], always with the point in the [000-1] direction; however some controversy exists as to whether these are hollow voids [5] or inversion domains [4]. Future TEM on our superlattice may provide the opportunity to compare these defects with the clusters observed in APT.

4. Conclusions

SIMS data indicated that a Mg-doped AlGaN/GaN superlattice had greater Mg incorporation in the AlGaN layers. A 3D APT reconstruction of the superlattice showed Mg clusters, which were found to be preferentially located in the AlGaN layers. Outside the clusters there was approximately the same Mg concentration in both types of layers. The greater density of clusters in the AlGaN layers therefore accounts for the difference in Mg concentration observed by SIMS. STEM-HAADF was then used to measure the layer thicknesses in the superlattice. These values were compared to those found by APT, showing that the AlGaN layers were reconstructed more accurately than the GaN layers.

Acknowledgements

SEB would like to acknowledge funding from the EPSRC, Sharp Laboratories of Europe and Pembroke College, Cambridge. RAO would like to acknowledge funding from the Royal Society. SIMS analysis was performed by Dr. A. Chew of Loughborough Surface Analysis, Ltd.

References

[1] Amano H, Kito M, Hiramatsu K and Akasaki I 1989 *Jap. J. Appl. Phys.* **28**, 2112

[2] Miller MK, Russell F and Thompson GB 2005 *Ultramicroscopy* **102**, 287

[3] Thompson K, Gorman B, Larson D J, van Leer B and Hong L 2006 *Microsc Microanal* **12**, 1736

[4] Vennegues P, Benaissa M, Beaumont B, Feltin E, De Mierry P, Dalmasso S, Leroux M and Gibart P 2000 *App. Phys. Lett.* **77**, 880

[5] Liliental-Weber Z, Benamara M, Swider W, Washburn J, Grzegory I, Porowski S, Lambert DJH, Eiting CJ and Dupuis RD 1999 *Appl. Phys. Lett.* **75**, 4159

Investigation of optimum growth conditions of InAlN for application in distributed Bragg reflectors

TC Sadler, MJ Kappers and RA Oliver

Department of Materials Science and Metallurgy, University of Cambridge, Pembroke Street, Cambridge, CB2 3QZ, UK

E-mail: tcs29@cam.ac.uk

Abstract. $In_xAl_{1-x}N$ is an alloy system which can be lattice matched to GaN, with many potential applications including distributed Bragg reflectors. This work investigated the effect of changing the ammonia flux during metal organic vapour phase epitaxy growth. The indium incorporation increases as the ammonia flux is increased from 44.6 mmol/min to 179 mmol/min, but remains constant for further increases up to 446 mmol/min. The surfaces across this range generally have low surface roughnesses suitable for use in distributed Bragg reflectors. However, some samples have significant surface pit densities, although these are not linked to similar rises in threading dislocations generated at the GaN / InAlN interface.

1. Introduction

The growth of InAlN lattice matched to GaN is an interesting recent development in the area of group-III nitrides. Lattice matching occurs at approximately 17.5 % In [1, 2]. One of the main applications for the material is for strainless distributed Bragg reflectors (DBRs) [1]. However InAlN has been perceived as difficult to grow using the two main techniques used in the group-III nitrides, metal organic vapour phase epitaxy (MOVPE) and molecular beam epitaxy (MBE), due to the very different growth conditions of the system end members. For example in MOVPE AlN is usually grown at a high temperature of approximately 1100 °C [3], but InN is usually grown at a low temperature of approximately 550 °C [4]. It has also been suggested that the system may have a spinode which would lead to spontaneous phase separation [5].

Unfortunately there are few systematic published studies of the effects of different growth conditions in MOVPE on the composition, morphology and defect structure of InAlN, although there has been some work exploring growth conditions in MBE [6]. Studies of this sort are needed not only because they shed light on an interesting alloy system with very different end members but more importantly because they will make it easier for others to optimise InAlN growth, allowing more work to be done on the material and increasing the likelihood that it will be incorporated into devices. In particular there has been little investigation of the material by transmission electron microscopy (TEM) or by atomic force microscopy (AFM).

Here we study the effect of ammonia (NH_3) gas flux rate in MOVPE on the system by investigating 6 different ~100 nm thick epilayers grown on two different occasions, using AFM, X-ray diffraction (XRD) and TEM.

2. Experimental methods

Our growth was carried out in a Thomas Swan 6 × 2" close-coupled showerhead MOVPE reactor, on pseudo-substrates consisting of ~6 μm of GaN grown on c-plane (0001) sapphire at 1020 °C, following deposition of a 30 nm GaN buffer at 560 °C. All the InAlN epilayers were deposited at 800 °C, employing a trimethylindium (TMI) flux of 16.1 μmol/min, a trimethylaluminum (TMA) flux of 15.8 μmol/min, and a pressure of 50 Torr using N_2 as the carrier gas. The ammonia (NH_3) flux was varied between 44.6 mmol/min (a flux of 1 slm) and 446.4 mmol/min (a flux of 10 slm).

XRD was carried out using a Philips P'analytical PW3050/65 X'Pert PRO HR horizontal diffractometer using Cu K_α radiation with an asymmetric Ge (220) 4-bounce primary monochromator and analysed using X'pert Epitaxy and Smoothfit 4.0 Philips Analytical B.V. It was used to measure the average lattice parameters of both the epilayers and the pseudo-substrates, using the positions of both 002 symmetric and 105 antisymmetric reflections. These were then used to calculate the epilayer compositions, assuming Vegard's law to hold for $In_xAl_{1-x}N$, isotropic strain in the c-plane of the samples [7] and using literature values of the lattice parameters and elastic constants of AlN and InN [8]. Whilst Vegard's law may not hold in this case, the precise nature of any deviations has yet to be studied in detail, so we cannot greatly improve on this approach. Fringe peak spacings were used to find the epilayer thicknesses, and defects were investigated using ω-scans of the 002 and 101 peaks.

Surface morphology was assessed by AFM using a Veeco Dimension 3100 with analysis using WSxM freeware [9].

TEM was carried out using a Philips CM30 operated at 300 kV, and samples were prepared by grinding and polishing to ~30 μm and then thinned to electron transparency using a Gatan precision ion polishing system.

3. Results and discussion

3.1. Overview of samples

The two sample series studied here were grown on separate occasions a few months apart, and each investigates a different range of NH_3 fluxes. The first, series A, used flows of 1 slm, 2 slm and 4 slm, and the second, series B flows of 4 slm, 7 slm and 10 slm. However due to process drift the growth temperature may not have been exactly 800 °C, and may have moved slightly between experiments. This is due to several practical reasons, for example the build up of material on the susceptor used and the difficulties of temperature calibration to an accuracy of a few degrees at 800 °C.

3.2. Initial investigation of epilayers: compositions and surface morphology

Initial experiments on all 6 wafers measured the epilayer composition by XRD, and also assessed the surface morphology by AFM using 1 μm × 1 μm scans. The distinguishing features of the surface are broad 'hillocks' ~50 nm across and 2-3 nm high, and pits which we initially hypothesized to be linked to dislocations due to similar features being linked to dislocations in other nitride systems [10, 11]. Table 1 gives the compositions, roughnesses and pit densities for all samples. Example AFM scans for each sample are shown in Fig. 1.

Table 1. The initial measurements for all samples investigated. All measurements from AFM scans (surface roughness and pit densities) come from the average of 4 scans with standard errors quoted.

Sample	NH_3 flux / slm	In incorporation / %	Roughness at 1 μm / nm	Pit density / μm^2 (from 1 μm scans)
A1	1	12.9 ± 1.4	0.29 ± 0.01	3.0 ± 1.0
A2	2	15.1 ± 1.4	0.44 ± 0.05	3.5 ± 0.3
A3	4	17 ± 1.45	0.77 ± 0.11	68.5 ± 9.3
B1	4	14.6 ± 1.4	0.70 ± 0.04	4.5 ± 0.9
B2	7	14.4 ± 1.35	0.87 ± 0.08	31.3 ± 2.6
B3	10	14 ± 1.4	0.64 ± 0.06	16.5 ± 2.9

The existence of process drift becomes apparent when comparing samples A3 and B1, which are grown under nominally identical conditions with a NH_3 flux of 4 slm. The difference in composition could be explained if our reactor was running slightly hotter for series B, by a difference of approximately 10 °C. Despite this some growth trends can be distinguished, the indium incorporation increases for smaller NH_3 fluxes between 1 slm and 4 slm, but there is no further increase as the flux is increased beyond 4 slm. The surface roughnesses are of a comparable order of magnitude for all samples, with a slight increase from 1 slm to 4 slm which levels out at higher flows. However other details do not have an obvious explanation, particularly the pit density. If the pits are linked to threading dislocations it seems odd that sample A3 should have an order of magnitude increase in dislocation density as it is the closest to lattice matching. It also seems strange that there is an increase in pit density from B1 to B3, as there is no change in composition or epilayer strain state and therefore no reason for extra threading dislocations to be generated in the epilayer. These questions warranted further investigation.

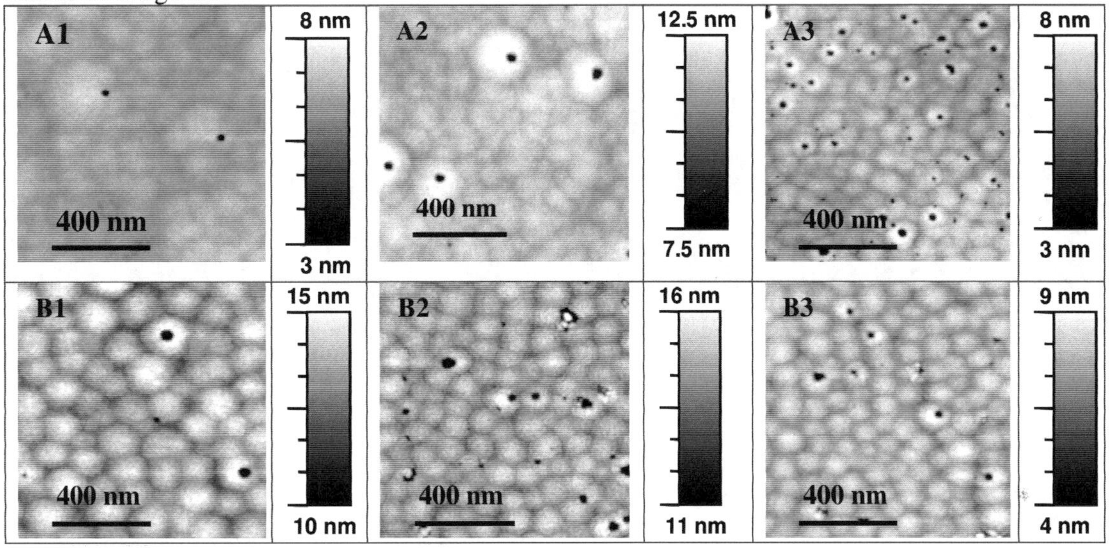

Figure 1. The surface morphology of all samples measured by AFM at 1 μm, showing nanoscale pits and hillocks.

3.3. The further investigation of defects by XRD and TEM

The first method used to see if there was additional dislocation generation in the InAlN epilayers was to compare the XRD peak widths of the 002 and 101 peaks for both the GaN pseudo-substrate and the InAlN epilayers using ω-scans. The 002 peak would be broadened by the introduction of screw and mixed type threading dislocations and the 101 by edge, mixed and screw. However, no significant difference was seen between the InAlN epilayer and the GaN peaks for any sample, and all samples had similar values (Tab. 2). Samples A3, B1 and B3 were then investigated by TEM using weak beam dark field (WBDF) imaging to view dislocations in cross section. However, no threading dislocations created at the GaN / InAlN interface were observed in B1 or B3, and only a few in A3 (Fig. 2). Our cross section samples had a volume corresponding approximately to an area of surface 100 nm by 5 μm, so if all the pits observed in AFM were created by such threading dislocations we would expect to see more in our sample.

Table 2. full width at half maximum for 002 and 101 peaks in XRD. Expressed: InAlN fwhm (GaN fwhm)

Sample	002 peak	101 peak
A1	228 ± 7 (244 ± 7)	399 ± 7 (383 ± 7)
A2	221 ± 7 (239 ± 7)	404 ± 7 (386 ± 7)
A3	227 ± 7 (242 ± 7)	394 ± 7 (381 ± 7)
B1	220 ± 7 (241 ± 7)	397 ± 7 (378 ± 7)
B2	210 ± 7 (240 ± 7)	366 ± 7 (376 ± 7)
B3	230 ± 7 (242 ± 7)	399 ± 7 (380 ± 7)

The evidence from X-ray peak widths and from TEM both strongly suggest that some of the additional pits observed in AFM are not caused by threading dislocations originating at the GaN / InAlN interface. While the threading dislocations observed do cause pits it may be the case that in some samples with large pit densities there are other pit generating mechanisms operating. It may also be the possible that dislocation loops are generated in the thickness of the film too close to the surface to be observed in our TEM studies to date.

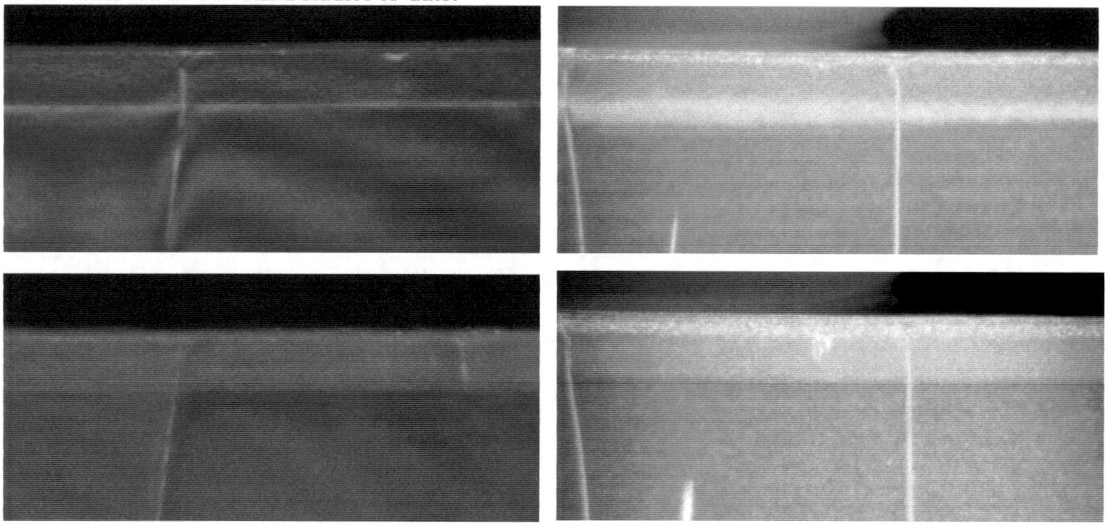

Figure 2. TEM of samples A3 and B1. WBDF images using ***g*** vectors of 0002 and $11\bar{2}0$ respectively leave the screw and edge components of a threading dislocation visible.

4. Conclusions

$In_xAl_{1-x}N$ is a useful and interesting alloy system with many applications. Here we have investigated the effect of NH_3 flux on 100 nm $In_xAl_{1-x}N$ epilayers. Indium incorporation increases between 1 slm (44.6 mmol/min) and 4 slm (179 mmol/min), but remains constant thereafter up to 10 slm (446 mmol/min). Under certain conditions the surfaces showed a significant number of pits similar to those caused by threading dislocations, but the mechanism for their generation remains an open question.

References

[1] Carlin J F, Zellweger C, Dorsaz J, Nicolay S, Christmann G, Feltin E, Butte R and Grandjean N 2005 *Phys. Status Solidi B* **242**, 2326
[2] Butte R et al. 2007 *J. Phys. D: Appl. Phys.* **40**, 6328
[3] Lobanova AV, Mazaev KM, Talalaev RA, Leys M, Boeykens S, Cheng K, and Degroote S 2006 *J. Cryst. Growth* **287**, 601
[4] Maleyre W, Briot O and Ruffenach S 2004 *J. Cryst. Growth* **269**, 15
[5] Matsuoka T 1998 *J. Cryst. Growth* **190**, 19
[6] Ive T, Brandt O, Kong X, Trampert A and Ploog KH 2008 *Phys. Rev. B* **78**, 035311
[7] Schuster M, Gervais PO, Jobst B, Hosler W, Averbeck R, Riechert H, Iberl A and Stommer R 1999 *J. Phys. D* **32**, A56
[8] Wright AF 1997 *J. Appl. Phys.* **82**, 2833
[9] Horcas I, Fernandez R, Gomez-Rodriguez JM, Colchero J, Gomez-Herrero J and Baro AM 2007 *Rev. Sci. Instrum.* **78**, 013705
[10] Oliver RA, Kappers MJ, Sumner J, Datta R and. Humphreys CJ 2006 *J. Cryst. Growth* **289**, 506
[11] Ponce FA et al. 2003 *Phys. Status Solidi B* **240**, 273

Journal of Physics: Conference Series **209** (2010) 012016 doi:10.1088/1742-6596/209/1/012016

Microstructural characterisation of a prototype layer structure for a GaN-based photonic crystal cavity

HWAR El-Ella, TC Sadler, MJ Kappers and RA Oliver

Department of Materials Science and Metallurgy, University of Cambridge, Pembroke St., Cambridge, CB2 3QZ, UK

E-mail: he234@cam.ac.uk

Abstract A semiconductor multilayer consisting of an n-type GaN/sapphire pseudo-substrate with a double sacrificial layer (20 nm of InGaN and 20 nm of InAlN) and a GaN cavity structure on top incorporating an InGaN quantum dot active layer was grown. Atomic force microscopy (AFM) was used to measure the morphology of the upper surface of each layer of similar test structures and to assess the morphological evolution of the full structure during its growth. Transmission electron microscopy (TEM) was also used to assess defect incorporation into the structure. AFM showed that the incorporation of a double sacrificial layer was observed to have no detrimental affect on either the formation of the quantum dots, or the morphology of the top GaN capping layer. TEM highlighted the occurrence of threading dislocations propagating through the sacrificial layers into the capping GaN layer, but did not resolve the occurrence of defect generation in the sacrificial layers.

1. Introduction

The group III nitride alloys provide a wide range of possible direct band-gaps making them an ideal choice for optical device applications. These materials are of particular interest in the context of single-photon sources (SPS). Current research into practical device formation for SPS has considered the use of two-dimensional photonic crystal (PC) designs with the hope of obtaining low volume, high quality factor (Q) cavities that can significantly enhance emission rates while retaining reliable single photon differentiation; simultaneously providing a versatile template for electronic integration [1]. However, little work has been performed on the nitride semiconductor system in this context, despite the potential advantages these materials offer over the more conventional arsenides in terms of available wavelengths, wavelength tunability and temperature stability of single photon emission [2].

Despite being ideal sources, the three-dimensionality of quantum dots (QDs) poses a difficulty for efficient photon extraction, as emission is probable in all directions. Furthermore epitaxial QDs form as embedded islands on, or within, a semiconductor matrix and therefore suffer from a high surrounding refractive index (n). As well as the surrounding strong photon refraction, the spontaneous emission lifetime within the QD is deeply related to the dipole moments of the immediate environment. QDs therefore need to be coupled with an optical cavity in such a way that can enhance the rate of spontaneous emission (by a quantity known as the Purcell factor, f_P [3]) while collecting and channelling the emitted photons into identical spatial modes [4]. The formation of high Q cavities with small modal volumes is therefore a key step in achieving a functional SPS. Recent published work on two dimensional PC membrane cavities have reported V_{eff} values down to ~$0.7(\lambda/n)^3$ and large

Q factors of the order of two million [5], with a potential theoretical increase to >10^7 with further optimization [6].

Fundamentally, the efficiency of a PC system depends on the quality of its internal and surface structure. The presence of internal structural defects and surface roughness disrupt the photonic band-gap incurred by the structures periodicity, and can severely degrade the *Q* of a promising design. One of the main priorities during fabrication is therefore in developing methods that can minimise the occurrence of structural defects, and ensure that surface roughness is kept to a minimum. The use of epitaxial deposition techniques (such as metalorganic vapour phase epitaxy – MOVPE) already prevents, to some extent, the scale of imperfections that would occur through other growth methods, as well as being vital to the formation of high quality QDs. The main source of imperfections therefore occurs during post-processing of epitaxial grown structures. One method of reducing defect formation during post-processing is by including sacrificial layers that react more vigorously with selective etchants compared to the rest of the structure. This allows for a 'clean' cleavage of the intended optical area from the rest of the structure, with minimal handling, and is clearly advantageous in comparison to other dry/wet etching methods. Such a process may be carried out through a photo-electro chemical (PEC) etch [7], in which a cathode is attached to an n-doped layer below the sacrificial layers and a light source with a wavelength corresponding to the sacrificial layers band-gap is directed upon them. The generation of a large concentration of holes in the sacrificial layers resulting from the abovementioned process promotes a higher etching rate, and can be tuned to selectively target other layers with differing band-gaps.

What naturally remains to be determined is what effect the inclusion of the sacrificial layer(s) has on the propagation of and/or generation of defects throughout the cavity and the roughness of interfaces within the structure. In this paper we have assessed and compared the surface morphology of the various layers by atomic force microscopy (AFM). Preliminary transmission electron microscopy has also been carried out in order assess defect propagation throughout the structure.

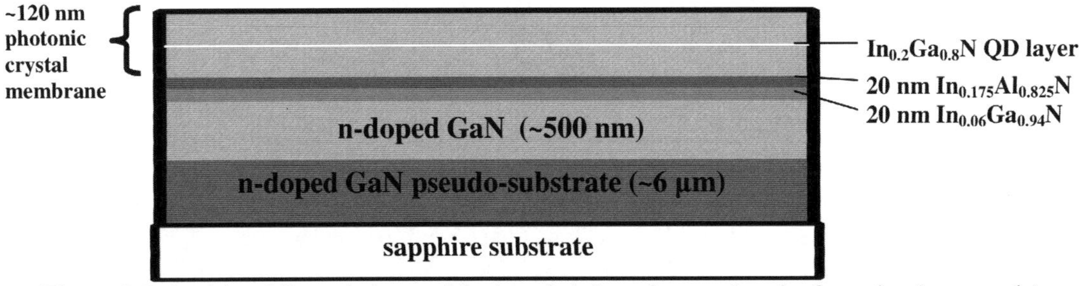

Figure 1. Epitaxial multilayer design used for intended photonic crystal cavity formation (not to scale).

2. Experimental

All samples were grown through MOVPE in a 6 x 2" Thomas Swan reactor close-coupled showerhead reactor, using trimethylgallium (TMG), trimethylindium (TMI) and ammonia (NH_3) as precursors. A SiH_4 flux was employed as a source of n-doping and H_2 was used as carrier gas for the growth of GaN unless otherwise stated, whilst N_2 was used as a carrier gas for all indium-containing layers. The complete structure involved the deposition of 6 layers on an n-doped GaN pseudo-substrate. A schematic of the grown multilayer is shown in Figure 1. Six other incomplete structures were grown to allow assessment of the morphology of the top surface of each of the layers shown in the figure 1 (apart from the pseudo-substrate). Two additional control samples were grown in order to compare and assess how the underlying sacrificial layers affected the growth of the QDs and the surface morphology of the final capping layer. These consisted of two samples of InGaN QDs on a GaN pseudo-substrate one of which was capped with ~10 nm of GaN. Surface morphologies were imaged using a Veeco Dimension 3100 atomic force microscope (AFM) in TappingMode™. Defect propagation throughout the epitaxial structure was probed with the use of a Philips CM30 transmission electron microscope (TEM) operating at 300 kV, through imaging close to the $(01\bar{1}0)$ zone axis using a ***g***,-4***g*** condition with ***g*** $=\{11\bar{2}0\}$. A cross-sectional TEM sample was prepared through polishing and dimpling to achieve a thickness of ca. 20 μm. The sample was then ion beam thinned until electron transparency was achieved using a Gatan precision ion polishing system (PIPS).

3. Results and Discussion

3.1. AFM Characterisation

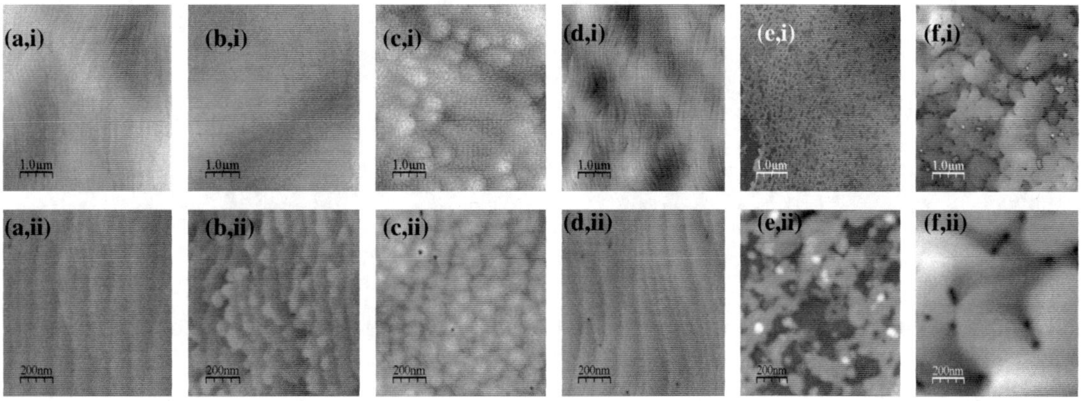

Figure 2. 5x5 and 1x1 μm^2 AFM scans of **(a i,ii)** GaN pseudo-substrate layer (image height = 7.7nm, 1.4nm) **(b i,ii)** InGaN sacrificial layer (h = 13.8nm, 1.9nm) **(c i,ii)** InAlN sacrificial layer (h = 6.2nm, 3.4nm) **(d i,ii)** Lower GaN layer (h = 3.7 nm, 4.7nm) **(e i,ii)** Single InGaN QD layer (h = 16.8nm, 9.3nm) **(f i,ii)** Capping GaN layer (h = 14.5nm, 9.5nm)

Surface morphologies were mapped and compared using AFM to scan the surfaces of each of the partial multilayers. The first layer to be grown was a ~500 nm n-doped (~ 1 x 10^{19} cm^{-3}) GaN connecting layer, the purpose of which was to cover the re-growth interface and provide a clean and flat surface for subsequent growth, as well as to provide an electrical connecting layer for a cathode necessary in the PEC etch process. Figure 2(ai) and 2(aii) show the surface morphology of the GaN pseudo-substrate with a terraced surface and a root mean square roughness (R_q) value of 0.69 nm over a 5 μm x 5 μm area. The subsequent growth of a double sacrificial layer above the n-doped GaN could potentially affect the morphology of the subsequent GaN layer, and thereby affecting the growth of the QD layer. 20 nm of $In_{0.06}Ga_{0.94}N$ and 20 nm of $In_{0.175}Al_{0.825}N$ were grown to provide the main sacrificial breaking point for the separation of the top three layers from the rest of the structure. These materials were chosen since they are expected to experience high etch rates in the PEC process. Figures 2(bi,bii) and 2(ci,cii) show the surface morphology of both the sacrificial layers, and figure 2(di,dii) shows the surface morphology of the subsequently deposited GaN layer to be very similar to that of the underlying GaN, with an R_q value of ~0.39 nm for a 5 μm x 5 μm area. The final three layers constituted the intended PC membrane and consist of a monolayer of $In_{0.2}Ga_{0.8}N$ QDs sandwiched in-between two ~60 nm GaN layers. The final capping GaN layer shown in figure 2(fi,fii)

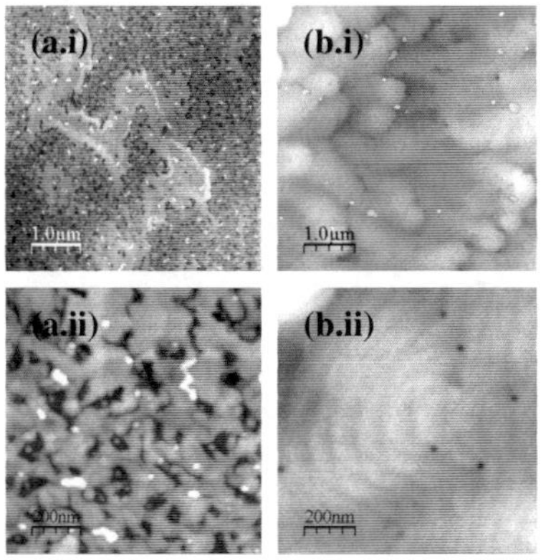

Figure 3. 5x5 and 1x1 μm^2 AFM scans of **(a i,ii)** InGaN QDs grown on a GaN pseudo-substrate layer (height = 7.7nm, 6.2nm) **(b i,ii)** InGaN QDs capped with a ~10nm GaN cap (h = 10.7nm, 2.8nm)

shows considerable surface roughness (R_q~1.73 nm over 5 μm × 5 μm), an effect which probably occurs since the final GaN cap was grown in N_2 to avoid exposing the QD layer (ei,eii) to an H_2 flux which might alter its properties.

The AFM scans of the control samples are show in Figure 3 and are similar to the corresponding QD and capped QD layer scans in Figure 2, with the capped QDs sample (figure 3(bi,ii)) having a slightly lower R_q value of ~1.26 nm over a 5 μm × 5 μm area.

3.2. TEM characterisation

Preliminary TEM images were taken along the $(01\bar{1}0)$ axis and two such micrographs are shown in figure 4. Although the stacked layers can be distinguished, the effect of the sacrificial layers on the GaN sandwiched QD layer is not observable. Threading dislocations can be seen propagating from the underlying n-type GaN through the sacrificial layers and the cavity structure in figure 4(a). However, no defect generation in the sacrificial layers is visible in these images. More detailed TEM characterisation will be required in order to check whether such defect generation occurs.

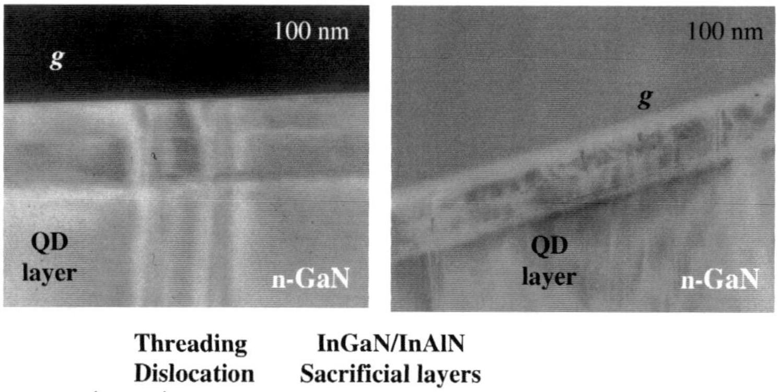

Figure 4. $(01\bar{1}0)$ bright field (a) and weak-beam dark-field (b) TEM images of the multilayer structure shown in figure 1

4. Conclusions

AFM scans of the various partial stacks show no significant deviations from similar morphology scans reported in the literature [8] [9]. With exception of the capping GaN layer, the various layers are observed to possess relatively smooth surfaces and the sacrificial layers are not observed to detrimentally affect the formation of the QDs. The preliminary TEM characterisation highlighted the occurrence of threading dislocations propagating through the sacrificial layers into the capping GaN layer. As the obtained TEM data was in no way conclusive, further studies will be carried out in order to clarify the roughness and abruptness of the various interfaces, as well as to more thoroughly assess defect incorporation throughout the stack and the influence of the sacrificial layers on the defect formation in the top GaN layers. Additionally, the effectiveness of PEC etching for the current sacrificial layers must still be assessed.

References

[1] Lounis B and Orrit M 2005 *Rep. Prog. Phys.* **68**, 1129
[2] Oliver RA, Jarjour AF, Taylor RA, Tahraoui A., Zhang Y., Kappers MJ and Humphreys CJ 2008 *Mat. Sci. Eng. B* **147**, 108.
[3] Purcell EM 1946 *Phys. Rev.* **69**, 681
[4] Vučković J, Pelton M, Scherer A and Yamamoto Y 2002 *Phys Rev A* **66**, 023808
[5] Asano T, Song BS, Akahane Y and Noda S 2006 *IEEE J. Sel. Top. Quant. Electron.* **12**, 1123
[6] Reithmaier JP 2008 *Semicond. Sci. Tech.* **23**, 123001
[7] Minsky MS, White M and Hu EL, 1996 *Appl. Phys. Lett.* **68**, 1531
[8] Oliver RA, Jarjour AF, Tahraoui A, Kappers MJ, Taylor RA and Humphreys CJ 2007 *J. Phys. C. Ser.* **61**, 889
[9] Oliver RA, Briggs AD, Kappers MJ, Humphreys CJ, Yasin S, Rice JH, Smith JD and Taylor RA 2003 *Appl. Phys. Lett.* **83**, 722

Dislocation reduction in MOVPE grown GaN layers on (111)Si using SiN_x and AlGaN layers

M Haeberlen, D Zhu, C McAleese, MJ Kappers and CJ Humphreys

Department of Materials Science and Metallurgy, University of Cambridge, Pembroke Street, Cambridge CB2 3QZ, UK

E-mail: mh544@cam.ac.uk

Abstract. Growth of GaN on Si(111) potentially enables cost efficient manufacturing of optoelectronic devices due to the possibility of using cheap large area substrates. However, GaN layers grown on Si(111) substrates suffer from high tensile stress that can lead to cracking at layer thicknesses exceeding 1 μm. Another challenge is the high dislocation density of GaN layers grown on Si(111) which is detrimental to device performance. In this paper we show that a step graded AlGaN buffer layer can compensate tensile stress, avoiding cracking, and at the same time reduce the dislocation density. An additional SiN_x interlayer in the GaN layer is shown to further reduce the dislocation density down to the high 10^8 /cm². Weak beam dark field TEM was used to study the dislocation reduction in cross sectional samples and for comparison of the step graded AlGaN buffer layer structure to a continuously graded one. STEM ADF was used to determine the exact location of dislocation bending with respect to the position of the interface.

1. Introduction

The potential cost reduction by successful growth of GaN based LEDs on large diameter (111) oriented silicon substrates is thought to give solid state lighting a boost towards replacing incandescent lighting [1]. However, in order to achieve this, a few obstacles still have to be overcome. The large mismatch in thermal expansion coefficients (46%) can lead to large tensile stresses resulting in cracking of the GaN layer. High dislocation densities in GaN grown on silicon of typically 10^{10} cm^{-2} originate from a large lattice mismatch of 17%. For comparison, LED structures on sapphire substrates are deposited on GaN layers which exhibit a typical dislocation density of ~$5x10^8$ cm^{-2}.

Build-up of tensile stress can be addressed by the integration of AlN interlayers deposited at low temperatures [2] or the use of an AlGaN buffer layer which results in compressive stress counteracting the tensile stress generated upon cooling in the top GaN layer [3]. In order to reduce dislocation densities, the in-situ deposition of SiN_x interlayers has been found to be effective [4].

In this paper we studied the propagation of dislocations through continuously or step graded AlGaN buffer layers by means of weak beam dark field TEM. Simultaneous imaging of compositional variations in the step graded AlGaN buffer together with residual contrast from dislocations in STEM ADF allows for exact determination of regions of dislocation interaction near interfaces. The results found in this study suggest that dislocation bending occurs at or very close to the interfaces between AlGaN layers of different composition. Dislocation bending in a continuously graded AlGaN buffer is much less pronounced indicating the presence of a threshold in strain for the bending over of dislocations as proposed for example in [5]. The combination of stress management layers with SiN_x interlayers is found to reduce the threading dislocation density significantly.

2. Experiment

2.1. Sample structures

Two different sample structures were grown on Si(111) for the study presented in this paper. Growth was carried out in a Thomas Swan close coupled showerhead MOVPE reactor. The sample structures are shown schematically in Figure 1. Sample A contains a continuously graded AlGaN strain management layer, whereas sample B employs a step graded AlGaN layer for compensation of tensile strain. A SiN_x interlayer for dislocation reduction was grown after 330 nm and 100 nm of GaN growth in sample A and sample B, respectively.

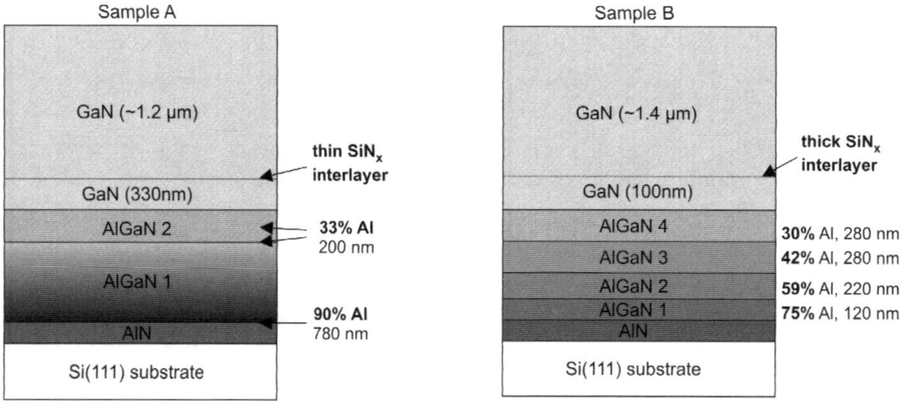

Figure 1. Schematic of sample structures grown for this study. Sample A: Continuously graded AlGaN strain management buffer with a thin SiN_x interlayer 330 nm above the AlGaN/GaN interface and Sample B: Step graded AlGaN strain management buffer with thicker SiN_x interlayer for dislocation reduction.

2.2. Experimental details

Cross sectional TEM samples were prepared so that both the [10-10] and [11-20] zone axes of the GaN film could be accessed in one sample. Sample preparation was carried out employing mechanical grinding followed by dimpling and ion beam thinning. Weak beam dark field images were taken in a Philips CM30 TEM operating at 300 kV along the [10-10] zone axis using (0002) and (11-20) reflections to reveal dislocations with a screw and edge component, respectively.

STEM ADF images were acquired along the [11-20] zone axis of sample B in a Tecnai F20 TEM operating at 200 kV equipped with a ADF detector with an inner diameter of 35 mrad and an outer diameter of 105 mrad. A camera length of 100 mm was chosen to enable Z contrast imaging together with imaging of residual diffraction contrast from dislocations as outlined for stacking faults in [6].

3. Results and Discussion

Figure 2 shows the WBDF images obtained from the samples containing a continuously graded (a) and, (b) a step graded strain management layer, respectively.

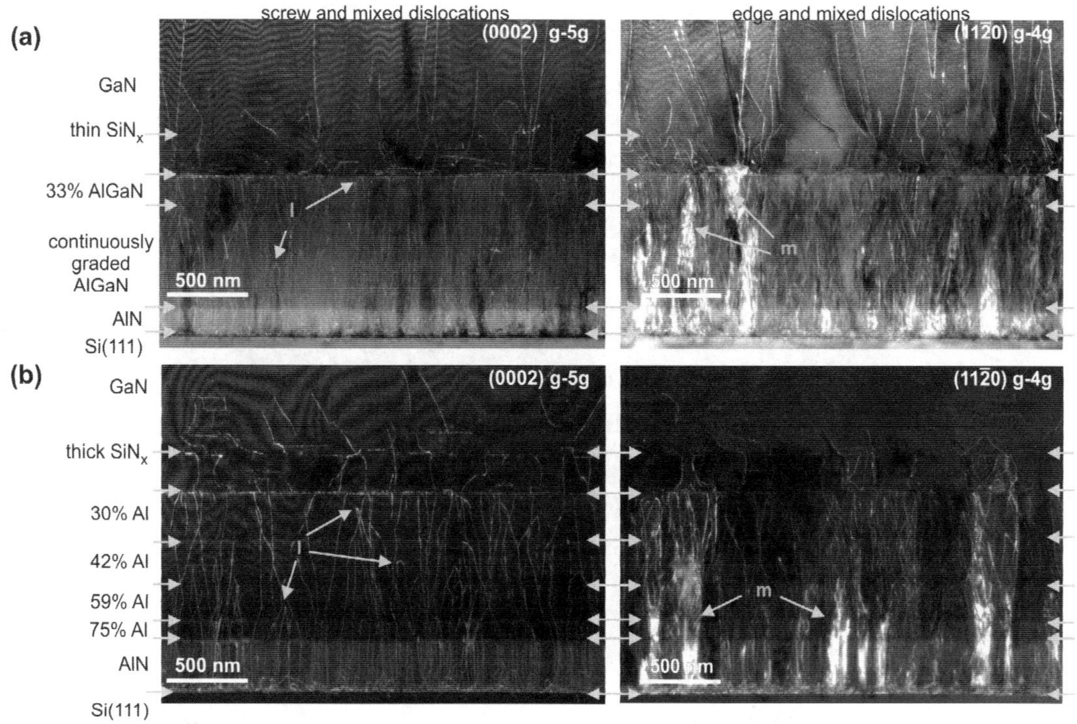

Figure 2. WBDF TEM images of sample (a) containing a continuously graded AlGaN layer and sample (b) with a step graded AlGaN layer. Screw dislocations annihilating by the formation of loops are marked with "l", misoriented grains visible in the (11-20) WBDF images are marked with "m", respectively.

It is apparent from the TEM images in figure 2 that significant differences in the dislocation structure exist in both samples. It can be seen that the formation of loops leading to annihilation of screw type dislocations is much more pronounced in sample B compared to sample A. Screw type dislocations are found to be bent slightly at each AlGaN/AlGaN interface leading to an increased likelihood of two dislocations annihilating. This way, the step graded AlGaN layer reduces screw type dislocations more efficiently than the continuously graded layer. The same observation can be made for edge type dislocations. Bending of edge type dislocations in sample A only occurs very close to the AlGaN/GaN interface, whereas dislocation bending can be noticed at each interface between AlGaN layers of different composition in sample B. The influence of the thickness of the in-situ deposited SiN_x layer can also clearly be seen in figure 2. It is found that the thicker SiN_x in sample B reduced the dislocations significantly, whereas the SiN_x layer in sample A hardly affects the dislocations at all. Misoriented grains, which can be seen in the (11-20) WBDF images of both samples, are efficiently reduced within both the continuously and step graded AlGaN layer. However, lower contrast from these misoriented grains after the AlGaN/AlGaN interfaces also indicates that dislocations are reduced in the step graded buffer layer structure.

In order to study dislocation bending in the step graded buffer layer in more detail STEM ADF images were acquired. A typical STEM ADF image of the uppermost portion of the step graded AlGaN layer is shown in Figure 3.

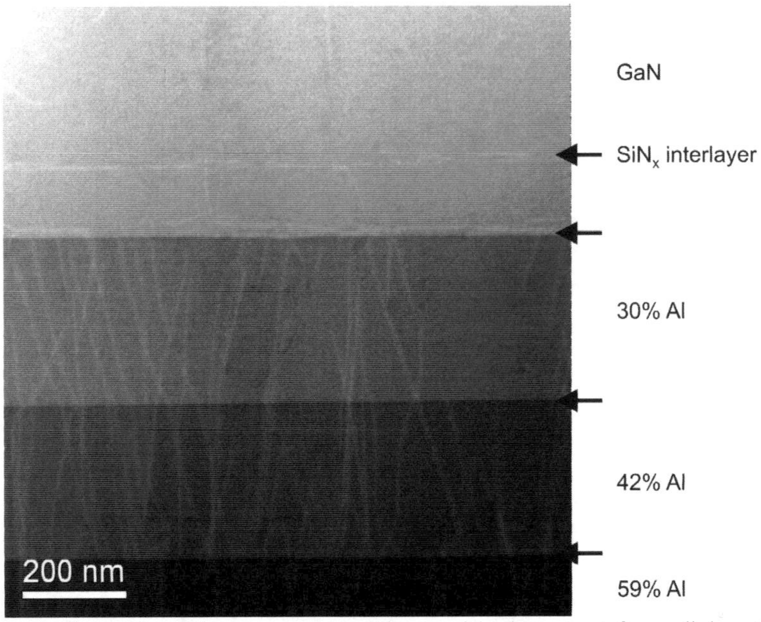

Figure 3. STEM ADF image of sample A. Z-contrast as well as residual contrast from dislocations is visible.

It can clearly be seen that the majority of dislocations (mainly edge type) bend over by angles of less than 10° at or very near the interfaces of the step graded AlGaN layer. Bending over of dislocations in AlGaN layer of different composition has already been reported in the literature [5,7] and occurs in order to reduce strain energy in the system. A threshold energy for the bending of dislocations of 10 eV has been proposed in [5]. Further experiments will be conducted to optimise the AlGaN buffer layer design for both dislocation reduction and strain management purposes.

4. Conclusions

MOVPE grown GaN layers on Si(111) with either a continuously or step graded AlGaN buffer layer for strain management have been investigated by means of weak beam dark field TEM and STEM ADF. It has been found that step graded AlGaN layers can lead to significant bending of edge type dislocations and that annihilation of screw type dislocations by loop formation is also favoured. Dislocation bending was only observed near the GaN/AlGaN interface of the continuously graded buffer layer sample.

References

[1] Zhu D, McAleese C, McLaughlin KK, Häberlen M, Salcianu CO, Thrush EJ, Kappers MJ, Phillips WA, Lane P, Wallis DJ, Martin T, Astles M, Thomas S, Pakes A, Heuken M and Humphreys CJ 2009 *Proc. SPIE* **7231**, 723118
[2] Dadgar A, Poschenrieder M, Bläsing J, Fehse K, Diez A and Krost A 2002 *Appl. Phys. Lett.* **30**, 3670
[3] Cheng K, Leys M, Degroote S, van Daele B, Boeykens B, Derluyn J, Germain M, van Tendeloo G, Engelen J and Borghs G 2006 *Journal of Electronic Materials* **35**, 592
[4] Cheng K, Leys M, Degroote S, Germain M and Borghs G 2008 *Appl. Phys. Lett.* **92**, 192111
[5] Romanov AE and Speck JS 2003 *Appl. Phys. Lett.* **83**, 2569
[6] Amali A, Rez P and Cowley JM 1997 *Micron* **28**, 89
[7] Follstaedt DM, Lee SR, Provencio PP, Allerman AA, Floro JA and Crawford MH 2005 *Appl. Phys. Lett.* **87**, 12111

Simulation supported analysis of the effect of SiN_x interlayers in AlGaN on the dislocation density reduction

O Klein[1], J Biskupek[1], U Kaiser[1], K Forghani[2], S B Thapa[2] and F Scholz[2]

[1] Central Facility of Electron Microscopy, Ulm University, 89081 Ulm, Germany
[2] Institute of Optoelectronics, Ulm University, 89081 Ulm, Germany

E-mail: oliver.klein@uni-ulm.de

Abstract. SiN_x interlayers can act as anti-surfactants and drastically reduce the dislocation density in pure GaN layers. In our work we could observe a very efficient dislocation annihilation of the **a**-type threading dislocations (TD) at a fractional SiN_x monolayer even in $Al_xGa_{1-x}N$ layers with relatively high Al content of x=0.2, grown on c-plane sapphire by MOVPE. The investigations were focused on the effect of the SiN_x interlayer on the dislocation density reduction of the **a**-type TDs. Weak-beam dark-field (WBDF) and high-resolution (HR) TEM analyses directly at the SiN_x interface indicate that the most frequently occurring effect for the reduction of the **a**-type TDs is the conversion of an **a**-type TD into an **a**-type basal dislocation due to lateral overgrowth of SiN_x by AlGaN. To confirm this effect, an appropriate dislocation model was developed for the **a**-type TD in AlGaN and its bending due to the SiN_x nano-mask. Corresponding image calculations were performed and compared with the experiments.

1. Introduction

In recent years aluminium nitride (AlN) has attracted much attention due to its extremely large direct band gap of approximately 6.0 eV and its impressive chemical and thermal stability. Thus AlN and $Al_xGa_{1-x}N$ ternary alloys are promising materials for high-power high temperature electronic applications and optoelectronic devices in the UV range. As group-III nitride wafers are still not available in sufficient amount and quality, AlN has to be grown on foreign substrates such as Al_2O_3 (sapphire). Unfortunately the large lattice mismatch between the $AlGaN/Al_2O_3$ interface of up to -14% leads to the formation of strain induced misfit dislocations at the interface, which are the cause of the generation of threading dislocations (TD), inducing a high dislocation density in the range of $10^{10}cm^{-2}$ and decreasing the crystal quality [1, 2]. As published among others in [3, 4], a promising possibility to reduce the dislocation density at the surface of an AlGaN layer system is the growth of intermediate SiN_x layers. SiN_x can act as an anti-surfactant and drastically reduce the dislocation density in AlGaN layers with high gallium content by lateral overgrowth. In our work we could observe a very efficient dislocation annihilation of the **a**-type TDs at a fractional SiN_x monolayer even in $Al_xGa_{1-x}N$ layers with a relatively high Al content of x=0.2. However, the annihilation effect of the pure edge **a**-type TDs does not occur homogenously over the whole SiN_x interface in our sample. Regions of the SiN_x layer at which the propagation of the **a**-type TDs is completely stopped alternate with regions where almost no annihilation effect was observed, which motivated further studies on the SiN_x monolayer and its effect on the dislocation reduction. Analyses regarding the annihilation process were carried out by exploiting the weak-beam dark-field (WBDF) method and by high-resolution TEM investigations of the a-type TDs directly at the SiN_x interface. The investigations were supported by an appropriate dislocation model, developed for comparative image calculations by the multislice algorithm.

2. Experimental methods

2.1. Investigated samples and layer growth

Our investigations centered on two different samples, denoted by S1 and S2 in this work. Sample S1 is composed of a 1.2 μm thick $Al_{0.2}Ga_{0.8}N$ layer with an in-situ deposited intermediate nominal SiN_x monolayer [5] using silane (SiH_4) as precursor with a deposition time of 4 minutes, located 150 nm above the substrate. It was grown by low pressure MOVPE at a temperature of 1120°C and a pressure of 80 hPa. Sample S2 consists of a pure 500 nm thick AlN layer grown at 1190°C and 35 hPa [2, 6, 7]. It was used for comparative WBDF and HRTEM pre-investigations, to prove the developed dislocation model for the pure edge **a**-type TD in wurtzite group-III nitrides. For both samples the actual layer growth was preceded by the deposition of a low temperature 20 nm thick oxygen doped AlN:O nucleation layer, to improve the crystal quality [8]. Trimethylaluminium/-gallium (TMAl/Ga) and NH_3 were used as group III and V precursors, respectively.

2.2. TEM measurements

Thin TEM foils were prepared for investigation in cross-sectional and plan-view geometries using standard techniques including mechanical polishing and low-angle argon thinning [9]. To image the dislocations and distinguish the different dislocation types, cross-sectional WBDF images close to the [01-10] zone were analysed by exploiting the **g·b** criterion. The WBDF method was carried out under the 3**g** condition. In wurtzite 2H group-III nitrides perfect dislocations can exist with three different types of Burgers vectors **b**: **a**-type TDs with **b** of type 1/3 [2-1-10], **c**-type TDs with **b** = [0001] and (**a**+**c**)-type TDs with **b** of type 1/3 [2-1-13]. Since most of the growth induced TDs in epitaxially deposited AlGaN layers propagate parallel to the [0001] direction, **a**-type TDs are pure edge dislocations, whereas **c**-type TDs are pure screw dislocations. The Burgers vector of a mixed TD propagating parallel to the c-axis is of type (**a+c**). According to the **g·b** criterion, only **c**-type and (**a+c**)-type dislocations are visible in the WBDF image when using the 0002 reflection. Exploiting the 2-1-10 reflection, only **a**-type and (**a+c**)-type TDs give rise to a contrast. To determine Burgers vectors directly, HRTEM images were recorded at the [01-10] zone as well as in the [0001] plan-view projection. The investigations were carried out with a Philips CM-20 microscope.

2.3. Model generation and image calculation

The atomic dislocation models for comparative image calculations were developed with the aid of the Hypercube Hyperchem™ software package [10]. The dislocations were placed manually into the bulk crystal system according to our experimental investigations by performing an appropriate Volterra construction [11]. The subsequent relaxation of the supercell was realised after a geometry optimization of up to 27000 atoms by using a conjugate gradient molecular mechanic algorithm with an MM+ type force field. The image calculations were carried out by performing multislice calculations using the Musli software under CM-20 conditions with a Cs value of 1.2 mm and a high-tension of 200 kV [12].

3. Results and discussion

3.1. Pre-investigations and general properties of SiN_x interlayers in AlGaN

The bright field (BF) image in Figure 1a shows an overview of the dislocation structure in sample S1, in which the TDs of all types are visible. There are regions at the SiN_x layer where the propagation of the TDs is completely stopped and the annihilation process is highly efficient (lower arrows). However this does not apply to all regions of the specimen. The WBDF images (b) and (c) are recorded from the same sample area by exploiting the 2-1-10 and the 0002 reflection, respectively.

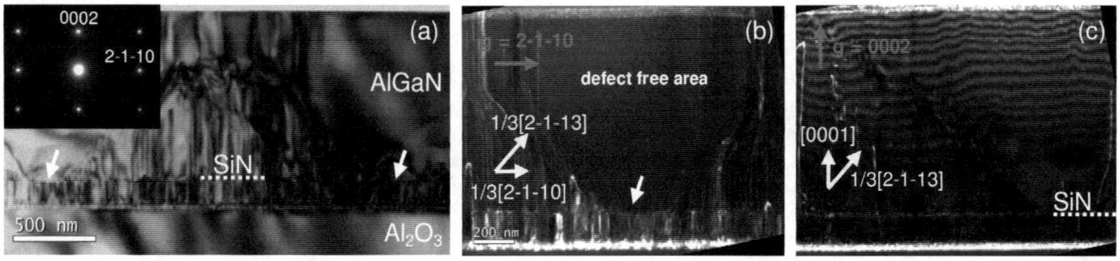

Figure 1. TEM cross-sectional views of sample S1. The BF image (a) from the [01-10] zone axis shows regions of the SiN_x layer at which the dislocation annihilation is highly efficient (bottom arrows in (a) and (b)), whereas the propagation of the TDs is not stopped in other regions. The WBDF images are recorded from the same sample area, exploiting the 2-1-10 (b) and the 0002 reflection (c). The Burgers vectors of the TDs visible in the respective WBDF image are also charted. In general, the sample shows a very high density of pure edge **a**-type TDs beneath the SiN_x layer (b), whereas the number of screw and mixed TDs is very small (c). Thus the appearance of large defect free areas is primarily based on the bending and annihilation of the pure edge **a**-type TDs at the SiN_x layer.

They illustrate a very high density of pure edge **a**-type TDs below the SiN_x interlayer, whereas the number of **c**- and (**a+c**)-type dislocations is very small. Thus the most frequently occurring effect of the SiN_x layer is the annihilation of the pure edge **a**-type TDs. A remarkable feature is the bending of the **a**-type TDs at regions close to the annihilation zone above the SiN_x interface, whereas the general bending direction always points away from the region with high annihilation efficiency. Furthermore the dislocation bending decreases with increasing distance from the zone of high annihilation grade. This bending effect and the annihilation of the **a**-type TDs at the SiN_x interface lead to the formation of large defect free areas.

In this work our investigations centered on the annihilation process at the SiN_x interlayer itself. As dislocation lines can not start or stop inside the crystal [11], they are always closed loops or end at surfaces. Thus, the TDs have to change the direction to become basal dislocations, propagating parallel to the surface and getting annihilated by interaction, to reduce the dislocation density at the surface of an epitaxially grown heterostructure. This effect is provided by the SiN_x interlayer. As we know from [4, 13], the growth of SiN_x on AlGaN is promoted at the dislocation cores due to free bonds. The treatment of the AlGaN layers with SiH_4 and NH_3 leads to a fractional coverage of the surface of the layers with SiN_x and causes the formation of a SiN_x nano-mask with SiN_x islands and exposed AlGaN regions (figure 2a). As SiN_x acts as an anti-surfactant, the SiN_x islands are laterally overgrown by the subsequent AlGaN layers. Thus, the two-dimensional epitaxial growth of AlGaN switches to a three-dimensional insular growth with lateral overgrowth of the SiN_x. The changing of the vertical growth direction to a lateral growth direction leads to the conversion of the a-type TDs into basal dislocations. The WBDF images of S1 in Figure 2b & c close to the SiN_x region show this effect for **a**-type and **c**-type TDs, respectively. The dislocation bending indicates the changing of the ratio between lateral and vertical growth at the SiN_x interface. If the dislocation density is sufficiently high, the dislocation bending leads to the annihilation of the TDs by the formation of dislocation

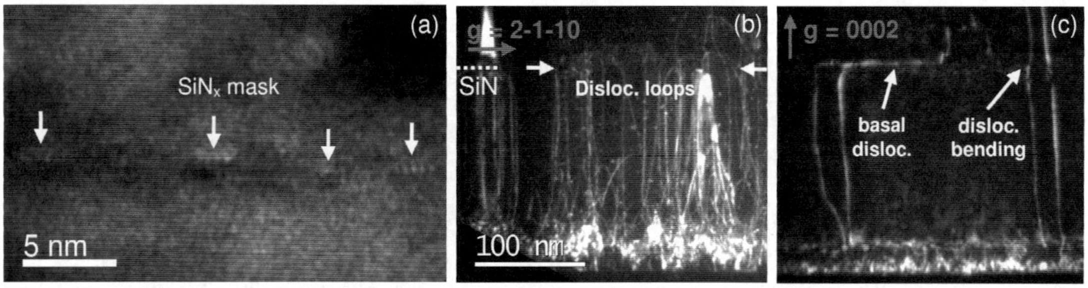

Figure 2. HRTEM (a) and WBDF micrographs of sample S1 in the area of the SiN_x interface at a zone of high annihilation grade. Image (a) illustrates the fractional coverage of AlGaN with SiN_x (Si deposition indicated by arrows). The bending of the **a**-type TDs and the annihilation by the formation of dislocation loops is clearly visible in image (b). Image (c) shows the propagation of the TDs with screw component. Even there, the conversion of the TDs into basal dislocations leads to an interaction of the TDs.

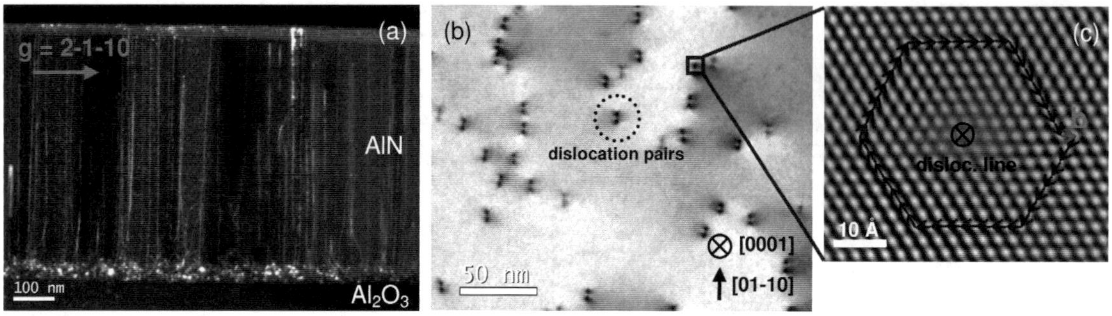

Figure 3. Cross-sectional WBDF image of S2 shows pure edge **a**-type TDs in AlN (a). In the [0001] plan-view image the arrangement of the **a**-type TDs in pairs along the [01-10] direction is clearly visible (b). This dislocation structure exactly corresponds to the Volterra cut along the [01-10] direction. The unclosed Burgers circulation around the dislocation core in the Fourier filtered HRTEM micrograph definitely identifies the TD as a pure edge **a**-type TD with Burgers vector **b** = 1/3 [2-1-10] (c).

loops. Otherwise the TD cannot interact after the bending and will continue to propagate parallel to the c-axis due to the increasing epitaxial growth with increasing distance to the SiN_x interface.

According to the annihilation process of the pure edge **a**-type TDs in 2H AlGaN, there is one possible mechanism for the conversion of such dislocations into basal dislocations at the SiN_x interface by lateral overgrowth. This is the bending of a pure edge **a**-type TD with $\mathbf{b}_1$ = -1/3 [2-1-10] to a pure edge **a**-type basal dislocation at the SiN_x interface, propagating parallel to the surface. This dislocation with Burgers vector $\mathbf{b}_3 = \mathbf{b}_1$ can interact with an adjacent pure edge **a**-type TD with $\mathbf{b}_2$ = 1/3 [2-1-10] and is therefore annihilated by the formation of a closed dislocation loop. Referring to the Volterra construction, this corresponds to the insertion of an additional (2-1-10) plane into the crystalline system. The developed dislocation model in our work is based on this relation and was verified by comparing experimental TEM studies with calculated images, exploiting the multislice algorithm.

3.2. Development and verification of the dislocation model for the pure edge a-type TD in Al(Ga)N

The dislocation model for the pure edge **a**-type TD in 2H Al(Ga)N was developed with the help of WBDF and HRTEM investigations on sample S2, by comparing the experimental images with the calculated ones, obtained by applying the multislice algorithm on the dislocation model. Beside the cross-sectional WBDF investigations (figure 3a), the plane-view BF and HRTEM micrographs indicate, that the formation of the pure edge **a**-type TDs in Al(Ga)N can be described as a cut along the [01-10] direction with a subsequent displacement at **b** = 1/3 [2-1-10] according to the Volterra construction. This construction rule is verified by the formation of dislocation pairs orientated along

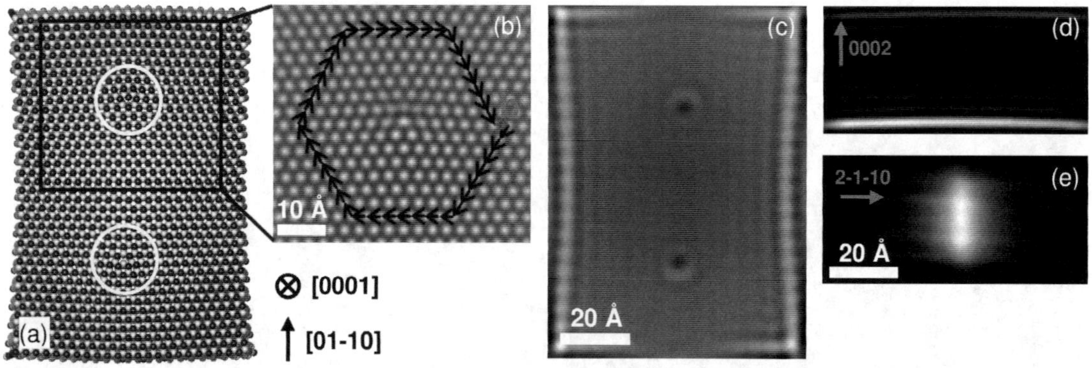

Figure 4. Atomic model for the edge **a**-type TD pair in Al(Ga)N (a) and calculated images (b-e) by applying the multislice algorithm on the model. In accordance with our TEM measurements, the HRTEM calculation identifies the dislocation as a pure edge **a**-type TD with **b** = 1/3 [2-1-10] (b). In the [0001] BF calculation the dislocation pair is clearly visible, as it was observed in our plane-view TEM investigations (c). Also the WBDF calculations and the verification of the **g·b** criterion are consistent with the experiment (d, e).

the [01-10] direction, clearly visible in the plan-view BF and HRTEM investigations (figure 3b & c).

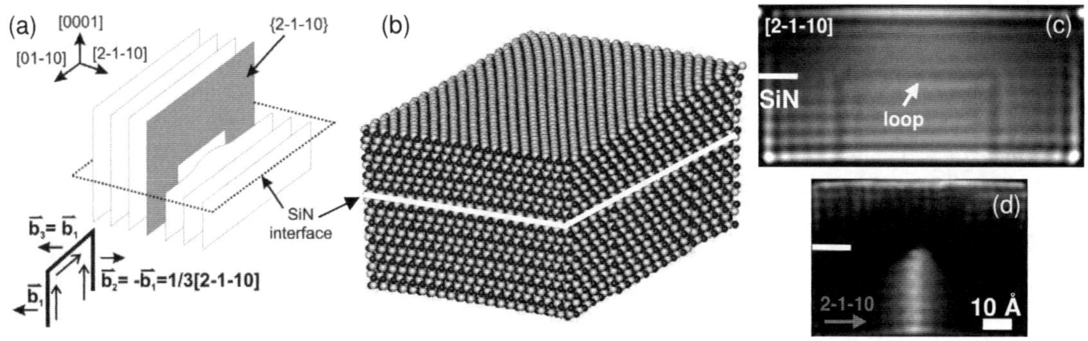

Figure 5. Extended atomic model for the annihilation process of **a**-type TDs in Al(Ga)N at the SiN_x interface by the formation of a dislocation loop (a, b). The BF calculation at the [2-1-10] zone clearly shows the dislocation loop created by an **a**-type basal dislocation (c). In the WBDF calculation close to the [01-10] zone the propagation of the dislocation line seems to end at the SiN_x interface as it was observed in figure 2b (d).

According to these results, it was now possible to construct the dislocation model for the pure edge **a**-type TD in Al(Ga)N. Figure 4 shows the relaxed model of the supercell after geometry optimization of around 17000 atoms and the image calculations applied to the model as described in chapter 2.3. We received the filled–core model for the pure edge **a**-type TD, representing a known core-type in 2H Al(Ga)N [14]. The WBDF calculations under the 3**g** condition close to the [01-10] zone verify the **g·b** criterion (figure 4d, e). The dislocation line is clearly visible by exploiting the 2-1-10 reflection, whereas the TD produces no contrast when using **g** = 0002. The consistence of the calculated images with our experimental investigations confirms the applicability of the construction rule for the formation of pure edge **a**-type TDs in Al(Ga)N and gives reason to extend the dislocation model in terms of the annihilation process at the SiN_x layer as we mentioned already in the previous chapter.

Figure 5a & b shows the atomic model for the annihilation process of the **a**-type TDs in Al(Ga)N at the SiN_x interface by the formation of a pure edge **a**-type basal dislocation due to lateral overgrowth, whereas SiN_x was not explicitly implemented in the model. The interaction of the basal dislocation with an **a**-type TD leads to the formation of a loop. Thus, the dislocation pair stops to propagate towards the surface. The relaxed model of the supercell contains around 27000 atoms. The calculated BF image at the [2-1-10] zone in figure 5c clearly shows the formation of the loop. Besides, the

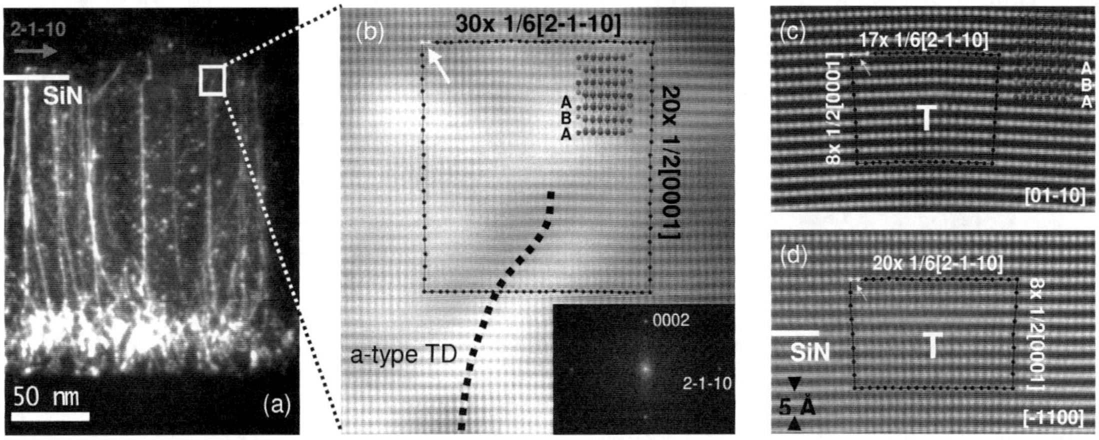

Figure 6. Burgers vector analysis of an annihilated pure edge **a**-type TD (a) at the SiN_x interlayer and comparison with HRTEM multislice calculations. The unclosed Burgers circulation in the filtered HRTEM image (b) results in a Burgers vector of 1/6[2-1-10] in the imaged projection. Thus, the dislocation line of the TD gets a component perpendicular to the image plane at the SiN_x interface. The unclosed circulation in the calculated image at the [01-10] zone (c) shows the Burgers vector 1/3[2-1-10] of the **a**-type basal dislocation, as provided in the model. In accordance to (b) we receive a Burgers vector of 1/6[2-1-10] in the calculated image from the [-1100] zone, corresponding to the imaging of the **a**-type Burgers vector from a 60° projection.

propagation of the TD is stopped as it was observed in our cross-sectional TEM investigations, clearly visible in the WBDF calculation by exploiting the 2-1-10 reflection (figure 5d).

The verification of the dislocation model was further carried out by comparing calculated and experimental HRTEM images at the SiN_x interface, where the dislocation bending and annihilation takes place due to lateral overgrowth. The experimental HRTEM investigations at the SiN_x nano-mask in the area of an annihilated pure edge **a**-type TD and the HRTEM image calculations on the developed model are in very good agreement according to the Burgers vector analysis. The simulation supported investigations in figure 6 show, that the annihilation of the pure edge **a**-type TDs in Al(Ga)N is realised by the conversion of an **a**-type TD into an **a**-type basal dislocation and subsequent interaction with an adjacent **a**-type TD due to lateral overgrowth of the SiN_x anti-surfactant.

4. Summary and conclusions

A detailed TEM study on the annihilation process of the **a**-type TDs at the SiN_x nano-mask in MOVPE grown Al(Ga)N layers on sapphire has been presented. We could partly observe a very efficient dislocation annihilation of the **a**-type TDs at the SiN_x interface even in $Al_xGa_{1-x}N$ layers with a relatively high Al content of x=0.2. Our supposed dislocation model explained the propagation of the **a**-type TD in the Al(Ga)N layer and at the SiN_x interface. It is shown by our simulation supported HRTEM and WBDF analyses, that the most frequently occurring effect for the reduction of the **a**-type TDs is the conversion of an **a**-type TD into an **a**-type basal dislocation at the SiN_x interface due to lateral overgrowth of the SiN_x by AlGaN. After bending, the basal dislocation can be annihilated by interacting with another **a**-type dislocation. The consistence of the calculated images with our experimental investigations confirms the applicability of the developed atomic dislocation model to the propagation of the **a**-type TDs at the SiN_x interface in the investigated crystal system.

Acknowledgements

We like to thank S. Grözinger for the sample preparation and J. Hertkorn, S. Schwaiger, F. Lipski for assistance in MOVPE growth. This work was partly financially supported by the Bundesministerium für Bildung und Forschung (BMBF) and the Deutsche Forschungsgemeinschaft (DFG).

References

[1] Kuwano N, Tsuruda T, Kida Y, Miyake H, Hiramatsu K and Shibata T 2003 *Physica Status Solidi (c)* **7**, 2444-7
[2] Thapa S B, Kirchner C, Scholz F, Prinz G M, Thonke K, Sauer R, Chuvilin A, Biskupek J, Kaiser U and Hofstetter D 2007 *Journal of Crystal Growth* **298**, 383
[3] Engl K, *et al.* 2006 *Journal of Crystal Growth* **289**, 6
[4] Tanaka S, Takeuchi M and Aoyagi Y 2000 *Jpn. J. Appl. Phys.* **39**, 831
[5] Hertkorn J, Lipski F, Brückner P, Wunderer T, Thapa S B, Scholz F, Chuvilin A, Kaiser U, Beer M and Zweck J 2008 *Journal of Crystal Growth* **310**, 4867
[6] Thapa S B, Hertkorn J, Scholz F, Prinz GM, Feneberg M, Schirra M, Thonke K, Sauer R, Biskupek J and Kaiser U 2008 *Physica Status Solidi (a)* **5**, 1774
[7] Thapa S B, *et al.* 2008 *Journal of Crystal Growth* **310**, 4939-41
[8] Hertkorn J, Bruckner P, Thapa S B, Wunderer T, Scholz F, Feneberg M, Thonke K, Sauer R, Beer M and Zweck J 2007 *Journal of Crystal Growth* **308**, 30
[9] Radnoczi G and Barna A 1996 *Surf. Coat. Technol.* **80**, 89
[10] Hypercube 2002 HyperChem™ Release 7 for Windows.
[11] Hirth JP and Lothe J 1982 *Theory of Dislocations* (New York: Wiley)
[12] Chuvilin A and Kaiser U 2005 *Ultramicroscopy* **104**, 73
[13] Elsner J, Jones R, Sitch P K, Porezag VD, Elstner M, Frauenheim T, Heggie MI, Öberg S and Briddon PR 1997 *Physical Review Letters* **79**, 3672
[14] Blumenau AT, Fall CJ, Elsner J, Jones R, Heggie M I and Frauenheim T 2003 *Physica Status Solidi (c)* **6**, 1684

The role of rough surfaces in quantitative ADF imaging of gallium nitride-based materials

JS Barnard, SE Bennett, RA Oliver, MJ Kappers and CJ Humphreys

Department of Materials Science & Metallurgy, University of Cambridge, Pembroke Street, Cambridge CB2 3QZ, United Kingdom.

E-mail: jsb43@cam.ac.uk

Abstract. We have studied the surface of GaN after preparations commonly used to thin cross sectional samples to electron transparency with atomic force microscopy (AFM) and scanning transmission electron microscopy annular dark-field (STEM-ADF) imaging. Tripod polishing and ion milling leave the GaN surface in varying degrees of roughness and strain. By comparing the roughness seen in the AFM with the 'roughness' observed in STEM-ADF imaging, we show most nano-scale variations in ADF can be attributed to surface roughness alone. However, some systematic differences are seen for very thin specimens.

1. Introduction

Unlike its phase-contrast imaging counterpart, scanning TEM annular dark field imaging enjoys good agreement between simulations and experiments [1,2]. Nowhere is this quantitative assessment more important than for measuring the extent of atomic ordering where the brighter (or darker) contrast of the sample can be interpreted as precipitates, chemical ordering and point defects [2]. However, the preparation of a thin TEM specimen is a grievous process and the sample surface can be left with surface strains, rough surfaces, as well an accompanying amorphous layer, all of which introduce varying levels of point-to-point contrast that may obscure the features of interest.

We have sought to examine the influence of genuine surface roughness on the standard deviation of the ADF signal when measured in a fully quantitative way. We have measured the roughness of the GaN surface, in cross section, using the atomic force microscope (AFM). We proceed to use these preparations to produce a cross-sectioned sample for examination in the STEM. We compare the surface height and ADF standard deviations to assess the level of agreement.

2. Experimental details

All surface preparations studied were applied to a standard MOVPE-grown GaN film with a thickness of 8.5 μm on c-plane sapphire. Tripod polishing followed standard practice: Diamond impregnated mats with abrasive sizes 30, 15, 9, 6, 3, 1 and 0.1 μm were used with a Southbay Technology hand-held tripod polisher. Mechanical polishing was concluded with a Syton 50 nm silica suspension on a cloth mat. A Gatan Precision Ion Polishing System (PIPS) was used for ion milling at 5 keV with a beam inclination angle of 7° at room temperature.

Figure 1. Cross-sectional AFM images of the GaN surface after tripod polishing (a), Ar ion milling to a dose of 4000 ions/nm^2 (b) and a dose of 30,000 ions/nm^2 (c) (height ranges for each image are indicated). The c-axis of GaN points left to right in all images.

A Veeco Dimension V atomic force microscope was used to measure the surface height profile of the prepared surfaces. It was operated in tapping mode using DESP SiN cantilevers; the fast scan direction was parallel to the c-axis of the GaN film. ADF imaging in STEM mode was performed on an FEI Tecnai F20 with an illumination angle of 12.9 mrad and collection angles of 45 to 230 mrad for high-angle ADF (HAADF) and 24 to 118 mrad for low-angle ADF (LAADF). The sampling rate and the defocus were adjusted to avoid lattice contrast that would artificially increase variance.

3. Surface roughness

Roughness measurements in AFM are based on the root mean square height difference from the mean: "RMS roughness". Taken across the whole image, it measures the *total variation*, and therefore low-spatial frequency components are included. Fine-scale roughness can be measured by decomposing the one-dimensional power spectrum function:

$$\langle \rho(k) \rangle = \frac{1}{N}\sum_{i=1}^{N}\left| \int_{-L/2}^{L/2} h_i(x)\exp(ikx)dx \right|^2$$

into narrow and broad spatial frequency bands; $h_i(x)$ are the set one-dimensional height profiles that were sampled N times across a square image of length L. Invoking Parseval's theorem, we calculated the total variance associated with each band by fitting polynomials to the band shape and integrating over all spatial frequencies. This allowed us to extract the *fine-scale* roughness associated with the broad high frequency band. All roughness values quoted below were measured over 1 μm^2 scan area.

Tripod polishing alone gave extremely smooth, flat surfaces over the whole GaN surface (figure 1(a)). The height range was typically less than 20Å and the total RMS roughness was small (2.2±0.4 Å). This roughness was dominated by polishing scratches in the surface, which left wide but shallow depressions on the surface. The fine-scale roughness was sub-ångström (0.9±0.02 Å).

Ion milling at a low dose of 4000 ions/nm^2 increased the total RMS roughness ten-fold to about 20±4 Å. However, most of this roughness was attributed to very large shallow pits (figure 1(b)). Spectral analysis showed that the fine-scale roughness was much more modest (1.1±0.03 Å), i.e. a small increase from the tripod-polished state.

A higher ion dose of 30000 ions/nm^2 lead to small increase in total RMS roughness (23±6 Å) as the shallow pits became channels (figure 1(c)). Spectral analysis showed that a much higher fraction of the total roughness was due to the fine scale roughness (17±8 Å). This was now seen as 'islands' that were typically 200 Å in width. Islands much smaller than this were difficult to detect because of the limited lateral resolution of the AFM, about 110 Å by our estimates.

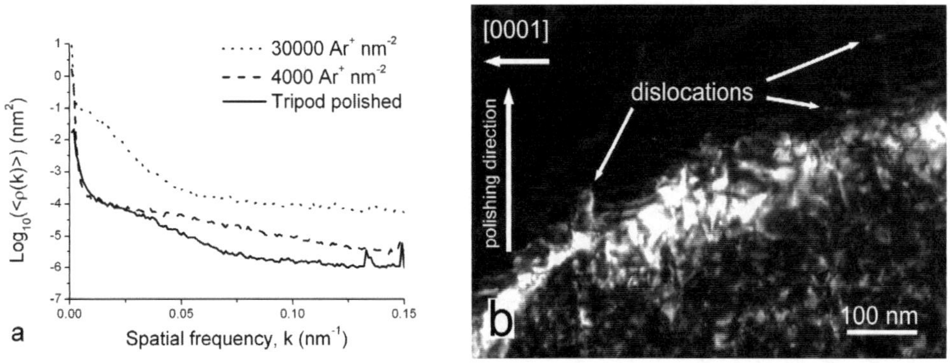

Figure 2. Normalized power spectra showing the effect of progressive ion milling on the fine scale roughness (high frequency band) as measured in the AFM (a). A (11-20) dark-field image of a tripod-polished only GaN sample shows significant presence of dislocations in the near surface region (b).

4. Surface roughness in STEM-ADF imaging: Theoretical values

Variance in the surface height contributes twice over to the variance of the specimen thickness. Thus thickness fluctuations can be converted into ADF intensity fluctuations if we know the dependence of the ADF intensity and its gradient on the specimen thickness. For this, we used Kirkland's STEMSLIC code, based on the multislice method, to calculate the total ADF signal at high and low collection angles corresponding to those on our microscope. [11-20]-oriented GaN super-cells were constructed with random displacements included to account for thermal vibrations (Einstein model, 25 configurations). This approximation is not accurate for the azimuthal distribution of thermal diffuse scattering, but is accurate for the total ADF intensity [4]. The theoretical standard deviations in the ADF signals were obtained by multiplying the ADF intensity gradient by the roughness measured by AFM (with a factor of $\sqrt{2}$ to convert the surface height standard deviation into a thickness deviation).

5. Surface roughness in STEM-ADF imaging: Measured values

To compare values of standard deviations, the absolute value of the current detected on the ADF detector must be known, as well as the incident probe current. We calibrated the ADF detector by placing an empty CBED disc with a small condenser aperture onto the detector and measured the output signal in the ADF image for a variety of spot sizes (C1 lens settings). The probe currents were then measured using a pico-ammeter connected to the Gatan Imaging Filter, which was configured for use as a Faraday cage. With fixed contrast and brightness settings, all ADF images could be converted into absolute intensity levels (fraction of the incident probe current).

We attempted to use a tripod-polished only thin foil for the ADF measurements. However, routine examination showed that the GaN is neither very thin, nor strain free. It is plagued with very high densities of sub-surface dislocations (figure 2(b)), leading to a sample in a highly strained state. Only by ion milling the tripod-polished sample with a low dose, could we obtain smooth, dislocation-free regions suitable for imaging in the STEM. Reasonably straight but tapered edges were analyzed in a similar way to the AFM, i.e. one-dimensional power spectra functions, but the thickness gradient at the edge had to be factored out. Thus, we compared the power spectra expected for several different specimen thicknesses (figure 3).

Figure 3 shows the standard deviations in both the high and low angle ADF signals (HAADF: 45-230 mrad and LAADF: 24-118 mrad) as measured by our approach (solid lines) and compared to the values expected for the low-dose ion-milled sample. The two broken lines correspond to the ADF variation one would expect if only fine-scale roughness occurred (dashed), or was rougher to the same extent as the RMS roughness (dotted). It is easy to see that the ADF variations concur only with very rough surfaces for the thicker regions while, for the thinnest regions (< 50 Å), the agreement is somewhere between the fine-scale and coarse scale roughness values.

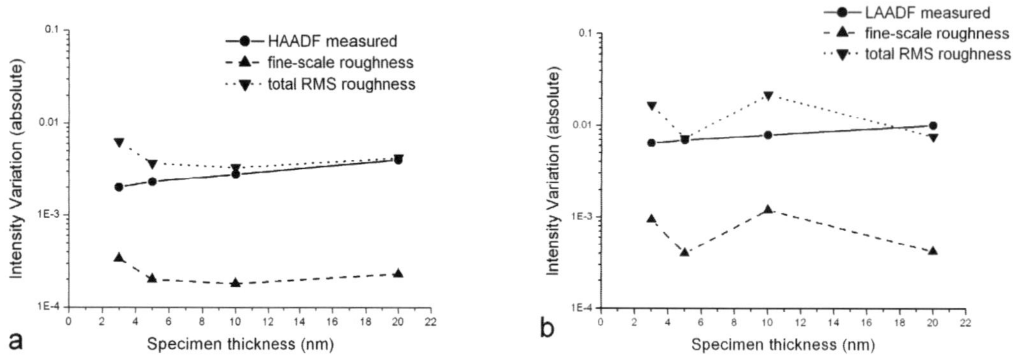

Figure 3. Measured standard deviations in the HAADF (a) and LAADF (b) at differing specimen thicknesses compared to that expected for the RMS surface roughness (20 Å, dotted line) and for fine-scale roughness (1.1 Å, dashed line). The ion-milling dose was 4000 ions/nm^2 (c.f. figure 1(b)).

6. Discussion and concluding remarks

It is surprising that the measured ADF variations cannot be explained by the fine-scale roughness alone. Several explanations are feasible: First, the presence of the amorphous layer and an ion-implanted surface region would lead to significant near-surface strain that relaxes when the sample is thin. Our comparison assumes that the amorphous layer is relatively thin and benign [5]. Given the sensitivity of the LAADF signal to strain, this might explain the excessive ADF variation seen here. Second, the roughness of top surface may not be statistically equivalent to the bottom because of electron channeling in the deeper parts of the crystal. However, given the thinness of the samples, this is likely to be small. Third, our use of the AFM to measure fine-scale roughness may be too challenging. The tip radius imposes a minimum lateral resolution of about 10 nm that effectively low-pass filters the power spectrum and removing the higher frequency components that might contribute significantly more to the total variance. However, significant quantities of energy are contained in the very low frequency components and these are still present in all spectra shown in figure 2(a). Perhaps by modeling and measuring the variance in the low-frequency band and subtracting that from the RMS roughness would give us a better estimate of the fine-scale roughness.

However, we have seen that reasonable agreement does exist between the theoretical and measured variances. Thus the RMS surface roughness can account for contrast variations seen in STEM ADF images on a scale from a nanometre to hundreds of nanometers.

References

[1] Boothroyd CB (1998) *J. Microscopy* **190**, 99
[2] Voyles PM, Chadi DJ, Citrin PH, Muller DA, Grazul JL, Northrup PA and Gossman H-JL (2003) *Phys. Rev. Lett.* **91**, 125505.
[3] Kirkland EJ (1998) *Advanced Computing in Electron Microscopy* Plenum Press, New York
[4] Muller DA, Edwards B, Kirkland EJ and Silcox J (2001) *Ultramicroscopy* **86**, 371
[5] Mkhoyan KM, Maccagnano-Zacher SE, Kirkland EJ and Silcox J (2008) *Ultramicroscopy* **108**, 791

TEM characterization of catalyst- and mask-free grown GaN nanorods

M Schowalter , T Aschenbrenner, C Kruse, D Hommel and A Rosenauer

University of Bremen, Institute for Solid State Physics, Bremen, Germany

E-mail: schowalter@ifp.uni-bremen.de

Abstract. Catalyst- and mask-free grown GaN nanorods have been investigated using transmission electron microscopy (TEM), scanning transmission electron microscopy (STEM) and energy filtered transmission electron microscopy (EFTEM). The nanorods were grown on nitridated r-plane sapphire substrates in a molecular beam epitaxy reactor. We investigated samples directly after the nitridation and after the overgrowth of the structure with GaN. High resolution transmission electron microscopy (HRTEM) and EFTEM revealed that AlN islands have formed due to nitridation. After overgrowth, the AlN islands could not be observed any more, neither by EFTEM nor by Z-contrast imaging. Instead, a smooth layer consisting of AlGaN was found. The investigation of the overgrown sample revealed that an a-plane GaN layer and GaN nanorods on top of the a-plane GaN have formed. The nanorods reduced from top of the a-plane GaN towards the a-plane GaN/sapphire interface suggesting that the nanorods originate at the AlN islands found after nitridation. However, this could not be shown unambiguously. The number of threading dislocations in the nanorods was very low. The analysis of the epitaxial relationship to the a-plane GaN showed that the nanorods grew along the [000-1] direction, and the [1-100] direction of the rods was parallel to the [0001] direction of the a-plane GaN.

1. Introduction

Despite intensive research in the last decade, heteroepitaxy of GaN-based light emitting diodes still suffers from the high density of defects such as threading dislocations and stacking faults that form due to the large lattice mismatch to the used substrate, as e.g. sapphire or SiC. On the other hand, GaN homoepitaxy is avoided due to the high costs of GaN substrates. Growing single crystalline and defect-free GaN nanorods could overcome such problems [1,2], since they could be grown almost strain-free. Usually, nanorods are grown on patterned or masked substrates, or growth may be initiated by a seed of metal droplets [3,4,5]. However, the seed metals often result in an incorporation of impurities and extended defects into the nanorods [6]. Catalyst- and mask-free grown nanorods exhibit the possibility to circumvent that, enabling fabrication of more effective optical devices. Catalyst-free growth of nanorods was reported for GaN on Si(111) and c-plane sapphire using AlN buffer layers [7,8]. However, the nanorods were only moderately ordered. In contrast, Aschenbrenner et al. [9] recently showed that highly ordered nanorods can be achieved by growth on nitridated r-plane sapphire substrates. In this paper, we report on the structural characterization of the catalyst-free grown GaN nanorods of Aschenbrenner et al. [9,10] using transmission electron microscopy (TEM) and related techniques. We determine the epitaxial relationship between the GaN nanorods with the substrate and investigate the stage of growth directly after the nitridation of the sapphire substrate and after the growth of the GaN nanorods.

2. Growth of the structures and experimental setup

The samples were grown using the following approach. The r-plane sapphire substrates were nitridated in a Thomas Swan vertical shower head reactor using a mixed NH_3/H_2 ambient with a ratio of 1:1 at 1050 °C [10] and overgrown with a thin GaN layer. In a second step the nitridated substrates were transferred to a molecular beam epitaxy (MBE) reactor and the nanorods were grown at a temperature of 830 °C with a background pressure of $1.3 \cdot 10^{-4}$ Pa and $6 \cdot 10^{-5}$ Pa of the N and Ga flux, respectively.

TEM specimens were prepared with the lift-out technique in a FEI focused ion beam (FIB) facility or by conventional thinning with a tripod holder to a thickness of about 15 µm and subsequent thinning in a GATAN precision ion polishing system (PIPS) using Xe ions with an energy of 5 keV under an angle of 5°. The specimens were investigated with a FEI Titan 80/300 microscope equipped with a Cs corrector for the objective lens, a Fischione high angle annular dark field detector (HAADF) and a GATAN post-column imaging filter. For all studies the microscope was operated at 300 kV.

3. Investigation of the nitridated sapphire surface

In a first step, we prepared a TEM specimen from a sample just after the nitridation of the sapphire substrate using the tripod technique. Figure 1 shows a high resolution TEM (HRTEM) micrograph of the specimen exhibiting the formation of islands on top of the substrate due to nitridation. To clarify the composition of the islands, energy filtered TEM (EFTEM) was performed using the K-edges of Al, N, O and C, respectively. From the energy filtered micrographs elemental maps were produced using the three-window technique. Figure 2a and 2b show N and C maps of the sample. In the N-map fluctuations can be observed close to the surface of the sapphire substrate indicating that the islands consist of AlN. The small signal above the N rich regions can be attributed to the presence of a thin layer of glue from the specimen preparation which can be seen from the C-map (Figure 2b). From the high resolution image of the islands we found that the [11-20] direction (the [0001] direction) of the AlN island are perpendicular (parallel) to the surface of the substrate.

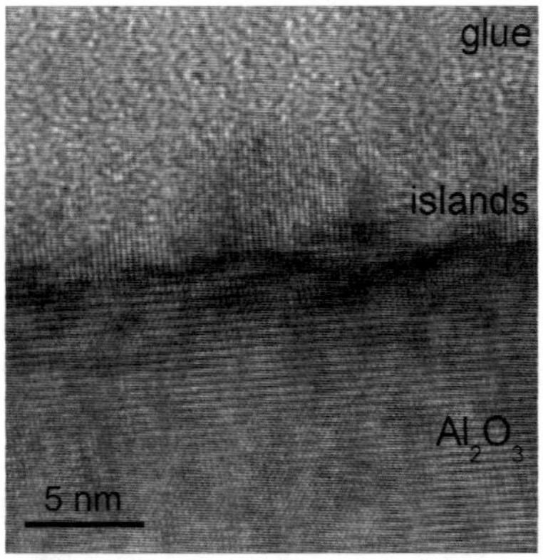

Figure 1. HRTEM micrograph of the sapphire substrate after nitridation. Islands with a height of about 5 nm have been formed on the surface of the sapphire substrate.

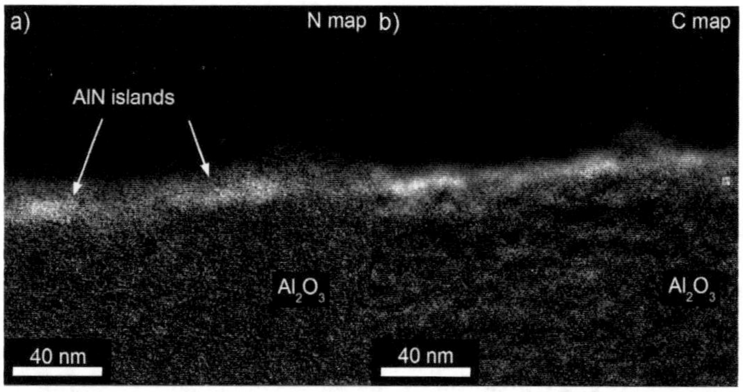

Figure 2. a) Nitrogen map and b) C map of the sapphire substrate after the nitridation. In the N map fluctuations are visible indicating the presence of AlN islands. The thin layer on top of the islands can be attributed to the presence of a thin layer of epoxy covering the islands.

4. Investigation of the overgrown sample

In a second step, we investigated the overgrown sample, which was prepared using FIB. Figure 3 shows a STEM micrograph of the structure after the overgrowth. From a first inspection one can see that the sapphire substrate is covered by a smooth layer of GaN [10], on which the nanorods are visible. From diffraction patterns we found that the layer was a-plane GaN, as one can expect from growth on r-plane sapphire. The Pt layer on top of the nanorods was deposited during specimen preparation in order to protect the nanorods from ion beam damage. With the 3 window technique elemental maps of Al, Ga and N were produced using the K-edge of Al and N and the L-edge of Ga. Figure 4 shows elemental maps of Ga, Al and N next to each other. In order to inspect closer the interface region between the sapphire and the a-plane GaN the same region was selected from each map and aligned according to the surface of the a-plane GaN layer. The inspection of the interface region in each map shows that a thin layer exists containing Al, Ga and N atoms. This was also done with an O map showing that the O atoms are confined to the sapphire substrate. HRTEM showed that only wurtzite type material is present on top of the sapphire substrate. Therefore, we conclude that the layer consists of wurtzite type AlGaN. In section 3 AlN islands were found at the surface of the sapphire but could not be revealed in the overgrown sample. This could be explained by a dissolution of the AlN islands or by the averaging effect through AlN islands and overgrown a-plane GaN.

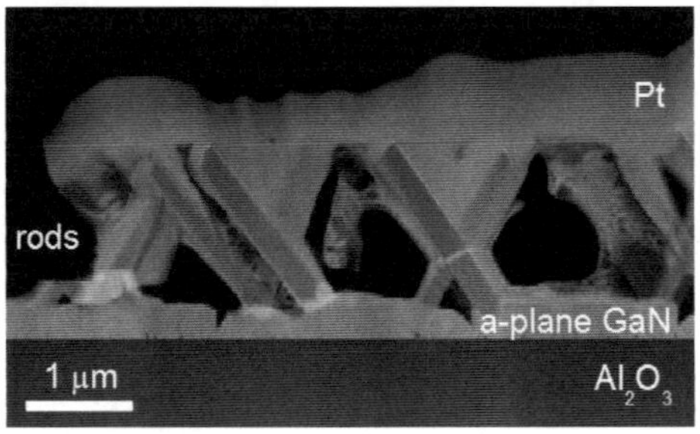

Figure 3. STEM micrograph of structure after the overgrowth. On the sapphire substrate an a-plane GaN layer has formed and nanorods are on top of the a-plane GaN. The Pt on top of the nanorods is due to specimen preparation.

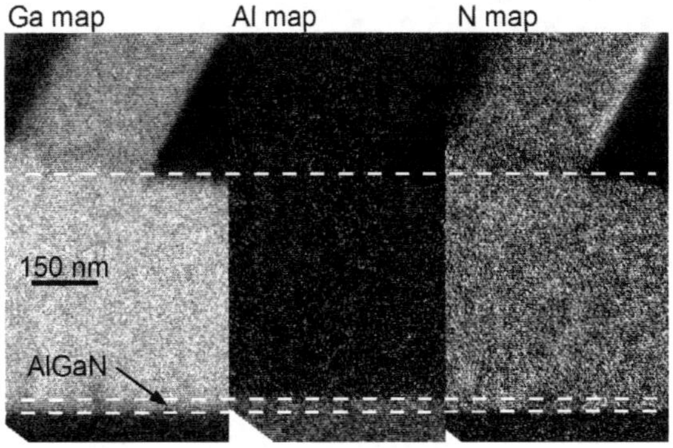

Figure 4. Elemental maps of Ga, Al and N. In order to inspect closer the interface region between the sapphire and the a-plane GaN the same region was selected from each map and aligned according to the surface of the a-plane GaN layer (upper dashed line). The inspection of the interface region in each map shows that a thin layer exists containing Al, Ga and N atoms (lower dashed lines). The layer seems to be laterally homogenous.

Figure 5 shows a scanning transmission electron microscopy (STEM) Z-contrast image of the structure. One can clearly distinguish between the sapphire substrate and the a-plane GaN layer on top of the substrate. In between the a-plane GaN and the sapphire substrate, a thin layer exhibiting a lower intensity is visible stemming from the AlGaN layer detected using EFTEM. However, also the HAADF image did not show significant fluctuations within the AlGaN to show the existence of islands at the interface after overgrowth. Figure 5 also shows the nanorods on the a-plane GaN layer, which reduce within the a-plane GaN towards the a-plane GaN/sapphire interface.

The origin of the nanorods appears to be within the a-plane GaN layer, but could also be at the a-plane GaN/sapphire interface as proposed by Aschenbrenner et al. [10]. Due to TEM specimen preparation parts of the nanorods may be truncated in such a way that the origin of the rods is outside of the TEM specimen. To clarify the position of the origin and the faceting of the nanorods STEM tomography and closer investigations of the early stages of the nanorod growth have to be done next. From thickness maps we could show that the darker regions close to the a-plane GaN/sapphire interface are due to voids within the a-plane GaN layer. We further investigated the epitaxial relationship between the nanorods and the a-plane GaN using selected area electron diffraction (SAD), convergent beam electron diffraction (CBED) and HRTEM. We found that the [1-100] direction of the rods was aligned parallel to the [0001] direction of the a-plane GaN and that the [11-20] directions of both materials were parallel to each other. We applied CBED in order to determine the polarity according to the method proposed by Mader and Recnik [11]. It was found that the growth direction of the rods was the [000-1] direction which was parallel to the [1-100] direction of a-plane GaN. This is in agreement with the common finding that GaN grows N-polar during MBE. Using different dark-field images we also investigated the defect structure in the nanorods. We only found one screw dislocation in one of the rods, suggesting that the defect density is very low. This agrees with the result from photoluminescence [9] where the nanorods showed no yellow luminescence.

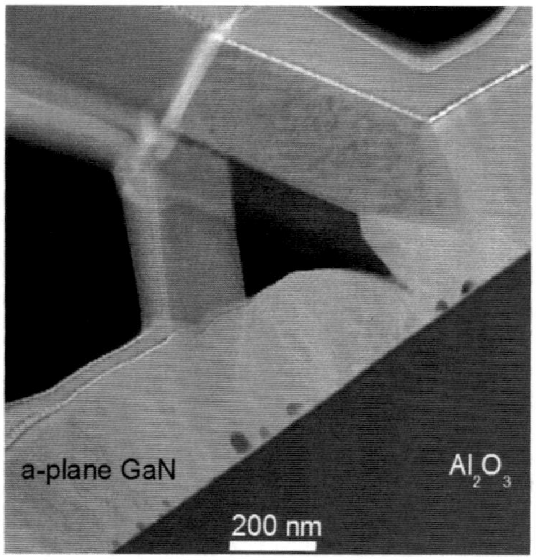

Figure 5. STEM Z-contrast image of the structure. The a-plane GaN layer and the nanorods are well distinguished from the sapphire substrate. The nanorods thin towards their origin, which seems to be within the a-plane GaN layer. However, Aschenbrenner et al. [9] proposed that the rods originate on top of the AlN islands. The Z-contrast images did not clearly show where the origin of the rods are, since the rods could be also truncated by specimen preparation and the origin could be outside of the TEM specimen.

5. Conclusion

In summary, we have determined the epitaxial relation between catalyst- and mask free grown GaN nanorods and the surrounding a-plane GaN. The rods exhibited an excellent crystalline quality. However, we could not yet completely clarify the origin of the nanorods.

References

[1] Kim H-M, Cho Y-H, Lee H, Kim S, Ryu S, Kim D, Kang T and Chung K 2004 *Nano Letters* **4**, 1059

[2] Sekiguchi H, Kishino K and Kikuchi A 2008 *Electronics Letters* **44**, 151

[3] Guha S, Bojarczuk NA, Johnsson MAL and Schetzin JF 1999 *Appl. Phys. Lett.* **75**, 463

[4] Lee S-K, Choi H-J, Pauzauskie P, Yang P, Cho N-K, Park H-D, Suh E-K, Lim K-Y and Lee H-J 2004 *phys. stat. sol. (b)* **241**, 2775

[5] Wang GT, Talin AA, Werder DJ, Creighton JR, Lai E, Anderson RJ and Arslan I 2006 *Nanotechnology* **17**, 5773

[6] Geelhaar L, Chèze C, Weber WM, Averbeck R Riechert H, Kehagias T, Komninou P, Dimitrakopulos GP and Karakostas T 2007 *Appl. Phys. Lett.* **91**, 093113

[7] Robins LH, Bertness KA, Barker JM, Sanford NA and Schlager JB 2007 *J. Appl. Phys.* **101**, 113505

[8] Tchernycheva M, Sartel C, Cirlin G, Travers L, Patriarche G, Harmand JC, Dang LS, Renard J, Gayral B, Nevou L and Julien F 2007 *Nanotechnology* **18**, 385306

[9] Aschenbrenner T, Kruse C, Kunert G, Figge S, Sebald K, Kalden J, Voss T, Gutowski J and Hommel D 2009 *Nanotechnology* **20**, 075604

[10] Aschenbrenner T, Goepel K, Kruse C, Figge and Hommel D 2008 *phys. stat. sol. (c)* **5** 1836

[11] Mader W and Recnik A 1998 *phys. stat. sol. (a)* **166**, 381

Journal of Physics: Conference Series **209** (2010) 012021 doi:10.1088/1742-6596/209/1/012021

Scanning probe microscopy studies on the growth of palladium and nickel on GaN(0001)

C Nörenberg[1,2], S Myhra[2] and PJ Dobson[1]

[1]Oxford University Begbroke Science Park, Sandy Lane, Yarnton, OX5 1PF, UK.
[2]Department of Materials, University of Oxford, Parks Road, Oxford OX1 3PH, UK

E-mail: christiane.norenberg@begbroke.ox.ac.uk

Abstract. The performance of GaN-based devices, such as blue LEDs and laser diodes, relies on good metal-GaN contacts. However, little is known about the formation and properties of these contacts on the atomic scale. We have studied the initial stages of growth of Pd and Ni on the GaN(0001) surface using variable temperature UHV-STM. The atomic-scale STM images have been compared with tapping-mode AFM images and point I-V measurements using conductive AFM in order to correlate performance on nano-scale and micro-scale. The growth mode of the metals and formation of alloys and interfacial compounds have important consequences for the electrical contact behaviour. Systematic control of the deposition conditions is therefore crucial for the performance of the contacts.

1. Introduction

Over the past decade, gallium nitride (GaN) technology and its device applications have developed rapidly. The performance of these devices (e.g. LEDs, laser diodes and high-power transistors) relies on high-quality metal-to-GaN contacts. Therefore, ohmic contacts and Schottky barriers to GaN have been studied extensively. Previous studies have focused on the improvement of ohmic contacts, such as Ti/Al and Pd/Al on *n*-type GaN, or Ni/Au, Ni/Pd/Au, Pd/Au on *p*-type GaN [1, 2]. Low-resistance, thermally stable ohmic contacts on *p*-type GaN have been reported using Pd/Ni and Pd/NiO based contacts [3, 4]. However, little is known about the formation and properties of these contacts on the atomic scale. So far, atomic-scale studies of metals on GaN(0001) using scanning tunnelling microscopy (STM) have only been reported for silver [5] and gold [6].

The formation of alloys and interfacial compounds between GaN and a metal contact influences the stability and electrical properties of the contact. For example, the formation of Pd gallides (Ga_2Pd_5 and Ga_5Pd) during annealing of Pd contacts on *p*-GaN results in enhanced thermal stability and reduced contact resistivity [3]. Control of the metal deposition conditions as well as the effective preparation of clean GaN surfaces is therefore crucial for the performance of the contacts. In a previous study, we have identified the effect of different surface treatments on surface morphology and stoichiometry of the GaN(0001) surface [7], whereas other authors have investigated the effect of specific surface treatments on the electrical properties of ohmic contacts to *n*- and *p*-type GaN [1, 8, 9].

Here, we have focused on atomic-scale studies of the initial stages of growth of palladium and nickel on the GaN(0001) surface using *in-situ* variable-temperature STM. Complementary meso-scale tapping-mode AFM imaging was carried out in order to characterize the density and distribution of structures as a function of processing parameters. Point I-V analysis in the contact mode (SECM) with a conductive Au-coated probe was also undertaken in order to elucidate the resultant point-contact electrical characteristics.

Journal of Physics: Conference Series **209** (2010) 012021 doi:10.1088/1742-6596/209/1/012021

2. Experimental

Pd and Ni were deposited in an ultra-high vacuum (UHV) chamber (base pressure 1×10^{-8} Pa) from an e-beam evaporator onto *in-situ* cleaned GaN(0001)/Al_2O_3 substrates, either at room temperature followed by annealing or at elevated temperature. The GaN/Al_2O_3(0001) templates consisted of atomically flat 1-5 μm thick *n*-doped (10^{18} cm^{-3}) GaN layers grown on Al_2O_3(0001) by MOCVD. *In-situ* cleaning involved either a 20 min anneal in an NH_3 atmosphere (2×10^{-5} Pa) or careful heating in UHV up to about 700 ºC to remove the surface oxide and produce Ga-rich GaN (0001) surfaces. The detailed cleaning procedures and the resulting GaN(0001) surface structures have been described elsewhere [7]. The growth experiments were carried out in the STM chamber of a JEOL 4500XT UHV STM. The growth was monitored *in-situ* using RHEED at 12 kV. Immediately after growth, the samples were investigated *in-vacuo* by STM using electrochemically etched tungsten tips.

AFM imaging was carried out *ex-situ* under ambient conditions with a Park Autoprobe instrument with a Multitask head and a 100 μm scanner. Tapping-mode *ex-situ* AFM topographic images were obtained with beam-shaped probes micro-fabricated from highly doped *n*-type Si with force constants in the range 3-10 N/m and resonance frequencies between 140 and 420 kHz. The radius of curvature of the probe at the apex was <10 nm. Point I-V analysis was carried out to elucidate point-contact conductivity. The measurements were carried out with a JEOL 4200D instrument operated in the contact mode with a conductive probe. I-V curves were obtained in contact mode by sweeping the tip-to-surface bias at fixed locations chosen from the topographical image.

3. Results and discussion

3.1. Palladium on GaN(0001)

When sub-monolayer amounts of palladium are deposited onto a Ga-rich GaN(0001) surface, followed by annealing at ca. 600 °C, Pd reacts with mobile excess Ga on the surface and is incorporated into the top layer. The result is the formation of a strained layer consisting of different domains oriented along the <1120> GaN directions (figure 1a). It is thought to be a monolayer (ML) of a Pd-rich alloy, as Ga and Pd can form a variety of intermetallic compounds [10] and the exact stoichiometry of the deposited layer is not known. Due to the large strain at the interface (16% lattice mismatch between Pd(111) and GaN(0001)) the surface consists of a patchwork of small domains rather than a complete monolayer.

Further Pd deposition followed by annealing at 600-650°C leads to the formation of three-dimensional (3-D) hexagonal flat-topped Pd nanocrystals (figure 1b). The crystallographic orientation of the Pd nanocrystals with respect to the substrate can be described as $(111)_{Pd}$ // $(0001)_{GaN}$, $[110]_{Pd}$ // $[1\bar{1}20]_{GaN}$. The (111) face of the Pd nanocrystals shows a periodicity of 5.8 ± 0.2 Å, which may correspond to a (2×2) adsorbate superstructure (figure 1c).

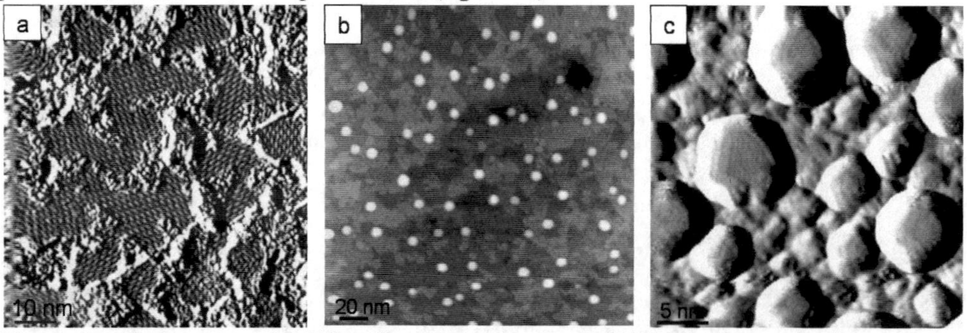

Figure 1. (a) Domains of Pd/Ga alloy strained onto GaN(0001), (b) nucleation of 3-D hexagonal Pd islands on GaN(0001) (sample bias V_s = 1.2 V, tunnelling current I_t = 0.8 nA), (c) higher resolution *log I* STM image of hexagonal Pd nanocrystals on GaN(0001) with $(111)_{Pd}$ // $(0001)_{GaN}$, $[110]_{Pd}$ // $[1120]_{GaN}$.

Journal of Physics: Conference Series **209** (2010) 012021 doi:10.1088/1742-6596/209/1/012021

3.1.1. I-V point analysis by C-AFM

Point I-V analysis was carried out at random locations on samples with 3-D Pd islands. Although particular features could not be targeted due to the small island size and high island density, three types of I-V curves were observed. Representative examples from a large data base are shown in figure 2. Type 1 (blue squares) suggests the presence of shallow electrically active interface states being accessed near zero voltage applied bias. One might speculate that the carrier transport is limited at higher bias by either the density of states and/or by life-time for charge transfer into or out of localized states. Type 2 (black dots) curves show diode behaviour, being forward biased at positive sample bias and reverse biased at negative sample bias. This characteristic is likely to be representative of conduction through the Pd islands, as Pd shows Schottky behaviour on *n*-type GaN. The effect is also observed in STM, where the best imaging conditions are obtained at large positive sample bias (>1.2 V), tunnelling into empty states, whereas imaging at negative sample bias is very difficult. Type 3 (red triangles) curves show an energy gap behaviour with a bandgap of ca. 4 eV, and then access to filled and unfilled states at ca. -2V and +2V sample bias, respectively. This may be representative of the GaN substrate, as GaN has a bandgap of ca. 3.5 eV, which may be slightly increased due to the presence of native surface oxide, and I-V curves measured on clean GaN look very similar.

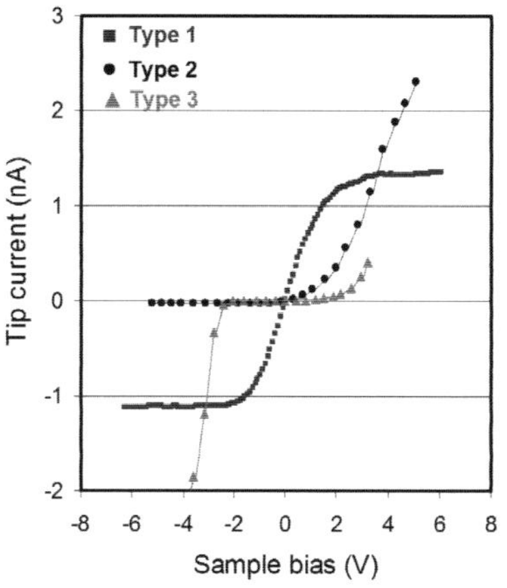

Figure 2. Point *I-V* curves of Pd/GaN(0001) sample measured using conductive AFM *Type 1* (blue squares) suggests the presence of shallow electrically active interface states being accessed near zero applied bias. *Type 2* (black dots) curves show diode behaviour, and are likely to be representative of conduction through the Pd islands. *Type 3* (red triangles) curves show a bandgap of ca. 4 eV and may be representative of the GaN substrate.

3.2. Nickel nanoislands on GaN(0001)

When nickel is deposited onto GaN(0001) followed by annealing at 600-650°C, it nucleates into mainly hexagonal Ni(111) islands without the formation of a wetting layer (figure 3). Tapping mode AFM (figure 3a, 1μm size) was used to confirm that the nm-scale STM images are representative of the larger surface area. The crystallographic relationship of the Ni islands and the underlying GaN substrate can be described as $(111)_{Ni}$ // $(0001)_{GaN}$, $[110]_{Ni}$ // $[1120]_{GaN}$. The island sizes are in the range 10-50 nm, with heights up to 2.4 nm, and some ripening has been observed during annealling. Structural detail on the flat island tops can be resolved in the logarithmic current (*log I*) imaging mode (figure 3c). The additional resolution emphasises the shape of the nanocrystals and the atomic corrugation on the flat tops. The measured periodicity along the $[110]_{Ni}$ direction (// $[1120]_{GaN}$) is 5 Å, which corresponds to twice the Ni [110] periodicity. This does not match the GaN [1120] periodicity (3.2Å), indicating that the Ni lattice has relaxed in this direction. The distance measured between the 'stripes' visible in figure 3(c) is 8.5±0.3 Å. For comparison, the Ni [112] periodicity is 2.16 Å, and the GaN [1100] periodicity is 2.76 Å. It may be possible that four Ni lattice spacings are strained onto three GaN lattice spacings in this direction.

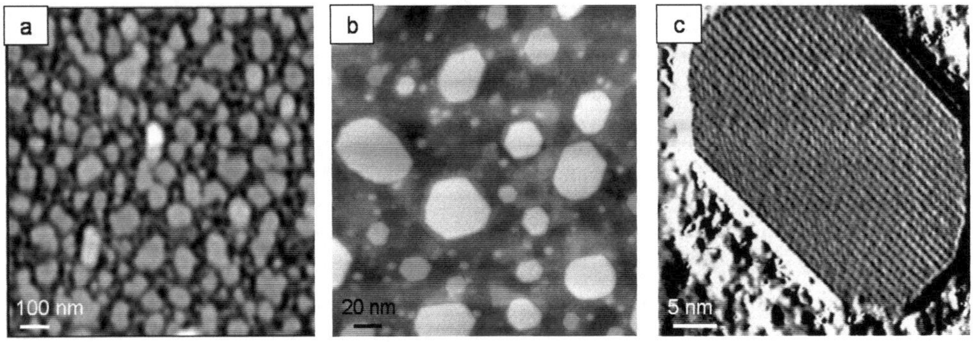

Figure 3. (a) AFM image of Ni islands on GaN (0001), (b) STM image of Ni islands on GaN(0001) (V_s= 1.8 V, I_t = 0.3 nA), (c) *Log I* STM image of Ni islands on GaN (32 nm)

4. Conclusions

We have investigated the deposition of Pd and Ni on GaN (0001) using *in-situ* elevated-temperature STM in order to elucidate growth modes and the formation of interfacial compounds, as this has been shown to influence the properties and performance of metal-GaN contacts.

The deposition of sub-monolayer amounts of Pd onto a Ga-rich GaN(0001) surface, either at elevated temperature or at room temperature followed by annealing, leads to the formation of a Pd-Ga surface alloy that wets the GaN surface. Further Pd deposition results in the nucleation and epitaxial growth of 3D hexagonal flat-topped Pd(111) nanoislands.

No wetting layer has been observed when nickel was deposited onto GaN(0001) and briefly annealed at 600-650 °C. Instead, nickel nucleates into hexagonal flat-topped Ni(111)islands. The lattice of the islands appears partially relaxed, however high-resolution TEM studies will be necessary to investigate the interface between the islands and the GaN substrate.

The work emphasises that a systematic control of the deposition conditions as well as the cleanliness and stoichiometry of the substrate is important for the control of the composition and morphology of the metal-GaN interface.

Acknowledgements

The authors thank Dr MJ Kappers (Cambridge University) for providing the GaN substrates, Dr RA Oliver and Dr MR Castell for useful discussions, and Prof. CRM Grovenor, Prof. GAD Briggs and Dr JA Crossley for support. An extended version of this paper is due to appear in *Microscopy and Analysis* (Wiley, 2009).

References

[1] Yan J, Kappers MJ et al. 2004 *phys. stat. sol. (b)* **241**,2820
[2] Pal S and Sugino T 2000 *Appl. Surf. Sci.* **161**, 263
[3] Jang HW, Kim KH et al. 2001 *Appl. Phys. Lett.* **79**, 1822
[4] Pan CC, Chen GT et al. 2006 *Appl. Phys. Lett.* **88**, 062113
[5] Wu K, Xue QZ et al. 2003 *Appl. Phys. Lett.* **82**, 1389
[6] Oreshkin AI, Maslova NS et al. 2004 *Phys. Low-Dim. Struct.* pp 165-172
[7] Oliver RA, Nörenberg C et al. 2003 *Appl. Surf. Sci.* **214**, 1
[8] Lee JL and Kim JK 2000 *J. Electrochem. Soc.* **147**, 2297
[9] Hartlieb PJ, Roskowski A et al. 2002 *J. Appl. Phys.* **91**, 732
[10] Massalski TB 1990 *Binary Alloy Phase Diagrams* 2nd ed. (Ohio: ASM International)

Lattice distortions in GaN thin films on (0001) sapphire

DV Sridhara Rao[1], R Beanland[1,2], MJ Kappers[1], D Zhu[1] and CJ Humphreys[1]

[1] Department of Materials Science and Metallurgy, University of Cambridge, Cambridge, CB2 3QZ, United Kingdom.
[2] Department of Physics, University of Warwick, Warwick, United Kingdom

E-mail: vsrd2@cam.ac.uk, dvsridhararao@gmail.com

Abstract. The GaN/sapphire microstructure and lattice distortions in GaN were investigated using convergent beam electron diffraction (CBED) and electron backscatter diffraction (EBSD). CBED studies showed small scale (<1°) triclinic strain in the electron microscope specimens with a tetragonal expansion (~0.001 nm) of the lattice along the GaN growth direction, which has been observed in continuous films as well as isolated or partly coalesced islands. The strains extend up to ~ 2 μm from the GaN/sapphire interface. EBSD studies show a clear mosaic structure. The largest component of the mosaic structure corresponds to a twist around the [0001] axis, perpendicular to the interface.

1. Introduction

Over the last decade, the group III-Nitrides AlInGaN have become an important materials system with optical applications in the visible and ultraviolet range of the spectrum. When heteroepitaxially grown on substrates such as the basal-plane (0001) of sapphire or silicon carbide, the resulting GaN film contains a high density of threading dislocations (TDs) ranging between 10^7 to 10^9 cm^{-2}. The large lattice mismatch stresses between GaN and the substrate are not accommodated by these TDs, but are believed to be relieved by an array of misfit dislocations at the GaN/substrate interface [1,2]. The origin of the TDs remains a topic of debate with basically two schools of thought: (a) the TDs are formed at the coalescence boundaries of slightly misoriented grains, i.e., the 'mosaic tilt-and-twist' model [1,2], and (b) the TDs originate from other defects in the highly imperfect region near the GaN/substrate interface, leading to spatially random TD arrangements with some (low-energy) linear array formation as a result of dislocation mobility at the GaN growth temperature (typically >1000°C) [3-5]. In view of these contrasting ideas, further work is needed to improve our understanding of the interplay of threading dislocations, strain distribution, low angle grain boundaries and other lattice distortions at the GaN/substrate interface. In this study, the GaN/sapphire microstructure and strain distribution were investigated in detail using convergent beam electron diffraction (CBED), and electron backscatter diffraction (EBSD) techniques. CBED is a TEM based technique offering the quantitative study of crystallographic parameters with high spatial resolution (~ few nm), whereas EBSD is a SEM based technique indicating the crystallographic tilts on a larger scale (~ several μm). In this paper, we discuss the findings of our studies.

2. Experimental

The GaN growth was carried out in a 6x2" close-coupled showerhead metal-organic vapour-phase epitaxy (MOVPE) reactor and the details are described elsewhere [6]. Four different samples were grown with a GaN nominal deposition thickness of ~0.75 μm (sample A), 1.2 μm (sample B), 1.65 μm (sample C), and 4.5 μm (sample D), on c-plane sapphire substrates with a nominal miscut of 0.25° to (11-20). Samples were grown with the GaN surface in different stages of coalescence, viz., islands (sample A), partly coalesced islands (samples B, C), and a fully coalesced continuous film (sample D). EBSD studies were performed using a FEI Helios Nanolab FIB/SEM equipped with an Oxford Instruments EBSD system. Orientation maps were acquired over regions approx. 30 μm square with a sample spacing of 75 nm and an accelerating voltage of 20 kV, a current of 1.4 nA and a sample tilt of 70°. Relatively slow scan speeds were used to obtain high data quality and indexing of close to 100% of the collected patterns. The data was processed using a self-written Matlab routine. Cross-sectional TEM specimens were prepared by argon ion beam thinning, and energy-filtered CBED experiments were performed using a Tecnai F20 TEM with a GIF 2000 at room temperature. The zone axis chosen was [12,1,-13,0], as this orientation was found to be sensitive to the changes in the lattice parameters (*c* and α, β, and γ) of GaN [7]. HOLZ simulations were performed using the JEMS [8] program and the conditions are described elsewhere [7]. In the calculations of lattice distortions, thin foil relaxation effects, due to having a thin TEM specimen, were not considered.

3. Results

3.1 Microstructure

TEM studies showed that isolated GaN islands exhibit {1-101} facets (Fig.1a). Mixed dislocations were bent towards the island coalescence boundary due to image forces at the free surface (TEM image, Fig.1b). Also, the AFM studies (Fig.1c) showed that the GaN island distribution appeared spatially random. The threading dislocation density at the surface of sample D was determined using TEM and AFM to be ~ $(3.7\pm0.3) \times 10^8$ cm^{-2} [9].

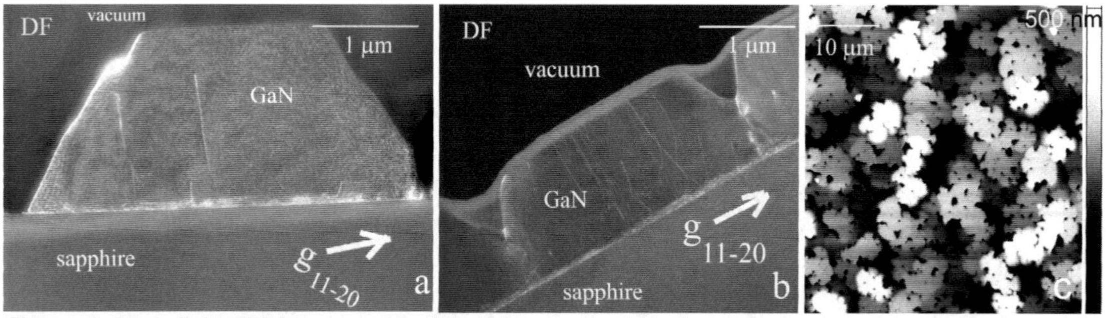

Figure 1. Dark field TEM images showing dislocations with edge component in (a) sample A, and (b) sample B. (c) An AFM image showing the surface morphology of sample B.

3.2 CBED

A simulated HOLZ pattern in the bright-field disk of a [12,1,-13,0] CBED pattern is shown in Fig.2(a), which exhibits mirror symmetry about the *a*-axis. This symmetry is broken by strains which change α and/or β; for example, a simulated pattern with a change in β is shown in Fig.2(b). In this case, β is the angle between [2-1-10] and [0001] of the GaN crystal. The experimental HOLZ patterns obtained from GaN islands in sample B showed breaking of this mirror symmetry, (representative image, Fig.2c), indicating strains of this type in the GaN electron microscope specimen. Detailed analysis of the positions of HOLZ lines in these patterns using the methodology described in [7] gives the distortions in α, β, γ, and *c*, for example, the respective estimated values are ~ $(90.0\pm0.03)°$ $(89.60\pm0.15)°$, $(119.9\pm0.01)°$ and (0.5191 ± 0.0002) nm for the HOLZ pattern in Fig.2(c). It is surprising to observe the presence of such large scale β-strains, however they were also visible in selected area diffraction patterns (Fig.2d), which showed that the angle between $\mathbf{g}_{0002}$ and $\mathbf{g}_{-2110}$ deviated by about 1° from 90°. Therefore, our results show that in our electron microscope specimen GaN is subject to a *triclinic* strain in the initial stages of growth. There is also an expansion along the [0001] growth direction. Similar phenomena were observed in samples A and C.

A careful study [7] of the HOLZ patterns recorded at different locations across the fully coalesced GaN film (sample D), as a function of distance from the GaN-sapphire interface towards the surface, indicated that: (a) lattice distortions are predominantly confined to the lower ~2 μm of the film, corresponding to growth where the GaN islands are not fully coalesced, and (b) the GaN lattice relaxes steadily further away from the substrate.

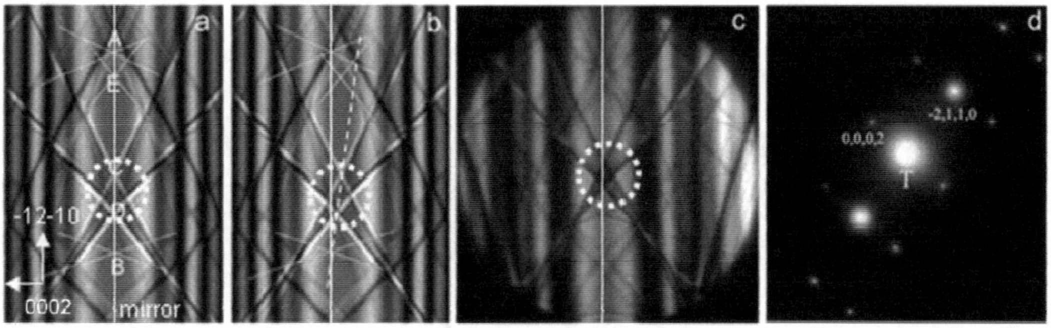

Figure 2. Simulated [12,1,-13,0] HOLZ patterns for (a) GaN and (b) *triclinic* distorted GaN (β=80°). (c) Experimental HOLZ pattern, (d) [01-10] SAD pattern of GaN near the GaN-sapphire interface.

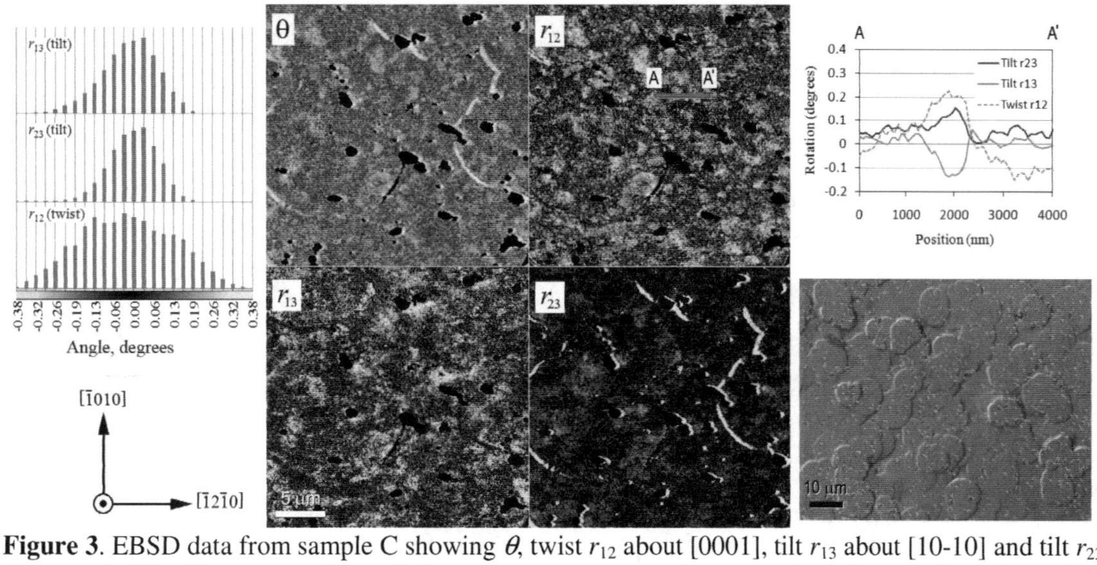

Figure 3. EBSD data from sample C showing θ, twist r_{12} about [0001], tilt r_{13} about [10-10] and tilt r_{23} about [-12-10]. Histograms of the different components are shown on the left, a section of the data taken along line A-A′ and an optical (Nomarski) image of the surface are shown on the right.

3.3 EBSD

The raw data, describing crystal orientation using Euler angles, were converted into matrix form and transformed by applying the inverse of the average rotation of the dataset. The resulting data show small rotations away from the average and can be displayed in several ways (Fig.3). The magnitude of the rotation θ can be obtained from the trace of the rotation matrix **R**, since tr[**R**] = $1+2\cos\theta$. Individual components, corresponding to a twist about [0001] (r_{12}); rotation (tilt) about [10-10] (r_{13}) or rotation (tilt) about [-12-10] (r_{23}) can also be obtained. Examination of Fig.3 and similar data from the other samples shows that (a) a clear mosaic structure can be seen, with regions – presumably corresponding to individual islands prior to coalescence – having more or less constant orientation distinct from surrounding material and (b) the component which shows the largest variation is the twist r_{12}. Note that the linear features with large tilts (~0.5°) present in the θ and r_{23} images are artefacts caused by surface relief, as can be seen in the Nomarski optical micrograph of Fig.3.

4. Discussion

The two techniques employed here sample complementary parts of the deformation of the material. CBED measures strain and is insensitive to a rigid-body rotation of the crystal, whereas EBSD measures rigid-body rotations and is not particularly sensitive to the strain in the material. These can be described as the symmetric (strain) and antisymmetric (rotation) parts of the full deformation tensor e_{ij}. The two are spatially related in that a strain (either elastic or plastic) must be present to accommodate rigid-body rotations in a continuous film. The initial GaN nucleation layer is in the form of islands each of which has a distinct orientation, and the EBSD data here show that the mosaic structure persists through the film. We found that the variation in tilt decreases with increasing layer thickness, similar to the EBSD study of Trager-Cowan [10], although in their study the r_{12} component was not extracted from the data. It is known that TDs are mainly mixed/edge-type rather than screw-type [11], which is consistent with the accommodation of a larger twist component r_{12}. The CBED studies show the strain in the material produces a triclinic distortion in the electron microscope specimens of the initial stages of GaN growth. This phenomenon could be due to a variety of parameters, such as substrate miscut, GaN-sapphire lattice misfit, thermal stresses and strain relaxation mechanisms in thin electron microscope specimens. The interplay of these parameters is complex, and modelling of the initial stages growth of GaN on sapphire is necessary to understand the origin and evolution of the lattice distortions. Interestingly, preliminary DFT calculations [12] indicate that a triclinic strain, of the scale observed, only results in a small lowering (~10-50 meV) of the energy band gap E_g of GaN. Cathodoluminescence (CL) studies are planned to verify these findings.

5. Summary

We have investigated the lattice distortions in *c*-plane GaN grown on sapphire using CBED and EBSD. The studies show a mosaic structure where twist is the major component and there is a triclinic distortion in our electron microscope specimens of the GaN in the initial stages of growth. The origin of these distortions is due to a complex interplay of several parameters, which requires modelling for a better understanding.

Acknowledgements

This research is funded by EPSRC (Grant No.EP/E035167/1). The authors are grateful to Dr. David Holec for the DFT calculations and Professor P. Stadelmann for the JEMS program.

References

[1] Ning XJ, Chien FR, Pirouz P, Yang JW and Asif Khan M 1996 *J. Mater. Res.* **11**, 580
[2] Wu XH, Fini P, Tarsa EJ, Heying B, Keller S, Mishra UK, Denbaars SP and Speck JS 1998 *J. Cryst. Growth* **189/190**, 231
[3] Narayanan V, Lorenz K, Kim W and Mahajan S 2002 *Philos. Mag. A* **82**, 885
[4] Oliver RA, Kappers MJ and Humphreys CJ 2006 *Appl. Phys. Lett.* **89**, 011914
[5] Moram MA, Oliver RA, Kappers MJ and Humphreys CJ, submitted to *Advanced Materials*
[6] Datta R, Kappers MJ, Vickers ME, Barnard JS and Humphreys CJ 2004 *Superlattices and Microstructures* **36**, 393
[7] Sridhara Rao DV, McLaughlin K, Kappers MJ and Humphreys CJ, submitted to Ultramicroscopy
[8] Stadelmann P, JEMS, version 3.2215U2007 (EPFL, Switzerland)
[9] Oliver R A, Kappers M J, Sumner J, Datta R, Humphreys C J, 2006 J. Crys. Growth, 289 506
[10] Trager Cowan C, Sweeney F, Trimby PW, Day AP, Gholinia A, Schmidt NH, Parbrook PJ, Wilkinson AJ and Watson IM 2007 *Phys. Rev. B* **75**, 085301
[11] Holec D, Zhang Y, Sridhara Rao DV, Kappers MJ, McAleese C and Humphreys CJ 2008 *J. Appl. Phys.* **104**, 123514
[12] Holec D, University of Leoben, Austria (private communication)

Atomic ordering in intermetallic CoAl alloys epitaxially grown on GaAs (001)

Q Wan, R Hey and A Trampert

Paul-Drude-Institut, Hausvogteiplatz 5-7, 10117 Berlin, Germany

Email: qianwan@pdi-berlin.de

Abstract. Epitaxially grown intermetallic CoAl alloys on GaAs (001) are investigated by transmission electron microscopy (TEM). Nanobeam diffraction proves the coexistence of the ordered B2 (CsCl) structure and its disordered BCC version, which is also verified by high-resolution TEM. From dark-field imaging, it is observed that the relative volume fractions of these two phases depend on the growth condition: the lower the growth temperature, the larger the proportion of B2 phase. The mechanism is discussed with respect to the diffusivity of vacancies and Co atoms.

1. Introduction

Intermetallic alloys, including FeAl, CoAl and NiAl, are of great importance since these materials not only have good strength-to-weight ratio but also excellent corrosion and oxidation resistance, which make them good candidates for high-temperature and soft magnetic applications [1]. In combination with semiconductor heterostructures, they are able to be integrated in a number of devices, e.g. spin injectors and mirrors. Furthermore, buried intermetallic film between two III-V structures could possibly be used in resonant tunneling diodes and metal base transistors [2]. Among these intermetallics, CoAl is especially attractive for epitaxial growth due to its low lattice mismatch of 1.2% with respect to GaAs. As a member of 3d transition metals, Co is well known for the formation of a stable B2 structure with Al over a wide composition range (46-58 at. % Co) [3]. This B2 structure can be considered as two interpenetrating primitive cubic sub-lattices [cf. Fig. 1(a)], where each Co atom has eight Al atoms as nearest neighbours and vice versa. On the other hand, a complete disorder between two sub-lattices results in a reduction of crystal symmetry to BCC as shown in Fig. 1(d).

Atomic ordering is closely related to the materials' electronic and magnetic properties. It is therefore highly desirable to achieve fully ordered structure of the film including a high quality interface to the substrate. However, overgrowth of semiconductors with metal layers is quite challenging because of the interfacial reactivity and the problem of combining polar with nonpolar crystals [4]. Here we study the microstructural properties of a CoAl layer grown on GaAs (001) substrate by molecular beam epitaxy (MBE). The atomic ordering and interface perfection were investigated by means of TEM methods.

2. Experiment

The CoAl thin films were deposited on GaAs (001) substrates by MBE. After growing a superlattice of $Al_{0.3}Ga_{0.7}As$ and GaAs and an AlAs layer as a buffer layer, which is used to suppress the interface reaction between the metal and the substrate, the template was then transferred into the As-free metal chamber, where a 40 nm CoAl layer was deposited with a very low growth rate of 0.147 nm/min. The growth temperature for CoAl was varied in a range of 100-200 °C.

Cross-sectional specimens for TEM were prepared conventionally by mechanical thinning and ion beam milling. Because the metal alloy is very sensitive to the ion bombardment, a cooling system operating at liquid nitrogen temperature is adopted to reduce the damage. Selected area electron diffraction patterns (SADP), high-resolution lattice images as well as dark-field images were recorded in a JEOL 3010 UHR microscope operating at 300kV. The EMS™ software package is employed to calculate the diffraction pattern and to simulate the high-resolution TEM contrast, which is based on the Bloch wave method. The spherical aberration, the focus spread, the semi-divergence angle and the objective aperture diameter used in the simulation were assumed to be 1.0 mm, 10 nm, 1.0 mrad and 20 nm^{-1}, respectively.

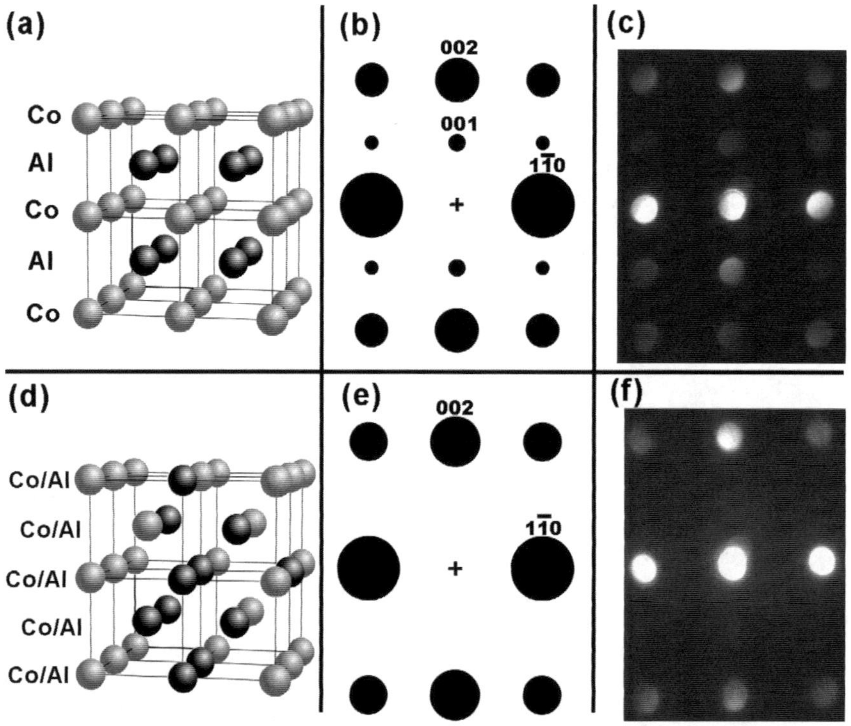

Figure 1. Crystal structure of the (a) B2 and (d) BCC phase and their respective simulated DP (b) and (e) along the [110] zone axis in kinematical approximation. The nano-beam DPs of neighbouring regions in the CoAl layer grown at 200 °C are also shown in (c) and (f) for comparison.

3. Results and Discussion

Electron diffraction is able to determine the crystal structure with high spatial resolution. Figures 1(b) and 1(e) represent the respective calculated electron DP of the B2 and BCC structure along the [110] zone axis in the kinematical approximation. As expected from their crystal symmetries, the DP of the B2 structure shows almost the same spot arrangement as the one of the BCC structure except for an extra (001) spot if one considers only fundamental reflections. The appearance of the reflections and their intensity are defined by their structure factor F . If CoAl is ordered, F_{001} is a nonzero value given by $f_{Co} - f_{Al}$, where f_{Co} and f_{Al} are the respective atom form factors of Co and Al. However, once Co and Al are mixed up, i.e., the ordering is destroyed, F_{001} equals zero given by

$f_{(Co,Al)} - f_{(Co,Al)}$, where $f_{(Co,Al)}$ is the average atom form factor of Co and Al. Spots like (001), which are present because of the ordering of the material, are generally called "superlattice" reflections.

For the detection of phase fluctuation on a nanometre scale, the conventional selected area diffraction method is not applicable because of its relatively large aperture size. We determine the local atomic ordering of the metal layer by the nanobeam diffraction method, which uses an electron beam with diameter of 2 nm. The nanobeam diffraction patterns of two neighbouring areas in the CoAl film are shown in Figs. 1(c) and 1(f), which are in agreement with the simulated DP of B2 and BCC structure. It is thus demonstrated that B2 and BCC phases are both present in the layer.

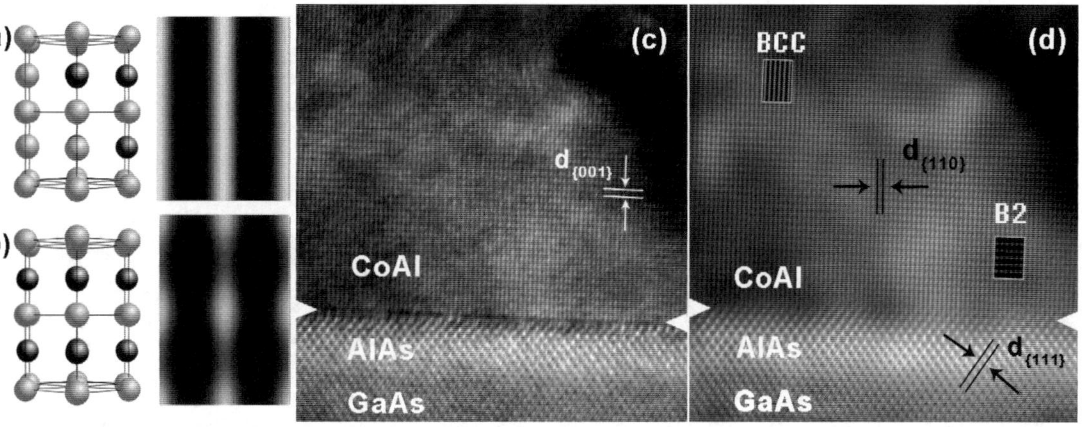

Figure 2. Crystal structure and its corresponding simulated lattice fringes for (a) BCC phase (b) B2 phase. (c) Cross-sectional High-resolution TEM image of the hetero-system along [110]-zone axis and its Fourier-filtered version (d). The simulated patterns for B2 and BCC are shown as insets.

The coexistence of both phases in the film is also demonstrated by high-resolution TEM. Figure 2(c) shows the cross-sectional lattice image of the hetero-system obtained along [110]-zone axis. The image has been Fourier filtered in order to reduce the noise and to obtain a more well-defined contrast as shown in Fig. 2(d). Besides the abrupt interface indicating no obvious interfacial reaction, a simple vertical line pattern as well as a rectangular network pattern are observed in the film. Determination of the crystal structure from a lattice fringe image requires a careful comparison between experimental and simulated images, because the contrast pattern is very sensitive to the specimen thickness (t) as well as the defocus (Δf) of objective lens. From a comparison of t-Δf maps (not shown here) with the high-resolution TEM image, it is found that the simple vertical line pattern coincides well with the simulated lattice image of BCC structure (t=5 nm, Δf=−76 nm), while the rectangular network pattern agrees with the B2 structure (t=6 nm, Δf=−76 nm), as shown in Fig. 2(a) and 2(b), respectively. Actually, since the interplanar spacing corresponding to the (002) spot is beyond the point resolution of our 300kV TEM, the lattice morphology of the BCC structure only displays the vertical lines, which originate from the interference between the direct beam and the (1-10) reflection [cf. Fig 1(f)]. The presence of the (001) spot of the B2 phase induces the visibility of lattice fringes parallel to the surface. Therefore, the observed high-resolution TEM contrast is a rectangular network pattern.

The spatial distribution of the B2 phase was analyzed by using dark-field imaging with the extra (001) as the g-vector. Figure 3(a) shows the cross-sectional TEM micrograph taken under such imaging conditions, where the bright and dark contrast in the metal layer is related to B2 and BCC phase, respectively. The fine speckle contrast confirms the coexistence of both phases. It is known from theoretical calculations that in a CoAl alloy system, the mobility of the Co atoms is quite high [5]. They can diffuse out of their original position leaving vacancies behind and occupy the available Al-type sites on the Al sub-lattice. These so-called anti-structure Co atoms cause local composition deviations of the CoAl layer from the equiatomic stoichiometry and lead to an increase of disorder. In order to intensify the atomic ordering of the intermetallic layer, a lower growth temperature of 100 °C was adopted, suppressing the diffusion of vacancies and Co atoms. Figure 3(b) shows the corresponding cross-sectional dark-field TEM image taken under the same imaging conditions as Figure 3(a). The proportion of bright areas is larger as compared to the sample grown at 200 °C indicating a higher fraction of ordered phase.

Another remarkable characteristic observed in both dark-field images is related to the continuous dark band running parallel to the film surface. The transition region occurs at about 30 nm from the interface independent of the growth temperatures used. Although the reason for this collective disorder transformation is not clarified yet, it can be attributed to the accumulated epitaxial strain in the layers [6]. Indeed, the lattice mismatch between CoAl and GaAs amounts to 1.2%. No indication for the plastic relaxation by misfit dislocation is observed at the interfaces even though the layer thickness reaches about 40 nm. It can be therefore concluded that the existing mismatch strain plays a role in the transition and leads to disorder close to surface.

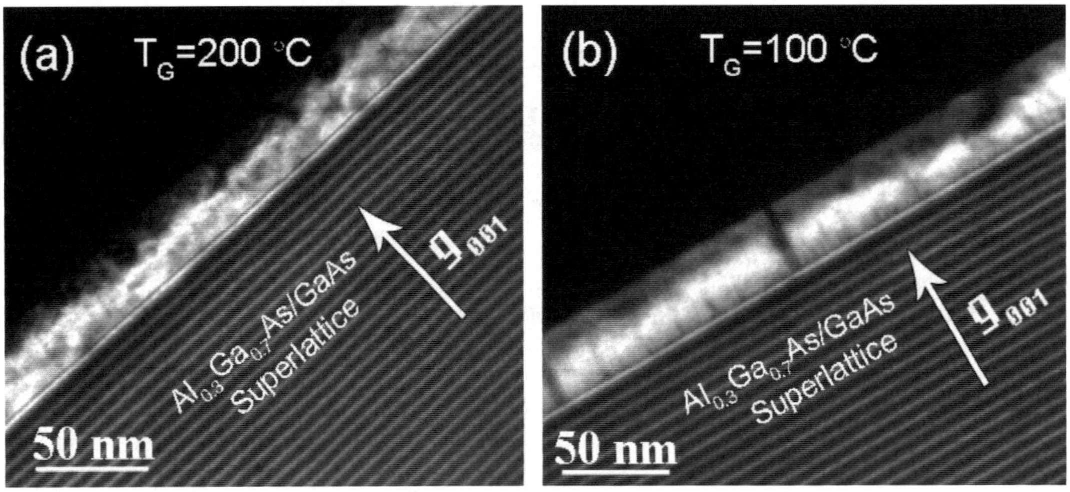

Figure 3. (110) Cross-sectional dark-field TEM image of the CoAl layer grown at (a) 200 °C and (b) 100 °C with ***g*** = 001. The bright contrast in the metal layer is related to the B2 structure.

4. Conclusion

In summary, the crystal structure, local ordering and its spatial distribution of CoAl grown on GaAs (001) is determined on nanometre scale. The coexistence of both BCC phase and ordered B2 phase is demonstrated. With lower growth temperature, a higher fraction of ordered phase is acquired.

Acknowledgement

The authors would like to thank M. Höricke for the sample preparation.

References

[1] Lee YP, Kim KW, Rhee JY, et al. 1999 *Phys. Rev. B* **59**, 546
[2] Sacks RN, Qin L, Jazwiecki M et al. 1999 *J. Vac. Sci. Technol. B* **17**, 1289
[3] Mayer D, Willerding B, Wandelt K et al. 1998 *J. Phys.: Condens. Matter* **10**, 10839
[4] Sands T et al. 1990 *Mater. Sci. Rep.* **5**, 99
[5] Stefanou N et al. 1987 *Phys. Rev. B* **35**, 2705
[6] Herman MA, Richter W, Sitter H *Epitaxy: physical principles and technical implementation* (Berlin: Springer)

Existence or absence of bandgap bowing in II-VI ternary alloys: comparison between common-anion and common-cation cases

Nacir Tit[1], IM Obaidat[1], AH Reshak[2] and H Alawadhi[3]

[1]Department of Physics, UAE University, Al-Ain, United Arab Emirates
[2] Institute of Physical Biology, South-Bohemia University, Czech Republic
[3]Department of Applied Physics, University of Sharjah, United Arab Emirates

Email: ntit@uaeu.ac.ae

Abstract. The common-anion and common-cation II-VI ternary alloys of the family Cd(Zn)Se(Te) are theoretically investigated based on two different methods. Within the virtual-crystal approximation (VCA), both the sp^3s^*-tight-binding method, with the inclusion of spin-orbit coupling, and the first-principle full-potential linear augmented plane waves (FP-LAPW) technique are employed to determine the alloy constituents' charge states (ionicities) and degree of bond polarity. The results show that: (i) in the common-cation ternary alloys (i.e., $CdSe_xTe_{1-x}$ and $ZnSe_xTe_{1-x}$), the anions alter a strong competition in trapping more charge. Such a competition builds up a compromised effect yielding the bowing behaviour. Whereas, (ii) in the common-anion ternary alloys (i.e., $Cd_yZn_{1-y}Te$ and $Cd_yZn_{1-y}Se$), the absence of such competition would cause the complete absence of bandgap bowing behaviour. The variation of the bandgap is found to be rather close to linear. The obtained good agreement between our theoretical results and the recently available photoluminescence data does further corroborate our claims.

1. Introduction

The recent advent in growth techniques of semiconductor nanostructures has paved the way for a technological revolution extending the field of applications from the Si-based microelectronics to the GaAs-based telecommunications. Among the produced nanostructures, the II-VI alloys find a broad field of applications in optoelectronics industry [1] (e.g., laser diodes emitting in the visible range, light detectors and emitters, and solar cells). There are already many new device structures which are commercially available and many more at the advanced research stage [1].

Understanding the chemical and physical properties of compound semiconductor alloys is important in making any future progress in their technological applications. One among the approaches is to give priority to focus on understanding the properties of the common-cation and common-anion ternary alloys. From there, one can proceed toward studying the properties of more complex alloys such as quaternary alloys and beyond [2]. However, despite decades of extensive studies, there are no commonly accepted explanations for the diverse observed behaviours of the bandgap versus composition (to name few, for instance, the bowing [2-4], linearity [2-4], band anti-crossing [2] and anomalous behaviours [5]).

2. Computational Methods

The present investigation focuses on finding the origins of the observed bowing in the common-cation and the linearity in the common-anion ternary alloys, members of the Zn(Cd)Se(Te) family. For such task we employ two different computational methods. The sp^3s^* tight-binding (TB) method [6], which includes the spin-orbit coupling, is used to evaluate the partial density of states (PDOS), the atomic electronic charge and alloy-bandgap energy to simulate the photoluminescence (PL) data. On the other hand, the first-principle full-potential linear augmented plane waves (FP-LAPW) technique [7], which is famous to be reliable in accurately predicting the ground-state properties, is used to calculate the electron charge-density contour plots aiming to corroborate our TB results.

3. Results and Discussions

3.1. Charge distribution

Figure 1 displays the TB-results of atomic charges versus composition for the 2 cases: (a) the common-cation $CdSe_xTe_{1-x}$ and (b) the common-anion $Cd_{1-x}Zn_xTe$ ternary alloys. In both panels, the charges of cations (Cd and Zn) vary around 2 electrons while the charges of anions (Se and Te) vary around 6 electrons as the compounds considered are all from the II-VI group. In Figure 1a, by focusing on the alloy's regime $0.25 \leq x \leq 0.75$, the anions Se and Te are shown to strongly compete with each other in trapping the charge. While in the binary states, x=0 and 1, one can clearly notice that $q_{Te} < q_{Se}$, in the alloy regime the charges of these two anions become comparable, and Te is even a bit more charged than Se. It is believed that this competition would make a compromise to cause the bandgap bowing behaviour (see below for further details). In Figure 1b, the cations Cd and Zn are shown to be robust to change their charge states (the variation is slow if not nil). It is shown that $q_{Cd} < q_{Zn}$ which is consistent with the fact that the former is more electro-positive than the latter. As a matter of fact, such charge behaviours reveal the lack of competition of cations in trapping/losing the charge which, in turn, is the main reason for a complete absence of the bowing character (see below for further details).

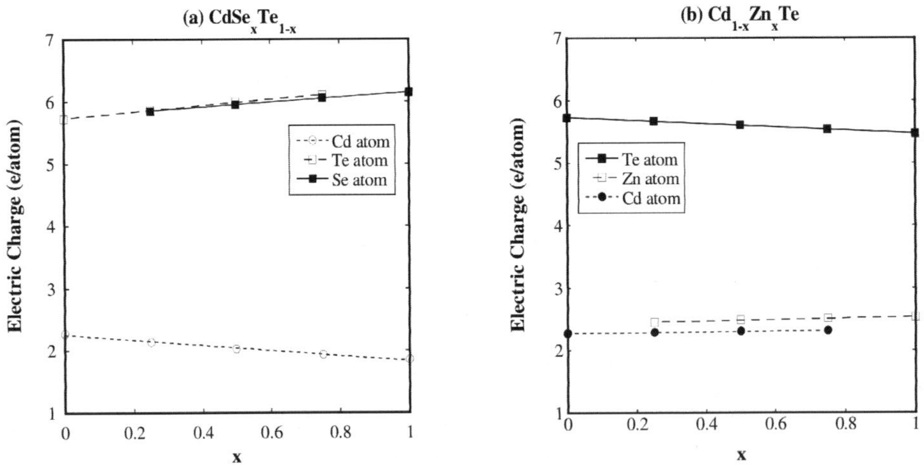

Figure 1. The TB-calculated atomic charge variation of alloy constituents versus composition in (a) the common-cation $CdSe_xTe_{1-x}$ and (b) the common-anion $Cd_{1-x}Zn_xTe$ ternary alloys

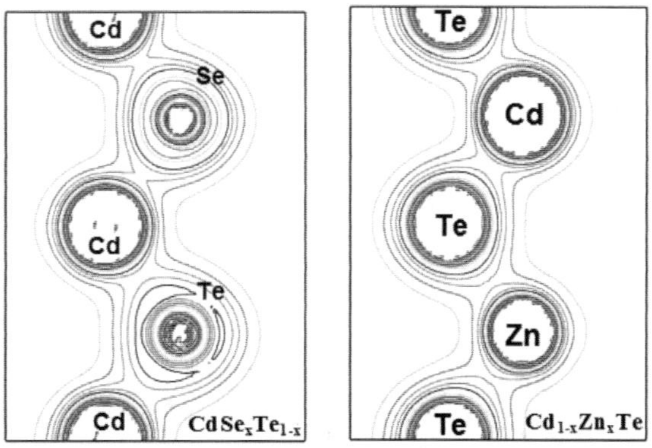

Figure 2. The electron charge density contour plots on the (110) plane obtained by the FP-LAPW as applied to $CdSe_xTe_{1-x}$ and $Cd_{1-x}Zn_xTe$ ternary alloys

Figure 2 shows the FP-LAPW-obtained electron charge density contour plots on the (110) plane for the two alloys: (a) $CdSe_xTe_{1-x}$ and (b) $Cd_{1-x}Zn_xTe$, with x=0.5. In both panels of Figure 2, the horizontal and vertical sides correspond to the $\langle 001 \rangle$ and $\langle 110 \rangle$ crystallographic directions, respectively. It is also important to mention that all four binary compounds (CdSe, CdTe, ZnSe and ZnTe) in their bulk zinc-blende (ZB) structures are found to be covalent with partial ionic character (i.e. polar materials). In the left panel of Figure 2, both Se and Te are shown to attract electrons as both are electronegative. In contrast to the TB predictions of Figure 1a, Se looks a bit stronger in electro-negativity than Te; which is consistent with their well-known chemical properties. However, in the case of common-anion ternary alloy (CdZnTe, which is shown on the right panel of Figure 2), the Zn-Te and Cd-Te bonds are shown to be similarly populated with charge. These bonds reveal the lack of competition between cations in changing their charge states in consistency to what was predicted by TB results in Figure 1b.

3.2. Modelling of photoluminescence data

Figure 3 shows our TB-theoretical modeling to various PL-data on the four alloys of interest. The figure shows the variation of the alloy bandgap versus lattice constant at low temperatures ($T \leq 15K$). The PL data shown correspond to: (a) PL1: on CdSeTe alloys from reference [8]; (b) PL2: on ZnSeTe alloys from reference [4]; (c) PL3: on CdZnTe alloys from reference [9]; and (d) PL4: on CdZnSe alloys from reference [10]. The curves correspond to our TB-results fitting using either a linear function for the common-anion or a quadratic function for the common-cation ternary alloys, respectively. The theoretical results (curves) are in excellent agreement with the various PL data, which in turn corroborate our theoretical findings mentioned above. The results of bowing parameters are listed in Table 1.

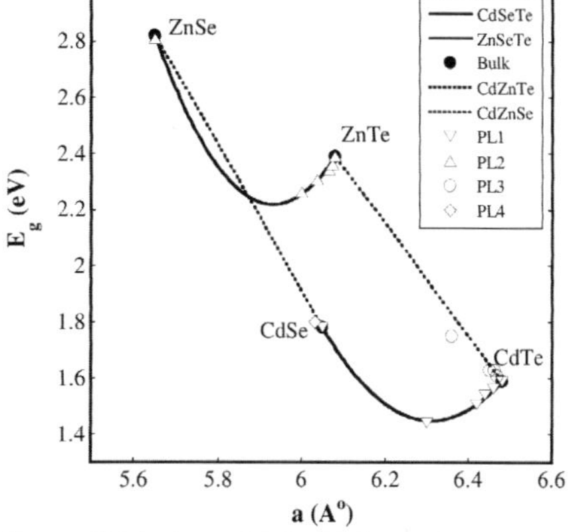

Figure 3. TB-theoretical modeling of PL data. PL1, PL2, PL3 and PL4 correspond to refs. [8], [4], [9] and [10], respectively.

Table 1 summarizes the theoretical and experimental results of the bowing parameter (*b* in eV units) for the two studied II-VI common-cation ternary alloys versus lattice mismatch $(\Delta a_0 / a_0)$, the valence-band offset (VBO) and the electro-negativity (χ in units of Pauling) of the elements. It gives a clear evidence that the bowing parameter is proportional to the following quantities: the lattice mismatch, VBO, anion electro-negativity and cation electro-positivity.

Table 1: Bowing parameter (*b* in *eV*) versus VBO, lattice mismatch, and electronegativity (χ in Pauling units) for the II-VI compound-semiconductor common-anion ternary alloys

alloy	$a_0(\Delta a_0 / a_0 \%)$	VBO *(eV)*	*b* (eV)	χ^{anion}_{ave} (Pauling)	χ^{cation} (Pauling)
ZnSeTe	5.88 (7.3%)	0.73^{A}	1.413^{B},1.239^{C}	2.325^{F}	1.69^{F}
CdSeTe	6.28 (6.8%)	0.57^{A}	0.904^{D},0.916^{E}	2.325^{F}	1.65^{F}

^{A}Reference [11]; ^{B}Reference [4]; ^{C}Reference [12]; ^{D}Present work; ^{E}Reference [8]; ^{F}Reference [13]

4. Conclusions

Within the VCA framework, we have used both the sp^3s^*-TB method, with inclusion of spin-orbit interactions, and the FP-LAPW technique to investigate the existence and origins of the bandgap bowing character. The results are summarized as follows:

(i) The common-cation ($CdSe_xTe_{1-x}$ and $ZnSe_xTe_{1-x}$) ternary alloys possess the bowing character as a result of a competition between the anions in trapping/losing the charge provided by the cation;

(ii) The common-anion ($Cd_{1-x}Zn_xSe$ and $Cd_{1-x}Zn_xTe$) ternary alloys possess rather a linear behaviour of bandgap versus composition. The bowing is absent because of the complete lack of competition between constituents to trapping/losing the charge;

(iii) Furthermore, in the case of the common-cation, the bowing is found to be proportional to the lattice mismatch, VBO, anion electro-negativity and cation electro-positivity;

(iv) The excellent agreement between our theoretical results and the PL data has corroborated our claims.

Acknowledgements

The authors are indebt to thank Dr. N. Jisrawi for fruitful discussions. This work was partially supported by the Research Affairs at the UAE University under Grant Number 08-02-2-11/09.

References

[1] Kolodziejski LA et al. 1995 *Annu. Rev. Mater. Sci.* **25**, 711
[2] Vurgaftman I, Meyerand JR and Ram-Mohan LR 2001 *J. Appl. Phys.* **89**, 5815
[3] El-Haj Hassan F, Hahemifar SJ and Akbarzadeh H 2006 *Phys. Rev. B* **73**, 195202
[4] Seong MJ et al. 1999 *Solid State Commun.* **112**, 329
[5] Wei SH and Zunger A 1989 *Phys. Rev. B* **39**, 6279
[6] Olguin D and Baquero R 1995 *Phys. Rev. B* **51**, 16891
[7] Reshak AH 2005 *Eur. Phys. J. B* **47**, 503
[8] Campo E, Hierl T and Hwang J 2004 *SPIE Proceedings* **5564**, 86
[9] Sanchez-Almazan FG et al. 1996 *J. Appl. Phys.* **79**, 7713
[10] Gupta L et al. 2003 *Pramana J. Phys.* **61**, 729
[11] Wei SH and Zunger A 1998 *Appl. Phys. Lett.* **72**, 2011
[12] Chang JH et al. 1999 *J. Korean Phys. Soc.* **34**, S4
[13] Wikipedia, the free encyclopedia: http://en.wikipedia.org/wiki/Main_Page

Refinement of chemically sensitive structure factors using parallel and convergent beam electron nanodiffraction

K Müller[1], M Schowalter[1], A Rosenauer[1], J Jansen[2], K Tsuda[3], J Titantah[4], D Lamoen[4]

[1] Institut für Festkörperphysik – Bereich Elektronenmikroskopie, Universität Bremen, Otto-Hahn-Allee 1, 28359 Bremen, Germany
[2] National Centre for HREM, Kavli Institute of Nanoscience, Delft University of Technology, 2628 Delft, The Netherlands
[3] Institute of Multidisciplinary Research for Advanced Materials, Tohoku University, 980-8577 Sendai, Japan
[4] EMAT, Universiteit Antwerpen, 2020 Antwerpen, Belgium

Email of K. Müller: mueller@ifp.uni-bremen.de

Abstract. We introduce a new method to measure structure factors from parallel beam electron diffraction (PBED) patterns. Bloch wave refinement routines were developed which can minimise the difference between simulated and experimental Bragg intensities via variation of structure factors, Debye parameters, specimen thickness and –orientation. Due to plane wave illumination, the PBED refinement is highly efficient not only in computational respect, but also concerning the experimental effort since energy filtering is shown to have no significant effect on the refinement results. The PBED method was applied to simulated GaAs diffraction patterns to derive systematic errors and rules for the identification of plausible refinement results. The evaluation of experimental GaAs PBED patterns yields a 200 X-ray structure factor of -6.33±0.14. Additionally, we obtained -6.35±0.13 from two-dimensional convergent beam electron diffraction refinements. Both results confirm density functional theory calculations published by Rosenauer et al. and indicate the inaccuracy of isolated atom scattering data, which is crucial e.g. for the composition evaluation by lattice fringe analysis.

1. Introduction

Development of efficient theoretical models during the past decade offers impressive insights into solid state properties for today's electron microscopists. In the field of transmission electron microscopy (TEM), the exact modeling of electron scattering properties, such as temperature and atomic scattering factors, has gained new interest. One should, however, bear in mind that the experimental verification should keep pace with new theoretical data. For example, DFT calculations of modified atomic scattering amplitudes (MASA [1]) have been published for the chemically sensitive 200 structure factor (SF) of $In_xGa_{1-x}As$ alloys. The MASA model accounts for the redistribution of electrons due to chemical bonding, leading to a significant improvement of accuracy for the composition evaluation by lattice fringe analysis (CELFA [2]) in the case of $In_xGa_{1-x}As$.

The intention of this article is threefold: First, the PBED method is introduced. An analysis of systematic errors of the refinement of SF and thickness will be given. The motivation to develop PBED techniques stems from the high computational efficiency due to plane wave illumination, as it has been demonstrated for structure refinements by Jansen et al [3]. Second, we report on measurements of the 200 SF for GaAs using our PBED method and two-dimensional convergent beam electron diffraction (CBED) refinements [4,5]. Third, PBED results for InAs will be given. Conclusions as to the precision of both methods and the difference between the MASA concept and isolated atom scattering data [6,7] will be drawn.

2. Introduction and test of the PBED method

2.1. Methodical basics

Plane wave illumination of a crystal leads to a diffraction pattern containing intense Bragg spots and thermal diffuse (TD) background. The Bragg intensities can easily be calculated using atomic scattering amplitudes and absorptive form factors e.g. from [6,7]. To profit from the plane wave illumination, the integrated spot intensities less the TD background are to be extracted accurately. The extracted intensities are then compared with those obtained from a single Bloch wave simulation per iteration in the refinement, making the PBED method much faster than e.g. two dimensional CBED. However, bending and thickness gradients in real specimens require the illuminated area to be only a few nanometers in diameter. For this purpose, we fabricated C2 apertures with diameters of less than 10μm which consequently lead to Bessel-like probe profiles.

The main steps of the PBED method are

1. Acquisition of PBED patterns from nanometre-sized areas,
2. Extraction of Bragg intensities and background subtraction,
3. Derivation of initial guesses for the refinement (thickness, orientation) and
4. Minimisation of the *R*-value defined by [3]

$$R = \sum_{m} \frac{(I_m^E - s \cdot I_m^S)^2}{\sum_{m} (I_m^E)^2} \tag{1}$$

in a nonlinear least-squares refinement procedure. In equation (1), R is the normalised summed difference between experimental and simulated Bragg intensities I_m^E and I_m^S, respectively. Our Bloch wave routines are written in MATLAB and the program can refine the specimen thickness, its orientation, Debye parameters, SF and the scaling factor s, which is used to scale the simulated intensities with respect to experimentally recorded counts. Different data sets may be used simultaneously for a constraint refinement of SF and Debye parameters. Except for the SF refinement the method is very similar to the multi slice least-squares program (MSLS [3]) written by Jansen et al.

2.2. Evaluation of simulated PBED patterns

Steps 2-4 listed in the last subsection have been applied to three simulated GaAs diffraction patterns in zone axis [053] containing TD background [8]. Parameters were chosen similar to the experimental situation in a Titan 80/300 used for the real measurements (sections 3, 4). A spherical aberration of 2mm was chosen for the condenser and a small beam convergence of 0.5mrad was applied to simulate a slight misalignment of the condenser. The probe diameter was 10nm and an acceleration voltage of 300kV was used. Specimen thicknesses were chosen to be 13.19nm, 23.08nm and 32.96nm.

After the background subtraction, the estimation of initial parameters is a serious problem. Whereas a guess of orientation can be derived from the intersection of Kikuchi bands, a guess of thickness might be ambiguous as shown in figure 1, where the *R*-value is plotted versus the thickness. Each minimum corresponds to one possible initial thickness and, if not measured separately, rules must be derived to identify the most plausible one. As we expect bonding effects to modify only low order SF, isolated atom potentials should be true for higher momentum transfers. We therefore performed potential scans for the real part $Re(U_{400})$ of the 400 Fourier component of the crystal potential as depicted in figure 2. Here, $Re(U_{400})$ was varied around the isolated atom value [6] (indicated by the arrow) for each of the seven thickness guesses in figure 1.

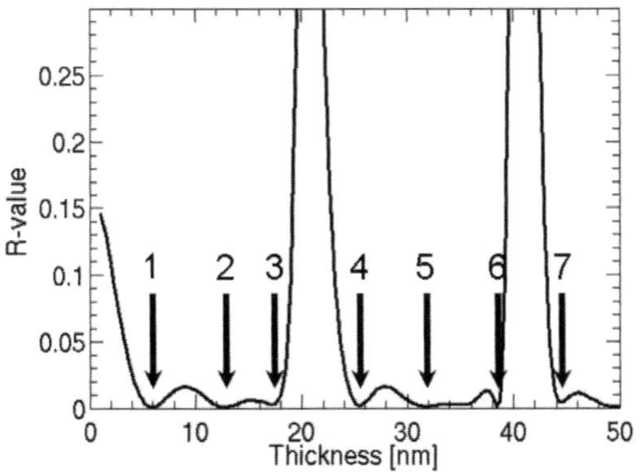

Figure 1. *R*-value as a function of thickness for the simulated diffraction pattern with a true thickness of 32.96nm. The arrows indicate possible initial thickness guesses.

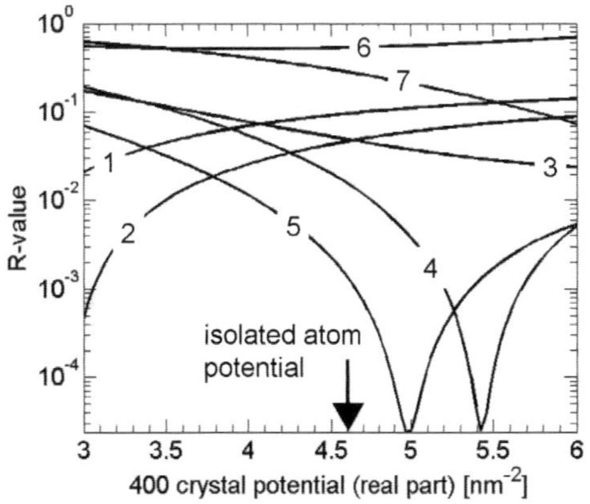

Figure 2. *R*-value as a function of the 400 potential (real part). Each curve corresponds to one thickness in figure 1, indicated by the numbers.

Obviously we trust that thickness most for which the 400 potential for isolated atoms is reproduced best, as it is the case for curve 5, corresponding to a thickness of 31.78nm. However, the isolated atom potential is not reproduced perfectly since we replaced the 200 SF by a 30% smaller value in all simulations of figures 1 and 2 to approach the real situation where the 200 SF is not known exactly.

Finally we refined $Re(U_{200})$, the thickness and the orientation for each of the three simulated PBED patterns to see how accurate our refinement result reproduces the true values. It turned out that the deviations of the 200 potential and the thickness from the input values were less than $0.03nm^{-2}$ and 0.16nm, respectively. Concerning the verification of the MASA model, this error is small enough, since $Re(U_{200})$ from [1] is 0.1 nm^{-2} larger than that from [6,7]. Furthermore, *R*-values well below 0.2% indicate that the integrated spot intensities of diffraction patterns acquired with aberration-manipulated, slightly convergent probes can be well approximated by plane wave illumination.

3. PBED and CBED measurements of the 200 structure factor for GaAs

3.1. PBED measurements

Three zero-loss energy filtered and six unfiltered GaAs PBED patterns have been recorded on a Titan 80/300 facility near the zone axis [053] at 300kV. All nine PBED patterns have been evaluated as described in section 2. The evaluation of experimental patterns turned out to be strictly analogous to that of the simulated ones, especially concerning the initial thickness guesses. The stability of each refinement has been checked by the inclusion of more beams, the reproduction of known properties (such as $Re(U_{400})$) and by assuring that the final thickness and the final orientation are close to the initial guesses.

The refinement results for the 200 SF (converted to X-ray SF) are shown in figure 3 as filled and empty black circles for unfiltered and zero-loss filtered measurements, respectively. Debye parameters of B_{Ga}=0.68Å^2 and B_{As}=0.59Å^2 were taken from [9]. From all PBED measurements, a mean 200 X-ray SF of -6.33±0.14 can be derived, whereas no significant influence of energy filtering can be observed in the investigated thickness range up to 25nm. Furthermore, we refined the Debye parameters for Ga and As, as well as the 200 SF constraint for all nine sets, allowing each set to have its individual thickness and orientation, which were also included in the refinement. The result for the 200 SF is shown as an encircled plus, which nearly falls together with the average over individual refinements.

3.2. CBED measurements

To compare our PBED results and the precision of our method with a traditional approach, the program MBFIT [4,5] was used to refine the 200 SF from seven zero-loss filtered CBED diffraction patterns. The results for the 200 SF (converted to X-ray SF) are depicted as crosses in figure 3. From these measurements, a 200 X-ray SF of -6.35±0.13 results, which agrees perfectly with the PBED value. However, the Debye parameters have been refined to $B_{Ga}=0.76Å^2$ and $B_{As}=0.62Å^2$, being larger than in the PBED case and yielding different theoretical values (grey triangles in figure 3).

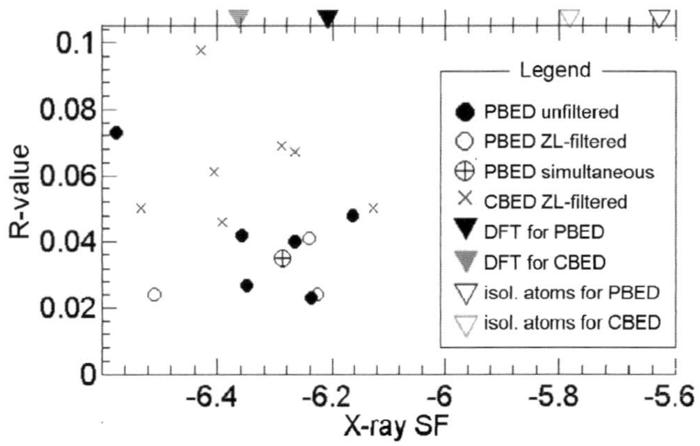

Figure 3. Refinement results for the PBED (circles) and the CBED (crosses) measurements of the GaAs 200 SF together with theoretical data for isolated atoms [6] (light triangles) and SF expected from the MASA concept [1] (filled triangles). Theoretical values for PBED and CBED differ slightly due to the different Debye parameters. The electron SF were converted to X-ray SF using the Mott-Bethe formula.

4. PBED measurements of the 200 structure factor for InAs

Analogous to section 3.1, seven unfiltered InAs PBED patterns have been recorded near the zone axis [053]. The study of bonding effects on the 200 SF for InAs is more difficult than in the case of GaAs since, in terms of X-ray SF, the DFT value of 53.68 is very close to the isolated atom X-ray 200 SF being 53.85. Thus the expected difference falls into the range of the precision observed in sections 3.1 and 3.2. However, the statistical evaluation of the seven refinement results yields an X-ray 200 SF of 53.69±0.11 with Debye parameters of $B_{In}=1.0Å^2$ and $B_{As}=0.86Å^2$ according to [9]. The constraint refinement using all seven sets simultaneously yields 53.71 for the X-ray 200 SF.

5. Conclusions

A method was developed and tested to measure structure factor amplitudes from PBED patterns. It was successfully applied to the chemically sensitive 200 SF of GaAs and InAs. Two-dimensional GaAs CBED patterns have also been refined to compare the precisions of both methods, which turned out to be comparable if derived from statistics. Methodologically, PBED simulations were about 100 times faster than two dimensional CBED simulations, making the method suitable for standard PC. Furthermore, no energy filtering is necessary. However, an advantage of the CBED method is that refinement results and initial guesses can be judged by visual control, whereas additional rules had to be derived for PBED refinements. All measurements for GaAs and InAs are in good agreement with the MASA model, whereas the isolated atom potentials cannot be verified. Using MASA for the quantitative compositional analysis of $In_xGa_{1-x}As$ alloys (e.g. with CELFA) is herewith shown to be essential also from the experimental point of view.

References

[1] Rosenauer A, Schowalter M, Glas F and Lamoen D 2005 *Phys. Rev.* B **72** 085326-1
[2] Rosenauer A, Fischer U, Gerthsen D and Förster A 1998 *Ultramicroscopy* **72** 121
[3] Jansen J, Tang D, Zandbergen H W and Schenk H 1998 *Act. Cryst.* A **54** 91
[4] Ogata Y, Tsuda K and Tanaka M 2008 *Act. Cryst.* A **64** 587
[5] Tsuda K and Tanaka M 1999 *Act. Cryst.* A **55** 939
[6] Weickenmeier A and Kohl H 1991 *Act. Cryst.* A **47** 590
[7] Bird D M and King Q A 1990 *Act. Cryst.* A **46** 202
[8] Rosenauer A, Schowalter M 2007 *Springer Proceedings in Physics 2007,* Cambridge (UK) 169
[9] Schowalter M, Rosenauer A, Titantah J T and Lamoen D 2009, Act. Cryst. A **65** 5

Nanoscale characterization of compound semiconductors using laser-pulsed atom probe tomography

M Müller, DW Saxey, A Cerezo and GDW Smith

EPSRC National Atom Probe Facility, Department of Materials, University of Oxford, Parks Road, Oxford OX1 3PH, UK

E-mail: michael.mueller@materials.ox.ac.uk

Abstract: Laser-pulsed atom probe tomography is introduced as a novel tomographic technique and its basic principles are explained. Atom probe provides 3-dimensional chemical maps with nanoscale resolution. For semiconductor research, needle-shaped atom probe samples are produced by focused ion beam lift-out and annular milling. An InAs buried quantum dot material was studied using laser-pulsed atom probe. The dot size, morphology and composition were evaluated as well as the wetting layers and interface composition. A laterally shifted In-rich core was visualised within the dots. Furthermore, a GeMn-based thin film was investigated to better-understand the mechanisms leading to the magnetic characteristics of this material system, which has potential applications as a magnetic semiconductor. We conclude that laser-pulsed atom probe has great potential for compound semiconductor research.

1. Principles of and recent developments in atom probe tomography

Three-dimensional nanoscale-analysis of semiconductor materials is an emerging and challenging area of research, and several advanced high-resolution methods like atom probe, electron and focused-ion beam tomography are now available. In terms of spatial resolution, atom probe tomography (APT) and electron tomography are the benchmark. In addition, atom probe tomography uniquely provides invaluable chemical information in 3D on the nanoscale.

1.1. Basic principles

APT relies on the successive removal of single surface atoms from a very sharp needle-shaped sample. Applying a high standing DC voltage in the range of 3-20 kV to this very sharp specimen ionizes surface atoms at sites of highest electrical field strength, which are then removed from the tip surface, by a process referred to as field evaporation [1]. In "conventional" APT, the standing DC voltage is held below the field evaporation threshold of the surface atoms, and nanosecond high-voltage pulses are applied to raise the field above the threshold and trigger field evaporation. The atoms removed subsequently pass through a time-of-flight (ToF) mass spectrometer equipped with a position-sensitive detector. The projection of atoms from the tip surface to this detector represents a point projection microscope and results in magnifications of up to 10^6 x. Based on the ion flight time and its position on the detector, the nature of the atom and its original position in the sample volume can be reconstructed [1].

Ultra short laser pulses can be used instead of voltage pulses to study materials with electrical conductivities lower than metals and alloys. For picosecond pulsing it is widely accepted that a short, spatially confined temperature rise at the specimen apex lowers the field evaporation threshold and leads to field evaporation at a given DC voltage [1,2]. This temperature rise just reaches from the tip base temperature (~ 50K) to about room temperature [1,3,4] and is therefore far below the threshold of thermal evaporation, sublimation or material breakdown.

With the implementation of laser pulsing "new" applications of APT have been progressing quickly. Various electronic materials have been successfully studied, addressing e.g. implantation profiling in silicon nanostructures, high-k dielectrics, compound semiconductors, photovoltaics and magnetic data storage materials [5-8].

For electronic materials, focused ion-beam lift-out and annular milling procedures [9,10] have become the predominant methods of sample preparation in order to achieve the very sharp needle-shaped specimen required. Generally, FIB-based preparation allows for flexible and reproducible tip preparation with limited ion-beam induced damage [11]. In particular, the combination of ion beam milling and high resolution SEM imaging (in dual-beam devices) has proven to provide maximal control throughout the entire preparation process.

Apart from laser pulsing and FIB sample preparation, further technological progress in recent years, e.g. the development of the local electrode technology [12], led to significantly increased data collection rates (reaching up to 5,000,000 ions/min), much larger field-of-views (FoV, up to 200 nm), improved mass resolution (better than 1/500 FWHM) and the use of "microtip arrays". This enables the evaluation of sample volumes in the order of 10^7 nm^3 in a fraction of the time needed by first-generation atom probes, resulting in more representative and comprehensive data sets [2]. For atom probe studies of semiconductor materials, laser assisted field evaporation (i.e. laser pulsing) has probably been the most significant progress in instrumentation pushing the frontiers of APT applications to this "new" class of materials.

2. Laser-pulsed APT applications in compound semiconductor research

2.1. Exploration of buried InAs quantum dots

Atom probe studies of buried quantum dots were conducted on a molecular beam epitaxy (MBE)-grown system of InAs quantum dots (material by courtesy of L. Chang, National Chiao Tung University, Hsinchu, Taiwan). The wetting layers with a thickness of about 2.6ML were grown at 485 °C with a growth rate of 0.28 Å/s. The capping with 30 nm of GaAs was realized at 600 °C and a growth rate of 3 Å/s. Microtip samples were prepared perpendicular to the (001) surface using the standard FIB preparation procedure introduced earlier, employing a Zeiss NVision40 dual-beam FIB. Atom probe experiments were carried using Imago's LEAP™ 3000X HR equipped with a picosecond laser system under optimized experimental conditions.

In the 3D atom map the quantum dots are represented by regions of locally higher indium density. These regions can be represented by isosurfaces, i.e. surfaces of either equal concentration or count density. Based on these isosurfaces, it was found that the dots possess on average (~ 70 quantum dots evaluated) a lens-shaped morphology with a base length aspect ratio of around 1:1.5 with X~(19.1±1.2) nm, Y~(13.1±1.0) nm and an average dot height of (4.1±0.2) nm.

Gallium was found to be substantially incorporated within the dots, the In-to-Ga ratio spatially averaged over the volume enclosed by the isosurfaces (~ 500 nm^3, typically confining 20,000-30,000 ion counts) was determined to be $In_{0.22\pm0.01}Ga_{0.78\pm0.01}$ (values derived from 70 quantum evaluated) in good agreement with kinetic modelling of intermixing during quantum dot growth [13] given the relatively high capping procedure (600 °C) for this material system. Two-dimensional indium concentration profiles derived from virtually projection-free cross-sections through the dots revealed the presence of a laterally shifted In-rich core within the dot with a maximum In-concentration of $In_{0.4}Ga_{0.6}$ [8]. Indium enrichment in the core was also observed by other authors [14-16], varying from e.g. x=0.62 [15] to x=1 [16] certainly at least partly depending on the different material systems studied and characterisation methods applied.

The comprehensive AP data also allows for wetting layer analysis. The as-reconstructed layer exhibit an asymmetrical In-concentration profile (Figure 1) steep at the InAs-on-GaAs interface but more gradual at the capping side, suggesting pronounced cation intermixing over around 3 nm. Given that this latter interface was grown at substantially higher temperatures, the stronger intermixing observed appears reasonable. Similarly, such asymmetry and its spatial extend were also indicated by various TEM studies [15,16]. The maximum indium concentration within the wetting layers reaches

$In_{0.2}Ga_{0.8}$. This maximum In-content in the wetting layer agrees remarkably well with other studies stating x to be between 0.15 [15] and 0.2 [16].

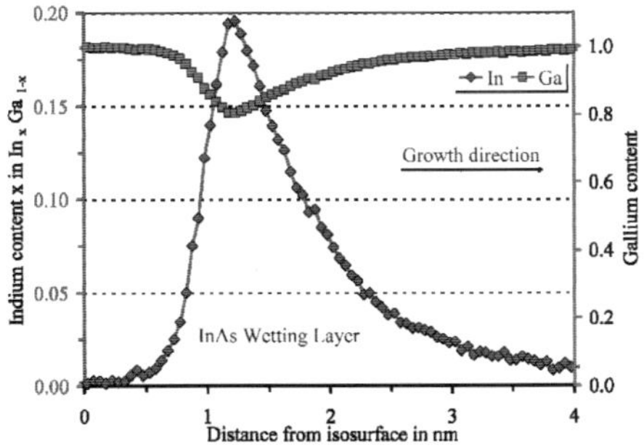

Figure 1: In-to-Ga-ratio across the interfaces of the InAs wetting layer. The In-concentration profile (green, -◆-) shows a clear asymmetry with a gradual InAs-on-GaAs interface, indicating stronger intermixing during the capping process.

For very well defined and consistent physical properties of QD-based devices, the quantum dots should possess highest uniformity in many aspects e.g. a narrow size distribution and similar compositions. The AP data derived, uniquely addresses the question of dot-by-dot variability providing firm statistics based on a large number of dots, but maintaining nanoscale accuracy. For example, variations in size and composition have been listed before [8]. In addition, Figure 2 illustrates the lateral positions of the In-rich cores in a stack of self-assembled dots. It is obvious that the core position varies from dot to dot as does the maximum indium concentration.

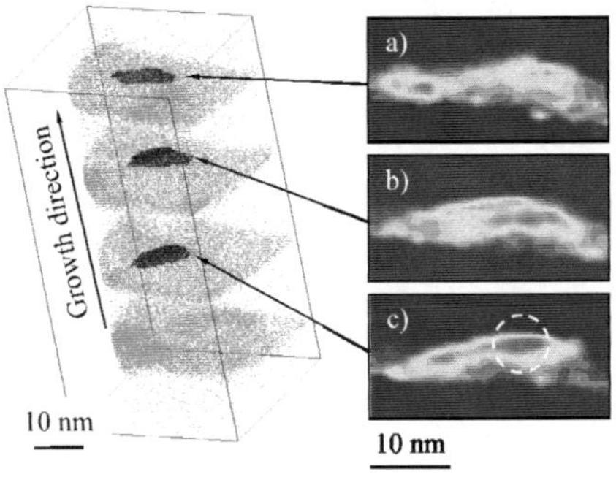

Figure 2: Illustration of the maximal, local In-distribution within consecutive quantum dots of a stack. On the left hand side the dots (blue isosurfaces) are shown in an atom map displaying In-atoms (green) of the wetting layers only. On the right hand side the corresponding 2D-concentration profiles based on 1 nm thick cross-sections along the major dot axis are presented. The red regions within the dots represent the highest In-concentration normalized to the maximum In-concentration in c) (see white dashed circle) of $In_{0.4}Ga_{0.6}$.

2.2. Nanostructural analysis of a thin-film GeMn magnetic semiconductor

The GeMn system is a promising candidate as a versatile magnetic semiconductor material, with potential applications in spintronics [17]. Current interest surrounds the structure-property relationship determining the magnetic/conductive characteristics of Ge thin-films containing a few percent Mn in a highly super-saturated solution [18,19]. An understanding of these material systems requires chemically sensitive analysis of the nanostructure with the highest possible resolution in three dimensions.

Laser-pulsed atom probe tomography was used to characterize a GeMn thin-film (by courtesy of D. Bougeard, Walter Schottky Institute, TU Munich, GER) containing 2-4% Mn, grown by MBE on a Ge(001) substrate to a thickness of approximately 100nm. String-like arrangements of Mn-rich clusters were found (Figure 3), oriented in the growth direction, with concentrations of 10-20% Mn at their centres. These observations are consistent with earlier characterization by TEM and X-ray diffraction [20], but provide new insights into the nanostructure that offer greater consistency with the magnetic and electrical characteristics.

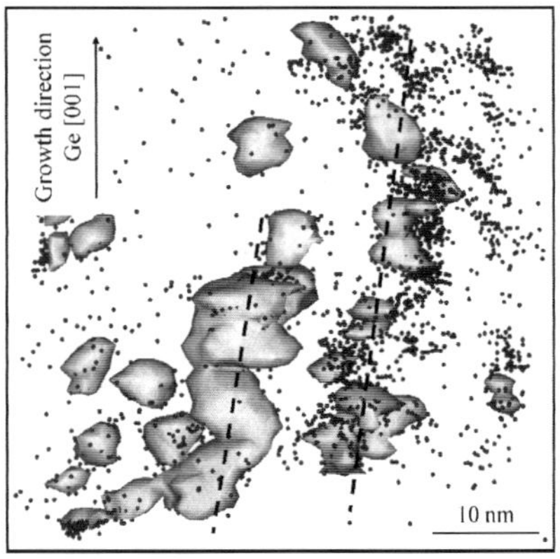

Figure 3. Iso-concentration surfaces (yellow) showing the distribution of Mn-rich regions in a GeMn thin film grown by MBE. Characterisation of the morphology and chemical composition of these structures in 3D is necessary for understanding the magnetic properties of the film. Individual detected Mn atoms are shown as blue dots. The vertical, dashed lines (black) indicate column-like structures of Mn-rich regions.

3. Discussion

Although small in number, early applications certainly prove the applicability of laser-pulsed atom probe tomography in compound semiconductor research and cast a positive light on future studies. Since research is still in its infancy, issues like optimising experimental conditions, chemical accuracy and homogeneity as well as data reconstruction for (compound) semiconductor nanostructures (quantum dots, multilayer systems) need to be addressed in due course. In particular, besides just focusing on new applications, fundamental studies on basic reference materials are indispensable for a better understanding and assessment of the technique itself.

References

[1] Miller MK, Cerezo A, Hetherington MG and Smith GDW. *Atom Probe Field Ion Microscopy.* Oxford Science Publications; 1996
[2] Cerezo A et al. 2007 *Mater. Today* **10**, 36
[3] Cerezo A, Smith GDW and Clifton PH 2006 *Appl. Phys. Lett.* **88**, 154103.
[4] Vurpillot F, Gault B, Vella A, Bouet M and Deconihout B 2006 *Appl. Phys. Lett.* **88**, 094105.
[5] Galtrey MJ, Oliver RA, Kappers MJ, Humphreys CJ, Stokes DJ, Clifton PH and Cerezo A 2007 *Appl. Phys. Lett.* **90**, 061903
[6] Gorman BP, Norman AG and Yan T 2007 *Microsc. Microanal.* **13**, 493
[7] Kelly TF, Larson DJ, Thompson K, Alvis RL, Bunton JH, Olson JD et al. 2007 *Annu. Rev. Mater. Res.* **37**, 681
[8] Müller M, Cerezo A, Smith GDW, Chang L and Gerstl SSA. 2008 *Appl. Phys. Lett.* **92**, 233115
[9] Miller MK and Russell KF 2007 *Ultramicroscopy* **107**, 761
[10] Thompson K, Lawrence D, Larson DJ, Olson JD, Kelly TF and Gorman B 2007 *Ultramicroscopy* **107**, 131
[11] Thompson K, Gorman B, Larson DJ, Van Leer B and Hong L 2006 *Microsc. Microanal.* **12**, 1736
[12] Kelly TF and Larson DJ 2000 *Mater. Charact.* **44**, 59
[13] Heyn C, Schramm A, Kipp T and Hansen W 2007 *J. Cryst. Growth* **301-302**, 692
[14] Liu N, Tersoff J, Baklenov O, Holmes Jr. AL and Shih CK 2000 *Phys. Rev. Lett.* **84**, 334
[15] Walther T, Cullis AG, Norris DJ and Hopkinson M 2001 *Phys. Rev. Lett.* **86**, 2381
[16] Wang P, Bleloch AL, Falke M, Goodhew PJ, Ng J and Missous M 2006 *Appl. Phys. Lett.* **89**, 072111.
[17] Jamet M, Barski A, Devillers T, Poydenot V, Dujardin R, Bayle-Guillemaud P et al. 2006 *Nat. .Mater.* **5**, 653
[18] Bougeard D, Ahlers S, Trampert A, Sircar N and Abstreiter G 2006 *Phys. Rev. Lett.* **97**, 237202
[19] Devillers T, Jamet M, Barski A, Poydenot V, Bayle-Guillemaud P, Bellet-Amalric E et al. 2007 *Phys. Rev. B* **76**, 205306
[20] Holý V, Lechner RT, Ahlers S, Horyk L, Metzger TH, Navarro-Quezada A et al. 2008 *Phys. Rev. B* **78**, 144401

Journal of Physics: Conference Series **209** (2010) 012027 doi:10.1088/1742-6596/209/1/012027

Off axis holography of doped and intrinsic silicon nanowires: interpretation and influence of fields in the vacuum

MI den Hertog[1], JL Rouviere[1], H Schmid[4], D Cooper[2], MT Björk[4], H Riel[4], F Dhalluin[2, 3], P Gentile[1], P Ferret[2], F Oehler[1], T Baron[3], P Rivallin[2], S Karg[4] and W Riess[4]

[1] INAC/SP2M, 17 rue des Martyrs, 38052 Grenoble Cedex 9, France
[2] LETI, 17 rue des Martyrs, 38052 Grenoble Cedex 9, France
[3] LTM, CNRS UMR-5129, 17 rue des Martyrs, 38052 Grenoble Cedex 9, France
[4] IBM Research GmbH, Säumerstrasse 4, 8803 Rüschlikon, Switzerland

E-mail: martien.den-hertog@cea.fr

Abstract. Intrinsic and axially modulated doped silicon nanowires (NWs) were studied by off-axis electron holography. Phase gradients in the vacuum were observed and compared to simulations of NWs with a varying charge density at the wire-oxide interface. It seems that intrinsic NWs are positively charged with a charge density around $2x10^{10}$ e.c. (electron charges) cm^{-2} and axially modulated doped NWs (with n-doped regions) are negatively charged with a charge density around $-1x10^{12}$ e.c. cm^{-2}. Expected fringing fields around the doped regions are incidentally observed but smaller than predicted by simulations. The effect of the surface charge on the reference wave is evaluated, and should not modify the obtained phase image.

1. Introduction

The scaling of transistors and integrated circuits for future technology nodes not only poses huge challenges on modelling, design and fabrication but increasingly also on the available tools required for structural and compositional analysis of individual devices. Many techniques exist that can probe the doping concentration such as Scanning Transmission Electron Microscopy [1], Atom Probe Tomography [2], Secondary Ion Mass Spectroscopy [3], Scanning Spreading Resistance Microscopy [4], Scanning Capacitance Microscopy [5], or Scanning Tunneling Microscopy [6], however, off-axis electron holography is the only technique that can measure active dopant atoms in the volume of the sample.

Electron holography has been successfully used to map active dopant distributions in semiconductor devices that are several hundreds of nanometers thick [7] and in a ~300 nm diameter core-shell nanowire [8]. To date, however, off-axis electron holography was not applied to sub-100 nm structures with modulated doping because it was generally accepted that the signal would be far too small. Off-axis electron holography is a TEM-based technique that uses a biprism (charged wire) to interfere an object wave that has passed through the sample and a reference wave that has traversed through only vacuum. From the interference pattern (known as the electron hologram) an amplitude and phase image can be reconstructed. In the absence of magnetic fields and diffraction contrast, the phase difference $\Delta\varphi$ between electrons that have passed through vacuum and electrons that have traversed the specimen is related to the crystal potential, or mean inner potential (MIP) $V(x, y, z)$, by

$$\Delta\varphi = C_E \int_0^t V(x, y, z) dz \qquad (1)$$

where C_E is an interaction constant (7.29×10^6 rad $V^{-1}m^{-1}$ at 200 kV) that depends on acceleration voltage and t is the thickness of the sample along the beam direction z [7].

Holography was performed on Vapour Liquid Solid grown silicon nanowires with axially modulated doping. For the first time doping contrast was observed in thin (60 nm) nanowires. Comparison of the experimental potential variations in the NWs with simulated potential maps reveals that a charge density of -1×10^{12} e.c. cm^{-2} is present at the wire-oxide interface (so-called sheet charge), which is in excellent agreement with the value obtained by transport measurements [9]. These results will be published elsewhere [10]. The aim of this paper is to evaluate the phase signal in the vacuum at the side of the NW, and compare this signal with simulations. Both NWs with modulated doping and homogeneous intrinsic NWs will be considered. This is both of fundamental and applied interest, as a correct interpretation of holography data for microelectronics purposes is possible only if the beam-specimen interaction and its effect on the measured potential is understood. Furthermore the potential variations around the specimen will influence the measurement [11]. A problem in electron holography is that according to potential simulations so-called fringing fields would be expected in the vacuum, for example at the side of a p-n junction [11-13]. In practice such fringing fields are only observed if the junction is reversed biased [12,13]. It was proposed by Fazzini et al. [13] that this is due to positive surface charges in the oxide created by the electron beam and they estimated a charge density around $1.5\text{-}3\times10^{13}$ e.c. cm^{-2}. It should be noted that simulations including such high positive charge densities on the surfaces of the studied NWs [10] would result in hardly any potential variations at the junctions, since the complete NW volume is depleted.

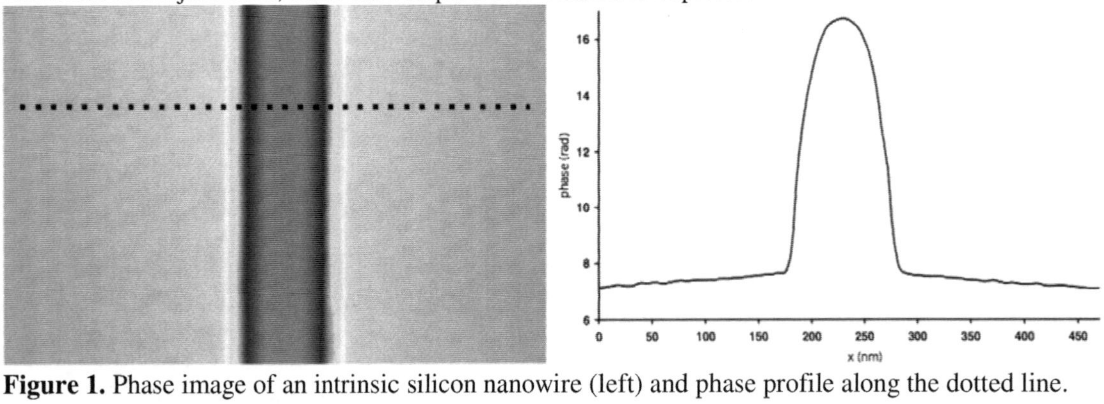

Figure 1. Phase image of an intrinsic silicon nanowire (left) and phase profile along the dotted line.

2. Experimental results

The nanowires were grown epitaxially on a lightly doped (10^{15} at. cm^{-3} p-type) (111) silicon wafer and contained three 150 nm long n-doped (phosphorus) regions with N_D of 10^{20}, 10^{19} and 10^{18} at. cm^{-3} separated by intrinsic segments. All doping concentrations were determined from electrical transport measurements on individual homogenously doped wires for each carrier concentration [9]. Growth conditions are given in [9,10]. The oxide shell on the nanowire surface was approximately 3 nm and grown under ambient conditions (native oxide).

Holography was performed on intrinsic NWs. Phase maps were extracted either by using the holoworks software or a home made Digital Micrograph script. A phase profile made perpendicular to the nanowire axis is shown in Fig. 1. In the vacuum the phase gradually decreases, indicating a positive charging of the NW. The phase gradient in the vacuum was compared to potential simulations, where an intrinsic NW was simulated with different charge quantities at the wire oxide interface, as shown in Fig. 2A. Simulations were performed with a TCAD simulation tool by Silvaco™ [14]. The potential simulations were converted to a phase signal as follows. A radial potential profile was made in the simulated potential map perpendicular to the NW axis. The radial profile was imported in Digital Micrograph (DM), the 2D cylindrical profile was reconstructed from this radial profile. A projection of the potential along the electron beam (*y*-axis) was realized using DM. The obtained projected potential (in V nm) was converted to radians (Equation (1)). A charge density of $2x10^{10}$ e.c. cm^{-2} at the wire-oxide interface seems to correspond well to the experimental phase gradient (Fig.2A). The shape of a phase profile made perpendicular to an intrinsic NW was always similar to Fig. 1, however, the gradient in the vacuum was sometimes larger, indicating a higher positive charge density at the wire surface (Fig. 2A).

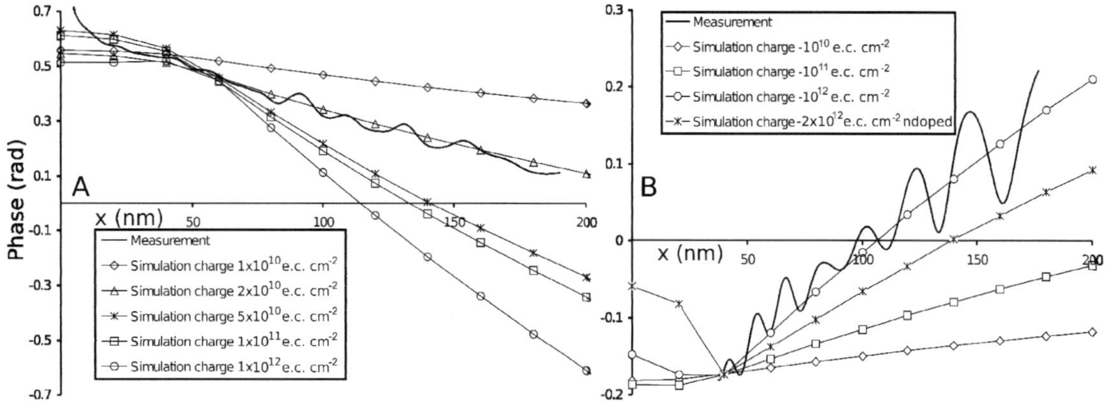

Figure 2. A: Phase gradient in the vacuum at the right of an intrinsic silicon NW (see Fig.1) compared with simulations including positive sheet charge values. B: Phase gradient at the right side of an axially modulated doped NW compared with simulations including negative sheet charge values.

A perpendicular profile made on NWs with axially modulated doping showed an increasing phase gradient in the vacuum going away from the wire, indicating negative charging of the wire. Such a phase gradient is compared to simulations in Fig. 2B. The simulations were performed for an entirely intrinsic NW and for an entirely 10^{19} cm^{-3} n-doped NW, with different negative sheet charge values. The experimental profile corresponds well to the simulation including a sheet charge of $-1x10^{12}$ e.c. cm^{-2}, which is in excellent agreement with [9,10]. From Fig. 2B it can be seen that phase modulations are expected around the doped regions, since the gradient depends also on doping concentration. In the case of a p-n junction even larger phase variations, of approximately 0.3-1 rad are expected in the vacuum around the doped regions, due to the space charge regions. However, phase variations around the doped regions were only incidentally observed and were around 0.1 rad, as shown in Fig. 3, smaller than predicted by simulations.

As shown in the foregoing the NWs are charged, which creates a linear phase gradient in the vacuum. If a charged NW creates a potential gradient in the vacuum, this will perturb the reference wave, as is shown schematically in Fig. 4. The potential simulations were used to describe the influence of the field in the vacuum. The potential $V(x)$ was converted to phase $\varphi(x)$, which was used in the object wave ψ_{obj} and the reference wave ψ_{ref} (Fig. 4). The intensity of the hologram is then given by [11,15]:

$$I_{holo}(x) = \left| \exp i\varphi(x+\frac{W}{2})\exp(i\gamma x\frac{2\pi}{\lambda}) + \exp i\varphi(x-\frac{W}{2})\exp(-i\gamma x\frac{2\pi}{\lambda})\right|^2 \qquad (2)$$

where W is the width of the hologram, and γ is the change in direction of the electron beam due to the birprism. The $W/2$ shift introduced in φ is a way to describe mathematically the interferences between the object and reference waves. In most experiments the biprism was almost parallel to the NW axis, therefore the influence of the field on the reference wave can be easily compared with simulations. The hologram simulations show that the influence of the potential on the reference wave is negligible. We speculate that this is due both to the potential variation in the vacuum that is linear in first approximation (see Fig 1,2) and the treatment of the hologram. The phase image is made symmetrical with respect to the NW axis by defining two reference regions symmetrically at both sides of the NW. A linear relation is fitted to the reference regions and the obtained gradients are subtracted from the phase image. Therefore the effect of any gradient that is not symmetric with respect to the NW axis is removed.

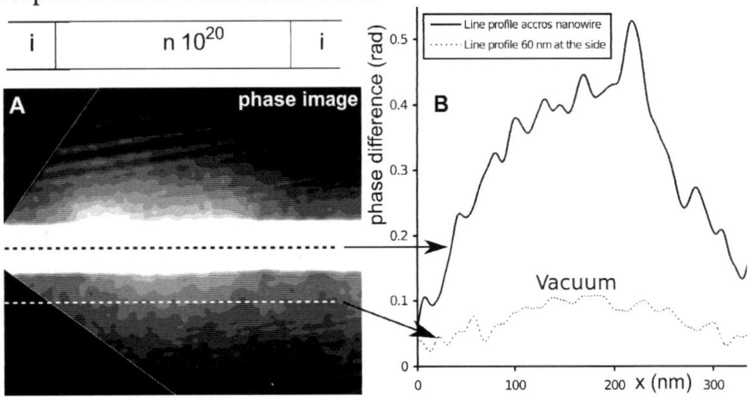

Figure 3. A: Phase image with adjusted grey scale to show phase variations in the vacuum at the side of an axially modulated doped nanowire, as indicated in the schematic (top). B: Phase profiles made on the wire and in the vacuum 60 nm at the side of the NW centre.

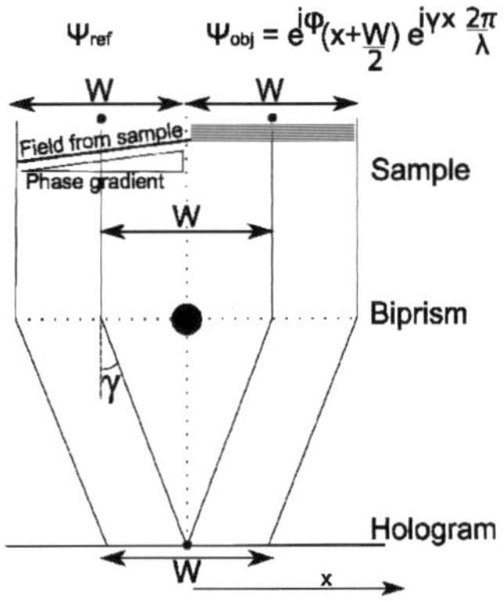

Figure 4. Schematic of the hologram formation. The points in the middle of the interfering waves are superposed in the hologram.

3. Conclusion

We have shown that electron holography can give valuable information on doping profiles in nanostructures [10] and phase gradients in the vacuum could be attributed to the sheet charge at the wire-oxide interface. However, phase variations in the vacuum around the i-n junction were only incidentally observed, and smaller than predicted by simulation. We speculate that this is due to an influence of the electron beam on the charge distribution. For this reason the effect of the electron beam on the charge distribution in the sample should be included in future potential simulations.

References

[1] Voyles P, Muller D. Grazul J, Citrin P and Gossmann H 2002 *Nature* **416**, 826
[2] Hoummada K et al. 2007 *Microelectronic Eng.* **84**, 2517
[3] Zelsacher R et al. 2007 *Microelectronics Reliability* **47**, 1585
[4] Eyben P, Janssens T and Vandervorst W 2005 *Mat. Sci. Eng. B*, **124**, 45
[5] Biberger R et al. 2008 *Microelectronics Reliability* **48**, 13391342
[6] Berthe M et al. 2008 *Science* **319**, 436
[7] Rau W et al. 1999 *Phys. Rev. Lett.* **82**, 2614
[8] Chung J and Rabenberg L 2006 *Appl. Phys. Lett.* **88**, 013106
[9] Björk M, Schmid H, Knoch J, Riel H and Riess W 2009 *Nature Nanotech.* **4**, 103
[10] den Hertog M et al. 2009 *Nanoletters* in print as DOI :*10.1021/nl902024h
[11] Matteucci G et al. 1991 *J. Appl. Phys.* **69**, 1835
[12] Twitchett A, Dunin-Borkowski R and Midgley P 2002 *Phys. Rev. Lett.* **88**, 238302
[13] Fazzini P, Merli P, Pozzi G and Ubaldi F 2005 *Phys. Rev. B* **72**, 085312.
[14] Silvaco International **2007** Atlas user manual. Santa Clara, CA, USA
[15] Yamamotoa K, Tanji T and Hibino M 2000 Ultramicroscopy **85**, 35

Observation of carrier distribution in GaAs semiconductors using phase-shifting electron holography

H Sasaki[1,2], K Yamamoto[2] and T Hirayama[2]

[1] Yokohama R&D Lab, The Furukawa Electric Ltd., 2-4-3 Okano, Nishi-ku, Yokohama, 220-0073, Japan
[2] Japan Fine Ceramics Center, 2-4-1 Mutsuno, Atsuta-ku, Nagoya, 456-8587 Japan

Abstract. Off-axis electron holography was used to map carrier distributions in GaAs semiconductors. After preparing cross-sectional transmission electron microscopy (TEM) specimens by a rotating-microprobe technique and a focused ion beam (FIB) system equipped with a low-voltage Ar ion beam for removing FIB damaged layers, we observed the different carrier concentration in GaAs specimens and evaluated the carrier diffusion at the interface between material with different carrier concentrations.

1. Introduction

The measurement of carrier distributions in semiconductor devices is very important in the development of new devices and for the analysis of device failure. The development of a technique for measuring the carrier distribution across a p-n junction in silicon has resulted in electron holography being widely used to measure the carrier distributions in silicon semiconductors [1–3]. This is generally done using transmission electron microscope (TEM) specimens prepared using focused ion beam (FIB) milling, which enables specific parts to be cut from a bulk sample. A high-voltage Ga ion beam, however, damages the surface layers of TEM specimens of compound semiconductors and thereby complicates the phase distributions measured by electron holography [4]. In this study we therefore used a low-voltage Ar ion beam in a FIB system to prepare GaAs specimens and used off-axis phase-shifting electron holography to map the GaAs carrier distribution with high spatial resolution and high phase-measurement sensitivity.

2. Experimental Datails

2.1. GaAs sample

We examined p-n thin films grown on a GaAs substrate by MOCVD. A structure of a TEM specimen is shown in Fig. 1. Silicon and zinc were used as n-type and p-type dopants, and the samples were designed for testing the sensitivity of electron holography to different carrier concentrations.

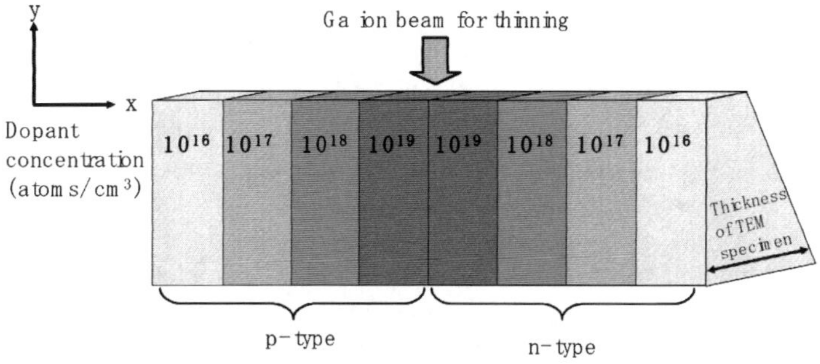

Figure 1. Structure of GaAs TEM specimen with p-type and n-type regions.

Journal of Physics: Conference Series **209** (2010) 012028 doi:10.1088/1742-6596/209/1/012028

2.2. FIB preparation

The procedure for preparing TEM specimens is shown in Fig. 2. The specimen was prepared using a SIINT SMI3050TB (Triple Beam) FIB system, which combined a Ga ion beam, a scanning electron microscope (SEM) for process monitoring, and an Ar ion beam to remove the FIB damaged layers [5]. We used a rotating microprobe so that a TEM specimen whose thickness is uniform along the *x*-direction shown in Fig. 1 could be prepared by irradiating it with Ga ions from the y-direction. First, a carbon layer was deposited on the surface of the sample to protect its surface from the Ga ion beam. As shown in Fig. 2(a), a small cross-sectional specimen with p-n junction was lifted out from the GaAs bulk sample using a microprobe. Figure 2(b) shows a scanning ion microscope (SIM) image of small specimen being held by the microprobe. As shown in Fig 2 (c), the microprobe was rotated by 98 degrees which resulted in changing the side surface to the top surface due to the configuration of microprobe in the FIB system. It could therefore be FIB-milled along the *y*-direction. Figures 2(d) and 2(e) show how the specimen was fixed to the thin Si grid by carbon deposition. The specimen suitable for TEM observation was thinned to 300 nm in the FIB system. Finally, the FIB damaged layer was removed by milling the sample surface for 5 minutes with a 1.0-kV Ar ion beam in the FIB system. Figure 2(f) shows a SEM image of a TEM specimen after Ga ion and Ar ion milling.

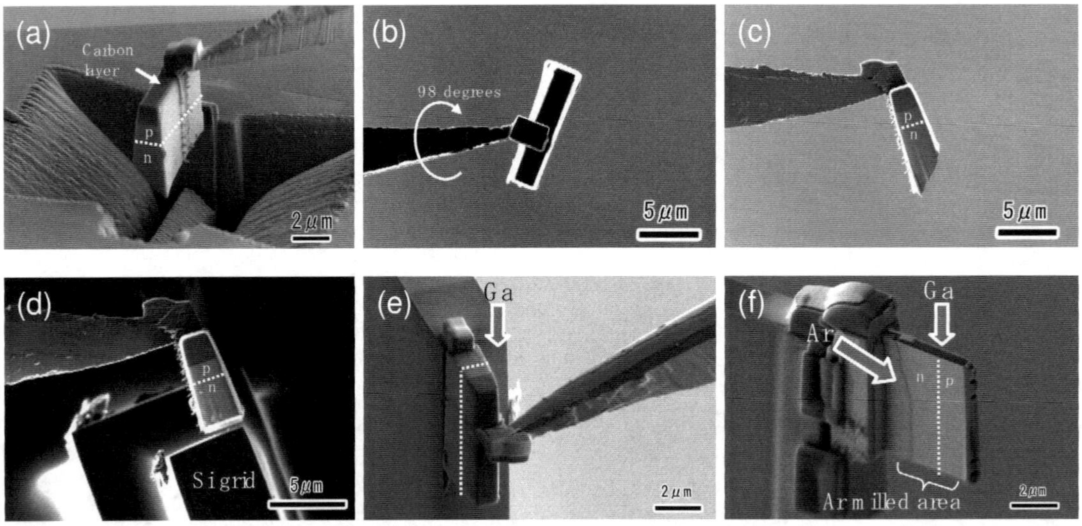

Figure 2. Preparation TEM specimens.a) SEM image of lifted TEM specimen. b) SIM images of specimen held by microprobe, c) microprobe and specimen after rotation, d) specimen close to Si grid. SEM images of e) specimen fixed to Si grid by carbon deposition, f) TEM specimen after Ga ion and Ar ion milling.

2.3. Phase reconstruction by phase-shifting method

Phase-shifting electron holography [6] was used to reconstruct the phase images because it produces a phase map that has higher spatial resolution and phase-measurement sensitivity than a phase map produced by the Fourier transformation method. In the conventional Fourier transformation method, the spatial resolution is determined by the radius of the low-pass filter used to get one of the sidebands and is consequently about three times the fringe spacing. The spatial resolution of an image reconstructed using the phase-shifting method, in contrast, is determined by the pixel size of the electron detector at a low to medium magnification. The spatial resolution in our experiment was approximately 2 nm, and the sensitivity of the phase measurement was estimated to be approximately 0.02 rad [6].

Electron holographic observations were made with a JEOL-3100F TEM operating at 300 kV, and equipped with a Schottky field-emission electron gun and an electron bi-prism. Digital holograms with 1024x1024 pixels were recorded using a GATAN 794 slow-scan CCD camera system. In this experiment, 13 holograms with different initial phases were used for phase reconstruction.

3. Results and Discussions

Figure 3 shows the reconstructed phase image. The p- and n-type regions are clearly seen as areas of dark and bright contrast, and some differences of changing dopant concentrations can also be seen.

The averaged phase profiles across the p-n junction were obtained from the phase image. The line profile of p-type regions is plotted in Fig. 4(a), and the corresponding profile of Zn dopant concentrations measured by secondary ion mass spectrometry (SIMS) is shown in Fig. 4(b). The interface between the 1×10^{19} cm^{-3} region and 1×10^{18} cm^{-3} regions is clearly seen in both profiles, but that between the 1×10^{17} cm^{-3} and 1×10^{16} cm^{-3} regions is not evident, because charges are trapped on specimen surfaces and the low-dopant-concentration regions are almost depleted in TEM specimen [7].

The averaged phase profile of n-type areas is shown Fig. 4(c), and the corresponding profile of Si dopant concentrations is shown in Fig. 4(d). In contrast to what is seen in the profiles for the p-type areas, the 1×10^{19} cm^{-3} and 1×10^{18} cm^{-3} regions are not clearly distinguished in the phase profile. This is attributed to dopant diffusion at this interface (indicated in the SIMS profile). In other words, carrier diffusion at the interface between 1×10^{19} cm^{-3} and 1×10^{18} cm^{-3} regions can be evaluated by these phase profiles. The contrast is almost the same in the regions with dopant concentrations below 1×10^{17} cm^{-3}, and this is also due to carrier depletion in thin TEM specimens.

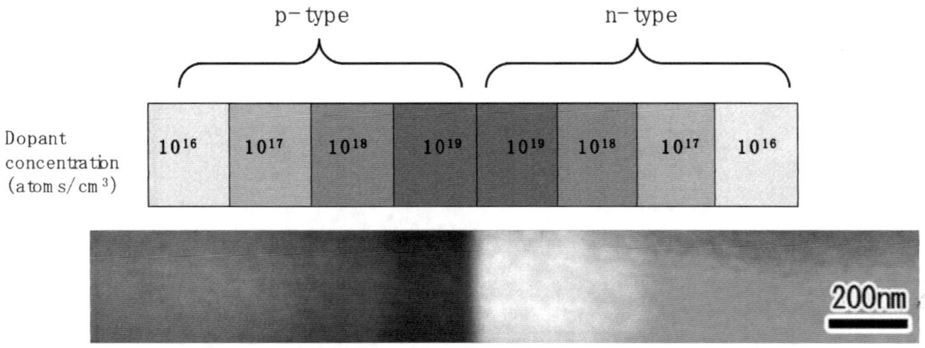

Figure 3. Phase image of GaAs specimen reconstructed by phase-shifting method.

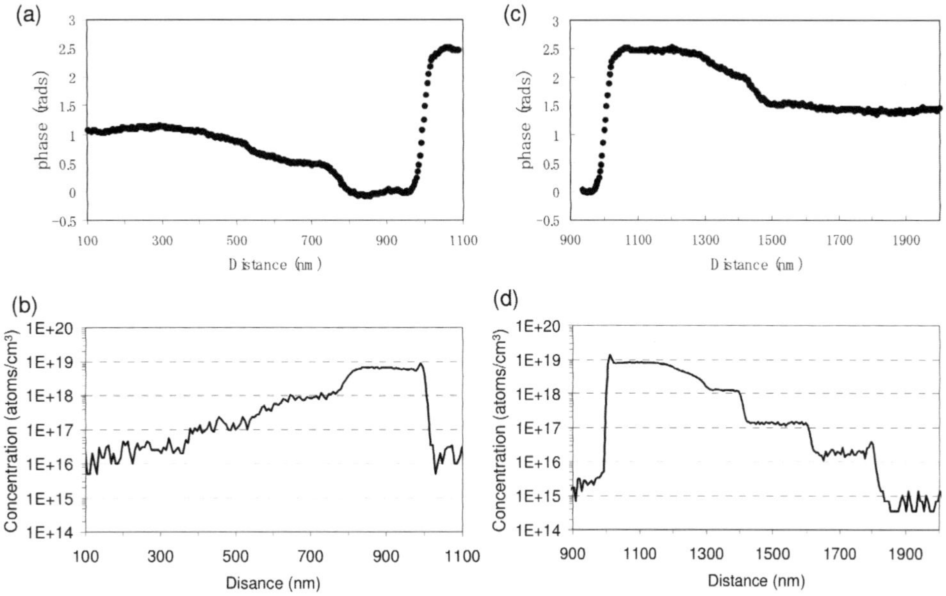

Figure 4.a) Line profile of phase image in p-type region. b) Zn SIMS profile. c) Line profile of phase image in n-type region. d) Si SIMS profile.

4. Conclusion

We prepared TEM specimens by a FIB system equipped with a rotating microprobe and a low-voltage Ar ion beam that makes it easy to remove FIB damaged layers on the specimen surface. We clearly observed the p-n junction and the difference in dopant concentrations between 1.0×10^{19} cm^{-3} and 1.0×10^{18} cm^{-3} regions and could also evaluate the carrier diffusion at the interfaces between regions with different carrier concentrations. This study indicates that these techniques will be highly useful for analyzing compound semiconductor devices.

References

[1] Rau WD, Schwander P, Baumann FH, Hoppner W and Ourmazd A 1999 *Phys. Rev. Lett.* **82**, 2614
[2] Wang Z, Hirayama T, Sasaki K, Saka H and Kato N 2002 *Appl. Phys. Lett.* **80**, 246
[3] Twitchett AC, Dunin-Borkowski RE and Midgley PA 2002 *Phys. Rev. Lett.* **88**, 238
[4] Cooper D, Twitchett AC, Somodi PK, Midgley PA, Dunin-Borkowski RE, Farrer I and Ritchie DA 2006 *Appl. Phys. Lett.* **88**, 063510
[5] Takahashi H, Sugiyama Y, Yamamoto Y, Sadayama S, Iwasaki K and Fujii T 2006 *Proc.16th Int Microscopy Cong* (Sapporo: Sapporo Convention Center) p 1104
[6] Yamamoto K, Kawajiri I, Tanji T, Hibino M and Hirayama T 2000 *J. Electron Microsc.* **49**, 31
[7] Sasaki H, Yamamoto K, Hirayama T, Ootomo S, Matsuda T, Iwase F, Nakasaki R and Ishii H 2006 *Appl. Phys. Lett.* **89**, 244101

Comparison of experimental and theoretical X-ray intensities from (In)GaAs specimens investigated by energy-dispersive X-ray spectroscopy in a transmission electron microscope

T Walther

Department of Electronic & Electrical Engineering, University of Sheffield, Sir Frederick Mappin Building, Mappin Street, Sheffield S1 3JD, UK

E-mail: t.walther@sheffield.ac.uk

Abstract. Experimental measurements of X-ray line intensity ratios in a transmission electron microscope are compared over several orders of magnitude of sample thicknesses, from the nm- to the mm- range, with Monte-Carlo simulations using two different software packages. It is shown that the form of the thickness dependence of the K/L ratio of characteristic X-ray lines for GaAs is reproduced qualitatively, but the numerical differences between software packages are large. A scheme is presented for improving the simple *k*-factor method, taking explicitly into account the thickness dependence that remains even after application of the usual absorption and fluorescence corrections. This is done in first-order approximation by linear regression. The improvement in determining the correct indium concentration in specimens of InGaAs is calculated to be 1at%.

1. Introduction

Microanalysis by energy-dispersive X-ray spectroscopy (EDXS) in a transmission electron microscope (TEM) offers much higher spatial resolution has than in a scanning electron microscope (SEM) but suffers from poorer counting statistics. The simplest approach for quantification is the use of *k*-factors, i.e. experimentally determined sensitivity factors [1]. More accurate quantification of elemental concentrations depends on the knowledge of the relative ionisation cross-sections of the corresponding elements (*Z*-effect), modelled already in 1930 by Bethe [2] and later refined by Duncumb and Reed [3], and proper corrections for absorption (*A*) [4] and for fluorescence (*F*) [5], which can be combined for a given specimen thickness and density in the so-called *ZAF*-correction. For SEM studies of bulk samples at moderate voltages full self-consistent spectrum modelling has become available which allows the user to model X-ray generation as a function of the product of mass and thickness and then integrate the result over the specimen thickness, e.g. with the *PROZA96* [6] program. For TEM, however, these approaches have not been implemented so far, as sample thicknesses are often not well known, while in SEM they can simply be set to ∞ for bulk. This study shows how the conventional *ZAF* approach for EDXS in a TEM can be improved for a given material system using re-calibration based on linear least-squares fitting data recorded over a wide range of sample thicknesses.

2. Comparison of Monte Carlo Simulations to Experiments

Monte Carlo simulations of electron scattering and X-ray generation as a function of foil thickness have been carried out using two different software codes. The freely available software code CASINO (version 2.4.2) [7], with default settings as described in the user manual, using tabulated values of the Mott scattering cross-section [8] and new calculations of the stopping power at low energies with a cut-off at 50eV [9] is a fast program to calculate L- and K-line intensities in non-relativistic approximation and is widely used for EDXS in a scanning electron microscope at voltages up to 30kV. The commercial HURRICANE code uses relativistic cross-sections as in [10], allows more complex geometries to be simulated and M-, L- and K-line intensities to be modelled, but it is much slower.

In order to study the effect of the depth at which a certain feature of interest is located within a medium-thick specimen (top-bottom effect) a model specimen for simulations was created that consisted of 2nm thin layers of a given element embedded in 200nm GaAs at a certain depth, *d*, with 1nm of carbon coating at the each surface. The acceleration voltage of the electron beam was set to U=200kV. The X-ray detector was assumed to form a take-off angle with the specimen of θ=90° (vertical take-off). Figures 1 and 2 show the decrease of the analytical signals for various X-ray lines, due to the increasing absorption effect as a function of depth in the GaAs specimen. While the curves obtained from the CASINO simulations can be fitted by simple exponentials [11], the corresponding curves calculated with the HURRICANE software are much noisier and reveal different trends, as can be seen from Figure 2. While soft X-rays such as N_K also decay exponentially, harder X-rays such as the Al_K do not attenuate appreciably or even increase in intensity with embedded depth. The latter trend is seen for the In_L line and probably due to the increased stopping power of electrons that have sufficiently slowed down when they approach the bottom half of the GaAs crystal.

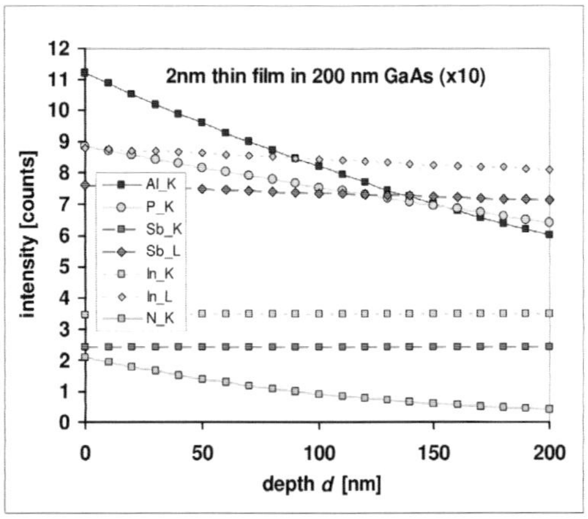

Figure 1. CASINO simulation of depth dependence of X-rays detected from a 2nm thin film embedded in 200nm thick GaAs with 1nm C on each surface. Voltage: U=200kV, take-off angle: θ=90°. From [11]

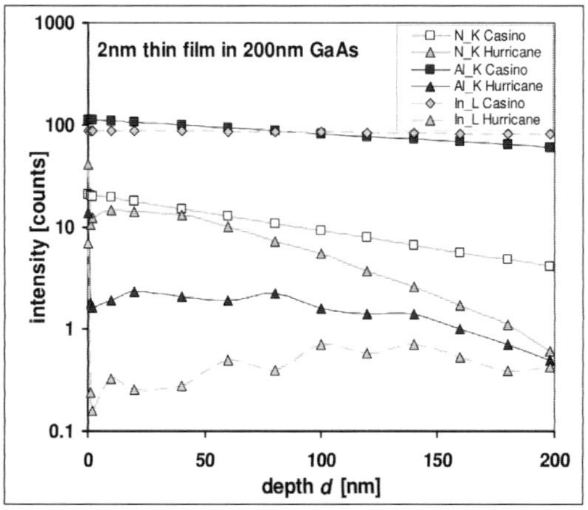

Figure 2. Comparison of CASINO and HURRICANE simulations of depth dependence of X-ray signals from 2nm thin layers of N, Al or In within 200nm thick GaAs (note logarithmic plot).U=200kV, θ=90°

If it is to be decided which software models electron scattering and X-ray production at 200kV more realistically, a comparison with experimental data has to be performed. Figure 3 compares the As_K/As_L ratio of GaAs modelled using both software codes, to experimental measurements using a JEOL 2010F field-emission TEM operated at U=197kV equipped with a Si:Li detector with ultrathin window and Oxford Instruments LINK ISIS300 software (rev. 3.2). The take-off angle was θ=20°. An aperture to suppress hard stray X-rays was used. The GaAs specimen was a wedge cleaved on {110} planes and mounted on a copper grid such that it could be imaged near (~4° off) a [001] zone axis so that, due to the geometry of the specimen, the sample thickness was always given as twice the distance from the specimen edge observed in scanning TEM mode. No 'hole count' correction was applied as the in-hole signal was very weak anyway. Secondary fluorescence by the copper support was taken into account by explicitly including the Cu lines into the quantification. Both software packages predict qualitatively similar curves, with a transition from a plateau at small thicknesses (hardly any absorption of both lines) towards another plateau at much higher values (where both lines are equally absorbed), which occurs at a specimen thickness of ~1μm where the absorption of the softer L-line becomes relevant. However, the ratio produced by the (non-relativistic) CASINO code is ~30 times higher than that by the HURRICANE code. Even if the difference in detector efficiency for both lines has not been taken into account here, this can be estimated from [12] to only account for a factor of ~0.8, so that the CASINO simulation looks more realistic than the HURRICANE simulation.

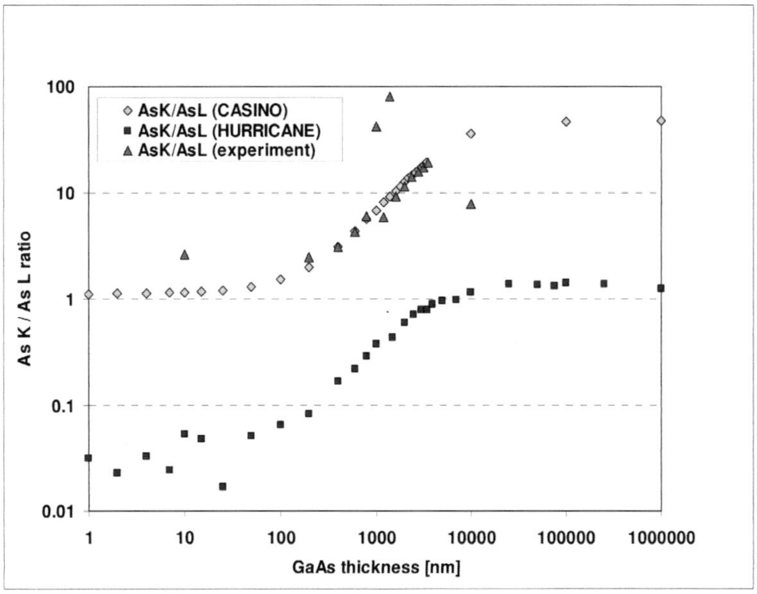

Figure 3. Comparison of As_K/As_L X-ray ratio in GaAs vs. thickness from experiments (red triangles) with simulations using CASINO (green diamonds) or HURRICANE code (blue squares).

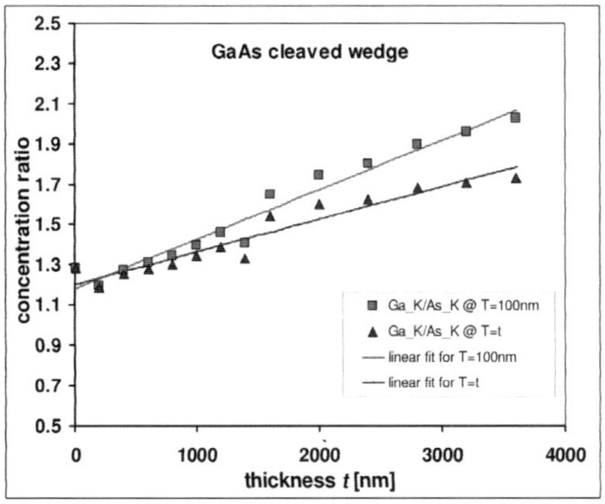

Figure 4. Experimental result of measuring the Ga_K/As_K ratio for cleaved GaAs using the LINK ISIS300 software, assuming a thickness of t=100nm throughout (red squares) or the true thickness (blue triangles).

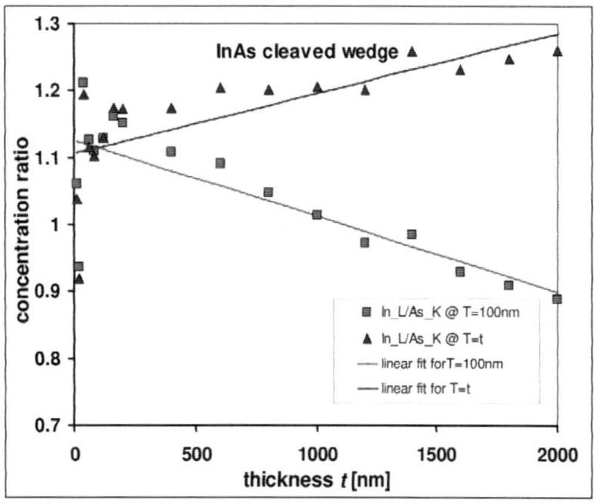

Figure 5. Experimental result of measuring the In_L/As_K ratio for cleaved InAs using the LINK ISIS300 software, assuming either a thickness of t=100nm throughout (red squares) or the true thickness (blue triangles).

Figures 4 and 5 show experimental concentration ratios of Ga/As for GaAs using both K-lines (Fig. 4) and In/As for InAs using the In L- and the As K-line (Fig. 5). The ratios were again determined for cleaved wedges. For the square data points in red a universal thickness of *T*=100nm was used for quantification using the LINK ISIS300 software, as could be used for cases where the true thickness is finite but not precisely known, while for the triangular data points in blue the true thickness was used for each measurement. Had the *k*-factors of the elements been correct, then the extrapolated curves should have intersected the vertical axes for zero thickness (*t*=0) at unity to reproduce the stoichiometry of the compounds. The fact the values extrapolated for *t*=0 are both too large (Ga_K/As_K=1.19 for GaAs and In_L/As_K=1.12 for InAs) confirms the corresponding *k*-factors for thin film approximation are incorrect by more than 10%. Also, had the correction for absorption and fluorescence worked properly, then the curves for the true crystal thicknesses should have produced identical and constant values, independent of thickness. The curves measured taken the thickness effects into account, however, clearly have remaining slopes and thus demonstrate that the absorption and fluorescence corrections by the LINK ISIS software are insufficient. For GaAs it reduces the slope marginally, while for InAs it overcompensates the negative slope, yielding a curve of positive slope.

Table 1. Result of linear regression fitting to the data shown in Figures 4 and 5

material	GaAs		InAs	
thickness used for fit	*T*=100nm	*T*=*t*	*T*=100nm	*T*=*t*
lin. correlation coefficient, *R*	+0.9798	+0.9697	−0.7989	+0.7136
vertical axis intercept	1.182±0.027	1.202±0.025	1.125±0.021	1.106±0.022
slope of fit [10^{-6}/nm]	245±14	162±14	−113±22	89±23

As a consequence, if the specimen thickness is roughly known, say to within ±100nm, the curves of Figures 4 and 5 allow the user to obtain the correct Ga/As and In/As ratios to 2% relative error, as given by the uncertainties in the intercepts and a small contribution by the slopes in Figures 4 and 5. For a ternary compound of $In_xGa_{1-x}As$ of unknown chemical composition *x* but known thickness *t* the indium concentration in the group-III sub-lattice can be determined to better than ±1at% (Δx=±0.01).

3. Summary

It has been shown that absorption/fluorescence corrections applying commercial software to experimental energy-dispersive X-ray microanalysis spectra can have relative errors larger than 10%. This can be improved by making systematic measurements of relative X-ray intensities as a function of known sample thickness and fitting the remaining thickness dependence to first order. For the $In_xGa_{1-x}As$ system this means the indium concentration thus calculated is accurate to better than 1at%.

References

[1] Goldstein JI, Costley JL, Lorimer GW and Reed SJB 1977 Quantitative X-ray Analysis in the Electron Microscope *Scanning Electr. Microsc.: Proc. Workshop Anal. Electron Microsc. (IIT Research Institute, Chicago, IL, USA, March 1977,* vol 1*)* ed O Johari, p 315
[2] Bethe HA 1930 *Ann. Phys.* **397**, 325
[3] Duncumb P and Reed SJB 1968 Quantitative Electron Probe Microanalysis *NBS Spec. Publ.* (vol 298) ed KFJ Heinrich, p 133
[4] Philibert J 1963 *Proc. Int. Symp. X-ray Optics and X-ray Microanalysis (Academic Press, New York, USA)* ed HH Pattee, VE Cosslett and E Engström, p 379
[5] Reed SJB 1965 *Brit. J. Appl. Phys.* **16**, 13
[6] Bastin GF, Dijkstra JM and Heijligers HJM 1998 *X-ray Spectrom.* **27**, 3
[7] Hovington P, Drouin D and Gauvin R 1997 *Scanning* **19**, 1
[8] Drouin D, Hovington P and Gauvin R 1997 *Scanning* **19**, 20
[9] Hovington P, Drouin D, Gauvin R, Joy DC and Evans N1997 *Scanning* **19**, 29
[10] Salvat F and Mayol R 1993 *Computer Physics Comm.* **74**, 358
[11] Walther T 2008 *J. Phys.: Conf. Ser.* **126**, 012090
[12] Williams DB and Carter CB 1996 Transmission Electron Microscopy *(Springer, New York)*, p 561

Theoretical and experimental factors affecting measurements of semiconductor mean inner potentials

RS Pennington[1], JJ Mortensen[2], T Kasama[1], CB Boothroyd[1] and RE Dunin-Borkowski[1]

[1] Center for Electron Nanoscopy, Technical University of Denmark, DK-2800 Kongens Lyngby, Denmark
[2] Center for Atomic-Scale Materials Design, Technical University of Denmark, DK-2800 Kongens Lyngby, Denmark

E-mail: robert.pennington@cen.dtu.dk

Abstract. We use density functional theory to explore the effect on calculations of semiconductor mean inner potentials of the presence of reconstructions, changes in lattice spacing and adsorbates on the surfaces of parallel-sided thin specimens. We also use electron holography to illustrate several factors that affect experimental measurements of mean inner potentials of semiconductor nanowires.

1. Introduction

Quantitative electron holography of semiconductors is of interest for measuring dopant distributions, specimen thickness profiles, compositions, and charge densities at interfaces and surfaces. For many applications, the mean inner potential V_0 of the semiconductor should be known accurately. However, measurements of V_0 are sensitive to many factors, including specimen charging, bonding, dynamical diffraction and the nature of the specimen surface [1]. In addition, the specimen orientation and thickness must be known accurately. Here, we assess some of the factors that affect predictions and measurements of V_0 by performing calculations of potentials in semiconductors with adsorbates and surface reconstructions and by using electron holography to record phase images of GaAs nanowires.

2. Calculation details and results

V_0 was calculated for selected semiconductors using the projector-augmented wavefunction approach within the density functional theory (DFT) formalism, implemented as GPAW [2]. The calculation volume contained the material of interest and a region with no material, as shown in Fig. 1 and also used by Kruse et al. [3]. The Hartree potential V_H was determined from the valence electrons, screened core electrons and nuclei for region B marked in Fig. 1 relative to that in region A. As will be detailed elsewhere, in order to take into account the unscreened core charges, a second potential V_{EC} was calculated by solving Poisson's equation for the unscreened charge according to the expression [2]

$$\rho(\mathbf{r}) = \tilde{\rho}(\mathbf{r}) + \sum_a \Delta\tilde{\rho}^a(\mathbf{r} - \mathbf{R}^a),$$

after which

$$V_0 = V_H + V_{EC}.$$

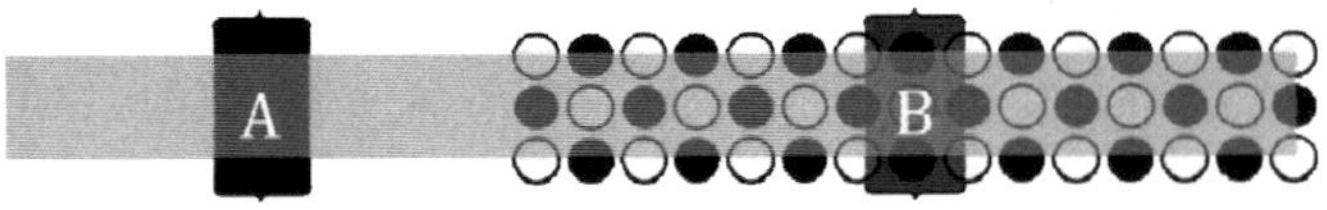

Fig. 1. Calculation geometry used in this work, with the material of interest and a region of vacuum.

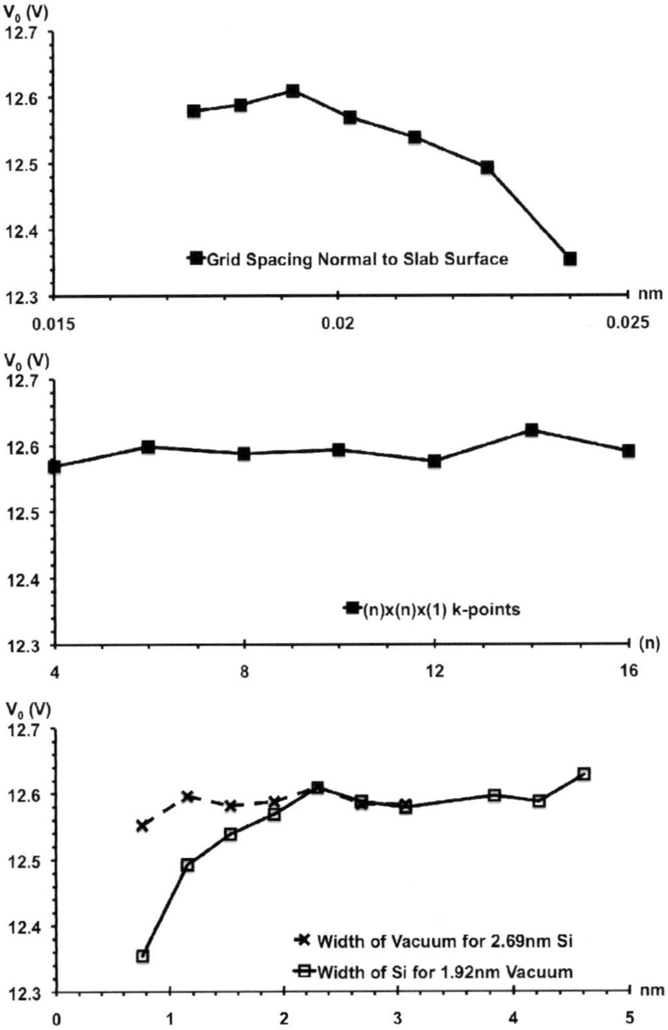

Fig. 2. Verification of parameter stability for calculations of V_0 for parallel-sided Si specimens with (110) surfaces. Grid spacings of $h_x = 0.0226$ nm and $h_y = 0.0192$ nm were used in the [001] and [1-10] directions parallel to the surface of the slab.

The sensitivity of V_0 to a range of input parameters is shown in Fig. 2 for parallel-sided Si specimens with (110) surfaces. As specimens containing only a few monolayers of material have decreased values of V_0, 2.69 nm of Si and 1.92 nm of vacuum were typically used in subsequent calculations for (110) surfaces. It was found that, in order to ensure stability, the atomic lattice of the crystal should be located on grid points in the direction normal to the surface of the slab. Unless otherwise stated, the calculation parameters used were $8 \times 8 \times 1$ k-points, 0.0192 nm per real-space grid point in the direction normal to the surface of the slab, and the PBE exchange-correlation functional [4], parallelized over 8 processor cores using domain decomposition.

2.1. Surface orientation and reconstruction

Table 1 shows the results of calculations of V_0 for parallel-sided Si slabs with different crystallographic orientations, with atom positions that are unchanged from those in a bulk Si crystal, and with the parameters optimized for each simulation. The difference between the calculated values for (110), (111) and (211) surfaces is below 0.05 V. However, the value for a (100) surface is 0.32 V above that for a (110) surface. Most significantly, the introduction of a (2×1) reconstruction on the (100) surface, with atom positions included from Ref. [5], changes V_0 by almost 0.6 V when compared to the unreconstructed (100) surface.

2.2. Adsorbates

Table 2 shows the results of calculations of V_0 for parallel-sided Si slabs with (110) surface orientations that have one atomic layer of either Si, C or Ge on their surfaces, both with the surface atoms located at positions that Si atoms would occupy on a perfect bulk Si lattice and then with the positions of either all of the atoms in the calculation (for a Si surface) or just the surface atoms (for C and Ge surfaces) relaxed until the force on the atoms was below 0.5 eV / nm. Although the arrangements of atoms in the surface layer may not reflect the positions of adsorbed atoms that would occur experimentally, the calculations illustrate the sensitivity of V_0 to the identities and positions of atoms in the surface layers, with a single relaxed adsorbed layer of C changing the calculated value of V_0 by more than 0.5 V.

2.3. Different semiconductors

Table 3 shows a comparison of DFT calculations of V_0 for different semiconductors with values that would be obtained from neutral atom electron scattering factors [6]. The fact that the present DFT values are close to those in Ref. [3] (also reproduced in Table 3) provides confidence in the applicability of the GPAW simulator for calculations of V_0 for a wide range of materials.

Table 1. V_0 for Si calculated as a function of surface orientation for optimized parameters.

Silicon surface	V_0 (V)
(110)	12.59
(111)	12.63
(211)	12.64
(100)	12.91
(100) (2×1) from [5].	12.32

Table 2. V_0 for Si calculated as a function of surface composition and relaxation.

Surface	V_0 (V)
Si (all-relaxed)	12.52
Si(C)	14.12
Si (C-relaxed)	12.01
Si (Ge)	12.46
Si (Ge-relaxed)	12.53

Table 3. V_0 calculated for different semiconductors with (110) surfaces using the present density functional theory (DFT) calculations, compared with values reproduced from the DFT calculations of Ref. [3] and obtained using neutral atom electron scattering factors (V_{fel}) taken from Ref. [6].

Material	V_0 (V) $_{DFT}$	V_0 (V) $_{Ref. [3]}$	V_0 (V) $_{fel}$
Si	12.59	12.57	13.76
Ge	14.69	14.67	15.60
AlP	11.40	11.39	13.42
AlAs	12.29	12.34	13.97
GaP	13.57	13.63	14.60
GaAs	14.15	14.19	15.38
InP	13.85	13.90	15.22
InAs	14.13	14.34	15.47
CdS	13.02	-	13.87

3. Electron holography results

Semiconductor nanowires are, in many ways, ideally suited for experimental measurements of V_0 as their orientations and cross-sections can be determined accurately and minimal specimen preparation is required. Fig. 3 shows part of a representative electron hologram of a GaAs nanowire that has a diameter of 160 nm, acquired at 300 kV using an FEI Titan 80-300 TEM operated in Lorentz mode. The figure also shows experimental phase profiles acquired from a wire oriented at the [110] zone axis and rotated about its long axis to several orientations. The phase profiles illustrate four of the difficulties of measuring V_0 experimentally, even from nanowires: the sensitivity of the line profiles to crystallographic orientation as a result of both changes in projected thickness and dynamical contributions to the phase shift, the difficulty of removing phase wraps at the edges of the wire where the sample thickness changes rapidly, the possibility of charging of the wire as a result of secondary electron emission during electron irradiation, resulting in a ramp in vacuum outside the wire for the 19.1° phase profile, and the unknown state of the nanowire surface.

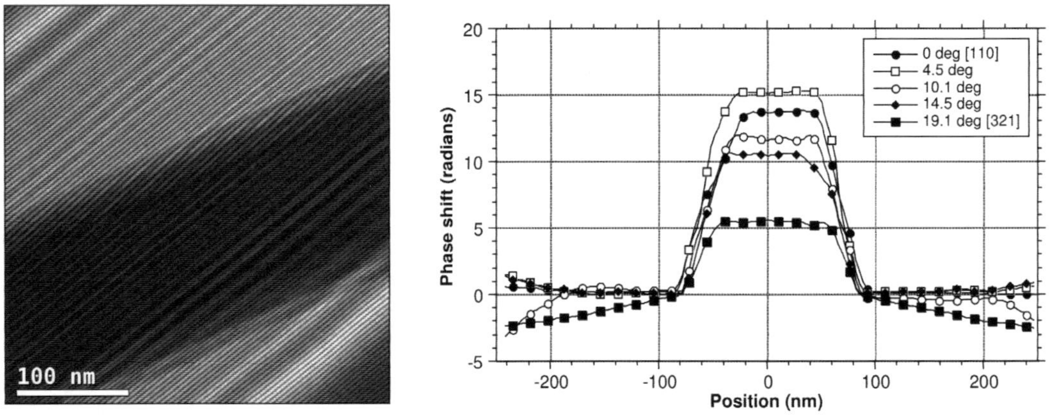

Fig. 3. Electron hologram and phase profiles for a GaAs nanowire examined at different orientations.

4. Conclusions

We have obtained preliminary density functional theory simulations and experimental measurements that illustrate several of the factors that may affect measurements of semiconductor mean inner potentials. The calculations illustrate the sensitivity of the mean inner potential to the presence of adsorbed or reconstructed layers on the specimen surface. The electron holography results illustrate several additional practical complications associated with experimental measurements of mean inner potentials, even for the relatively simple GaAs nanowire system.

Janusz Sadowski of Lund University is thanked for provision of the GaAs nanowire samples.

References

[1] Völkl E, Allard LF and Joy D (Eds) 1999 *Introduction to Electron Holography* (Springer)
[2] Mortensen JJ, Hansen LB and Jacobsen KW 2005 *Phys. Rev. B* **71**, 035109
[3] Kruse P, Schowalter, Lamoen D, Rosenauer A and Gerthsen D 2006 *Ultramicroscopy* **106**, 105
[4] Perdew JP, Burke K and Ernzerhof M 1996 *Phys. Rev. Lett.* **77**, 3865
[5] Over H, Wasserfall J, Ranke W, Ambiatello C, Sawitzki R, Moritz W1997 *Phys. Rev. B* **55,** 4731
[6] Rez D, Rez P and Grant I 1994 *Acta Cryst. A* **50**, 481

Advantages of low beam energies in a TEM for valence EELS

M Stöger-Pollach[1] and P Pongratz[2]

[1]University Service Centre for TEM, Vienna University of Technology,
Wiedner Hauptstr. 8-10/052 A-1040 Vienna, AUSTRIA
[2]Institut für Festkörperphysik, Vienna University of Technology,
Wiedner Hauptstr. 8-10/138 A-1040 Vienna, AUSTRIA

Email: stoeger@ustem.tuwien.ac.at

Abstract. Since the availability of monochromators in transmission electron microscopes (TEMs), electron energy loss spectrometry (EELS) is widely used to determine band gaps and the dielectric properties of semiconductors on a nano-metre scale. Nevertheless, three physical effects hamper straightforward analysis: (a) relativistic energy losses, (b) the delocalization of the energy loss which is in the 10 nano-metreer range for valence losses, and (c) the presence of interface plasmons. When reducing the operation voltage of the TEM one can kill two birds with one stone: (a) the relativistic losses will disappear as soon as $v_e<c_0/n$ (with v_e as the speed of the electron, c_0 as the vacuum speed of light and n as the refractive index of the investigated sample) and (b) the delocalization will decrease, because it also depends on the energy of the incident electron probe. The determination of the optical properties of quantum structures is discussed in the case of GaP/GaAs interface at 200 keV and 20 keV beam energy, respectively. Further, the influence of the delocalization of the energy loss signal is discussed theoretically and experimentally.

1. Introduction

Using low beam energies in transmission electron microscopy (TEM) has a long tradition in biological samples. This has a few reasons: i) beam damage is reduced because the energy of the probe electron is reduced – and usually biological samples are very beam sensitive –, ii) due to the fact that biological samples have a large Carbon content, the mean free path for inelastic scattering is very low, this means that even samples of a thickness in the range of 70-100 nm can be easily trespassed by electrons accelerated with 80 kV and iii) when the incident electron energy is reduced and the mean free path for inelastically scattered electrons is decreased, the contrast in the image is enhanced. Beside advantages like reducing beam damage and enhancing contrast also two big disadvantages are coming up, when reducing the beam energy. First, the resolution of the microscope is reduced. This is shown in Fig. 1, where the contrast transfer function (CTF) of a TECNAI G20 microscope equipped with a super twin lens and a LaB_6 electron source is plotted. In all cases the extended Scherzer defocus was selected. For biological samples the CTF seems to be still good enough, but in materials science efforts need to be done in order to bring it back to the high resolution regime. First successful results were presented on the European Microscopy Conference 2008 (EMC 2008), where a C_s corrected high resolution images recorded at 30 kV was shown [1]. The second disadvantage is the decrease of brightness when reducing the operation voltage of the TEM. Little brightness automatically means that image recording times increase and drift and other instabilities also need to be considered. Again a way out of this dilemma was shown on the EMC 2008: a high brightness source [2]. As a consequence next generation TEMs will be able to compensate all disadvantages of lowering the high tension and give back the operator the full power of a modern transmission electron microscope.

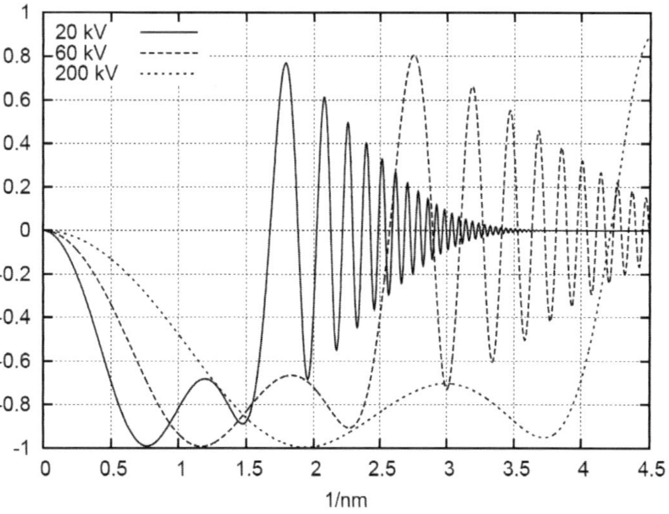

Fig. 1. Contrast transfer function of a TECNAI microscope operated at 20, 60 and 200 kV.

beam energy	energy spread	ext. Scherzer defocus
20 kV	0.35 eV	-124.3 nm
60 kV	0.7 eV	-93.6 nm
200 kV	1.1 eV	-67.2 nm

Table 1. Instrumental conditions for the TECNAI G20 ST. Further: C_s = 1.2 mm, C_c = 1.1 mm, d*V*/*V* = 2ppm, d*I*/*I* = 5ppm, convergence angle = 0.1 mrad

Although such next generation machines are not available commercially at the moment, this work was done on a conventional TEM and shows that lowering the beam energy has also positive effects on valence energy loss spectrometry (VEELS). One effect has an instrumental reason: the improvement of the energy resolution. Another one was first described by N. Bohr [3] in 1913: the decrease of the delocalization of the energy loss. A third advantage is the prevention of the excitation of relativistic energy losses in semiconductors as described elsewhere [4]. Due to the fact that the third positive aspect was described in detail in [4] this work is focused on the improvement of the spectral energy resolution and on the shrinking of the delocalization.

2. Improving the spectral resolution

Conventionally, the spectral resolution is improved by using a monochromator in the condenser system of a field emission microscope (FEG-TEM). Due to the construction of conventional gun systems used in FEG-TEMs a reduction of the high tension below 60 kV reduces the intensity of the beam tremendously. With the construction of the new X-FEG, this problem is overcome.

On the other hand, the construction of a LaB_6 operated electron gun allows reducing the high tension down to 20 kV without losing too much intensity. The loss of intensity can be compensated by the Wehnelt bias. In contrast to a FEG-TEM, using a monochromator on a LaB_6-TEM is not useful, because a LaB_6 gun has much less brightness which would then be further reduced. An improvement of the energy resolution in the EELS spectrum can therefore only be obtained by a reduction of the filament temperature and an improvement of the point spread function (PSF) of the detector. Unfortunately, a reduction of the filament temperature is counterproductive, because brightness is reduced too. This means that all the improvement of the energy resolution in the loss spectrum is reached by the improvement of the PSF. Figure 2 shows the measured PSF at 200 kV and 20 kV of our GIF 2001 system. For its determination a part of

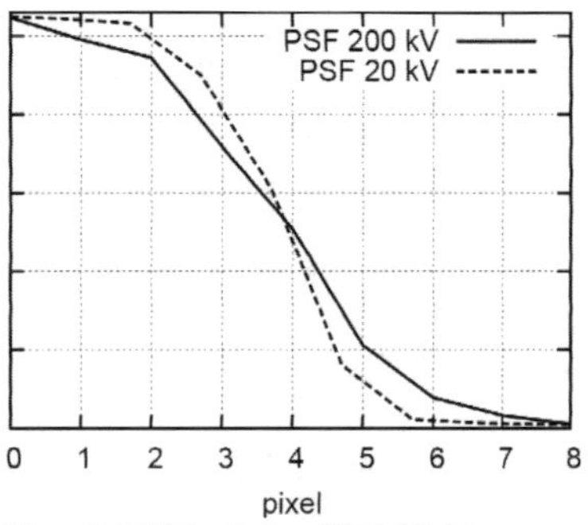

Fig. 2. PSF of the GIF 2001 at 200 keV and 20 keV beam energy.

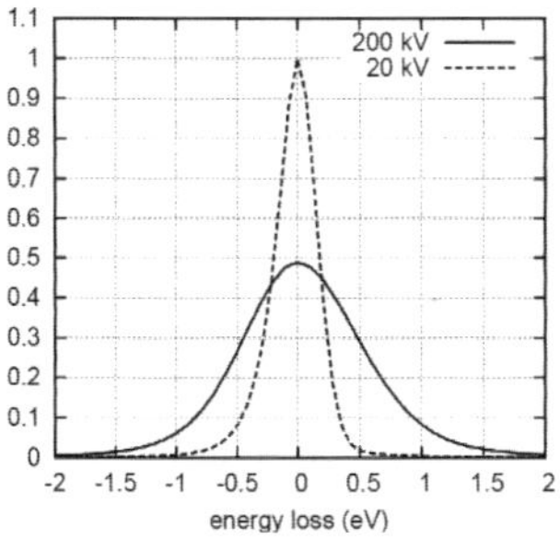

Fig. 3. Zero loss peaks recorded with 200 keV and 20 keV beam energy.

the CCD was blocked in order to have a knife-edge in the energy dispersive direction. Then the CCD was illuminated homogeneously and the PSF was measured. The positive effect of the improved PSF on the energy resolution can be directly measured when the full width at half maximum (FWHM) of the zero loss peak (ZLP) is measured. Figure 3 shows the ZLPs recorded at 200 kV and 20 kV, respectively. The FWHM of the ZLP is reduced to 0.35 eV instead of 1.1 eV when the TEM is operated at 200 kV.

3. Delocalization in EELS at low beam energies

Besides an improvement of the energy resolution in the EELS spectrum, a reduction of the delocalization of the energy loss is induced when the beam energy is decreased. Figure 4 shows the reduction of delocalization at low beam energies in comparison to 100 keV and 200 keV electron beams using once Bohr's cut-off criterion [3] and once the Rayleigh criterion [5]. The Rayleigh criterion seems to be very crude, because it does not include Coulomb interaction. The resulting values are about five times larger than those obtained by using Bohr's cut-off criterion. Common for both models is that the delocalization is reduced by a factor of approximately 3 when using 20 keV instead of 200 keV electrons. Comparing Bohr's cut-off criterion with a quantum mechanical model [6] shows an excellent agreement.

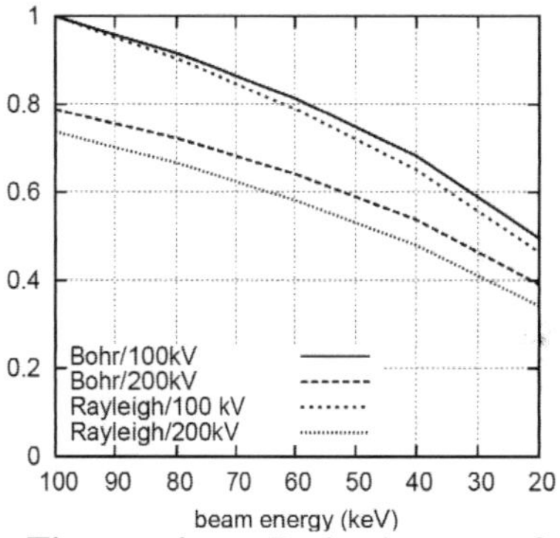

Fig. 4. Reduction of delocalization compared to 100 keV and 200 keV electrons for an energy loss of 16.4 eV.

4. Experimental results

For the experimental determination of the optical properties of GaP and GaAs, respectively, a valence EELS (VEELS) spectrum image was recorded across the GaP/GaAs interface. This has the advantage, that the phenomena of i) delocalization and ii) interface plasmon excitation can be studied simultaneously.

4.1. Delocalization

For the VEELS experiments an FEI TECNAI G20 equipped with a GIF 2001 system was operated at 200 kV and 20 kV. The GaP/GaAs interface was rotated parallel to the energy dispersive axis of the spectrometer using a double tilt rotation holder allowing recording a ΔE-$\boldsymbol{r}$ pattern (Figure 5, left side).

On the right hand-side of Figure 5 it can clearly be seen that for the 200 kV experiment the delocalization is in the range of 10 nm as given by Bohr's cut-off criterion. Although the signal is much more noisy for the 20 kV experiment, the delocalization can be estimated to be roughly 4 nm, which is also in good agreement with calculations. This experimental test has a fundamental

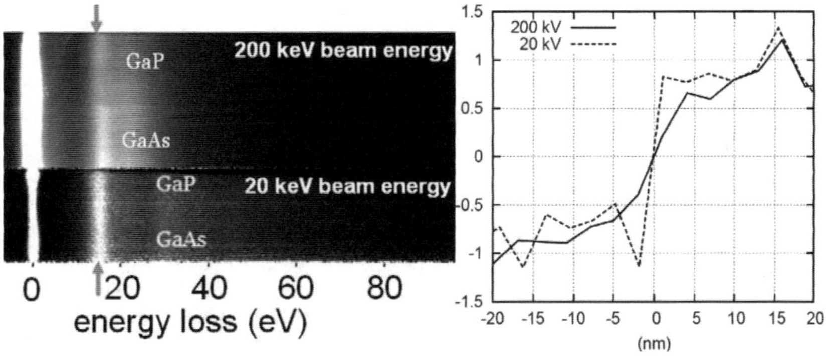

Fig. 5. left: Spectrum image of the GaP/GaAs interface at 200 kV and 20 kV at 16.4 eV energy loss (see arrow). right: Normalized intensity profiles at the GaP plasmon maximum for the determination of the delocalization of the energy loss.

importance for the determination of the optical properties of non-isolated quantum structures: using 200 keV electrons, a simple Kramers-Kronig Analysis of VEELS experiments is failing. Besides the influence of relativistic effects [5], delocalization also hampers VEELS analysis of nano-objects using highly energetic electrons.

4.2. Interface plasmons

Besides delocalization and relativistic energy losses a third feature is altering the VEELS signal: the interface plasmon. Due to the fact that interface plasmons (IPs) are at lower energies than volume plasmons, the delocalization of these collective oscillations is even stronger. Figure 6 (left) shows the calculated position of the IP at the GaAs/GaP interface. In the non-relativistic formalism for a flat interface between materials A and B, with the complex dielectric functions ε_A and ε_B, the contribution of the IP is proportional to $\mathrm{Im}[(\varepsilon_A - \varepsilon_B)/(\varepsilon_A + \varepsilon_B)]$. Strong IPs are expected when $\mathrm{Re}(\varepsilon_A + \varepsilon_B) = 0$, and may produce loss signals below the band gap energy. For the present interface the maximum of the IP is found to be 3.4 eV (Figure 6, left hand-side). Bohr's cut-off criterion gives a delocalization of this energy loss of 55 nm for 200 keV and still 20 nm for 20 keV probe electrons. This is shown in Figure 6 on the right hand-side.

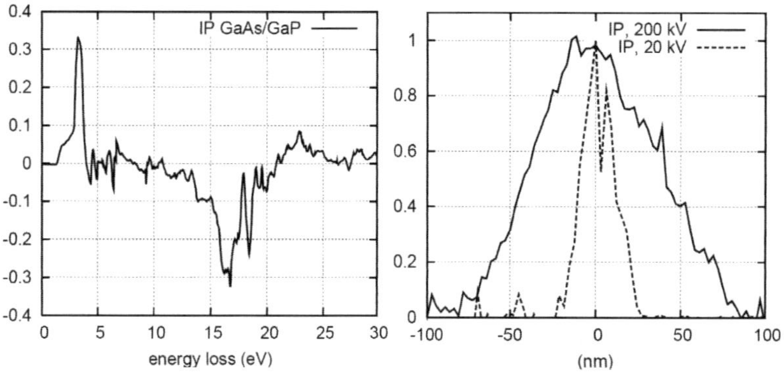

Fig. 6. left: Energy loss contribution to the IP of the GaAs/GaP interface. The negative values at the volume plasmon position are due to the *begrenzungs-effekt*.
right: Intensity profile of the spectrum images at 3.4±0.3 eV showing the delocalization of the IP.

5. Conclusion

The recent experiment demonstrated that the determination of band gaps and dielectric properties of buried quantum structures is still erroneous when using TEM-EELS applications. Although the delocalization can be decreased by decreasing the beam energy and also relativistic effect can be avoided completely, the appearance of IPs hampers straightforward analyses. Even in the case of an interface between two materials with similar dielectric functions – as the GaAs/GaP interface –, IPs can be detected.

Acknowledgements

The Hochschuljubiläumsstiftung der Stadt Wien, contract Nbr. H-01585/2007, is kindly acknowledged for funding.

References

[1] Freitag B 2008 EMC Technical lecture from FEI company: *Where are we on abberation correction and where do we go next?*
[2] Freitag B et al. 2008 *Proc. EMC 2008, Vol. 1: Instrumentation and Methods,* pp. 55-56, M. Luysberg, K. Tillmann, T. Weirich (Eds.), Springer Verlag Berlin-Heidelberg 2008
[3] Bohr N 1913 *Phil. Mag.* **25**, 10
[4] Stöger-Pollach M 2008 *Micron* **39**, 1092;
Stöger-Pollach M et al. 2007 *Springer Proc. in Phys.* **120**: Microscopy of Semiconducting Materials 2007, pp. 345-348, A.G. Cullis, P. Midgley (Eds.)
[5] Lord Rayleigh 1896 *Phil. Mag.* **5**, 167
[6] Muller DA and Silcox J 1995 *Ultramicroscopy* **59**, 195

Through-Focal HAADF-STEM of buried nanostructures

MP Guerrero-Lebrero[1], J Pizarro[1], E Guerrero[1], PL Galindo[1], A Yánez[1] and SI Molina[2]

[1] Dept. Lenguajes y Sist. Informáticos, Universidad de Cádiz, Spain
[2] Dept. Ciencia de los Materiales e I.M. y Q.I. Universidad de Cádiz, Spain

E-mail: maria.guerrero@uca.es

Abstract. High-Angle Annular Dark-Field Scanning Transmission Electron Microscopy (HAADF-STEM) in combination with strain mapping techniques provides a powerful tool for quantitative analysis of crystalline semiconductor materials. Due to the complex interaction of a focused probe and a sample in HAADF, the calculation of each pixel in a simulation process requires a complete multislice iteration, making the overall computing process a rather demanding task in time and memory. SICSTEM is a parallel software code recently developed for running on the University of Cadiz Supercomputer (3.75 Tflops) that allows the simulation of images from large nanostructures containing more than one million atoms. The software has been designed to be able to generate not only one dimensional line scans or two dimensional images, but also to perform optical sectioning in the STEM simulation process, providing an easy way to simulate 3D HAADF-STEM images. In this work we consider GaAs capped GaSb nanostructures epitaxially oriented on a GaAs substrate. A methodology has been developed by combining the through-focal series STEM imaging and image analysis to estimate shape and position of buried GaSb nanostructures.

1. Introduction

The control of the shape, position and strain of nano-objects constituting self-assembled semiconductor nanostructures is a key aspect to improve their functionality. To improve the design of these nanostructures, it is important to have tools for quantitative interpretation. HAADF STEM image simulation tools can reveal information on individual atoms, but due to the high computational requirement needed, especially memory and time, nowadays can be only applied to a few unit cells. To improve these computational requirements, parallel HAADF STEM software has been developed. SICSTEM [1] is a HAADF STEM image simulation tool which is able to perform optical sectioning and to simulate aberration corrected microscopy, providing depth resolution at the nanometre scale in reasonable time.

The aim of this work is to estimate the 3D location of GaAs capped GaSb nanostructures epitaxially oriented on a GaAs substrate, from simulated images. Direct localization of buried nanostructures in three dimensions can be carried out by the intensity analysis of a through-focal series of images. A through-focal series is based on an acquiring process with the electron beam focused in a particular plane, then focus is updated to a new value by adding a fixed increment. Thus, the process consists of an iterative task of image acquisition and focus variation in order to get a set of frames from which information in 3D can be extracted.

Previous work demonstrated that the through-focal series acquisition provides the possibility of a 3D reconstruction of the sample investigated. In [2] it was shown that location of isolated heavy atoms in a matrix with a low atomic number was possible, although only third-order aberration correctors were used.

2. Experiment and results

In this work a square prism shaped GaSb quantum dot of dimensions 1×1×2 nm^3 buried in GaAs (001) of dimensions 3x3x25.4 nm^3 has been considered. The GaSb nanostructure was centred in the *x-y* plane, while in *z* direction three different locations were considered, namely at 4.24, 12.72 and 21.2 nm from the surface. Figure 1 shows the three models considered in this study where the position of each GaSb quantum dot can be clearly appreciated.

The corresponding supercells consisting of around 11800 atom coordinates, occupancy and Debye-Waller factors were created and applied as inputs to the SICSTEM software, using parameters for a 300 kV dedicated VG603 STEM (*Cs* = -0.037 mm, *C*5=100mm, inner detector angle = 58 mrad, outer detector angle = 200 mrad and objective aperture = 22 mrad). In about 65 hours, a through-focal series containing 51 images per model of 512x512 pixels, with a focal step size of 0.5 nm was generated.

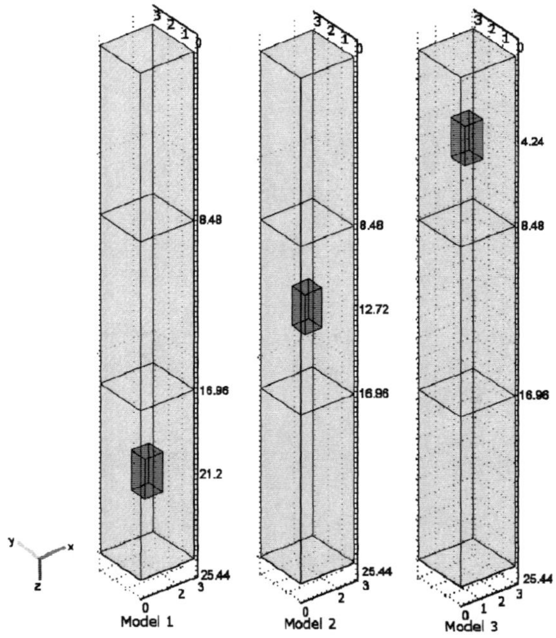

Figure 1. Three different models (1, 2 and 3) of GaSb quantum dot positions within a GaAs matrix.

Each *Z*-contrast image along the <110> zone axis consists of atomic dumbbells formed by GaAs or GaSb atoms. Each dumbbell comprises two atomic columns, one containing Ga atoms throughout the whole material while each neighbouring atomic column is either pure As or pure Sb atoms. The intensity of *Z*-contrast images is strongly dependent on the atomic number of the elements that constitute each probed atomic column. Those peaks of a dumbbell with similar intensities correspond to GaAs (Z_{Ga}=31, Z_{As}=33), whereas higher intensities are more likely associated with Sb (Z_{Sb}=51).

We focus on the anion peaks where Ga atoms can be neglected. Figure 2 shows a profile taken at the centre of the image (inside the buried nanoprism) and the other one near to the border (outside the buried nanoprism). From a given simulated image the integrated intensities around anion peaks [3] have been calculated and the minimum value is assumed to be the As atom reference in the image.

A significant relative difference between each integrated intensity and the reference value will determine the existence of Sb atoms, their x and y positions, and hence, the 2D shape of the buried nanostructure. Figure 3 (a) shows the determined shape at 22.5 nm depth for model 1 and (b) is a plot of normalised intensity differences for every anion position in the frame.

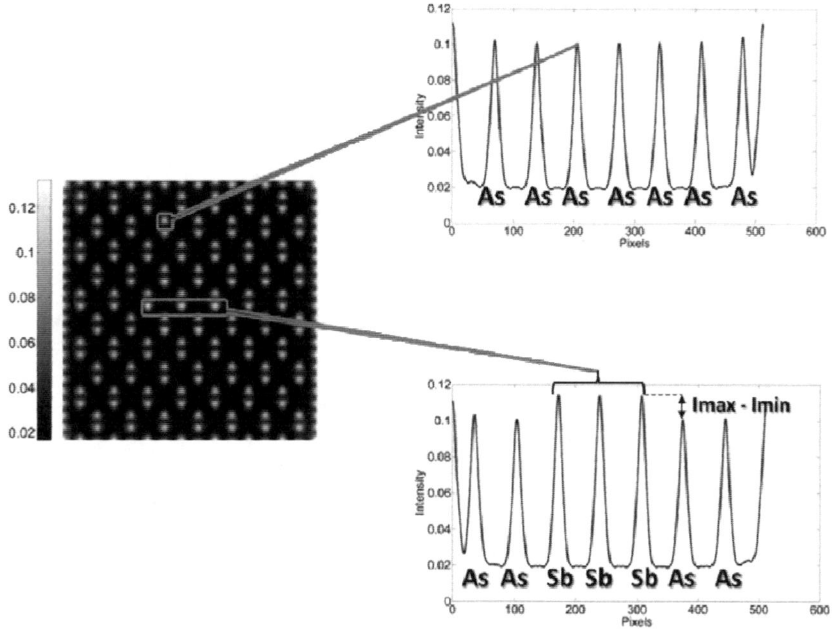

Figure 2. Simulated intensity profiles taken at the centre of an image (inside the buried nanoprism) and near to the border (outside).

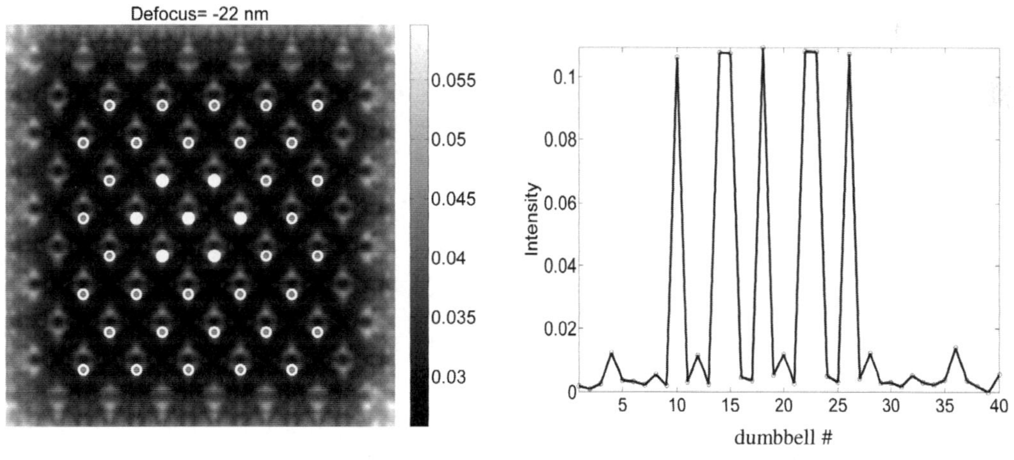

Figure 3. Simulated image for model 1 acquired at 22.5 nm underfocus. The shape of the nanoprism is indicated by white full circles that represent Sb atoms; empty circles represent As atoms (left). Relative difference between each integrated intensity value and the reference value for every anion position marked in the frame (right).

In order to estimate the z position of Sb atoms, let's define ΔI_f as the relative difference between maximum (I_{max}) and minimum (I_{min}) anion integrated intensity at a given defocus value of f, $(I_{max}-I_{min})/I_{min}$. This operation is carried out through the focal series of images. The highest value of the ΔI_f set could be interpreted as the z position where the Sb atoms come into focus and are clearly visible.

Figures 4a, 4b and 4c show ΔI_f as function of the defocus values considered in the acquisition process for geometric models 1, 2 and 3. Maxima of these plots are found at around 4.5, 14 and 22.5 nm thickness, respectively, and in the corresponding geometric models these positions would correspond to the centres of the nanoprisms: 4.22, 12.5 and 21.2 nm. This discrepancy could be due to channeling effects. In order to determine the exact nanoprism length, new simulations considering different nanoprism sizes and positions will be necessary.

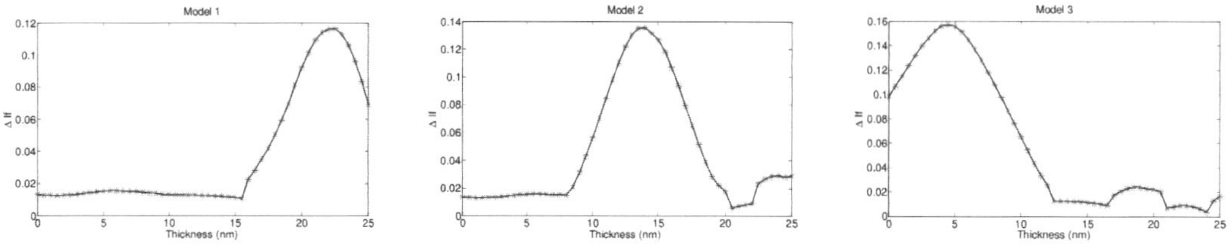

(a) (b) (c)

Figure 4. ΔI_f as function of thickness values considered in the acquisition process for geometric models 1, 2 and 3.

3. Conclusions

In this paper the shape and position of a GaSb nanoprism inside GaAs using HAADF-STEM focal series simulated images has been estimated. The proposed methodology can be applied to 3D reconstruction when nanostructures with heavy atoms are embedded in supercells with light atoms. Further research will be focused on applying this approach to experimental images. Problems may arise when segregation is present and/or due to the resolution in z-direction due to the channeling effect.

References

[1] Pizarro J, Galindo PL, Guerrero E, Yáñez A, Guerrero MP, Rosenauer A, Sales DL and Molina SI 2008 *Applied Physics Letters* **93**, 153107

[2] van Benthem K, Lupini AR, Oxley MP, Findlay SD, Allen LJ and Pennycook SJ 2006 *Ultramicroscopy* **106**, 1062

[3] Molina SI, Sales DL, Galindo PL, Fuster D, González Y, Alén B, González L, Varela M and Pennycook SJ 2009 *Ultramicroscopy* **109**, 172

Journal of Physics: Conference Series **209** (2010) 012033 doi:10.1088/1742-6596/209/1/012033

Characterization of core-shell GaAs/AlGaAs nanowire heterostructures using advanced electron microscopy

MJ Tambe[1], LF Allard[2] and S Gradečak[1]

[1] Department of Materials Science and Engineering, Massachusetts Institute of Technology, Cambridge MA 02139, USA
[2] High Temperature Materials Laboratory, Oak Ridge National Laboratory, Oak Ridge TN 37831, USA

E-mail: gradecak@mit.edu

Abstract. To explore the unique properties of the nanoscale, advanced fabrication and characterization techniques are required. Specifically analyses in two orthogonal directions, plan-view and cross-section, were used to prove the core-shell morphology of GaAs/AlGaAs nanowires and determine their cross-section to be hexagonal. High-resolution transmission electron microscopy and high angle annular dark field scanning transmission electron microscopy confirmed the core-shell interface to be defect-free, coherent, and sharp (<1nm). Energy dispersive X-ray spectroscopy determined the shell composition to be $Al_{0.9}Ga_{0.1}As$ uniformly along the length of the nanowire. These results demonstrate the power of electron microscopy to aid the development of semiconductor nanotechnology.

1. Background and motivation

Semiconductor nanowires have emerged as a promising new platform for nanoscale electronics, optoelectronics, and sensors. Core-shell nanowire heterostructures possess three distinct advantages over thin film systems: defect-free integration of highly misfit materials [1], the potential to study one dimensional physics [2] and the potential to create integrated nanowire devices [3].

The GaAs/AlGaAs system is of significant interest for high mobility electronics because it possesses intrinsically high electron mobility that can be further enhanced in two dimensional GaAs/AlGaAs heterostructures. GaAs/AlGaAs core-shell nanowire heterostructures would represent a unique model-system to achieve one dimensional carrier confinement. Initial studies [4,5] have demonstrated that core-shell nanowire fabrication may be realized in GaAs/AlGaAs, and in-depth structural analysis of GaAs/AlGaAs [6] core-shell nanowires is needed to develop nanowire devices.

In this work, the controlled growth of core-shell GaAs/AlGaAs nanowires is demonstrated. Chemical analyses by energy dispersive X-ray spectroscopy (EDX) confirm the core-shell morphology with a shell composed of $Al_{0.9}Ga_{0.1}As$. Structural analysis by transmission electron microscopy (TEM) and lattice-resolved high angle annular dark field (HAADF) scanning transmission electron microscopy (STEM) confirms the epitaxial shell deposition and the entire nanowire heterostructure to be defect-free and single crystalline.

Journal of Physics: Conference Series **209** (2010) 012033 doi:10.1088/1742-6596/209/1/012033

2. Experimental

Core-shell GaAs/AlGaAs nanowires were fabricated by the vapour-liquid-solid mechanism [7] in a metal-organic chemical vapour deposition reactor as described previously [6]. Plan-view samples were prepared by ultrasonically removing nanowires in ethanol suspension and then depositing onto lacey formvar/carbon TEM grids. Cross-sectional samples were prepared as shown in Figure 1a, a plastic coverslip was rubbed on the surface of the as-grown substrates to collect the nanowires. The plastic coverslip was then embedded in epoxy and sliced approximately 75nm thick by microtomy (Leica and RMX using a Diatome Histo diamond blade).

The morphologies of as-grown nanowires were studied using a JEOL 6320FV scanning electron microscope (SEM). The structure and chemical composition of individual nanowires were characterized using a JEOL 2010F TEM, a VG HB603 STEM equipped with EDX, as well as a JEOL 2200FS aberration-corrected STEM. During high angular annular dark field (HAADF) STEM imaging of cross-sectional samples without carbon coating, the *Z*-contrast was clear but strong sample charging effects were observed, as shown in the fast Fourier transforms (FFT) shown in Figure 1b. After the samples were carbon coated, the sample charging was minimized, but this had considerable effects on the contrast at the core-shell interface. Therefore cross-sectional sample were imaged without carbon coating, coated with carbon, and then imaged again to determine both crystal structure and compositional profile.

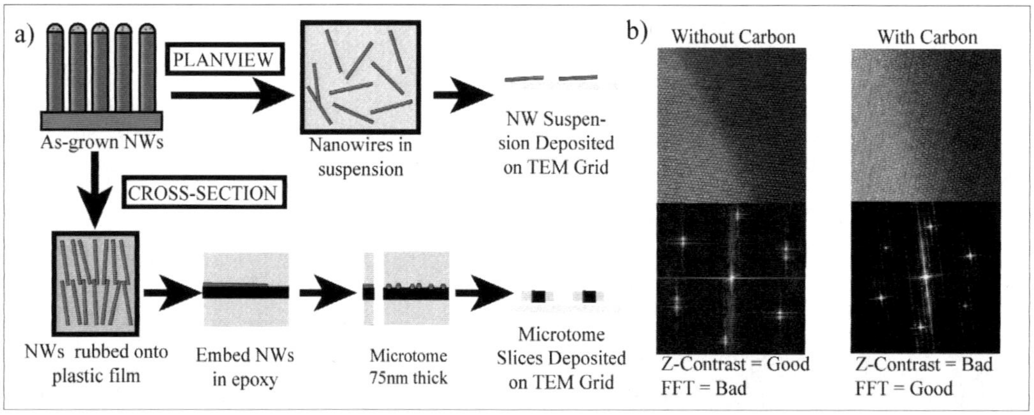

Figure 1. a) A schematic of the TEM sample preparation procedures. b) HAADF STEM images with corresponding FFT for cross-sectional samples with and without carbon coating.

3. EDX chemical analysis

The compositions of the nanowires were determined by EDX analysis in both plan-view and cross-sectional view. EDX profiles were then fitted to a numerical model with Gaussian beam broadening:

$$\Phi_Z(x) = \frac{1}{\sigma\sqrt{2\pi}} \int_{-\infty}^{\infty} h_Z(x') e^{-\frac{(x-x')^2}{2\sigma}} dx' \qquad (1)$$

Here $h(x')$ is the thickness profile integrated over the sample height of element Z and σ is the Gaussian beam broadening width. The results from both plan-view and cross-sectional analysis were consistent with a GaAs core and an $Al_xGa_{1-x}As$ shell with $x = 0.89 \pm 0.03$. For spectra acquired on the VG HB603 $\sigma = 0.5$ nm. For results taken on the JEOL 2010F σ = 3-4 nm.

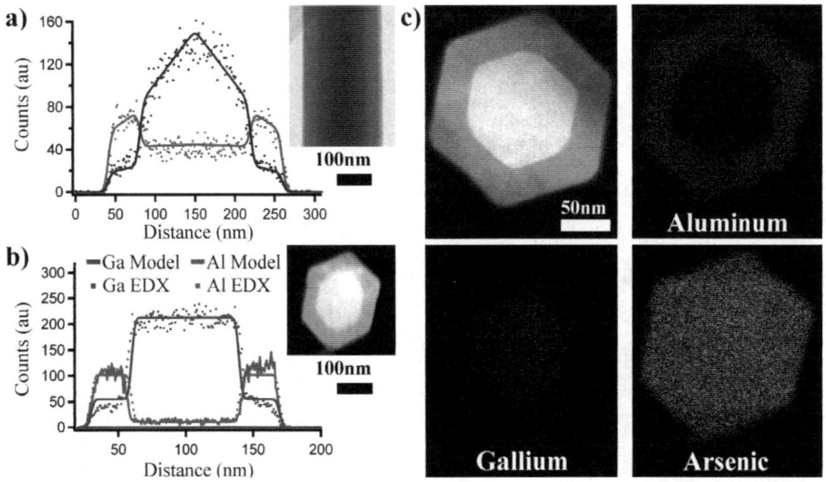

Figure 2. a) EDX linescan data from a core-shell nanowire collected in plan-view and fitted to the model shown in equation 1. Inset is BF-STEM image of the nanowire. b) EDX linescan data from a core-shell nanowire collected in cross-section and fitted to the model shown in equation 1. Inset is DF-STEM image of the nanowire c) HAADF STEM image and corresponding EDX chemical maps.

4. Structural analysis

The structures of the nanowires were analyzed using SEM, TEM, and lattice resolved HAADF STEM. SEM inspection, as shown in Figure 3a, reveals that the nanowires grow vertical to the substrate. Bright-field (BF) TEM analysis, as shown in Figure 3b, reveals that the nanowires are zinc-blende single crystals growing along the <111> direction. The nanowires were imaged at multiple tilt angles and no dislocations or twin planes were observed.

For more detailed information about the core-shell interface, lattice-resolved HAADF STEM images, as shown in Figure 3c with <111> zone axis, of the GaAs/AlGaAs interface were taken. From these images it can be seen that the core-shell interface is epitaxial and coherent. Comparison of the HAADF-STEM with the corresponding BF-STEM, as shown in Figure 3d, demonstrates that the contrast in the HAADF image is caused by compositional variation. FFT of the lattice-resolved images determines the facets to be {110}.

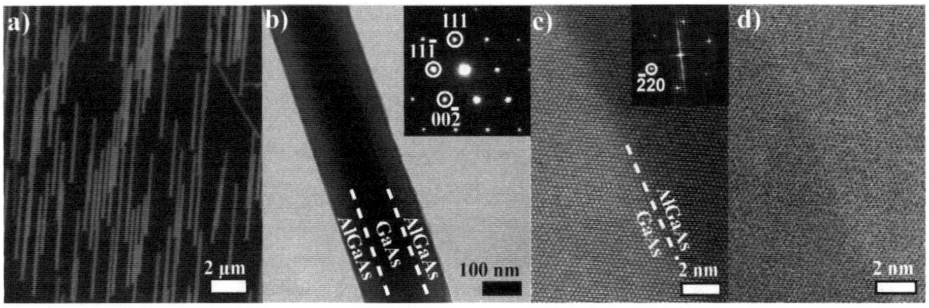

Figure 3. a) SEM image of as-grown core-shell nanowires. b) BF-TEM image of core-shell nanowire. Inset is a corresponding SAD pattern. c) Lattice-resolved HAADF STEM image of the core-shell interface in cross-sectional view. Inset is FFT. d) The BF-STEM image corresponding to c).

5. Anisotropic shell deposition

In addition to the commonly observed regular hexagonal cross-section of GaAs/AlGaAs nanowire heterostructures [5], anisotropic shell deposition has also been observed. Plan-view STEM, as shown in Figure 4a, shows a core-shell morphology and appears similar to core-shell nanowires with regular hexagonal cross-section (Fig 3b). However, the anisotropic nature of the shell is clearly revealed in cross-sectional HAADF STEM, as shown in Figure 4b. This anisotropy is related to crystallographic facets of the nanowire core. If all facets of the nanowire core belong to one family of planes ({110} or {112}), then shell deposition is uniform. However, if the facets of the nanowire core belong to multiple families of planes, the shell deposition will be anisotropic. From FFTs of lattice-resolved HAADF STEM images, the shell thickness is seen to be greater on {110} facets than on {112} facets.

Lattice-resolved HAADF STEM images, as shown in Figure 4c, also reveal the shell to be amorphous. This is further demonstrated by a STEM shadow image, as shown in Figure 4d, in which the crystalline shell shows clear Kikuchi lines whereas the shell does not. The amorphous nature of the shell is unclear, but is likely due to oxidation of the Al-rich AlGaAs shells. Further investigation is underway to investigate growth on different nanowire facets.

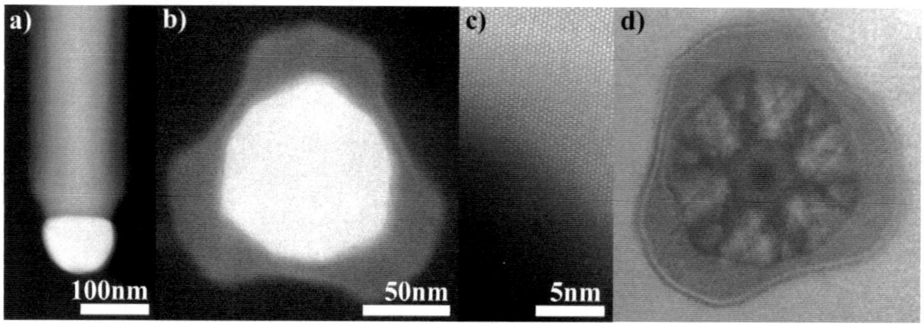

Figure 4. a) DF-STEM image of anisotropic shell nanowire. b) HAADF STEM image of anisotropic core-shell nanowire in cross-sectional view. c) Lattice-resolved HAADF STEM image of core-shell interface. d) "Shadow" image of anisotropic shell nanowire in cross-sectional view.

6. Summary

In summary core-shell GaAs/AlGaAs nanowire heterostructures have been characterized using multiple electron microscopy techniques. HAADF STEM imaging gives information about both the structure and composition profile at the GaAs/AlGaAs interface. EDX analyses confirm the core-shell morphology and consistently determine the shell to be $Al_xGa_{1-x}As$ with $x = 0.89 \pm 0.03$. BF-TEM and lattice resolved HAADF STEM structural analysis demonstrates the core-shell interface to be epitaxial, coherent, and sharp. As evidenced by anisotropic shell nanowires, both cross-sectional and plan-view analysis are necessary to determine the morphology of core-shell nanowires.

Acknowledgements

The authors thank EA Fitzgerald for access to MOCVD facilities, the MIT Center for Materials Science and Engineering, an NSF-funded MRSEC, for use of electron microscopy facilities, and ST Boles for MOCVD assistance and helpful discussions. SG acknowledges 3M, the Interconnect Focus Center, MIT startup funds, and the NSF CAREER Award (DMR-074555) for financial support. A portion of this research was conducted at the SHaRE User Facility, which is sponsored by the Division of Scientific User Facilities, Office of Basic Energy Sciences, U.S. Department of Energy.

References

[1] Liang Y, Nix WD, Griffin PB and Plummer JD 2005 *J. Appl. Phys.* **97**, 043519
[2] Liang G et al. (2007) *Nano Lett.* **7**, 642
[3] Ng HT et al. (2004) *Nano Lett.* **4**, 1247
[4] Noborisaka J, Motohisa J, Hara S and Fukui T 2005 *Appl. Phys. Lett.* **87**, 093109
[5] Ouattara L et al. 2007 *Nano Lett.* **7**, 2859
[6] Tambe MJ et al. 2008 *Appl. Phys. Lett.* **93**, 151917
[7] Wagner RS and Ellis WC 1964 *Appl. Phys. Lett.* **4**, 89

MOCVD growth mechanisms of ZnO nanorods

G Perillat-Merceroz[1, 2], PH Jouneau[2], G Feuillet[1], R Thierry[1], M Rosina[1] and P Ferret[1]

[1] CEA - LETI, Minatec, 17 rue des Martyrs, 38054 Grenoble cedex 9, France
[2] CEA - INAC, 17 rue des Martyrs, 38054 Grenoble cedex 9, France

E-mail: guillaume.perillat-merceroz@cea.fr

Abstract. ZnO is a promising material for the fabrication of light emitting devices. One approach to achieve this goal is to use ZnO nanorods because of their expected high crystalline and optical quality. Catalyst free growth of nanorods by metalorganic chemical vapour deposition (MOCVD) was carried out on (0001) sapphire substrates. Arrays of well-aligned, vertical nanorods were obtained with uniform lengths and diameters. A thin wetting layer in epitaxy with the sapphire substrate is formed first, followed by pyramids and nanorods. The nucleation of nanorods occurs either directly at the interface, or later on top of some of the pyramids, suggesting various nucleation mechanisms. It is shown that crystal polarity plays a critical role in the growth mechanism with nanorods of Zn polarity and their surrounding pyramids with O polarity. A growth mechanism is proposed to explain that most threading dislocations lie in the wetting layer, with only a few in the pyramids and none in the nanorods.

1. Introduction

ZnO has been studied for decades, but recently generated renewed interest due to its potential applications for optoelectronic devices [1]. Its large band gap (around 3.4 eV, similar to GaN) makes ZnO a candidate for short wave length light emitting diodes (LEDs) and laser diodes. The large exciton binding energy (60 meV compared with 25 meV for GaN) is promising for light emission at room temperature, presuming that the difficult problem of p-type doping is solved. Growth of ZnO nanowires or nanorods is widely reported because of possible uses as field emitters, solar cells, gas sensors or light emitting diodes (LEDs). Their high crystalline quality and purity, due to growth without a catalyst, are adequate for optoelectronic applications. MOCVD growth [2] enables fast, large area deposition at an industrial scale. Understanding of the growth mechanism of nanorods is needed for better control of their density and size and should give an indication as to why the nanorods are defect-free.

2. Experimental details

ZnO nanorods were grown using catalyst-free MOCVD in a horizontal hot-wall Epigress reactor. The c-plane sapphire substrates were annealed at 1200°C in oxygen. Diethylzinc (DEZn) was used as Zn precursor, N_2O as the oxidiser, with a VI/II ratio between 100 and 1000. The ZnO nanorods were grown at 750-850°C, with different deposition times in order to follow the formation of the nanorods. The morphology of the samples was investigated by scanning electron microscopy (SEM) using a Zeiss Ultra 55 microscope. Structural studies were carried out by transmission electron microscopy (TEM) on two microscopes: a 300 kV FEI Titan and a 400 kV JEOL 4000 EX.

3. Results and discussion

3.1. Morphology

After a growth of 20 seconds to 1 minute, a thin 2-dimensional ZnO wetting layer is observed on the sapphire substrate (as shown by cross sectional TEM in Figure 5b), together with nanorods and pyramids (Figure 1a and b). The higher density seen in the 20 seconds growth sample is due to different growth conditions. After one hour of deposition, nanorods have grown in diameter and length. At this stage, the nanorods have a typical length of 4 μm and a diameter of 200 nm (Figure 1c). Nanorods grow along the c-direction as demonstrated by X-ray diffraction [3]. Some pyramids exhibit nanorods on their top and some do not.

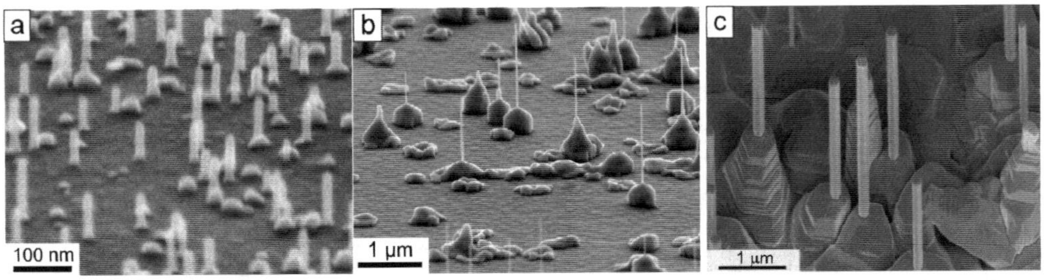

Figure 1. Tilted SEM images of ZnO nanorods: (a) after 20 s of growth, (b) after 1 min of growth, (c) after 1 hour of growth.

3.2. Role of crystal polarity

We used convergent beam electron diffraction (CBED) to determine the crystal polarity of nanorods and pyramids. Nanorods were found to be of Zn polarity while pyramids had O polarity (Figure 2). This is in agreement with a recent result by Sun et al. [4] who reported that nanorods grown by pulsed laser deposition exhibited Zn polarity. Contrary to that the buffer layer shows O polarity.

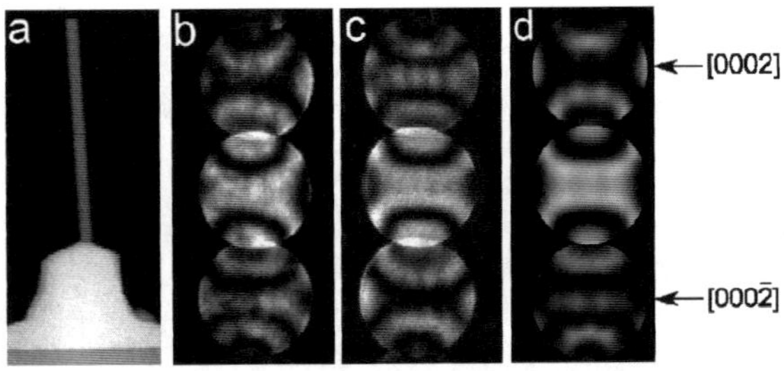

Figure 2. Determination of the polarity of both the nanorod and the pyramid. (a) STEM image of the zone analyzed; (b) and (c) respective CBED patterns of the nanorod and the pyramid. Growth direction upwards. (d) Simulated CBED pattern with the JEMS software.

Facets of the hexagonal nanorods have been indexed based on intensity profiles of nanorods viewed along the <1-100> or <11-20> zone axis. It was found that they are of {1-100} type (Figure 3). Laudise et al. have also reported the higher growth rate along the [0001] direction (Zn polarity) and the presence of {1-100} facets in hydrothermal growth [5]. Growth of nanorods requires Zn polarity nuclei which have a high growth rate in the [0001] direction, partly because of the low free energy of the {1-100} facets.

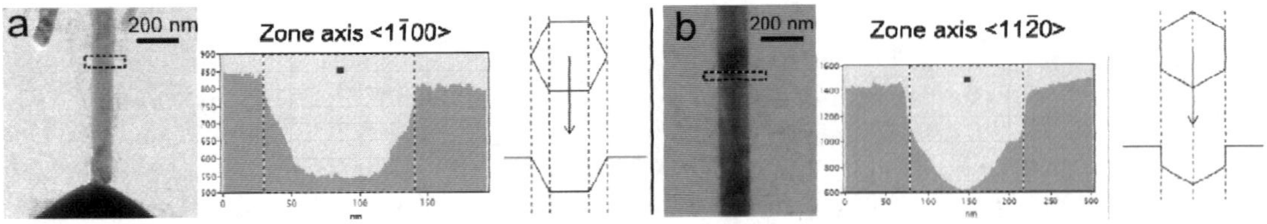

Figure 3. Identification of the side facets of the nanorods by intensity profiling of TEM bright-field images: (a) <1-100> zone axis (b) <11-20> zone axis.

3.3. Nanorod nucleation

Nanorods originate either from the ZnO / sapphire interface (Figure 4a), or from the top of some pyramids (Figure 4b). Contrasts in the nanorod are bend contours resulting from the glue-induced strain. No defects are visible in nanorods dispersed on a carbon film (Figure 5d). For one sample, pyramids and nanorods were observed to be aligned in a <11-20> direction (Figure 4c). In this case, nucleation of the pyramids and of the nanorods would be linked to atomic steps at the substrate surface [6]. However, as the alignment along a particular direction is not always seen and as some nanorods nucleate on top of pyramids, atomic steps could not explain the nucleation of all nanorods. The exact nucleation mechanisms remain under investigation.

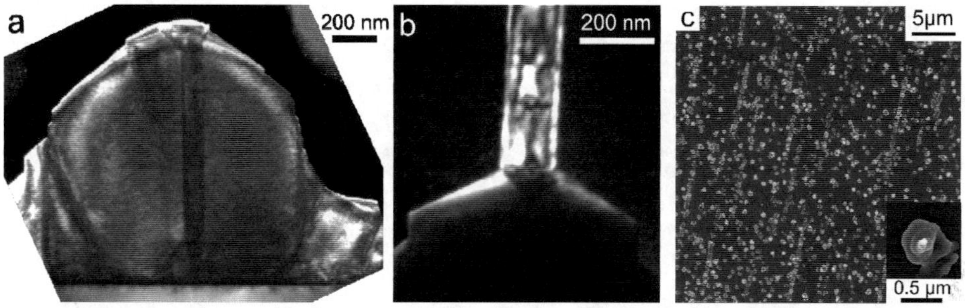

Figure 4. Nucleation of ZnO nanorods. (a) Nucleation at the sapphire / ZnO interface. The nanorod pointing out of the pyramid was removed during TEM sample preparation (b) Nucleation on top of a pyramid. (c) Top view SEM image showing lateral alignment of pyramids and nanorods along a <11-20> direction.

3.4. Growth mechanism and defects

As it is usually observed, the orientation relationship is (Figure 5a):

$$[0001]\ Al_2O_3\ //\ [0001]\ ZnO \text{ and } [-1010]\ Al_2O_3\ //\ [11\text{-}20]\ ZnO,$$

leading to an in-plane mismatch of 18.4% between the sapphire substrate and ZnO. A regular network of misfit dislocations is formed at the interface in order to relax the heteroepitaxial stress (Figure 5b and c): with a thickness of 8 mono-atomic layers, the wetting layer is almost totally relaxed. But no dislocations are observed in the nanorods (Figure 5d). This high crystalline quality of nanorods is confirmed by photoluminescence studies [7].

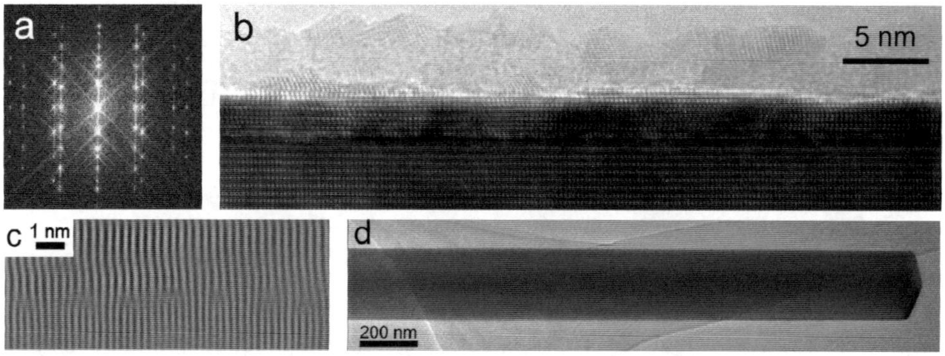

Figure 5. Stress relaxation and defect formation: (a) Fourier transform at the ZnO / sapphire interface, (b) high resolution TEM image of the ZnO wetting layer on sapphire after 1 min of growth, (c) [01-10] filtered image showing the network of misfit dislocations at the sapphire / ZnO interface and (d) bright field TEM image of a nanorod with no dislocations.

The nanorods nucleating at the ZnO / sapphire interface are very narrow (3 to 20 nm, Figure 6a) suggesting that few misfit dislocations are present at the base of the nanorod. Furthermore, the ratio between the nanorod length and the diameter is such that the probability for a dislocation to thread into the nanorods is low. When a nanorod has grown vertically, a pyramid nucleates around it. The pyramid grows both, laterally and vertically, and dislocations follow the growth front, threading from the nanorod base to the sides of the pyramid (Figure 6b, c and d). In the case of a pyramid with no nanorod inside, growth on the wetting layer is similar. Most dislocations in the wetting layer bend over, due to the lateral growth of the pyramid (Figure 7).

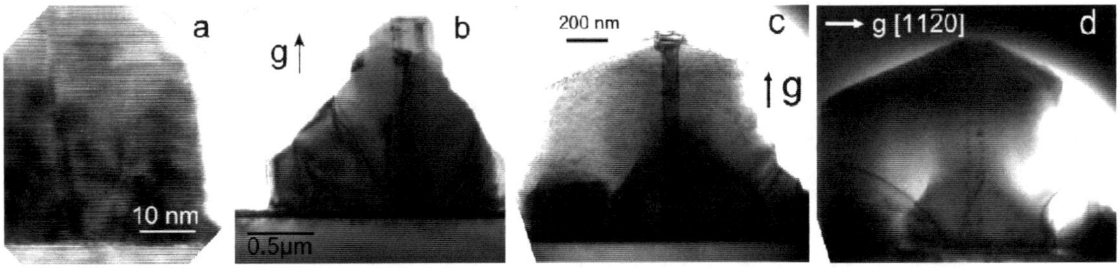

Figure 6. Nanorods surrounded by pyramids: (a) High resolution TEM image (1 min growth). (b), (c), (d) Two-beam bright field TEM images (30 min growth). The nanorods pointing out of the pyramids were removed during TEM sample preparation. Images (c) and (d) are taken on the same pyramid and with the same magnification.

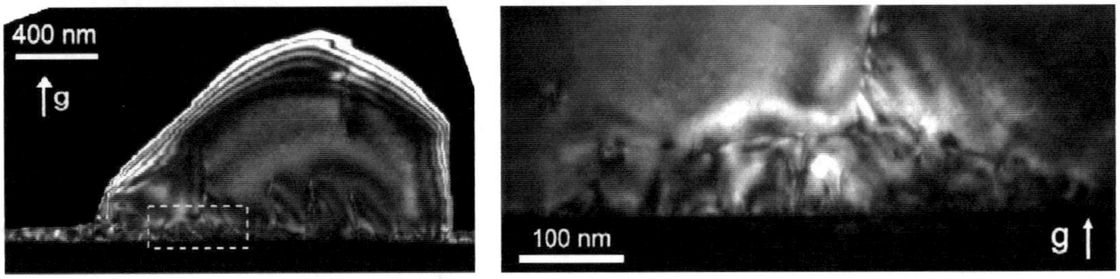

Figure 7. Defect structure of nanorod free ZnO pyramid: (0002) two beam dark-field image of a pyramid on the wetting layer (left) and zoom into the interface region (right).

4. Conclusion

We have proposed a growth mechanism of ZnO nanorods. A 2-dimensional wetting layer covers the sapphire substrate, and nuclei appear locally to form pyramids and nanorods. Nanorods grow because some nuclei are of Zn polarity and because of the large growth rate anisotropy between [0001] and [000-1] directions. No dislocations are observed in the nanorods. Threading of dislocations from the substrate are not very probable because of the nanorod narrowness. The dislocations are not favoured energetically because of the nanorods length. Relatively few dislocations are observed inside the pyramids because of their lateral overgrowth on top of the wetting layer which bends dislocations under the pyramids.

References

[1] Ozgur U, Alivov YI, Liu C, Teke A, Reshchikov MA, Dogan S, Avrutin V, Cho S and Morkoc H 2005 *J. Appl. Phys.* **98**, 041301
[2] Park WI, Kim DH, Jung S and Yi G 2002 *Appl. Phys. Lett.* **80**, 4232
[3] Rosina M, Ferret P, Jouneau P, Robin I, Levy F, Feuillet G and Lafossas M 2009 *Microelectron. J.* **40**, 242
[4] Sun Y, Cherns D, Doherty RP, Warren JL and Heard PJ 2008 *J. Appl. Phys.* **104**, 023533
[5] Laudise R and Ballman A 1960 *J. Phys. Chem.* **64**, 688
[6] Munuera C, Zuniga-Perez J, Rommeluere JF, Sallet V, Triboulet R, Soria F, Munoz-Sanjose V and Ocal C 2004 *J. Cryst. Growth* **264**, 70
[7] Robin IC, Gauron B, Ferret P, Tavares C, Feuillet G, Dang LS, Gayral B and Gerard JM 2007 *Appl. Phys. Lett.* **91**, 143120

Quantitative investigation of the onset of islanding in strained layer epitaxy of InAs/GaAs by X-ray mapping in STEM

T Walther and M Hopkinson

Department of Electronic & Electrical Engineering, University of Sheffield, Sir Frederick Mappin Building, Mappin Street, Sheffield S1 3JD, UK

E-mail: t.walther@sheffield.ac.uk

Abstract. The Stranski-Krastanow transition from two- to three-dimensional growth describes the onset of roughening and the formation of islands. For InAs/GaAs, islanding occurs abruptly for a deposition of about one unit cell of InAs. We combine scanning transmission electron microscopy and X-ray elemental mapping to measure the layer thicknesses and chemical compositions of thin layers of InAs deposited on GaAs(001) for nominal thicknesses of 1.6, 1.8 and 2.0 monolayers of InAs. Different methods to analyse ratios of such elemental maps have been compared to measure quantitatively the indium content within these ultra-thin layers. The most accurate and reliable of the three methods investigated is shown to be the analysis of the inverse slope of plots of As/In ratio maps versus scan window size perpendicular to the layers.

1. Introduction

The Stranski-Krastanov transition describes the sudden onset of islanding during the growth of a previously perfectly flat film. For the InAs/GaAs(001) system this has been reported from surface analytical techniques to occur at a deposition of 1.8 monolayers of indium, i.e. just before the completion of a single unit cell of InAs [1, 2]. We investigate this transition by X-ray mapping combined with scanning transmission electron microscopy (STEM), with the aim to quantify the amount of indium in layers grown at nominal InAs thicknesses just around that value and to correlate this directly with the onset of islanding observed.

2. Experimental

Three layers of InAs of nominal thicknesses of 1.6, 1.8 and 2.0 monolayers indium (defined as $1ML=d_{002}=a/2=0.283$nm for GaAs) were deposited upon GaAs(001) with 70nm GaAs barrier layers. The samples were grown by molecular beam epitaxy at 505°C, using elemental gallium, indium and arsenic sources. The InAs growth rate was set as 0.10 ML/s, a value derived from calibrations based on indium induced reflection high-energy electron diffraction (RHEED) oscillations on InAs substrates, taking into account the differences in lattice constant. The nominal thicknesses are subject to errors of ± 0.1ML due to the nature of the RHEED measurement and the possibility of growth rate drift. A <100> cross-sectional sample was prepared using standard cutting, grinding, polishing and Ar^+ ion milling techniques and studied in a JEOL2010F field-emission transmission electron microscope operated at 197kV in STEM mode and equipped with a Gatan Imaging Filter, an Oxford Instruments Si:Li X-ray detector with ultrathin window and the Oxford Instruments ISIS300 software for quantification. Figs. 1 and 2 depict bright-field and annular dark-field images recorded with the

Journal of Physics: Conference Series **209** (2010) 012035 doi:10.1088/1742-6596/209/1/012035

specimen oriented edge-on, close to a <100> zone axis. 128×100 pixel X-ray maps at a sampling of 2.2nm/pixel were recorded for all X-ray lines (K and L) of all elements (In, Ga and As) using signal integration from successive scans during a long (~2 hrs) time where drift was corrected to 1-2 pixels precision. The sample was tilted 19° towards the detector, with a 4° in-plane component that blurred the layers in projection to apparent widths of ~6nm (3 pixels). The sample thickness was estimated to be ~80nm. The maps are shown in Figure 3.

Figure 1. Edge-on bright-field (BF) STEM image of InAs/GaAs layer structure. Growth towards the top right corner.

Figure 2. Annular dark-field (ADF) STEM image of the same area. Quantum dots are only visible in the top of the three InAs layers.

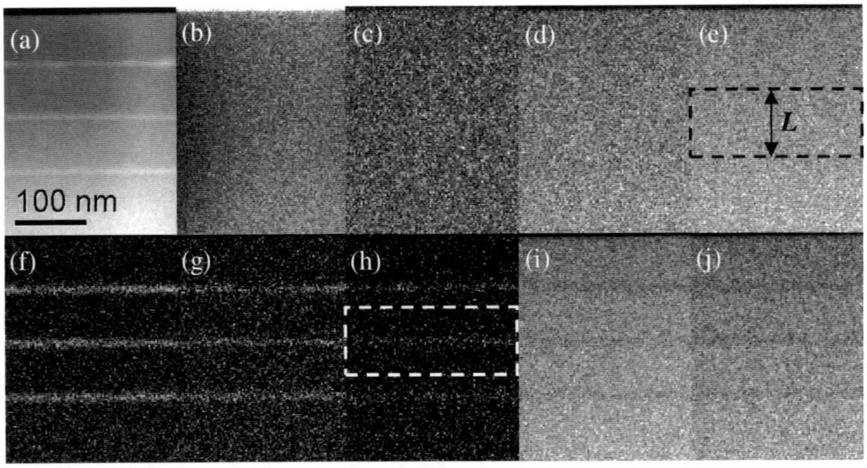

Figure 3. (a) ADF image and X-ray elemental maps of (b) C_K (26-184 counts), (c) background of S_K (0-29 counts), (d) As_L (30-205 counts), (e) As_K (29-240 counts), (f) $In_{L\alpha}$ (1-33 counts), (g) $In_{L\beta}$ (0-29 counts), (h) $In_{K\alpha}$ (0-20 counts), (i) Ga_L (50-336 counts) and (j) Ga_K (32-246 counts).

We then compared three different techniques of evaluating the indium content of the layers quantitatively, namely by measuring (and for 1. and 2., integrating the signals along growth direction):

1. the relative decrease of gallium concentration from line profiles of the Ga/As ratio (Fig. 4);
2. the relative increase of indium concentration from line profiles of the In/(Ga+In) ratio (Fig. 5); and
3. the inverse slope of plots of the As/In ratio as a function of the length L of the scan window perpendicular to the layers (Fig. 6), as indicated schematically in Figs. 3e and 3h for K-lines.

The last method is equivalent to the concept of concentric electron probes in TEM, originally introduced for X-ray analysis of atoms segregated to planar faults studied by convergent beam illumination [3] and later extended to STEM illumination [4]. Its principle is that the number of matrix atoms (here: As) within an irradiated volume is proportional to the size of the area illuminated times the specimen thickness, while the number of segregated atoms (here: In) in the illuminated volume is proportional to the lateral extent of the illumination times the specimen thickness. Hence, the ratio of the first to the second should be proportional to the extension of the illuminated area and independent of the specimen thickness. If beam broadening were negligible, the slope would be inversely related to the effective chemical thickness of the planar fault or thin layer analysed. With beam broadening taken into account, deviations from linearity will occur which can be quantified by linear regression analysis and thus give a measure of the expected accuracy [3, 4]. Using Monte Carlo simulations of electron scattering and X-ray generation this method called *conceptEM* was previously shown to be more reliable and more accurate for the measurement of the chemistry of thin layers than standard profiling methods [5], given the same illumination conditions. In this experimental study, the biggest problem was to find appropriate *k*-factors because the ISIS300 software provides *k*-factors only for specific X-ray lines within individual spectra but not for X-ray maps generated from integrating spectral intensity over windows of certain widths that will depend on the individual energies of X-ray lines chosen. As a compromise, for the corresponding windows we determined self-consistent effective *k*-factors based on the assumption that the stoichiometry of the group III/V ratio is unity (i.e. 50at% of the InGaAs material is comprised of As atoms) and that In replaces Ga in InGaAs (i.e. the sum of In and Ga atoms must also account for 50at%). This implies that maps of the As signals (group-V sub-lattice) should be of the same height as the sum of maps ($k_{In,As}$ In + $k_{Ga,As}$ Ga) for the group-III sub-lattice.

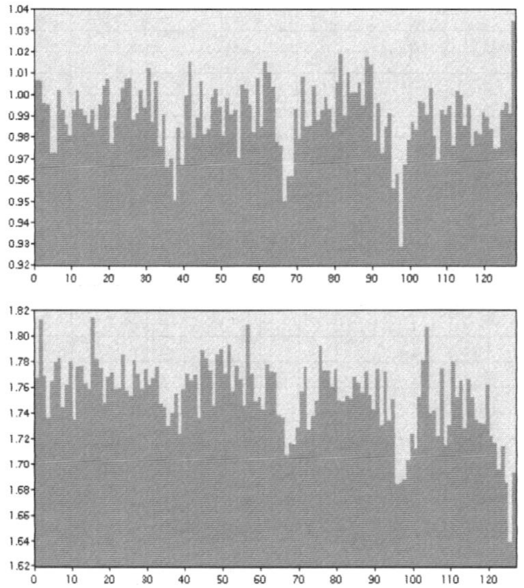

Figure 4. Relative decrease of gallium signal, measured by profiles of the Ga_K/As_K ratio (above) and the Ga_L/As_L ratio (below) vs. distance along [001] in nm. The max. decrease of the Ga signal amounts to ~4, 7 and 9% of the Ga signal in GaAs (i.e. to be multiplied by 50at% for occupancy of group III sub-lattice).

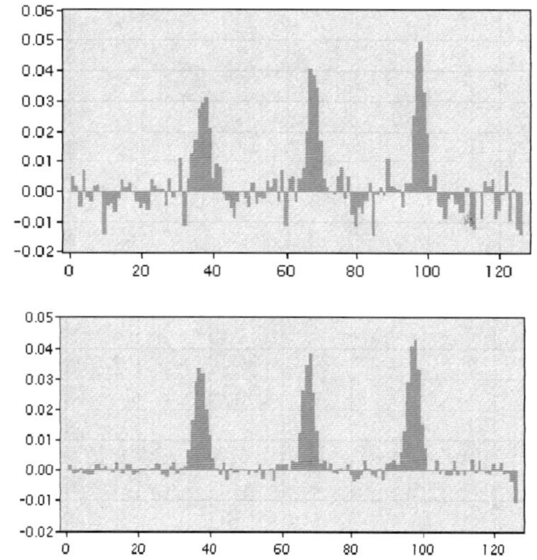

Figure 5. Relative increase of the indium signal as measured by profiles of the ratio $In_K/[In_K+k_{GaK,InK}\,Ga_K]$ (above) or of the ratio $In_L/[In_L+k_{GaL,InL}\,Ga_L]$ (below) vs. distance in nm. The peak increase of In amounts to ~3, 4 and 5% of the group III sub-lattice.

Figures 4-6 show results from the three quantification methods, and Table 1 compares the numerical results obtained for the total amount of indium in the layers. It can be seen that the first method based on the relative decrease of the Ga signal is simple but suffers from the largest statistical errors (due to the noisiness of the line scans in Figure 4) and the results from K-line quantification are systematically larger (by 0.38±0.12ML) than those from L-line quantification. The second method measuring the increase in the In signal from the profiles yields consistent values for K- and L-line quantification (the difference of –0.04±0.11ML is within statistics) but all values are still much smaller than the nominal values expected from the calibrated deposition rates. The third method quantifying the slopes of the plots of the As/In ratio vs. scan window length in Figure 6 yields generally larger indium concentration values closer to the nominal ones and smaller statistical error bars due to linear regression analysis. The data for K- and L-line quantification are, however, only consistent within these small error bars of ~±0.1ML for the measurement of the 3rd layer, while the values obtained for the 1st and 2nd layer from K- and L-lines disagree by more than this statistical error, which we attribute to these regions of the specimen being somewhat further from the specimen edge, hence a little thicker so that the softer X-rays are more strongly absorbed. This trend can, however, only be noticed when the error bars are sufficiently small. It should be noted that the correlation coefficients of the linear fitting were 0.994±0.005 for the K-lines and 0.998±0.001 for the L-lines, which from previous simulations [3] would indicate relative errors of a few %, in agreement with our observations.

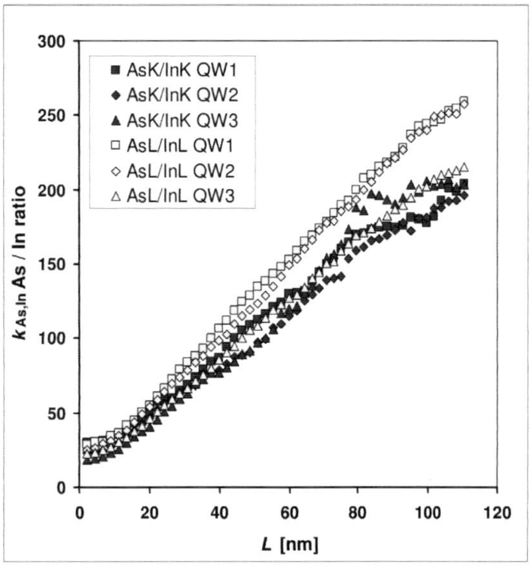

Figure 6. Plot of *k*-factor corrected As/In ratio as function of window size around In(Ga)As layers.

Table 1. List of amount of indium in the three layers measured by the various techniques.

method	lines	1st layer	2nd layer	3rd layer
RHEED	-	1.6±0.1	1.8±0.1	2.0±0.1
decrease of $k_{Ga,As}$ Ga/As	K	0.90 ±0.3	0.95 ±0.3	1.48 ±0.3
	L	0.47 ±0.3	0.71 ±0.3	1.01 ±0.3
increase of In/(In+$k_{Ga,In}$Ga)	K	1.32 ±0.25	1.34 ±0.25	1.49 ±0.27
	L	1.25 ±0.21	1.39 ±0.25	1.63 ±0.27
inverse slope of $k_{As,In}$ As/In	K	1.77 ±0.13	1.90 ±0.11	1.67 ±0.13
	L	1.46 ±0.07	1.48 ±0.05	1.76 ±0.07

Values in the table are given in equivalents of full InAs monolayers.

3. Summary

The most accurate and reliable method to quantify the indium content of InGaAs thin layers from X-ray microanalysis is to perform linear regression analysis to plots of As/In ratio measurements as a function of area size analysed around these layers. In thin regions quantification from K- and L-lines can yield consistent values to better than ±0.1ML.

References

[1] Leonard D, Pond K and Petroff PM 1994 *Phys. Rev. B* **50**, 11687
[2] Nabetani Y, Yamamoto N, Tokuda T and Sasaki A 1995 *J. Cryst. Growth* **146** 363
[3] Walther T 2004 *J. Microsc.* **215**, 191
[4] Walther T 2006 *J. Microsc.* **223**, 165
[5] Walther T 2008 *J. Phys.: Conf. Ser.* (Proc. EMAG 2007, Glasgow) **126**, 012091

Study of annealed InAs/GaAs quantum dot structures

Y Qiu, Z Y Zhang, R A Hogg, A G Cullis and T Walther

University of Sheffield, Department of Electronic and Electrical Engineering, Mappin Street, Sheffield, S1 3JD, UK

E-mail: Y.Qiu@shef.ac.uk

Abstract. Two similar samples of InAs/GaAs quantum dots and their response to annealing at 700ºC have been studied using scanning transmission electron microscopy. The only difference in the growth of both samples is the inclusion of p-doping by carbon in the GaAs barriers between the quantum dots in one of the wafers. The size and line density of and the change of indium concentration within the quantum dots have been investigated to understand the effect of carbon doping on the diffusion of indium atoms within the structures during the anneal.

1. Introduction

In(Ga)As quantum dots (QDs) emit light at ~1300nm, which can be used for infrared applications such as light emitting diodes (LEDs) and laser diodes. Interdiffusion inside the device due to internal heating can affect the device performance and life time [1]. Recently, carbon doping has been used to reduce the temperature sensitivity of such devices [2-4] but it is unclear how this mechanism works exactly. Therefore, we have studied the change of the microstructure in C-doped and undoped InAs/GaAs QD samples upon annealing.

2. Experimental

Two samples with eight layers of InAs/InGaAs QDs were grown in a molecular beam epitaxy (MBE) system on Si-doped GaAs (100) substrates. Nominally 0.8nm InAs was deposited to form QD layers which was then followed by 5nm InGaAs deposition to reduce the strain. Then, 33nm GaAs barrier layers were deposited (cf. Figure 1). The only difference between the samples is that in one of them the GaAs ~14nm above the QD layers has been p-doped by carbon to a concentration of ~$3x10^{17}/cm^3$. A post-growth annealing process (PAP) was applied to both samples for 5mins at 700ºC. In photoluminescence (PL), the carbon-doped sample shows a smaller blue shift of only ~13nm compared to over 50nm for the undoped sample and a stronger PL intensity after annealing (cf. Figure 1). To study the microstructure of these samples, specimens for cross-sectional transmission electron microscopy (TEM) were prepared by standard methods involving glueing, cutting, grinding, polishing and Ar^+ ion milling until perforation. We applied annular dark-field (ADF) STEM using a JEOL 2010F field-emission transmission electron microscope equipped with a scanning unit providing an electron probe of ~0.25nm in size at ~10mrad semi-angle of beam convergence and 0.5mm spherical aberration.

3. Results

Figure 1 sketches the details of the layers and compares the photoluminescence of InAs/InGaAs QDs (with C-doped and undoped GaAs barriers) before and after the annealing process. ADF imaging has been used at medium magnification (120kX) to determine the spacings between neighbouring QDs and at high magnification (1500kX) to measure the sizes of the QDs (Fig. 3) and the thickness of the InGaAs strain relaxing layer for both samples.

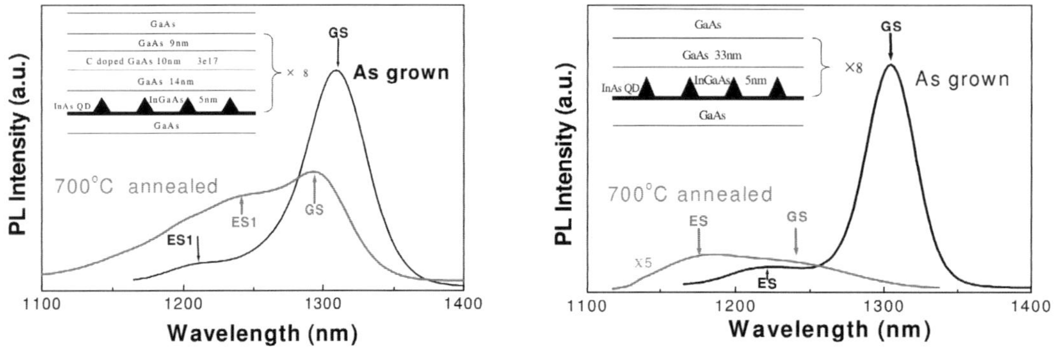

Figure 1. Photoluminescence of InAs/InGaAs quantum dots in the C-doped (left) and the undoped sample (right) before and after annealing

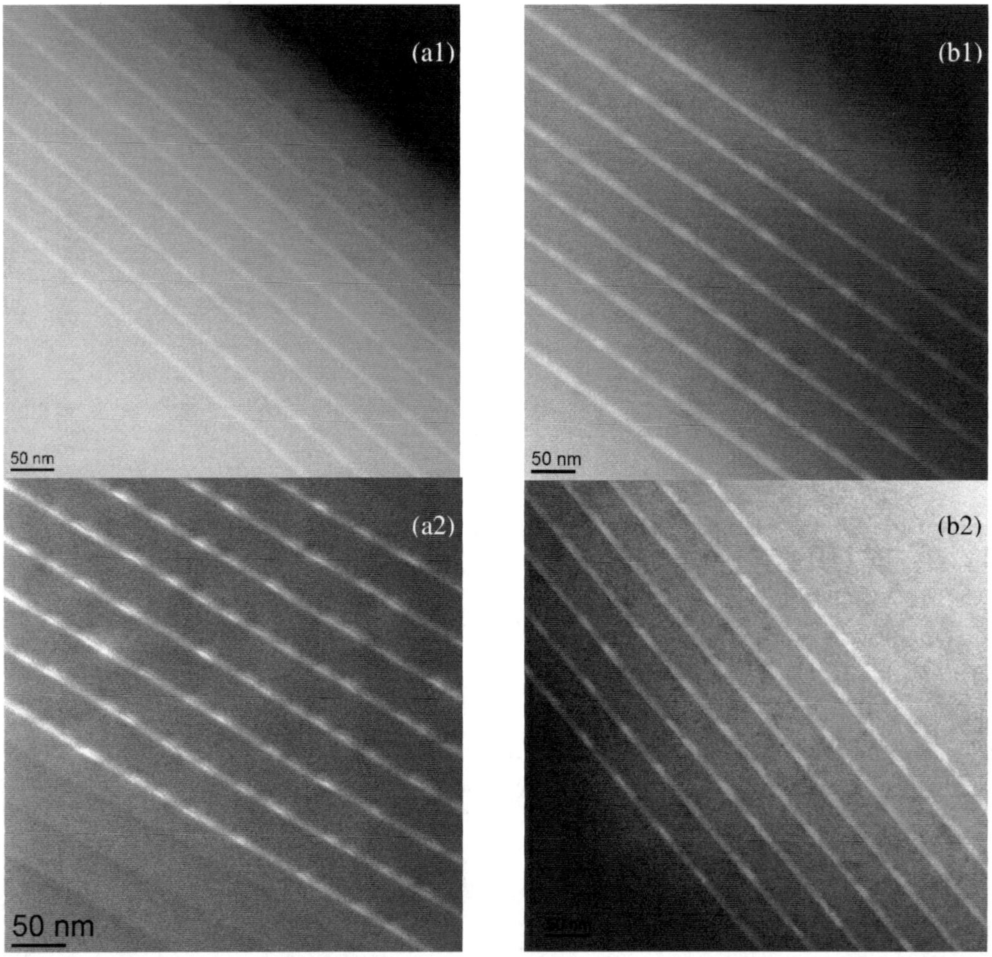

Figure 2. Overview of the samples before and after anneal: (a1) C-doped sample, (a2) C-doped sample after anneal, (b1) undoped sample, (b2) undoped sample after anneal. The growth direction points towards the top right in all images. Note the high contrast in (a2).

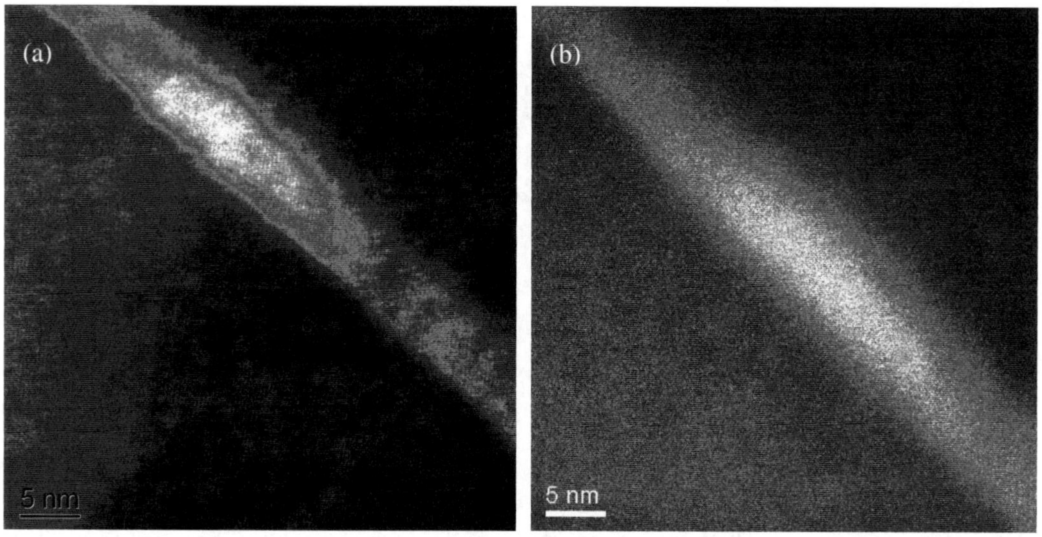

Figure 3. QD after anneal in the C-doped sample (a) and after anneal in the undoped sample (b). The pseudo colour of the images has been obtained after calculating the square root of the annular dark-field intensity.

4. Discussion

4.1. Study of quantum dots

The InAs/InGaAs QD sizes and their spacings in C-doped (wafer # 1466) and undoped (wafer # 1467) as-grown samples (1466ag and 1467ag, respectively) and annealed C-doped and undoped samples (1466a and 1467a, respectively) have been measured and compared.

4.1.1. Size of Quantum Dots

Around 50 QDs have been investigated for each sample using ADF imaging at 1500kX magnification to determine the height and length of the QDs. The intensity in ADF imaging is approximately proportional to the square of the average atomic number ("Z-contrast") if the inner collection angle is very large (high-angle ADF) and the specimen uniformly flat over the field of view. Hence, maps of the chemistry can be obtained for systems such as InGaAs where only two types of atomic species can interdiffuse (here: In and Ga on the group-III sub-lattice). Neglecting contrast reduction due to finite collection angles [5] we can approximate a chemical map by calculating the square root of the ADF image intensity. Then the heights and the lengths of the QDs have been measured as full widths at half maximum (FWHM) of intensity line profiles across these maps. Figures 4 and 5 display the heights and lengths of QDs after annealing for both C-doped and undoped samples.

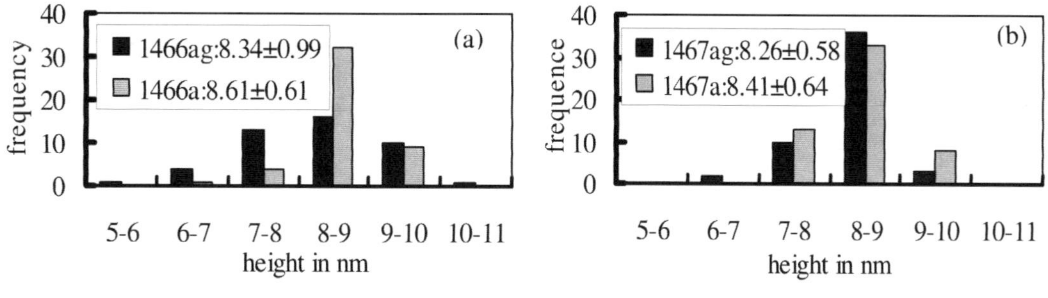

Figure 4. QD heights as-grown and after anneal for C-doped (a) and undoped (b) samples

Journal of Physics: Conference Series **209** (2010) 012036 doi:10.1088/1742-6596/209/1/012036

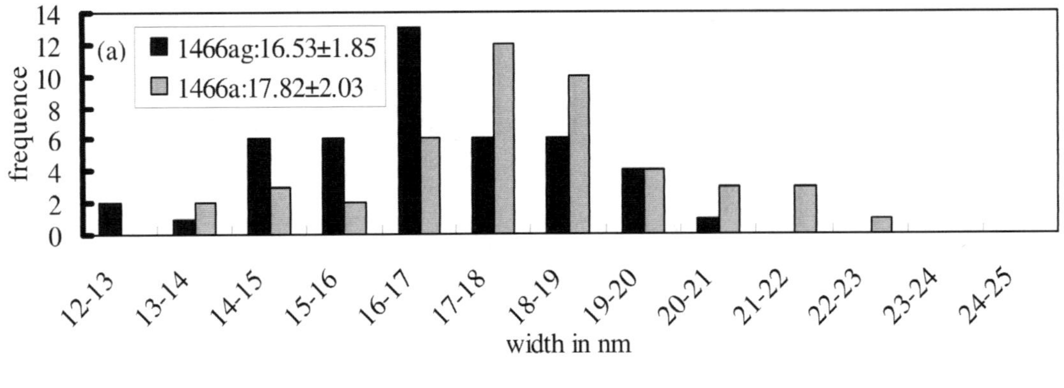

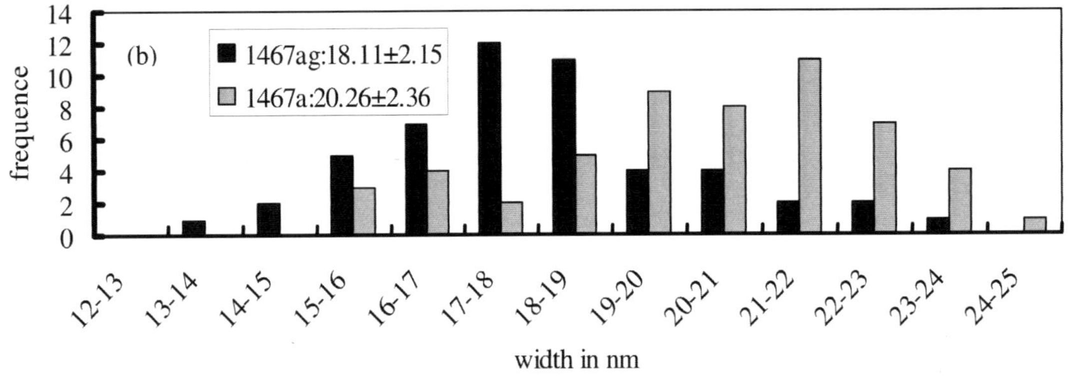

Figure 5. QD widths as-grown and after annealing for C doped (a) and undoped (b) samples

The heights and lengths of the QDs in the C-doped sample increase during the anneal by 3.2% and 7.8%, respectively. On the other hand, the heights and lengths of the QDs in the undoped sample increase by 1.8% and 11.9%, respectively. This means that the QDs in both samples become larger but undergo shape changes upon annealing, with the QDs in the undoped sample becoming flatter.

The general increase in QD size upon annealing is due to interdiffusion of indium between the InAs QDs and the surrounding GaAs barriers. This must reduce the In concentration in the QDs. The drop of In concentration due to interdiffusion increases the bandgap of the QDs from that of InAs towards that of a ternary InGaAs alloy, consequently a blue shift of the ground state (GS) transitions in PL intensity is observed. On the other hand, the slight increase of the heights of the QDs upon anneal reduces the quantum confinement, which would result in a weaker red shift of GS transitions in both samples.

4.1.2. Line density: spacings between nearest quantum dots
More than 300 distances between neighbouring quantum dots have been measured by ADF overview images taken at 50kX, 60kX, 120kX, 150kX, and 200kX magnifications (examples in Fig. 2). Figure 6 compares the histograms of the statistics of both samples before and after the anneal.

Fig. 6(a) shows that new QDs seem to form between existing ones only in the C-doped sample.

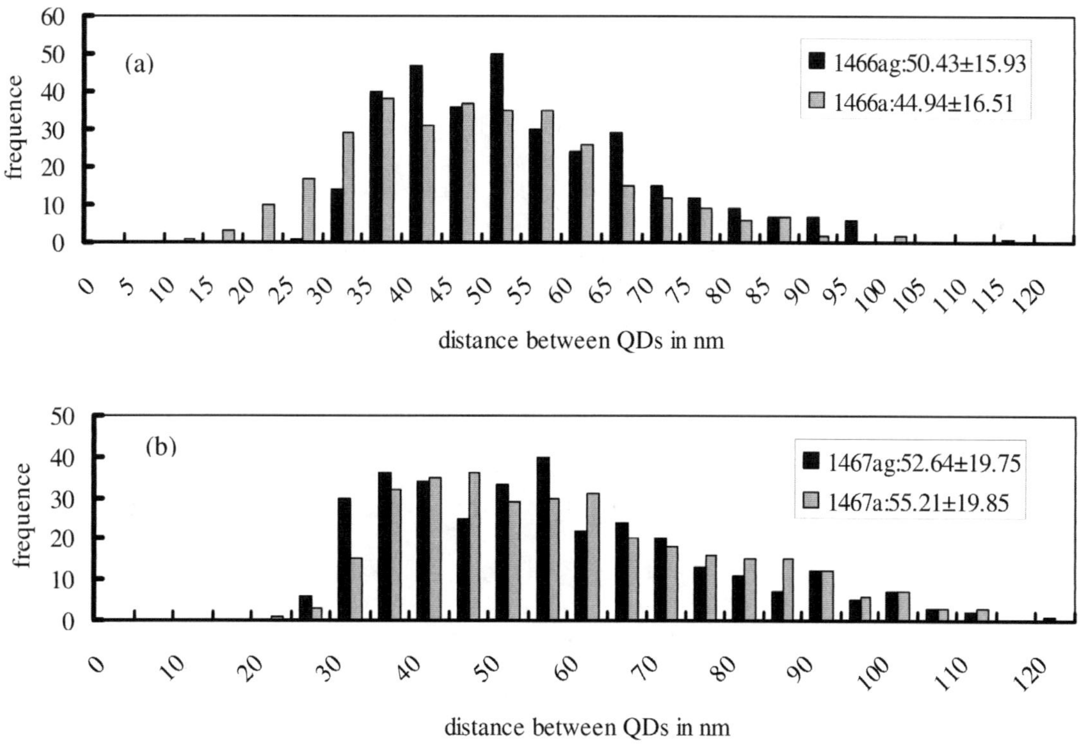

Figure 6. Histograms of spacings between nearest QDs for C-doped (a) and undoped (b) samples

4.2. Study of InGaAs barrier layer thickness

We studied the changes in indium concentration in the barrier layers for C-doped and undoped samples before and after anneal qualitatively by comparing the image contrasts, and we used ADF imaging to monitor thickness changes of the InGaAs strain relaxing layer. Table 1 shows the results.

Table 1. Comparison of InGaAs layer thicknesses (all values in nm)

doping	C-doped		undoped	
sample	1466ag	1466a	1467ag	1467a
measurement 1	5.59	5.09	5.61	5.55
measurement 2	5.66	5.10	5.59	5.60
measurement 3	5.34	5.25	5.35	5.49
measurement 4	5.58	5.11	5.70	5.60
average	5.54	5.14	5.56	5.56
standard deviation	0.14	0.08	0.15	0.05
mean error of average	0.07	0.04	0.08	0.02

The thickness of the InGaAs barriers above the QDs has been measured by ADF-STEM at 1000kX, 1200kX and 1500Kx magnification at four different locations between the quantum dots. According to Table 1, the barrier layer thickness decreases significantly upon anneal, by about 7%, only for the C-doped sample, whereas the thickness stays roughly the same for the undoped sample. Due to the decrease of the thickness of the InGaAs barriers and the conservation of the number of indium atoms, in the C-doped sample In atoms must have moved from the barriers into the QDs, thereby increasing the indium concentration in the QDs upon anneal by lateral segregation, similar to what has been observed for strained SiGe/Si layers [6, 7]. This also agrees with the much higher ADF contrast in the annealed C-doped sample compared to the undoped annealed sample (cf. figure 3).

5. Conclusion

The annealing process encourages interdiffusion between the InAs QDs and the surrounding (In)GaAs material, resulting in a decrease in indium composition and hence an increase of the bandgap, which shifts GS transitions to the blue. On the other hand, the slightly increased heights of the QDs upon annealing reduce the quantum confinement, leading to a slight red-shift of GS transitions in both samples. The latter effect is, however, much less pronounced than the change in bandgap due to alloying.

The increase of the In concentration in the QDs in the C-doped sample causes a reduction of the bandgap by increased quantum confinement in this material only and thereby counteracts the dominant effect of interdiffusion. As a result, the PL intensity the GS energy of the C-doped sample is less affected and blue-shifted by only 13nm while the undoped sample is blue-shifted by more than 50nm.

Finally, the strong interdiffusion causes the integrated PL intensity of the undoped sample to drop significantly compared to the C-doped sample, while in the C-doped sample the loss of indium due to interdiffusion is partly compensated by the lateral segregation of indium atoms from the InGaAs capping into the QDs.

In effect, C-doping stabilises both the intensity and the energy of optical emission from the quantum dots, however, this may probably be achieved not by directly suppressing interdiffusion but by compensating out-diffusion of indium from the quantum dots by lateral segregation of indium from the InGaAs layers to the quantum dots. This means C-doping influences the lateral diffusivity of indium atoms in such structures. The main question remaining is whether this is due to some of the doping atoms diffusing down to the QD layers rather than attracting Ga vacancies from there. In this case C acceptor atoms on As sites, which should be the most stable configuration in the GaAs lattice [8], could neutralise the charges on Ga vacancies, which would tend to increase further the already large local lattice distortions at the strained layer interfaces, thereby lowering even further the Peierls stress required to be overcome by diffusing atoms.

References

[1] Mowbray D et al 2007 *International Journal of Nanoscience* **6**, 291
[2] Deppe DG and Holonyak N 1988 *J. Appl. Phys.* **64**, R93
[3] Ooi BS, Bryce AC, Marsh JH and Roberts JS 1997 *Semicond. Sci. Technol.* **12**, 121
[4] Zhang ZY, Jiang Q and Hogg RA 2008 *Applied Physics Letters* **93**, 07111
[5] Walther T 2006 *J. Microsc.* **221**, 137
[6] Walther T, Humphreys CJ and Robbins DJ 1997 *Defect and Diffusion Forum* **143**, 1135
[7] Walther T, Humphreys CJ and Cullis AG 1997 *Appl. Phys. Lett.* **71**, 809
[8] Cheong B-H and Chang KJ 1994 *Phys. Rev. B* **49**, 17436

Journal of Physics: Conference Series **209** (2010) 012037 doi:10.1088/1742-6596/209/1/012037

TEM studies of multilayered $In_{0.33}Ga_{0.67}As$ quantum dots

DV Sridhara Rao[1], K Muraleedharan[1], R Balamuralikrishnan[1], R Muralidharan[2], T Srinivasan[2] and D Banerjee[3]

[1] Defence Metallurgical Research Laboratory, Kanchanbagh Po, Hyderabad-58, India.
[2] Solidstate Physics Laboratory, Timarpur, Lucknow Road, Delhi-54, India
[3] Defence Research and Development Organisation, New Delhi, India.

E-mail: vsrd2@cam.ac.uk, dvsridhararao@gmail.com

Abstract. We have employed TEM techniques to investigate vertical alignment in multilayered $In_{0.33}Ga_{0.67}As$ quantum dots (QDs) as a function of deposition thickness. Our experimental results are discussed in the context of an analytical model reported in the literature, in terms of the so-called pairing probability, which can assess the quality of vertical alignment.

1. Introduction

$In_xGa_{1-x}As$ quantum dots (QDs) find applications in optoelectronic devices. $In_xGa_{1-x}As$ grows on GaAs as a two-dimensional (2D) layer up to a transition thickness (t_c), and further growth results in a 2D-3D (island) morphological transition with a wetting layer. This is known as Stranski-Krastanov growth [1]. This growth mode is attractive as it results in self-assembled islands. However, many applications require the growth of QDs with large areal density, which is usually not realizable in the case of single-layered QDs. This can be overcome by the growth of multilayered dots, i.e. by repeated growth of a coherent layer of $In_xGa_{1-x}As$ QDs followed by a thin GaAs spacer layer. The tensile strain field due to the buried dots transmits through the spacer layers, resulting in preferential growth of dots on top of the buried dots, thereby leading to their vertical alignment [2].

In this work, we have used TEM techniques to investigate vertical alignment in multilayered $In_{0.33}Ga_{0.67}As$ quantum dots (QDs) as a function of deposition thickness. Our results suggest that the dots are neither prominent nor vertically aligned when the deposition thickness is just above t_c. However, for deposition thickness ~ $1.5t_c$, the dots are well formed with good vertical alignment. Our experimental results are discussed in the context of an analytical model reported in the literature, in terms of the pairing probability (P) [3], which can be used to determine the quality of vertical alignment.

2. Experimental

The (001) epi-ready GaAs substrates were loaded in the growth chamber of a RIBER 2300 R&D molecular beam epitaxy (MBE) system. After desorption of the native oxide at 580°C, a buffer structure consisting of a semi-insulating GaAs (0.5 μm thick) layer and an $Al_{0.24}Ga_{0.76}As$/GaAs superlattice (20 periods, thickness of one period is 10 nm) was grown by using Ga, In and As solid sources and under arsenic overpressure (or beam equivalent pressure, BEP ratio of about 20). The multilayered QD structure consisting of 20 periods of $In_{0.33}Ga_{0.67}As$ layers with 20 nm thick GaAs spacers was grown at 500°C, without growth interruption. During growth, in-situ reflection-high energy electron diffraction (RHEED) was used to establish that the transition thickness of $In_{0.33}Ga_{0.67}As$ layers was 12 monolayers (MLs) (~3.47 nm). Two different samples were grown with a nominal thickness of 13 MLs (just above t_c, case-A) and another with 17 MLs (~ 1.5 t_c, case-B) respectively. The latter structure was capped with a 5 nm thick GaAs, whereas the 13 MLs sample was left uncapped. The samples were cooled to room temperature under arsenic overpressure.

Cross-sectional TEM specimens were prepared by argon ion beam thinning using a Gatan Duo Mill (Model 600) with a liquid nitrogen cold stage. The samples were investigated using a Philips EM430T analytical TEM (operated at 250 kV) and a Tecnai 20 UT high-resolution TEM (HRTEM, 200 kV). The specimen preparation procedure ensured that the foils were near <110> zone axis at zero tilt conditions. This zone contained the direction of growth (i.e., [001]). As a result, the multilayers could be investigated in the edge-on orientation with respect to the electron beam. Also, (002) composition-sensitive two-beam dark field (DF) imaging was used for the estimation of the structural parameters (height and width) of dots [4].

3. Results and Discussion

In case-A, the cross-sectional TEM studies (Fig. 1A) revealed that the buried dots are not well formed. However, in case-B, the QDs are formed prominently (Fig. 1B), with a wetting layer, confirming the S-K growth mode. The wetting layer thickness is 3.5±0.3 nm, which is, within measurement errors, close to the transition thickness of 3.47 nm (corresponding to 12 MLs). The extra material deposited beyond the transition thickness has contributed to the growth of the dots to a bigger size. Also, the QDs are coherent with the GaAs spacers (Fig. 1C). Further, the QDs are well aligned along the growth direction, and they resemble the shape of a plano-convex lens. The strain effects are dominant in the <110> on-zone imaging condition; hence the structural parameters of the QDs are estimated by using the (002) two-beam dark (DF) field imaging condition. Due to the composition sensitive nature of this imaging condition, $In_{0.33}Ga_{0.67}As$ appears darker than GaAs (Figs. 1A,1B). In addition, a contrast variation has been observed within each dot. This can be attributed to a shape factor [5] caused by the plano-convex morphology of the dot, due to which the electron beam passes through varying thicknesses of $In_{0.33}Ga_{0.67}As$ and the embedding GaAs, and also from one phase to the other, as we move from the top of the dot to its base. Beanland [5] has also shown that the actual size of the dots is the region enclosed within the black contrast periphery and its outer edge has to be used for the estimation of the height (h) and diameter (D) of the dots.

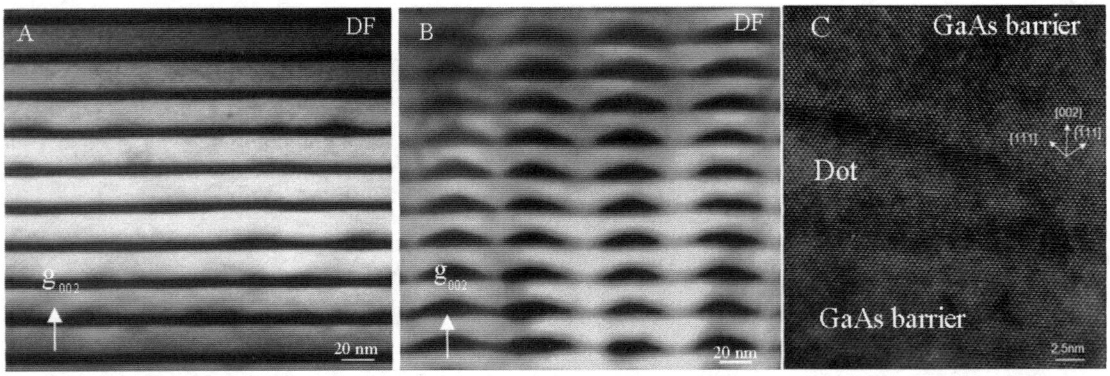

Figure 1. (002) Dark field TEM micrographs of multilayered quantum dots corresponding to a $In_{0.33}Ga_{0.67}As$ deposition thickness of (A) 13 MLs (case-A) and (B) 17 MLs (case-B). The lattice resolution image of a buried dot is shown in figure C.

We have quantified the vertical alignment of the QDs along the growth direction, using the analytical model proposed by Xie Qianghua et al. [3]. In this model, developed for the InAs/GaAs system, the authors quantified the vertical alignment using the concept of pairing probability (*P*). The factors considered are: dot size (*d*), inter-dot spacing (*l*), thickness of spacer layer (Z_s), and the elastic properties (such as misfit and elastic constants) of the InAs and GaAs layers. The average dot size is given by the diameter of an equivalent sphere that has the same volume as that of the plano-convex lens of height *h* and diameter *D*. For an ideal case (i.e. perfect vertical alignment), $P = 1$. It decreases with the deterioration of the alignment. Based on the value of *P*, the authors have classified the alignment as being strongly correlated (*P* ~ 0.9 or higher), moderately correlated (*P* ~ 0.4–0.9), or uncorrelated ($P < 0.4$).

Table 1. Material parameters used for estimation of pairing probability. For ternary alloy, the respective parameters were estimated using Vegard's law.

Parameter	GaAs[a]	InAs[a]	$In_{0.33}Ga_{0.67}As$
Young's Modulus (*E*, GPa)	86.96	51.42	75.23
C_{11} (GPa)	118.8	83.4	107.12
C_{12} (GPa)	53.8	45.4	51.03
C_{44} (GPa)	59.4	39.5	52.83
Bulk Modulus[b] (*B*, GPa)	76.28	57.13	69.96
Poisson's ratio (γ)	0.31	0.35	0.323
Lattice mismatch (with GaAs)	-	7.2%	2.3%
Lattice parameter (*a*, nm) (Rounded-off to fourth decimal)	0.6058	0.5653	0.5787
[a] Reference [6] [b] Bulk Modulus=$E/[3(1-2\gamma)]$			

We have calculated the pairing probability for QD structures (cases A, B) using the structural parameters obtained from our experiments. The QDs in the top layer were not considered in the estimations, as the capping affects the structural parameters. In case-A, though the buried dots are not well formed, we have nevertheless attempted to estimate the required structural parameters. The average dot height (*h*) and diameter (*D*) are estimated to be about 3 nm and 17 nm leading to an average dot size of 8.78 nm. The average interdot spacing (*l*) has been measured to be 70 nm. In case-B, the average dot height (*h*) and diameter (*D*) of the buried dots are estimated to be 9 nm and 42 nm leading to an average dot size of 23.3 nm. The average interdot spacing (*l*) has been measured to be 58 nm. Ternary material parameters, such as lattice parameter and bulk modulus, were calculated using Vegard's law from the reported data for InAs and GaAs (Table 1). Similar to the analysis of Qianghua Xie et al. [3], the pairing probability as a function of the spacer thickness has been plotted for the two cases (Fig. 2). At a spacer thickness of 20 nm (our experiments), *P* has been estimated to be about 0.25 and 0.92 respectively, indicating that uncorrelated and strongly correlated conditions are prevalent in the respective cases. These predictions are in good agreement with our experimental observations (Figs. 1A and 1B), indicating that the model continues to hold for QDs with lower indium content.

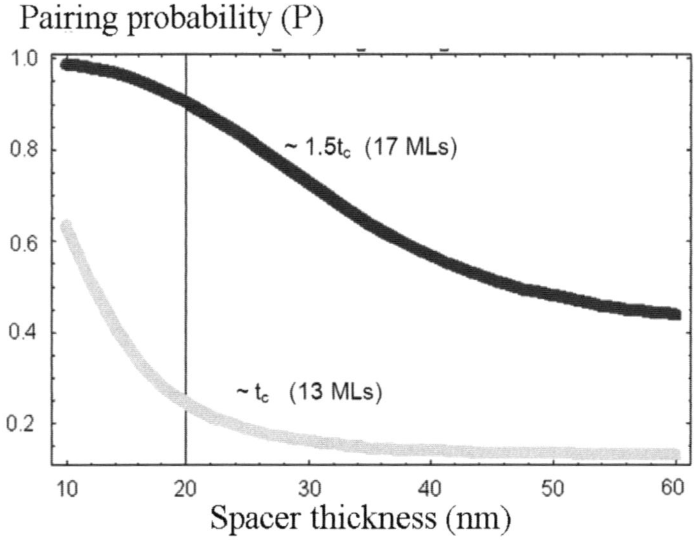

Figure 2. Pairing probability (*P*) as a function of spacer thickness of $In_{0.33}Ga_{0.67}As$ dots at a deposition thickness: ~t_c (13 MLs, corresponding to figure 1A) and ~1.5t_c (17 MLs, corresponding to figure 1B).

4. Summary

We have investigated the formation of $In_{0.33}Ga_{0.67}As$ multilayered dots with GaAs spacers. The dots are well formed with good vertical alignment for a deposition thickness close to 1.5 times the transition thickness. In contrast, the dots are not prominently formed and relatively poorly aligned for the deposition thickness just above the transition thickness. In both cases, the dots are predominantly coherent. Further, considerable agreement on the vertical alignment of dots has been found between our experiments on $In_{0.33}Ga_{0.67}As$ quantum dots and the analytical model of Qianghua Xie et al. [3] developed for InAs quantum dots.

Acknowledgements

This research is funded by Defence Research and Development Organisation. The authors are grateful to Director, Defence Metallurgical Research Laboratory, Hyderabad and Director, Solid State Physics Laboratory, Delhi, for their support and permission to communicate this work.

References

[1] Bimberg D, Grundmann M and Ledentsov NN, 1999 *Quantum Dot Heterostructures* (John Wiley and Sons, Chichester)
[2] Priester C and Grenet G 2001 *Phys. Rev.* B **64**, 125312
[3] Xie Q, Madhukar A, Chen P and Kobayashi NP 1995 *Phys. Rev. Lett.*, **75(13)** 2542
[4] McCaffrey JP, Washewski ZR, Robertson MD and Corbett JM, 1997 *Phil. Mag. A* **75**, 803
[5] Beanland R 2005*Ultramicroscopy* **102**, 115
[6] Liu GR and Jerry SSQ 2002 *Sem. Sci. and Techol.* **17**, 630

Journal of Physics: Conference Series **209** (2010) 012038 doi:10.1088/1742-6596/209/1/012038

Study of InGaN/GaN quantum dot systems by TEM techniques and photoluminescence spectroscopy

RJ Kashtiban[1], U Bangert[1], B Sherliker[2], M P Halsall[2] and AJ Harvey[1]

[1] Nanostructured Materials Research Group, School of Materials, The University of Manchester, P.O. Box 88, Manchester, M1 7HS, UK
[2] Microelectronics and Nanostructures Research Group, School of Electrical and Electronic Engineering, University of Manchester, M60 1QD, UK

E-mail:Reza.jalilikashtiban@postgrad.manchester.ac.uk

Abstract: InGaN/GaN multilayer quantum dot structures produced by MOCVD techniques on c-plane sapphire were studied by transmission electron microscopy (TEM) and photoluminescence (PL) techniques. Indium fluctuations ranging from 1-4 nm were observed with both energy filtered TEM (EFTEM) and high angle annular dark field (HAADF) scanning TEM. The existence of V-shaped defects with nucleation centres at the termination of threading dislocation were observed in HAADF images. There was also evidence of the formation of large quantum dots at low densities from lattice HRTEM images. This was further confirmed by PL measurements through the observation of a single sharp line at low power with the typical saturation behaviour at higher power excitation.

1. Introduction

Self-organized quantum dot structures (SOQDs) are currently being studied for device applications. Ternary III-V alloys employing indium are the most promising materials for applications spanning the frequency range from Terahertz to visible/near UV. InGaN SOQDs have recently been used in the active region of laser diodes to improve their performance [1]. Despite the wealth of potential applications the growth of desirable In-rich quantum structures is still a challenge because of the high lattice mismatches, and therefore it is important to obtain a detailed understanding of the formation process and the atomic arrangements of the group III elements. In this work high angle annular dark field (HAADF) (where contrast arises from the sensitivity of the scattering intensity to the atomic number), energy filtered transmission electron microscopy (EFTEM) (which shows up compositional changes in the sample structure and provides concentration maps) as well as high resolution transmission electron microscopy (HRTEM) and photoluminescence (PL), are combined to study quantum dots in the InGaN/GaN system.

2. Experimental

Samples were grown on c-plane sapphire by metal-organic chemical vapour deposition (MOCVD): a thick GaN buffer-layer was grown at 1120 °C followed by 5 repeats of 15nm GaN at 765 °C and 2 nm InGaN at 745 °C with nominally 15% In (top layer was uncapped). HRTEM, EFTEM and HAADF images were acquired with a Tecnai F30 FEG-TEM microscope operating at 300 kV. EFTEM maps were produced by passing electrons which have lost at least 443 eV, corresponding to the In M4,5 absorption edge, through a Gatan imaging filter. The images were taken under short periods of irradiation to prevent electron beam induced artefacts [2], and the electron current density was kept as low as possible to further reduce any electron beam damage. Care was also taken to capture images from a fresh area within 2 minutes to reduce damage due to prolonged exposure [3]. The specimens for cross-sectional TEM analysis were prepared by mechanical polishing followed by ion beam milling

A 100 nm thick Al film was evaporated on to the sample surface which was covered by specific diameter polystyrene beads (~650nm). The sample was then placed in liquid nitrogen to remove the beads. This produced sub-micron apertures on the sample surface, through which microscopic photoluminescence (μ-PL) spectra were taken using the 405 nm line of a cube diode laser as the excitation power source. The laser beam was focused on the sample surface by an objective lens. The PL signal was collected by the same objective lens, and detected by a liquid-nitrogen-cooled charge coupled device (CCD) detector through a monochromator. The laser spot and its position on the sample surface were monitored by a CCD camera. The sample was mounted in a cold finger cryostat cooled to about 3.8K.

3. Results and Discussion

Figure 1(a) shows a HRTEM image of the bottom 3 quantum dot layers. There is evidence for a large quantum dot on the bottom layer (dark pyramidal region). The dot diameter and height are estimated to be 15 nm and 2.5 nm respectively. Similar quantum dots were observed by other groups [4] The dark region is due to strain caused by lattice mismatch between the InGaN region and the GaN barrier. Figure 1(b) shows a HRTEM image of all five GaN/InGaN layers and part of the thick GaN buffer (bottom left). From AFM images the density of the surface dots was evaluated to be about 6×10^{8} cm^{-2} [5].

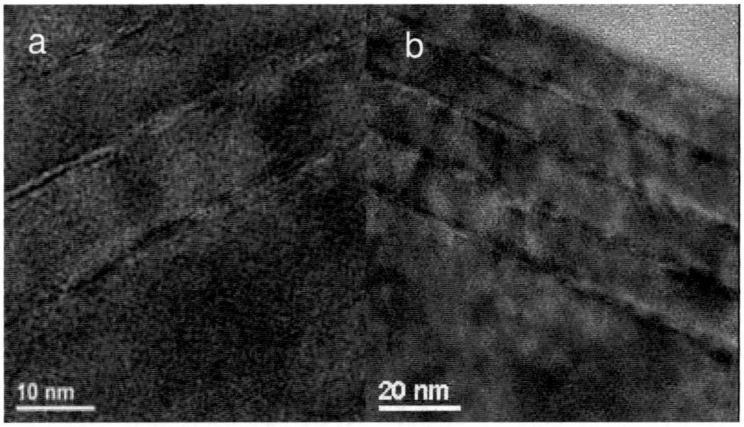

Figure 1. HRTEM image of (a) the bottom 3 quantum dot layers, (b) all five repeats of GaN/InGaN layers and part of the thick GaN buffer layer (bottom left)

Figure 2(a) shows an HAADF scan in which four quantum dot layers can be identified. The image was acquired from a relatively thick part of the sample; therefore the dot layers appear very uniform. The dark area at the bottom left of the image is the vacuum. The InGaN layer width is measured to be approximately 2 nm, which agrees well with the nominal growth thickness. The thickness of the GaN barrier is measured slightly less than the nominal thickness of 15 nm. Figure 2(b) is a HAADF scan from an area of the sample which is thinner than in (a). Indium fluctuations within the quantum dot layers can be seen here, which range from 1 to 4 nm in length. Bright stripes parallel to the basal plane are InGaN layers (*Z* of In = 49) and dark ones are GaN layers (*Z* of Ga = 31). Figure 2(c) is an HAADF scan from a thicker area of the sample. Two V- shaped defects [6] can be seen at the surface, one of which nucleates at the termination of a threading dislocation.

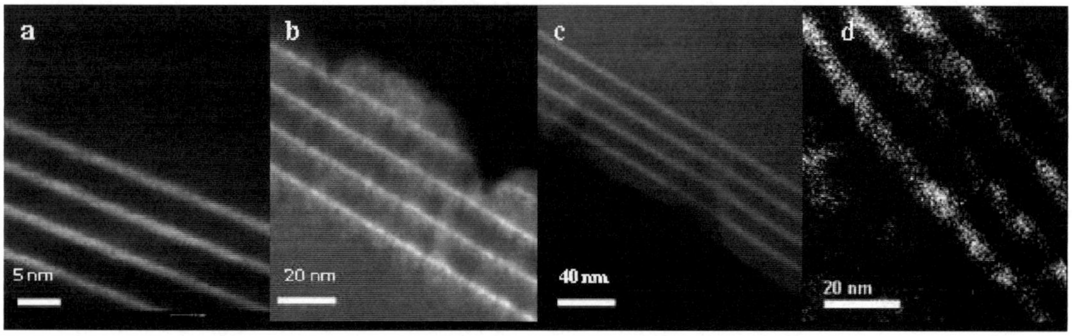

Figure 2. HAADF images of (a) four quantum dot layers, (b) an area thinner than in (a) and (c) an area with V-shaped defects and threading dislocation. (d) EFTEM map of In distribution in the quantum dot layers.

In order to understand the compositional information of the sample EFTEM was carried out. Figure 2(d) represents an EFTEM map produced by passing electrons through the filter that have lost 443 eV corresponding to the In M4,5 absorption, to show the indium distribution in the structure. It can be seen that the distribution in the quantum dot layers is not uniform, with fluctuations leading to small-scale clustering of indium.

In order to see manifestation of quantum dot formation in the optical behaviour of the sample PL experiments were carried out. Figure 3 represents PL spectra for two different samples at 4K as function of excitation power taken through a sub-micron size aperture (~550nm); the upper spectrum in (a) is from an unmasked sample (macro-PL). The usual excitonic emission (the large band) in (a) consists of several sharp lines while the macro-PL spectrum exhibits a continuum (broad peak). The discrete structure of the emission indicates that the epilayer is strongly disordered due to the immiscibility of indium in GaN within the wetting layer and to well width fluctuations, suggesting that there are localized exciton centres. Figure 3 (b) shows a series of spectra from a sample with a longer deposition time. At low excitation powers there is a single sharp line. This line reaches saturation point as the excitation power increases, as is typical of a quantum dot. We attribute this line to a single quantum dot. The AFM studies by Davies et al. [5] suggest a low dot density of about 6×10^{8} cm^{-2} and statistical studies of the number of apertures (~550 nm apertures) containing at least one dot (using PL) agree with this AFM density.

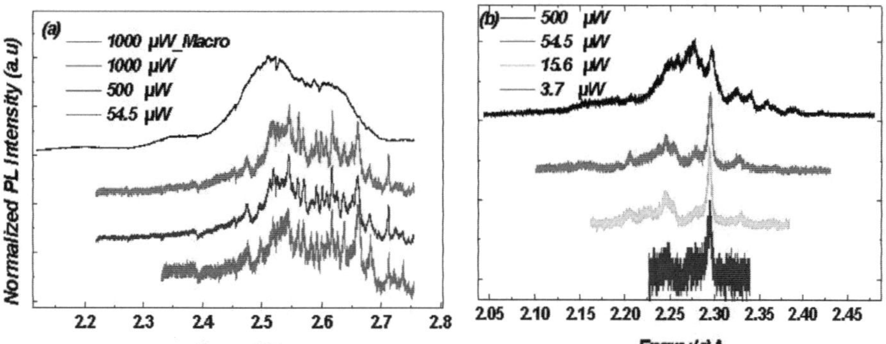

Figure 3. PL spectra at 4K as function of excitation power, taken through a submicron size aperture (~550 nm) for two samples with InN deposition times of 12s and 17.5s. Spectra are displaced vertically upwards with increasing excitation power.

4. Conclusion

EFTEM images using the In M4,5 absorption edge showed that In distribution is not uniform in the quantum wells and that there are fluctuations leading to clustering of In. There is agreement between EFTEM and HAADF showing the indium clusters within the quantum dot layers range from 1 - 4 nm in length. V-shaped defects were observed, some of which nucleated at the termination of a threading dislocation. There is evidence of large quantum dots in HRTEM images with estimated diameter and height of 15 nm and 2.5 nm respectively. This is confirmed by micro PL results which show a single sharp PL line at lower excitation power similar to a typical quantum dot emission which saturates with increasing power. AFM suggests a low density of 6×10^8 cm^{-2} surface dots.

Acknowledgment

Cooperation of Dr. P J Parbrook and Dr. T Wang from the ESPRC National Centre for III-V Technologies, Department of Electronic and Electrical Engineering, University of Sheffield for growing the samples is greatly acknowledged.

References

[1] Arakawa Y 2002 *IEEE J.* **8**, 823
[2] Smeeton TM, Kappers MJ, Barnard JS, Vickers ME and Humphreys CJ 2003 *Appl. Phys. Lett.* **83**, 5419
[3] O'Neill JP, Ross IM, Cullis AG, Wang T and Parbrook PJ 2003 *Appl. Phys. Lett.* **83**, 1965
[4] Wang Q, Wang T, Bai J, Cullis AG, Parbrook PJ and Ranalli F 2008 *Appl. Phys. Lett.* **93**, 081915
[5] Davies SC, Mowbray DJ, Ranalli F, Parbrook PJ, Wang T, Yea B, Sherliker B, Halsall H, Kashtiban RJ and Bangert U, to be published in *Appl. Phys. Lett.*
[6] Shiojiri M, Chuo CC, Hsu JT, Yang JR and Saijo H 2006 *J. Appl. Phys.* **99**, 073505

Structure and luminescence of sol-gel synthesized anatase nanoparticles

U Hörmann[1], U Kaiser[1], M Albrecht[2], J Geserick[3] and N Hüsing[3]

[1] Ulm University, Electron Microscopy Group of Materials Science, Albert-Einstein-Allee 11, 89081 Ulm, Germany
[2] Leibniz Institute for Crystal Growth, Max-Born-Str. 6, 12489 Berlin, Germany
[3] Ulm University, Institute of Inorganic Chemistry I, Albert-Einstein-Allee 11, 89081 Ulm, Germany

E-mail: ute.hoermann@uni-ulm.de

Abstract. Two samples of mesoporous anatase nanoparticles, prepared by the sol-gel method, were characterised by Cs-corrected high resolution transmission electron microscopy (HRTEM), X-ray powder diffraction (XRD) and Raman spectroscopy. Statistical evaluation of TEM data showed an average diameter of these crystallites of 8.8 nm and 11.1 nm, respectively. Optical spectroscopy by cathodoluminescence (CL) in a scanning electron microscope (SEM) showed free exciton transitions related to the direct and the indirect band gap of anatase TiO_2. From the analysis of the excited states of the free excitons an exciton binding energy of 10 meV and a Bohr radius of 2.35 nm is obtained. The small Bohr radius could explain the absence of quantum confinement in the particles presented in this study.

1. Introduction

Anatase is a wide band gap semiconductor with indirect band gap [1]. The band gap of bulk anatase was calculated by Wunderlich [2] to 3.20 eV, which is in accordance with experimental results.

Applications of mesostructured nanocrystalline titania are based on photoexcitation. In particular, titania is an important material in photo catalysis [3] due to its high catalytic activity. Other applications using photon excitations are the Grätzel type solar cell [4] and electrochromic devices [5].

It is therefore of great interest to get access to basic materials properties, such as the Bohr radius, effective masses and band structure to understand the size dependence of the electronic properties. While size dependence has been extensively studied and measured in terms of size dependent Raman shifts or photo luminescence transition lines [1,6], the existence of a quantum confinement effect in anatase and rutile nanoparticles is still under debate, e.g. questioned by Monticone et al. [7] who performed absorption measurements of anatase TiO_2 with sizes ranging from 2-8nm and found no indications for the existence of quantum confinement at all. These authors explained their findings by a size dependence of the effective hole mass in this material system.

2. Experiment

The anatase nanoparticles were synthesised in a sol-gel process at room temperature at pH=2. The sample denoted SDS was prepared from ethylene glycol modified titanate (EGMT) in the presence of

Journal of Physics: Conference Series **209** (2010) 012039 doi:10.1088/1742-6596/209/1/012039

the surfactant sodium dodecyl sulphate (SDS) in a molar ratio of SDS:Ti = 0.2:1. For the sample denoted as Brij56 glycol modified titanate (GMT) was converted in the presence of the surfactant Brij56 in a molar ratio of Brij56:Ti=0.03:1. After drying the samples they were calcined for 4 h at 400 °C in air in order to remove the surfactant.

Nitrogen sorption measurements were performed on a NOVA 4000e and Autosorp MP1 instrument (Quantachrome). The surface area was calculated according to Brunauer, Emmett and Teller (BET) in the p/p_0 range of 0.05-0.3 and the pore size distribution was determined according to Barrett, Joyner and Halenda (BJH) from the desorption branch of the isotherms.

XRD measurements were performed by using Cu-K_α radiation ($\lambda = 0.154$ nm) on a PANalytical MPD PRO diffractometer.

For the TEM sample preparation the powders were dispersed in ethanol and 5 minutes treated in an ultrasonic bath. Afterwards the dispersion was dripped onto a holey carbon film coated copper grid. The samples were structurally characterised by Cs-corrected high resolution transmission electron microscopy using a FEI Titan 80-300 operated at 300 kV.

Raman measurements were carried out on a Jobin-Yvon LabRam using a 630 nm laser for excitation. Cathodoluminescence spectroscopy has been performed using a Zeiss DSM microscope with a Gatan MonoCL3 cathodoluminescence system attached. Light was dispersed by a 2400 l/mm grating blazed at 300 nm. A Peltier cooled charge coupled device camera was used for detection.

3. Results and Discussion

3.1 HRTEM

The HRTM images show that in both samples the anatase is nanocrystalline and mesoporous. The nanoparticles in both samples are either facetted or exhibit an irregular roundish shape. Particles with bad crystallinity or amorphous particles, not shown here, were observed more often in the Brij56 sample than in SDS. However, it has to be considered that the anatase is unstable against irradiation under the electron beam. Thus, no quantification can be made on the relative amount of amorphous anatase in the samples. Defects are observed in both samples, e.g. twins, see figure 1.

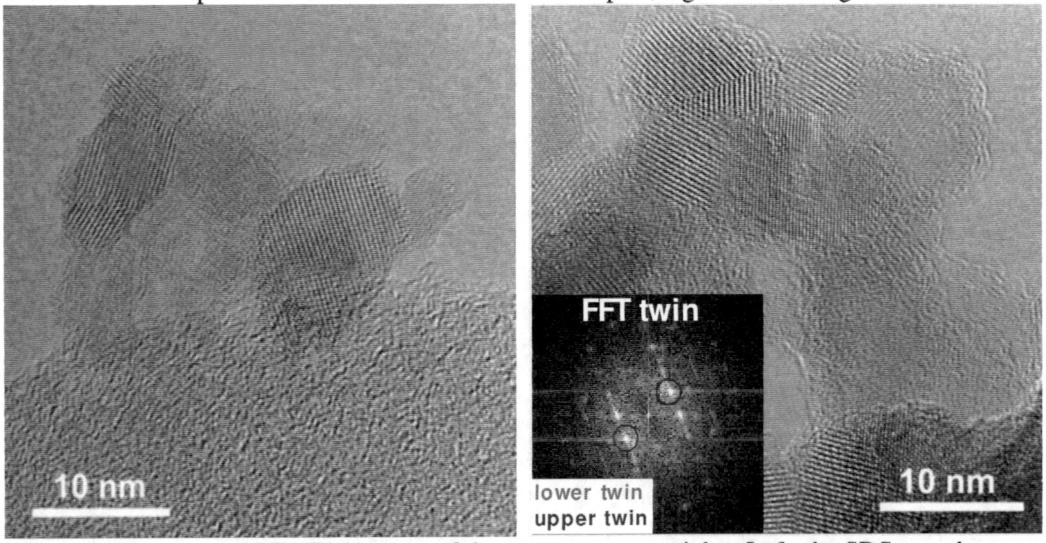

Figure 1. Cs-corrected HRTEM images of the anatase nanoparticles. Left: the SDS sample. Right: the Brij56 sample with crystalline nanoparticles and a twin. Inset: power spectrum.

Particle statistics was performed on both samples. 495 particles were measured on SDS and lead to an average particle size of 8.8±2.7 nm in diameter, ranging from 3 to 19 nm. On the Brij56 sample 485 particles were evaluated, resulting in an average diameter of 11.1±3.2 nm ranging from 5 to 23 nm.

3.2 X-Ray Powder Diffractometry

Both samples were found to consist of pure anatase, see figure 2. The particle sizes were determined from the FWHM of the {101} reflection, using the Scherrer equation with a form factor of K=0.9. The resulting particle diameters were 7.0 nm for the SDS sample and 6.2 nm for the Brij56 sample. These values are smaller than the diameters determined from the TEM micrographs. A possible explanation for this discrepancy may be the following [8]: (i) in very small particles the lattice constant depends on the crystallite size. This variation causes peak broadening. (ii) twin domains contribute twice, independently, to the spectrum while for the TEM analysis a twinned particle was counted only once. (iii) oxygen saturation in the surface area causes peak broadening. In all these cases peak broadening leads to an underestimate of the particle diameter from XRD. Nevertheless, the quantitative effect for these powders is a matter for further investigation.

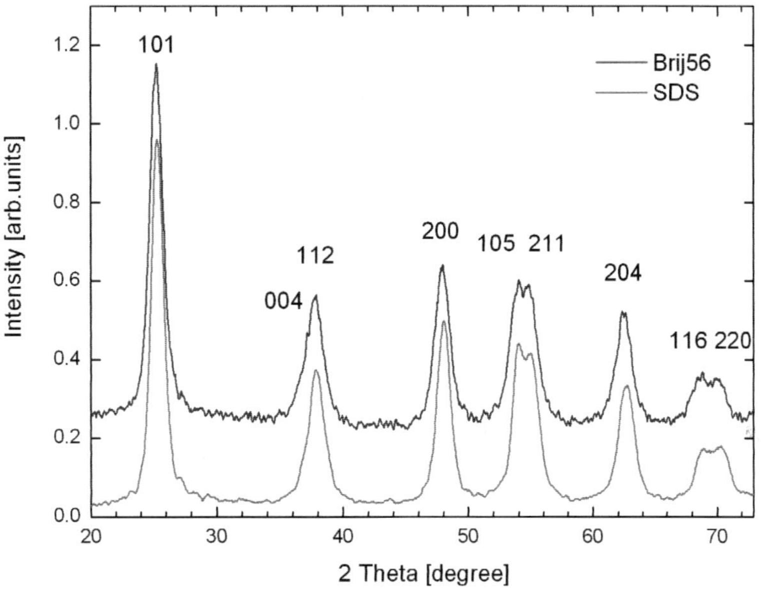

Figure 2. X-ray diffraction pattern of both samples with anatase reflections indicated.

3.3 Raman Spectroscopy

The Raman measurements show the phonon excitations for anatase. The crystallite size can be determined from the peak positions and from the full widths at half maximum (FWHM) of the E_g peak, see figure 3 [9]. The E_g peak of SDS has the maximum at 147.4±0.2 cm^{-1}. The maximum in the Brij56 sample was found at 145.2±0.1 cm^{-1}. The resulting particle diameters are 8 nm for SDS and 18 nm for Brij56. The FWHM measures 16.3±0.5 cm^{-1} for SDS and 12.4±0.1 cm^{-1} for Brij56. This leads to 6 nm particle diameter in SDS and 8 nm in Brij56. Again, in accordance with the TEM measurements the nanoparticles are larger in the Brij56 sample.

3.4 Nitrogen sorption

From the nitrogen sorption the BET surface area was determined to 184 m^2/g for the SDS sample and 140 m^2/g for the Brij56 sample. The pore diameter was determined to 4.9 nm for SDS and 6.9 nm for Brij56. The lower surface area and larger pore size for the Brij56 are in agreement with the results from TEM particle statistics and Raman spectroscopy, demonstrating larger particle size in Brij56.

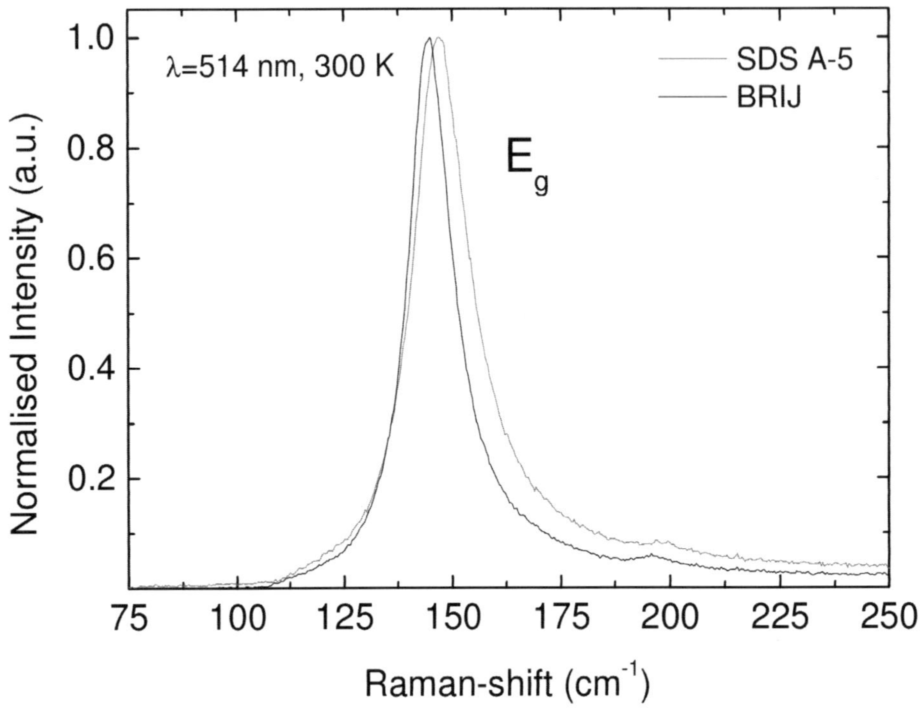

Figure 3. Raman spectrum of both samples. Shift of the E_g peak is due to phonon confinement.

3.5 Scanning electron microscopy cathodoluminescence investigation

Cathodoluminescence (CL) measurements have been performed on both samples. Optical transitions of the samples were independent of the average crystal size of the analysed samples; spatially integrated spectra at 7 K are dominated by a broad luminescence band centred at 2.26 eV. This broad band has been reported to appear in anatase phase TiO_2 single crystals, epitaxial layers and nano-crystals and has been attributed to self-trapped excitons at the TiO_6 octahedron [10]. Site selective spectra taken with a focused electron beam revealed transitions at 3.35 eV, with a line width of about 4 meV. Spectra taken at different sites indicated spatial anti-correlation of areas showing broad band luminescence at 2.26 eV and of those showing sharp UV transitions. Figure 4 shows a typical CL spectrum of the sharp transitions at higher energy resolution. At the high energy site two main peaks at 3.3962 and 3.3535 eV are dominating the spectrum. Both lines can be fitted by Lorentzian curves. The lines are separated by 42 meV. The higher energy transition has an intensity that is about 1% of that at lower energies. Temperature dependent measurements show a shift to lower energies according to that of the band gap of anatase TiO_2 reported by Tang [11]. Beside the main transition a number of phonon replica are observed. While the lower energy line at ~3.35eV has exclusively replica that correspond to the B_{1g} mode, the higher energy transition at ~3.40eV has additional replica that are higher in intensity than the main peak. The latter is characteristic for the zero phonon line and its phonon replica of indirect transitions in semiconductors.

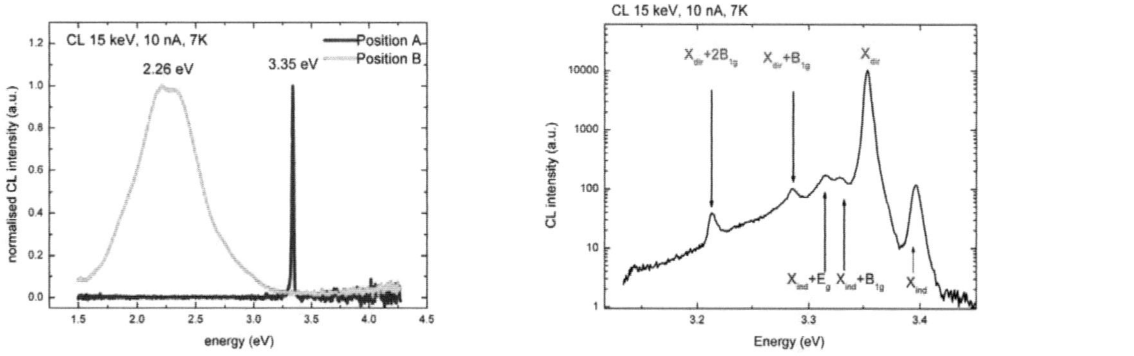

Figure 4. CL spectra typical for both samples. Left: overview at different locations. Right: high resolution spectrum of excitonic transitions.

Tang et al. [11] report on an indirect and direct gap that was found at 3.46 and 3.42 eV, respectively from absorption measurements in anatase single crystals. From the energy separation of the optical transitions, their intensity relation and their specific phonon replica it is tempting to assign the transitions found in our anatase nanocrystals to free excitonic transitions at the direct and indirect gap of anatase TiO_2. To confirm this we may in the following focus on the direct transition and its phonon replica. As can be seen in Figure 4 the 1- and 2- phonon replica have an asymmetric shape, with a steep edge at the low energy side and a high energy wing. This shape, if retained after background subtraction, could be characteristic for free excitons and due to the kinetic energy of the excitons that is described by a Maxwell distribution. Further support for the hypothesis of free excitons comes from the almost linear increase of the FWHM of the luminescence peaks with increasing temperature. The transition at 3.3535 eV in Figure 4 shows a high energy shoulder. A spectrum of that transition at higher resolution is shown in Figure 5. A second peak is revealed with an energy separation of 7.48 meV to the direct free exciton transition. We attribute this transition to the excited state of the direct free exciton and estimate the exciton binding energy and the Bohr radius from this separation. To do so, we need the basic parameters such as effective masses of electrons and holes and the static dielectric constant. There are a number of different values that have been reported especially for effective masses of electrons and holes and for the static dielectric constant. Taking $me=10m_0$ and $m_h=0.8m_0$ from [12] we obtain from

$$E(n) = E_g - (\mu e^4 / 2\hbar^2 \varepsilon^2 n^2) \tag{1}$$

and

$$\mu = m_e m_0 / (m_e + m_0) \tag{2}$$

a dielectric constant of 3.18 which is in good agreement with literature data ($\varepsilon = 31$ [13]). We can also estimate a Bohr radius for the free exciton of $a_B^x = 2.35$ nm and an exciton binding energy of 10 meV. In fact we have no direct correlation between crystal size and optical transitions, however, taking our value of the Bohr radius we do not expect quantum confinement effects for the particles studied in this work since their diameter exceeds the Bohr radius by a factor of 3. Comparing our data to that from the literature it is interesting to note that Monticone et al. [7], analysing the absorption edge of anatase nanoparticles, found quantum confinement effects for anatase nanoparticles only for with sizes in the range of 0.8 nm, which supports our experimental findings of the bulk behaviour of larger nanoparticles investigated in this study.

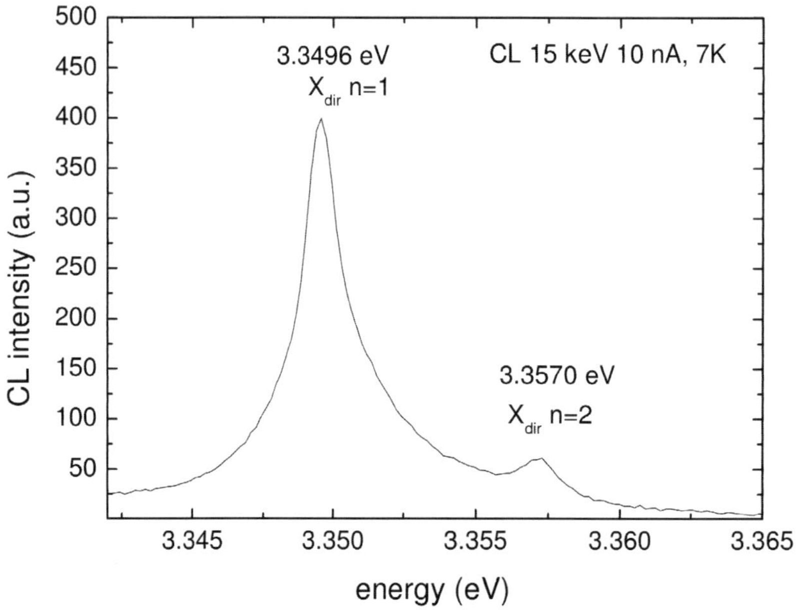

Figure 5. High resolution CL spectrum showing the ground state and the first excited state of the free excitonic transitions.

4. Conclusion

In this study we analyse optical and structural properties of anatase nanoparticles by means of HRTEM, Raman spectroscopy and CL. Cs-corrected HRTEM images show mesoporous nanoparticles with numerous defects. Additionally, small amounts of amorphous particles in the same size range as the anatase nanoparticles are being formed. The nanoparticles with an average diameter of 8.8 and 11.1 nm, respectively, show strong free exciton transitions independent of their size. From the excited state of these transitions we determine the Bohr radius of anatase transitions to be 2.35 nm. According to the best of our knowledge this the first report on free excitonic transitions in anatase TiO_2.

References

[1] Serpone N, Lawless D and Khairutdinov 1995 *J. Phys. Chem.* **99,** 16646
[2] Wunderlich W, Miao L, Tanemura M, Tanemura S, Jin P, Kaneko K, Terai A, Nabatova-Gabin N and Belkada R 2004 *Int. J. Nanosci.* **3**, 439
[3] Linsebigler AL, Lu G and Yates JT 1995 *Chem. Rev.* **95**, 735
[4] Hagfeldt A and Grätzel M 1995 *Chem. Rev.* **95**, 49
[5] Chen X and Mao SS 2007 *Chem. Rev.* **107**, 2891
[6] Bersani D, Lottici PP and Ding XZ 1998 *Appl. Phys. Lett.* **72**, 73
[7] Monticone S, Tufeu R, Kanaev AV, Scolan E and Sanchez C 2000 *Appl. Surf. Sci.* **162**, 565
[8] Kaszkur Z 2006 *Z. Kristallogr. Suppl.* **23**, 147
[9] Zhang WF, He YL, Zhang MS, Yin Z and Chen Q 2000 *J. Phys. D* **3**, 912
[10] Tang H, Berger H, Schmid PE and Lévy F 1993 *Solid State Comm.* **87**, 847
[11] Tang H, Lévy F, Berger H and Schmid PE 1995 *Phys. Rev. B* **52**, 7771
[12] Enright B and Fitzmaurice D 1996 *J. Phys. Chem.* **100,** 1027
[13] Roberts S 1949 *Phys. Rev.* **76**, 1215

Ab initio based atomic scattering amplitudes and {002} electron structure factors of $In_xGa_{1-x}As/GaAs$ quantum wells

JT Titantah[1], D Lamoen[1], M Schowalter[2] and A Rosenauer[2]

[1] EMAT, Universiteit Antwerpen, Groenenborgerlaan 171, 2020 Antwerpen, Belgium
[2] Institut für Festkörperphysik, Universität Bremen, D 28359 Bremen, Germany

E-mail: john.titantah@ua.ac.be

Abstract. The atomic scattering amplitudes of the various atoms of the systems $Ga_{1-x}In_xAs$, $GaAs_{1-x}N_x$ and $InAs_{1-x}N_x$ are calculated using the density functional theory (DFT) approach. The scattering amplitudes of N, Ga, As and In in the model systems are compared with the frequently used Doyle and Turner values. Deviation from the latter values is found for small scattering vectors ($s<0.3Å^{-1}$) and for these scattering vectors dependence on the orientation of the scattering vector and the chemical environment is reported. We suggest a parametrization of these modified scattering amplitudes (MASAs) for small scattering vectors ($s<1.0Å^{-1}$). The MASAs are exploited within zero pressure classical Metropolis Monte Carlo (MC), finite temperature calculations to investigate the effect of quantum well size on the electron {002} structure factor (SF) of $Ga_{1-x}In_xAs$ quantum wells.

1. Introduction

Accurate chemical composition analysis of technologically important III-V nanostructures is of importance. It is known that minute compositional variations can induce large changes in the photoluminescence properties of these materials. Using transmission electron microscopy (TEM), chemical composition quantification can be carried out with methods based on the composition sensitive {002} reflection [1] obtained from the atomic positions and the atomic scattering amplitudes (ASAs) of the atoms of the system. The most commonly used set of ASAs is that based on the relativistic Hartree-Fock (HF) calculations of Doyle and Turner (DT) on isolated atoms [2]. Using density functional theory (DFT) calculations on the binary crystals, Rosenauer et al. [3] demonstrated the importance of using DFT to obtain more accurate modified atomic scattering amplitudes (MASAs). The latter set of parameters is more accurate since it accounts for charge redistribution in the system as a result of bonding. However, the MASAs that have so far been used are those obtained from DFT calculations based on the binary zinc-blende crystals. The effect on the MASAs of the static atomic displacements present in the ternary and quaternary systems has never been investigated. This effect should become very important in technologically important $Ga_{1-x}In_xAs_{1-y}N_y$ where the insertion of nitrogen can introduce atomic displacements of more than 15% of the GaAs bond length. The study of the effect of static displacements on the structure factor (SF) [3-5] often ignores this important issue. An accurate composition analysis based on the structure factor should incorporate these effects at all levels. Another issue of interest is the system size dependence of the {002} electron structure factor of nanostructured III-V semiconductors.

In this work we have performed DFT calculations on a series of $Ga_{1-x}In_xAs$, $GaAs_{1-x}N_x$, $InAs_{1-x}N_x$ and the binary counterparts, all in the zinc-blende phase, with unrelaxed and relaxed atomic coordinates. Within the formulation of ref. [3] we have calculated the MASA of Ga, In, As and N in various chemical environments. The MASAs of all these atoms are compiled as a function of the magnitude of the scattering vector $s=\sin\theta/\lambda$ (where θ is the scattering angle and λ is the wavelength) and compared with the results of Doyle and Turner [2]. Then special attention is devoted to the relaxation of the multi quantum wells (MQWs) and the calculation of the {002} structure factors.

2. Theoretical considerations

The DFT calculations of the MASAs are performed using the all-electron full-potential-linearized augmented-plane-wave (LAPW) DFT code WIEN2k [6]. The MASAs are calculated using the method developed in ref [3]. These MASAs will be used to evaluate the SFs of the strained MQWs. QWs with In fractions of 20, 40, 60 and 80% are considered. InGaAs/GaAs QWs with In content as high as 80% have not been synthesized due to a dislocation-induced transition from two-dimensional (2D) to 3D growth mechanism for increasing well size. In fractions as high as 80% were considered in our simulations to investigate the effect of the inherent high strain present in such systems on the use of the electron structure factor for chemical composition quantification. The InGaAs/GaAs heterostructures have been relaxed as follows. The unit cell used in our simulation consists of a slab of InGaAs QW of thickness t_w on a GaAs barrier of thickness t_b. While the lateral widths along the [100] and [010] directions of the box are kept fixed in order to simulate a pseudomorphic growth condition on the GaAs barrier, the atomic positions and the length $L= t_w+t_b$ of of the simulation box in the growth direction (i.e. along the [001] direction) is relaxed using a zero pressure Metropolis Monte Carlo Scheme. Periodic boundary conditions were considered in all three dimensions, whereby if an atom exited the box at one end, its periodic image in the other end is retained. We use the Tersoff potential [7] that was recently parametrized by us [8] using accurate ab initio energies and the experimental elastic constants of the binary crystals.

3. MASA

DFT calculations were performed on systems containing Ga, In, As and N either as binary crystals (GaAs, InAs, GaN and InN all in the zinc-blende phase) or on supercells made of 64 atoms mixed ternary $III_{1-x}III_xV$ or $V_{1-x}V_xIII$ crystals. The mixed systems were relaxed using the DFT approach and the average MASA of each atom type was calculated. In Fig. 1 we plot the MASAs for various reflections and for all the model systems as a function of the magnitude of the scattering vector s. An acceleration voltage of U=200 kV is used. The spread of the MASAs demonstrate their chemical environment sensitivity. As Fig. 1 shows, for Ga, As and In, the HF calculations accurately reproduces the DFT results for $s>0.3Å^{-1}$. For smaller s-values deviations (towards larger values) from the HF calculations become quite important. The agreement for N is poor, thus discrediting the use of the DT ASAs on nitrogen containing systems. However the agreement of the HF results with our DFT calculations for scattering vectors larger than $s=0.3Å^{-1}$ encourages the use of the DT results for higher order reflections like the {220} and {004} reflections. Another important aspect of the DFT result is the orientation sensitivity at small scattering vectors which is exemplified by {111} spots. We have also found that for small scattering vectors, atomic relaxation increases the MASA (see Fig.2).

In order to account for all these effects on the MASA, we performed a fit of the MASAs of Ga, In, As and N as a function of s. For these atoms we consider the binary zinc-blende GaAs, InAs, GaN, InN and the relaxed ternary $Ga_{1-x}In_xAs$, $GaAs_{1-x}N_x$ and $InAs_{1-x}N_x$ systems. In analogy with the work of Doyle and Turner [2], who needed parameter tables for economic data storage, we retained $f(s)$ (in nm) as the sum of four exponential functions as follows:

$$f(s)=\sum_{i=1}^{4} a_{\alpha i}\, e^{-b_{\alpha i}s^2} \qquad (1)$$

where $a_{\alpha i}$ and $b_{\alpha i}$ are fitting parameters and α is the atom type as reproduced in Table 1.

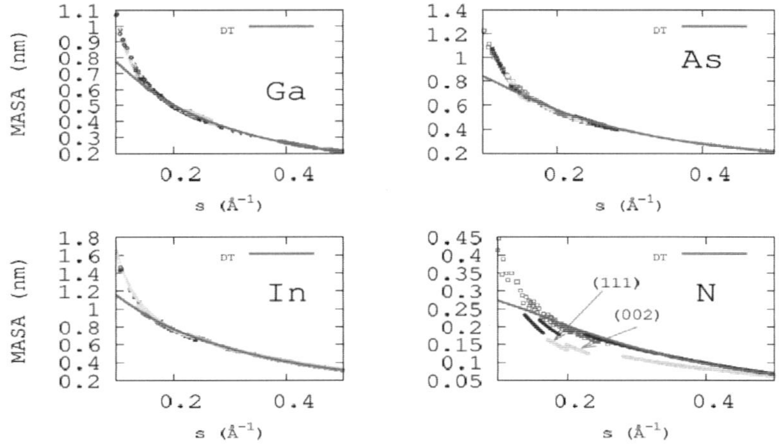

Figure 1. Electron atomic form factors as a function of *s* compared with the Doyle and Turner (DT) results.

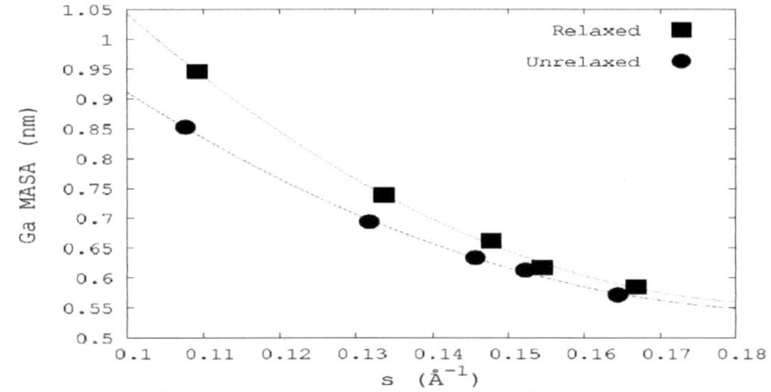

Figure 2. The MASA of Ga for relaxed and unrelaxed $Ga_{24}In_8As_{32}$ system (64 atoms) showing the effect of atomic relaxation on the MASA.

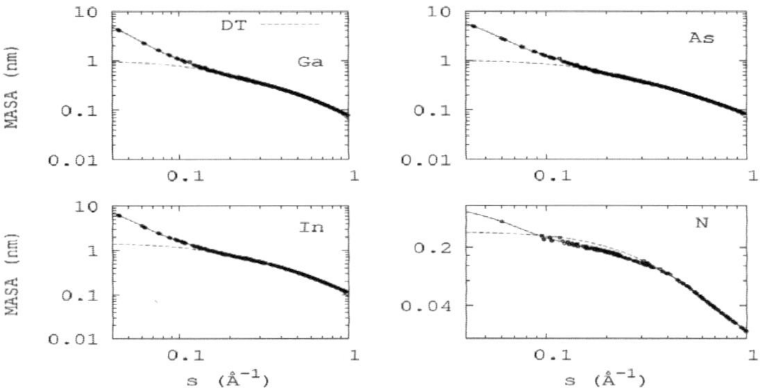

Figure 3. Fitted MASAs for Ga, In, As and N in zinc-blende binary and ternary materials compared with the Doyle and Turner results (DT). The points correspond to DFT calculations for several binary and ternary alloys and the full lines are the fitted expressions.

Figure 3 compares the fitted functions with the DT results for Ga, In, As and N. Table 1 gives the parameters of the fit, which yield very accurate form factors for small scattering vectors. The coefficients $a_{\alpha i}$ must be corrected by the factor $\left(1+ {eU}/{m_0c^2}\right)$ for the appropriate electron energy.

Table 1. The parameters of the fit of the MASAs of Ga, In, As and N.

Atom α	a_1	b_1	a_2	b_2	a_3	b_3	a_4	b_4
Ga	5.5541	632.8870	1.0827	119.4029	0.2713	14.5702	0.2060	1.4976
In	8.0200	626.1903	1.5765	118.4147	0.4236	13.4353	0.3083	1.5710
As	6.6263	628.0908	1.2804	116.9245	0.3183	13.7264	0.2030	1.4090
N	0.3121	295.5509	0.0544	38.6426	0.1071	6.2606	0.03323	0.8909

4. $In_xGa_{1-x}As$ / GaAs multi quantum well simulations

Typical layer by layer changes in the bond lengths of the atoms in a MQW are shown in Fig. 4 for a ~10nm wide $In_{0.4}Ga_{0.6}As$/ GaAs quantum well. The pair correlation function shows a bimodal bond length distribution with Ga-As bonds at about 2.4 Å and In-As bonds at about 2.6 Å. The Ga-As bond length is seen to shrink within the well.

In the absence of static displacements and at low temperatures we can write the modulus square of the 'average' {002} SF as:

$$S_0^{002}(x) = \left|(1-x)f_{Ga} + xf_{In} - f_{As}\right|^2 \tag{2}$$

It is commonplace to use the SF of the (200) diffraction spot for composition analysis. In $Ga_{1-x}In_xAs$ heterostructures the (200) TEM dark field measurement is based on the intensity of the (200) electron SF [9]. For sufficiently thin specimens and off-zone axis imaging conditions, this intensity is proportional to the modulus square of the SF. If we neglect static atomic displacements, strain effects and temperature effects, this intensity can be written as the modulus square of the 'average' {002} SF. Any deviation from this dependence could result from one or more of these effects. Because of tetragonal distortion the (200) and (002) spots become inequivalent. Typical profiles of the modulus square of the (002) and (200) SF across a MQW are depicted in Fig. 5 together with the prediction of Eq. (2). Note that the deviations from a constant value within the InGaAs MQWs are caused by the statistical fluctuations of the In-concentration. The 'average' SF deviates slightly from the calculations. The figure shows the GaAs value at 0.04 nm^2, an interface region of about 1 nm thick for large x, where the SF is lower than that of the GaAs region, and the well region in which the intensity is found to be lower than that of the barrier for the 20% In systems. The low intensity interface region is often seen in quantum wells of $Ga_{1-x}In_xAs$ and dilute nitrogen $Ga_{1-x}In_xAs_{1-y}N_y$ [10-12] and has been attributed to the lower In content at these interfaces. Our calculations confirm this.

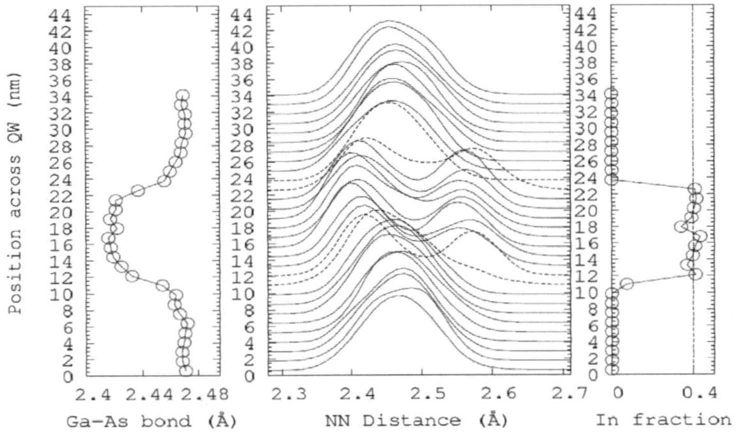

Figure 4. Left: Variation of the Ga-As bond length across the well of a multi-well $Ga_{1-x}In_xAs$ system: Nearest neighbour pair correlation showing the bimodal bond length distribution within the well. Ga-As bond length shows up at about 2.4 Å and that of In-As at about 2.6 Å. Right: Variation of In content for the 40% In containing $Ga_{1-x}In_xAs$ quantum well.

The quantum wells considered in this work are highly strained, the strain level being dependent on the size of both the QW and the barrier. In a recent work we showed that as the well size increases beyond 5 nm for the In contents considered in this work, continuum elasticity theory fails to explain the nature of the strained state of the QWs [13]. Such large wells were shown to be characterized by a strained interface layer of about 2-4 monolayers thick and a strain-relaxed interior. We have analysed the effect of the strain on the (002) structure factor. For the large QWs, in which strain relaxation occurs, the real and imaginary parts of the (002) SF are found to be sensitive to the strain level even though the modulus square is only weakly sensitive as Figs. 5 shows.

The dependence of the square modulus of the (002) vs. (002) structure factors on the In fraction and well size is shown in Fig. 6. While the well sizes vary from 12 monolayers (MLs) to 40 MLs, the sum of the well and the barrier sizes is fixed at 120 MLs (about 34 nm). Comparison with the prediction of Eq. (2) shows that for thin layers the use of this expression for In content quantification based on the (002) spot can lead to an overestimation of the In fraction especially for high In containing QWs meanwhile the use of the (200) spot may lead to a slight underestimation of this fraction. The reason for the deviation of the (200) spot from the 'average' profile is the fact that the in-plane lattice parameter is constrained at the GaAs bulk value and the MASAs used for the (200) SF calculation are those obtained at this lattice parameter. Strain and/or static atomic displacements are responsible for the deviation of the (002) intensities from the average values.

5. Summary

We have computed modified atomic scattering amplitudes of $Ga_{1-x}In_xAs$, $GaAs_{1-x}N_x$ and $InAs_{1-x}N_x$ ternary systems. For small scattering vectors (s<0.3Å^{-1}), major deviations from Hartree-Fock calculations are recorded in addition to remarkable chemical environment sensitivity. For such vectors orientation sensitivity of the reflection spots are highlighted and atomic relaxation is also found to increase the MASA. These findings call for caution to be taken when using various sets of MASAs for quantification purposes. We have investigated the effect of well size on the (002) and (200) structure factors of multi quantum wells using the Metropolis Monte Carlo approach. The (002) and (200) SFs of the MQWs differ from each other and both show deviation from the 'average' values. We have found that for thin layers, the 'average' intensity underestimates the (002) SF. This means that quantifications of the In content exploiting the (002) spot will overestimate the In fraction in thin GaInAs MQWs.

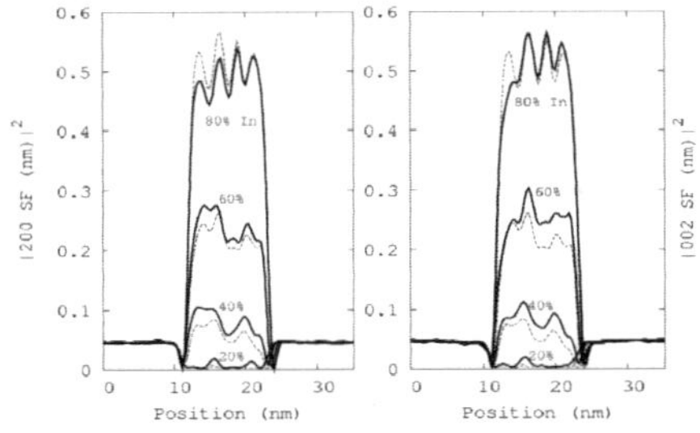

Figure 5. Variation of the modulus square of the (200) and (002) SFs of the MQW along the growth direction.

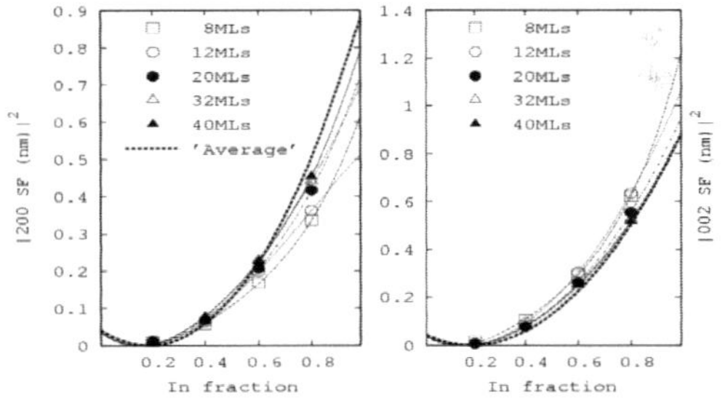

Figure 6. Variation of the square modulus of the (200) and (002) SFs with In content for various well sizes ranging from 12 to 40 MLs.

Acknowledgement: We acknowledge financial support from the FWO-Vlaanderen under contract G.0425.05 and the Deutsche Forschungsgemeinschaft under contract number RO2057/4-1.

References

[1] Rosenauer A, Fischer U, Gerthsen D and Forster A 1988 *Ultramicroscopy* **72**, 121
[2] Doyle PA and Turner PS 1968 *Acta Cryst. A* **24**, 390
[3] Rosenauer A, Schowalter M, Glas F and Lamoen D 2005 *Phys. Rev. B* **72**, 085326
[4] Glas F 2005 *Philos. Mag.* **11**, 2055
[5] Volz K, Rubel O, Torunski T, Baranovskii SD and Stolz W 2006 *Appl. Phys. Lett.* **88,** 081910

[6] Blaha P, Schwarz K, Madsen GKH, Kvasnicka D and Luitz J: *WIEN2k, An Augmented Plane Wave + Local Orbitals Program for Calculating Crystal Properties* (Karlheinz Schwarz, Techn. Universität Wien, Austria), 2001 ISBN 3-9501031-1-2.
[7] Tersoff J 1986 *Phys. Rev. Lett.* **56**, 632
[8] Titantah JT, Lamoen D, Schowalter M and Rosenauer A 2007 *J. Appl. Phys.* **101**, 123508
[9] Bithell EG and Stobbs WM 1989 *Philos. Mag. A* **60**, 39
[10] Herrera M, González D, Hopkinson M, Gutiérrez M, Navaretti P, Liu HY and Garciá R 2005 *J. Appl. Phys.* **97**, 073705
[11] Largeau L, Bondoux C and Patriarche G 2001 *Appl. Phys. Lett.* **79**, 1795
[12] Luna E, Trampert A, Pavelescu EM and Pessa M 2007 *New Journal Phys.* **9**, 405
[13] Titantah JT, Lamoen D, Schowalter M and Rosenauer A 2008 *Phys. Rev. B* **78**, 165326

Journal of Physics: Conference Series **209** (2010) 012041 doi:10.1088/1742-6596/209/1/012041

Structure of an incommensurate 90° Si grain boundary resolved with the help of a Cs-corrector for illumination

JL Rouviere[1], F Lançon[2], K Rousseau[1,4], D Caliste[2], PH Jouneau[1] and F Fournel[3]

[1] CEA, INAC/SP2M/LEMMA, Minatec, 17 rue des Martyrs, 38054 Grenoble, France
[2] CEA Grenoble, INAC/SP2M/L_SIM
[3] CEA Grenoble, LETI – Minatec
[4] now at: Serma Technologies, 7 parvis Louis Néel, 38040 Grenoble, France

Email : jean-luc.rouviere@cea.fr

Abstract. The atomic structure of an incommensurate (001)/(110) Si grain boundary (GB) or 90° Si GB has been studied by transmission electron microscopy (TEM) and refined by atomistic simulations (Stillinger-Weber potential). Samples were made by bonding one (001) Si wafer with one (110) Si wafer and carefully orienting the 2 wafers in order that they have a common $[1\bar{1}0]$ direction. In the interfacial direction perpendicular to $[1\bar{1}0]$, the $[110]_I$ direction of grain I is parallel to the $[001]_{II}$ direction of grain II and, as the ratio of this 2 vectors is $\sqrt{2}$, it is impossible to find 2 integers n and m such that $n[110]_I=m[001]_{II}$. The structure is incommensurate in this direction. *Z*-contrast images obtained in an FEI-Titan microscope equipped with a probe Cs-corrector easily resolve the Si dumb-bells in the two grains and allow us to determine the complex atomic structures of the interface. On the other hand, near on-axis high resolution TEM images obtained in a JEOL 4000EX microscope are very efficient to analyse the long range order of the interface.

1. Introduction

In many situations, energy is lowered by creating periodic structures. For instance, in Si polycrystalline materials, grains are generally not at all randomly distributed. Grain boundaries (GBs) with low energies tend to be created, and these low energy GBs correspond to particular orientations of the grains that produce periodic interfaces with short interfacial periods [1]. But what happen if the two crystals are bonded together in such an orientation that no periodic lattice is geometrically possible (what we call an incommensurate case)? Will the orientation of the grains be slightly changed in order to create a periodic lattice? Such a crystal has already been created in gold and its structure has been described as quasicrystalline [2]. In this work the fabrication, characterization and simulation of an incommensurate grain boundary in Si is reported.

2. Experimental details

The incommensurate grain boundary in silicon was made by bonding together one (001) 4" Si wafer, with one (110) 4" Si wafer. The two mirror-like wafers were first cleaned to remove any organic contamination or particles. Secondly, both surfaces were hydrophilic recovered by a native oxide with rms roughness of ~1.5nm.Thirdly, the surfaces were passivated by hydrogen using HF vapour. Then the two wafers were placed face to face and carefully oriented in such a way that they had a common

$[1\bar{1}0]$ direction. The bonding was initiated by pressing gently one edge of the wafer. From the pressed point, a wave front propagated and started to bond the two wafers together [3]. The bonding was then fixed by annealing at high temperature (about 1000°C).

$[1\bar{1}0]$ cross-sections were then classically realized by mechanical thinning and ion milling. Samples were observed either with a JEOL 4000EX microscope equipped with a LaB_6 filament or with a FEI-TITAN microscope fitted with a CEOS probe Cs-corrector.

Atomic models of specific configurations were manually built starting from models determined from experimental images by using the V_Sim software [4]. The energy of these configurations was finally minimized by using a Stillinger-Weber potential and by periodising the structure [5-6].

3. Experimental results

3.1. HRTEM images

Figure 1 is a high resolution transmission electron microscopy (HR-TEM) image of the grain boundary (GB) obtained with the JEOL 4000EX. The GB is observed along the common $[1\bar{1}0]$ direction. Perpendicular to $[1\bar{1}0]$, the [110] direction of the first grain ($[110]_1$) is parallel to the [001] direction of the second grain ($[001]_2$) (fig. 1). As the ratio of the norms of these two vectors is equal to $\sqrt{2}$, it is mathematically impossible to have a common period along these directions and the GB is said to be incommensurate along that direction. Grain 2 can be deduced from grain 1 by a rotation of 90° around the common $[1\bar{1}0]$ axis. From images such as fig. 1, it is very difficult to solve the atomic structure of the interface because, due to the limited resolution of the microscope (0.17 nm at Scherzer defocus), not all the atomic columns are resolved and because the structure is not really periodic although some features are repeated. Indeed it is rather difficult to determine what is repeated and what is not. These periodicities are best seen by slightly disorienting (2 or 3°) the sample away from the $[1\bar{1}0]$ common direction. The contrasts in the two grains are then slightly different and the two grains are easily distinguished (fig. 2). Long range order is then analyzed more easily.

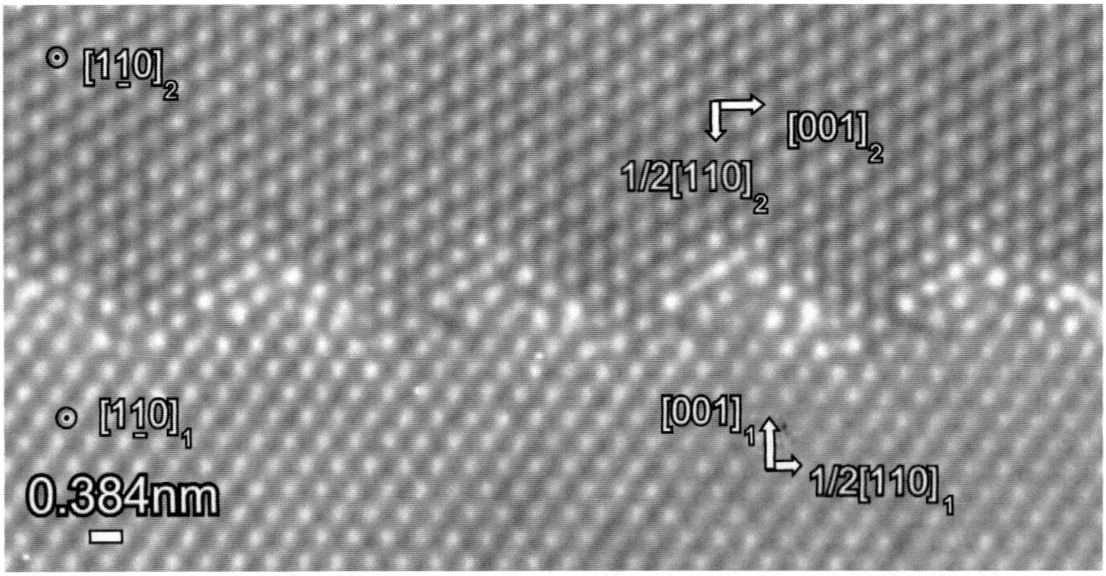

Figure 1. $[1\bar{1}0]$ HRTEM image of the (001)/(110) incommensurate Si GB. Defocus is about 50 nm, and atoms are black. The Si dumb-bells separated by 0.135 nm are not resolved.

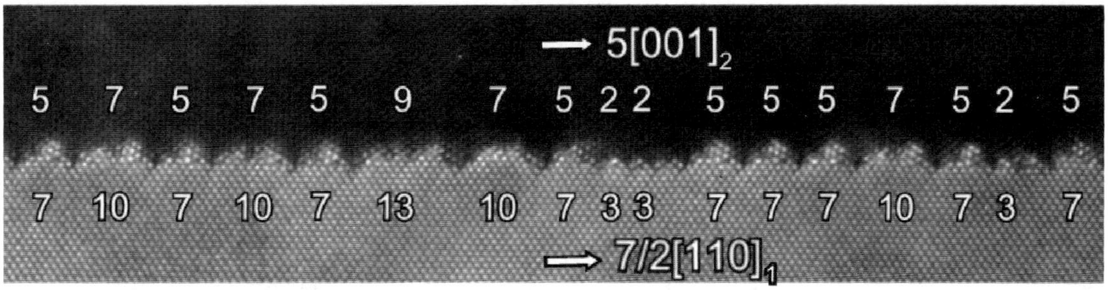

Figure 2. Slightly off-axis HRTEM image of the (001)/(110) incommensurate Si GB (defocus is about 50 nm, atoms are black).

The interface is not flat but has a zigzag shape. The interface is not periodic, but composed of a limited number of segments distributed at first glance with no apparent order. These segments are characterised by their respectively vectors $\boldsymbol{p}_1=k_1/2[110]_1$ and $\boldsymbol{p}_2=k_2[001]_2$ in the lower and upper crystals (convention of figure 2). Figure 2 shows the more frequent (k_1,k_2)-segments : (7,5), (10,7), (13,9) and (3,2). It can be checked that the k_1/k_2 ratios are good approximants to $\sqrt{2}$ although they do not all belong to the diophantian series [2]: (1,1), (3,2), (7,5), (17,12). However, it was impossible to solve the atomic structure of these segments by analysing HR-TEM images because the structures are too complex. In Figures 1 and 2, atoms are black. The Si dumb-bells separated by 0.135 nm are not resolved. Tunnels are white and it is impossible to really separate where the hexagonal, pentagonal and heptagonal tunnels are. We had to use Z-contrast images to do it.

3.2. STEM images

Figure 3 is a high resolution scanning TEM (HR-STEM) image of the GB. This image was acquired on the TITAN microscope (300 kV) with a Fishione HAADF detector by using a spot size of 6, a gun lens of 6, an extraction voltage of 4300 V, a convergence angle of 20 mrad, a collection angle of 40 mrad, and a scanning time of 20 s. To enhance the contrast the image was filtered using a 7x7 pixels mask with weights of 32, 4, 2 and 1 for the central, first, second and third adjacent pixels, respectively. The atomic structure can be directly deduced from figure 3: each white spot corresponds to an atomic column and Si dumb-bells, composed of two columns separated by 0.135 nm, are mostly resolved.

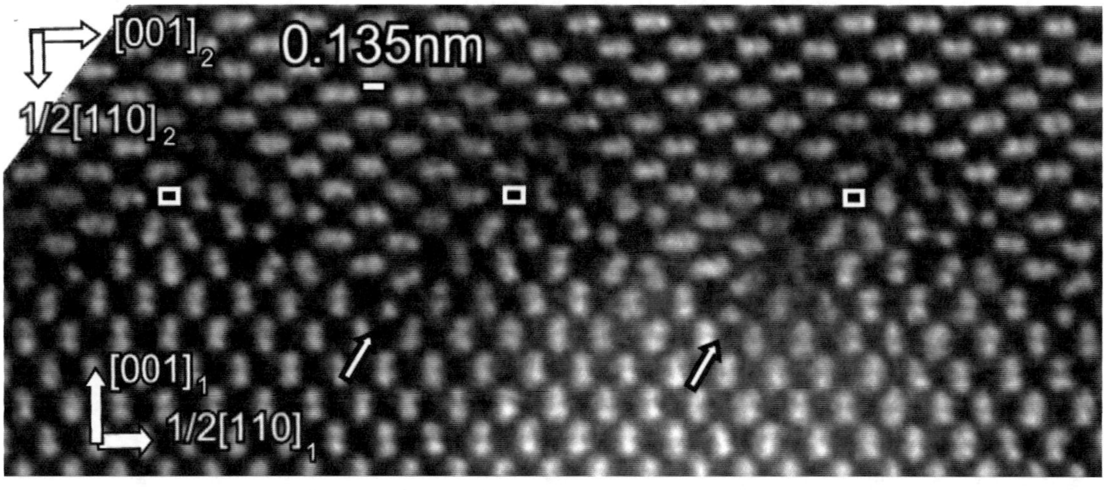

Figure 3. High resolution STEM image showing several periods of the (7,5) segment. The common $[1\bar{1}0]$ direction of the two grains is the direction of observation. Arrows points towards atomic columns that have special reconstructions (no dumb-bell) doubling the period along $[1\bar{1}0]$. The white squares outline specific tunnels composed of rings of 7 atoms (heptagonal tunnels).

Figure 3 contains several consecutive (7,5) segments. Several isolated columns can be seen and correspond to a special reconstruction doubling the period along $[1\bar{1}0]$. The atomic structure of others segments have been solved but are not reported here.

3.3. Atomistic simulations

The atomic structure of a (7,5) segment is shown in figure 4. All the Si atoms have 4 neighbours. Some columns (pointed by arrows in fig. 4) have special reconstructions (no dumb-bell) along $[1\bar{1}0]$. On these special columns, one bond is parallel to $[1\bar{1}0]$ and the 3 others are nearly in the $(1\bar{1}0)$ plane. These special reconstructions lead to a doubling of the period along the $[1\bar{1}0]$ direction: the period is $[1\bar{1}0]$ instead of ½$[1\bar{1}0]$ (fig. 4b).

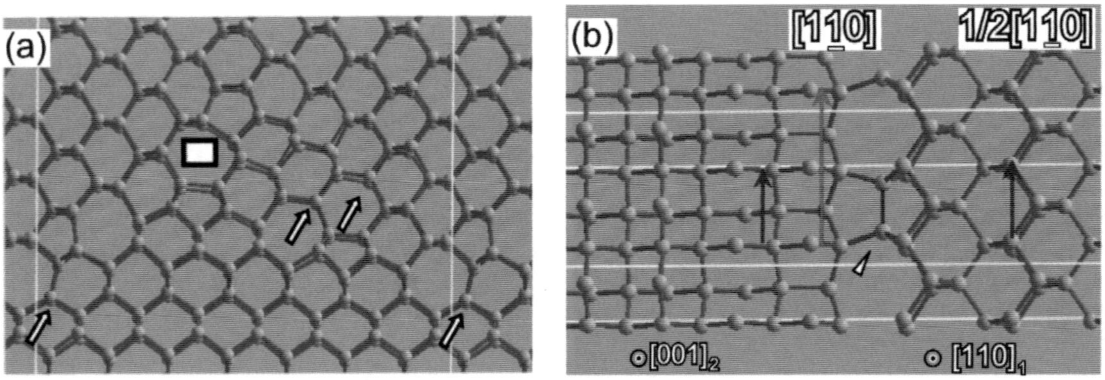

Figure 4. (a) Atomic model of the (7,5) segment projected on to the $(1\bar{1}0)$ plane showing atoms and bonds between neighbours. Vertical white line determine the size of the periodic box used for simulations, of size $7/2[110]_1$ in the lower grain 1 and $5[001]_2$ in the upper grain 2) The white square outlines the heptagonal tunnels of Fig. 3. Arrows points towards atomic columns that have special reconstructions along $[1\bar{1}0]$. (b) Section of the GB view along the directions $[001]_2$ // $[110]_1$ and taken near the atomic column corresponding to the arrow situated at the bottom left of fig. 4a. The periodicity in grains 1 and 2 is outlined by the two small ½$[1\bar{1}0]$ vectors. The $[1\bar{1}0]$ period of the interface is outlined by a vertical vector.

4. Conclusion

Thanks to the combination of electron microscopy techniques and atomistic simulation, the atomic structure of a complex non-periodic grain boundary has been resolved. The Si tetracoordination (4 neighbours) is kept at the interface. The interface is formed of quasi-period (k_1,k_2) segments.
Cs-corrected STEM was an efficient technique to solve the atomic projected structure of these segments. However, it is easier to determine the long range (k_1,k_2) segment ordering in slightly off-axis HR-TEM image because the two grains have slightly different contrast and because larger fields of view are more easily obtained.

References

[1] Sutton AP and Ballufi RW, *Interfaces in Crystalline Materials*, Oxford University Press 1995
[2] Lancon F, Penisson JM and Damen U 2000 *Europhys. Lett.* **49**, 603
[3] Fournel F, Moriceau H, Aspar B, Rousseau K, Eymery J, Rouviere JL and Magnea N 2002 *Appl. Phys. Lett.* **80**, 793
[4] Caliste D, http://inac.cea.fr/L_Sim/V_Sim
[5] Stillinger FH and Weber TA 1985 *Phys. Rev. B* **31**, 5262
[6] Vink RLC, Barkema GT, van der Weg WF and Mousseau N 2001 *J. non-Cryst. Sol.* **282**, 248

Structural properties of annealed SiO_x

G Nicotra[1], C Bongiorno[1], M Miritello[2], F Priolo[2] and C Spinella[1]

[1]CNR-IMM, Stradale Primosole 50, 95121 Catania, Italy
[2]MATIS CNR-INFM and Dipartimento di Fisica, Università di Catania, Via S. Sofia 64, 95123 Catania, Italy

E-mail: giuseppe.nicotra@imm.cnr.it

Abstract. On the basis of recent results, concerning the quantitative determination of the clustered silicon concentration in substoichiometric silicon oxide layer and based on electron energy loss spectroscopy (EELS) and energy-filtered transmission electron microscopy, in this paper we demonstrate that the structural properties of Si nanoclusters (Si ncs) strongly depend on the technique used for specimen preparation. In particular here we report on annealed SiO_x films prepared by plasma enhanced chemical vapour deposition (PECVD) and magnetron sputtering, showing that their structural properties are very different. It is shown that in PECVD films only around 30% of the Si excess agglomerates in clusters while an almost complete agglomeration occurs in sputtered films. These data are explained on the basis of the different initial structural properties of the as-deposited films which become crucial for the subsequent evolution.

1. Introduction

Silicon oxide films are present in almost all microelectronic and optoelectronic integrated circuits. The synthesis and the precise control of Si nanostructures embedded in a silicon oxide host one important capability, given the possibility of exploiting this system to obtain new functions in novel devices for microelectronics and for optoelectronics [1,8].

There are several techniques to synthesize a Si-rich silicon oxide (SRSO), among them Si ion implantation in silica [2,4], electron-beam deposition [5-6], PECVD [1,8] and magnetron sputtering [7] are the most used. Si nanoclusters (ncs) embedded into a SiO_2 matrix are achieved after thermal annealing thus producing a phase separation between Si and silica. The size distribution properties depend both on the Si excess and on the thermal budget of the subsequent anneal [1]. Energy filtered transmission electron microscopy (EFTEM) studies have shown that clusters start in the amorphous phase and became progressively crystalline only above 1000°C [8]. In spite of the fact that SRSO is obtained by several different techniques, an accurate study defining the similarities and differences among materials produced by different methods is indeed lacking. In this paper we investigate and compare the structural properties of SRSO layers obtained by PECVD and magnetron sputtering. We demonstrate that, in contrast to what is generally believed, properties of films grown by different methods are indeed very different as a result of the agglomeration processes. Implications of these results will be discussed.

2. Experimental

SiO_x thin films (with a total Si concentration in the range 35–55 at.%, corresponding to Si excesses ranging from $1.8x10^{21}$ to $2.8x10^{22}$ at./cm^3) have been grown by using a parallel plate PECVD system and an ultra-high vacuum magnetron sputtering system. The substrates have been heated at 300-400 °C during the deposition [1,10]. After deposition, the SiO_x films have been annealed for 1h at temperatures between 1000 and 1300°C in ultra-pure N_2 atmosphere. The structural properties of both as-deposited and annealed SiO_x films have been studied by EFTEM carried out in cross-sectional configuration by using a JEOL JEM 2010F TEM/STEM equipped with a 200-kV Schottky field-emission electron gun and an ultra high-resolution objective lens pole piece. The energy filter is a post-column Gatan Image Filter (GIF) with a 1k x 1k CCD. Samples were thinned by mechanical lapping, followed by ion milling using a 4 keV Ar^+ beam incident at an angle of 8° and 6° in the final step. Under these conditions the thinning procedure does not induce appreciable extra heating of the sample and consequent modifications of his structure. EFTEM allows to detect Si nc (both crystalline and amorphous and independent of the crystal orientation) dispersed in a SiO_2 matrix, due to the different plasmon loss energies in Si and in silica. All the analyses have been performed in regions thin enough to avoid projection effect by the superposition of ncs.

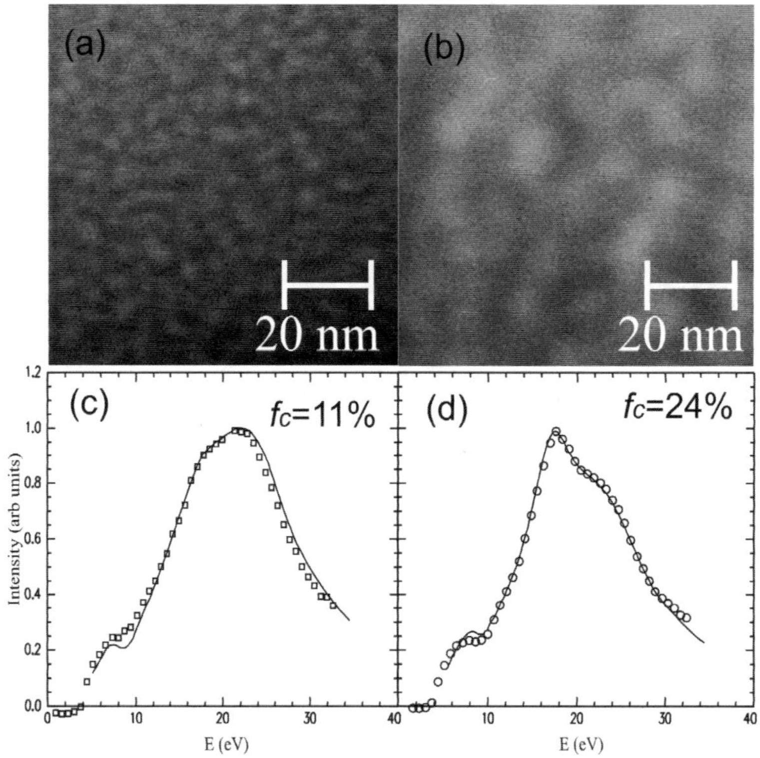

Figure 1. EFTEM images of the SiO_x films at the same Si excess and prepared by PECVD (a) and sputtering (b). Bright zones are associated with the presence of Si ncs. EELS low-loss spectra, open squares, for as-deposited PECVD (c) and sputtered (d) SiO_x films respectively. Continuous lines are the fits to the spectra, performed according to the Barrera-Fuchs model.

3. Result and discussion

Samples have been analyzed by EFTEM. In figure 1 the analyses of a PECVD sample (a), and a sputtered sample (b) both with a Si excess of about $1.6x10^{22}$ at./cm^3 and annealed at 1100 °C are shown. It is evident that the Si ncs, appearing as bright regions on a dark background, are much larger in sputtered samples (the mean radius is 4 nm for the sputtered sample and 2 nm for the PECVD one). We have hence performed a quantitative analysis of the EFTEM data, by estimating the amount of Si clustered within the nanostructures.

This task has been performed by using previously established analytical methodologies, based on the measurement of the Si ncs mean size and density [8], or on the analysis of the shape of the electron energy loss spectra [11], based on the Barrera-Fuchs model [12]. From this fitting procedure we find a unique parameter which is the clustered Si volume fraction f_c. The values f_c=11% and f_c=24% have been obtained in figure 1 (c) for the PECVD sample, and (d) for the sputtered one respectively. This study has been performed on many other samples and the results for both sputtered and PECVD samples having different amounts of Si, all annealed at 1100 °C, are reported in figure 2.

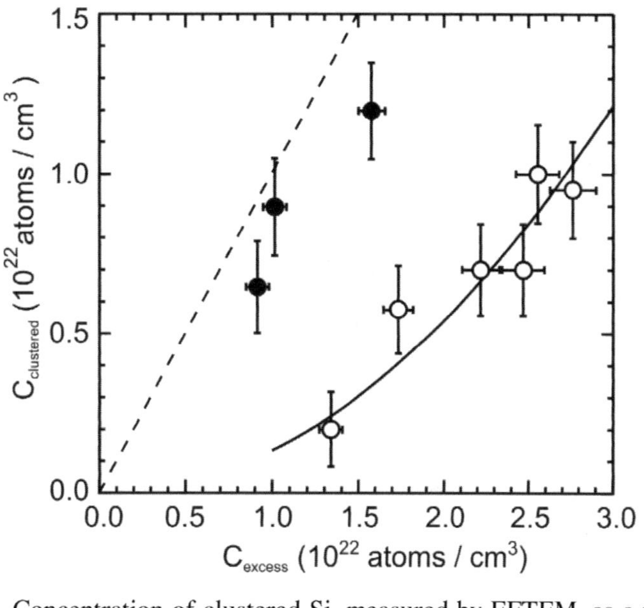

Figure 2. Concentration of clustered Si, measured by EFTEM, as a function of the Si excess and measured by RBS. Open symbols refer to PECVD samples, solid symbols to sputtered samples. The dashed line denotes complete phase separation between Si and SiO2, the continuous one is a guide for the eyes.

The figure reports the Si present in form of Si nc as a function of the amount of Si excess measured by Rutherford Back-scattering (RBS). In this graph the dashed line denotes complete phase separation between Si and SiO_2, while the continuous one is a guide for the eyes. It is clear that in PECVD samples (open circles) only about 30% of the Si in excess agglomerates while the rest remains dispersed in the matrix. This behaviour had indeed been observed before [8,11,13]. In contrast, sputtered samples (solid circles) show an almost complete agglomeration of the excess Si in ncs, with the matrix remaining as almost stoichiometric SiO_2. A question arises on why there is this strong difference in agglomeration properties of the Si ncs in samples prepared by different methods. The presence of N contamination in the former samples coming from the growth method employing N_2O as one of the gaseous precursors [1] is not believed to be the main reason, since the presence of Si-N bonds is expected to lead to an easier phase separation with respect to Si-O bonds [14]. In fact, phase separation and Si ncs formation require the nucleation of Si clusters and their growth through atomic mass transport. It is the starting properties of the as-deposited samples that should be profoundly different and give rise to a different evolution of the two systems. Indeed, the structure of SiO_x can be described as a simple Si/SiO_2 mixture (random mixture model (RMM) [15]), so proposing phase separation and neglecting the existence of Si intermediate oxidation states, or, alternatively, as a mixture of all Si oxidation states able to respect the overall film stoichiometry (random bonding model (RBM) [16]). In general the two models represent two different extremes and both "pure" models have never been verified. Films can therefore be "more" similar to a RMM or to a RBM and this should affect their thermal evolution. We have hence studied the as-deposited samples by EELS in the electron microscope. Figure 3 reports the EFTEM images and the low-loss EELS spectra for as-deposited PECVD ((a) and (c)) and sputtered ((b) and (d)) SiO_x films, having the same Si excess of the samples shown in figure 1. The EELS spectra are background subtracted by power law model. It is quite clear that the energy loss of the sputtered sample (d) has a marked component at lower energies, around the bulk Si plasmon (17 eV), whilst the one coming from the PECVD sample (c) is shifted to higher energies, around the bulk SiO_2 plasmon (at 23 eV). It should be emphasized that EFTEM imaging at 17 eV has not evidenced the presence of Si ncs in both samples. However, the circumstance that the energy loss spectrum of the sputtered sample is peaked at 17 eV is a strong indication that the $Si\text{-}SiO_2$ phase separation has already occurred in that as-deposited sample with Si domains whose size is possibly below the detectable limit of the EFTEM imaging technique for the used energies (cluster radii lower than 0.7 nm).

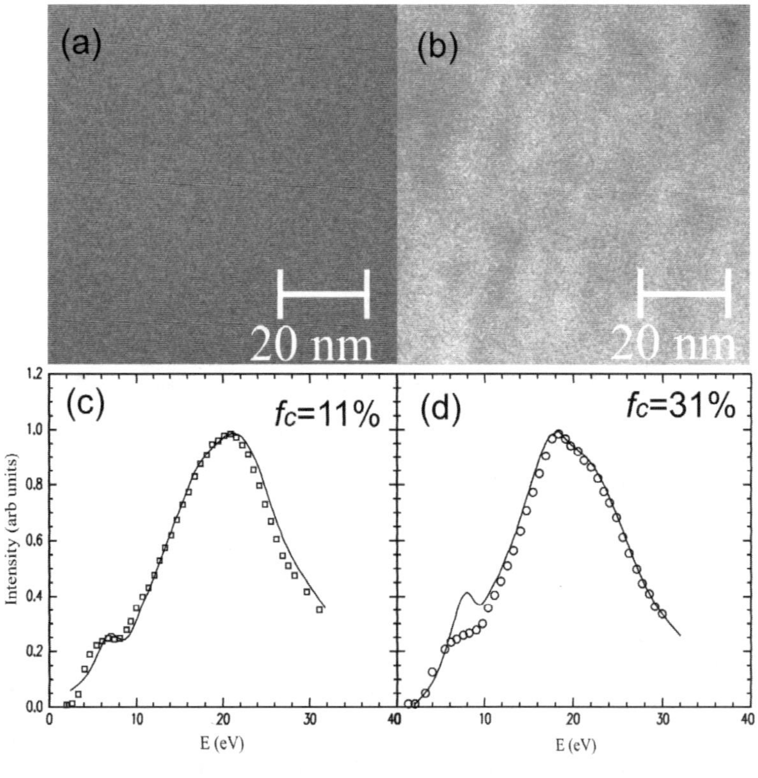

Figure 3. EFTEM images of the as-deposited SiO_x films having the same Si excess and prepared by PECVD (a) and sputtering (b). EELS low-loss spectra, open squares, for as-deposited PECVD (c) and sputtered (d) SiO_x films respectively. Continuous lines are the fits to the spectra, performed according to the Barrera-Fuchs model. Both films have the same Si excess as those shown in figure 1.

Then we tried to fit the spectra shown in figure 3 by using the Barrera-Fuchs model [12], i.e. by assuming that the electron energy loss probability, in the plasmonic region (5–30 eV), is given by the sum of three contributions coming from pure SiO_2 (the host), pure spherical Si ncs having a well defined size, and surface plasmon losses coming from the Si-SiO_2 interfaces, weighted by the unique parameter f_c. We imposed a Si nc radius of 0.7 nm consistent with the experimental evidence that the clusters are not visible by EFTEM at 16eV. The best fit to the spectrum coming from the as-deposited samples (continuous line in figure 3 (c,d)) was found for f_c = 11% and f_c = 31% of the Si excess, respectively, which is roughly equal to the Si excess present in the film. These values are roughly equal to the fraction of clustered Si detected in the same samples after thermal annealing. Therefore, virtually invisible Si nc are already present in the as-deposited SiO_x layers prepared by both the investigated techniques; these clusters account for almost all the Si excess for sputtered samples, which therefore have a structure closely resembling that predicted by a pure RMM. In contrast a PECVD film presents a structure more similar to the RBM. Small deviations from pure RBM [17] are commonly observed in CVD films [18]; the energy transferred to the incoming atoms by the plasma, and/or the surface mobility due to the substrate heating allow for a partial, chemistry-driven rearrangement of the growing film. The same argument holds also for sputtered films, where the very high energy associated to the sputtered atoms leads to the formation of the most stable phases (Si and SiO_2) and therefore to an almost complete phase separation. As a cross check we also investigated the matrix surrounding the Si ncs by electron energy loss spectroscopy (EELS) in a scanning transmission electron microscopy (STEM) configuration. The incident electron beam was focused to a 0.5 nm spot radius and scanned along a straight line connecting two adjacent Si clusters.

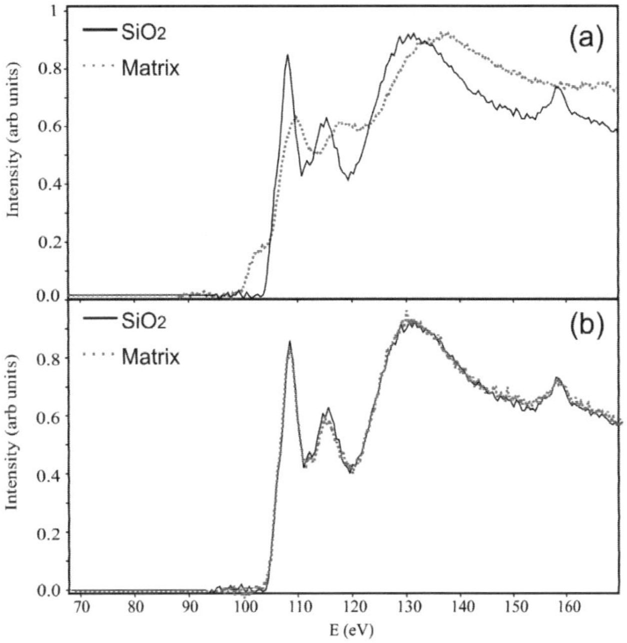

Figure 4. EELS core-loss spectra, dashed line, of the matrix surrounding Si ncs taken by STEM in between two adjacent ncs for PECVD (a) and sputtered (b) films. The continuous line in both panels is for pure SiO_2.

Dashed lines in figure 4 show the correspondent core-loss spectra of the PECVD (a) and of the sputtered sample (b) taken at a point in between the two examined clusters, i.e., in a region where only the matrix is visible in the micrographs of figure 1. We can observe that the core-loss spectrum taken in the sputtered sample is identical to the one taken on a reference SiO_2 sample (continuous lines in figure 4). This is not true for the spectrum coming from the PECVD sample since it is significantly different from the SiO_2 one, thus confirming that in this latter case the matrix is only partially depleted from all the silicon excess initially dispersed in it. In the same time, EELS spectra recorded in the energy region between 99-104 eV, and acquired in the matrix among two adjacent Si clusters in the annealed PECVD sample, exhibit a complex shape indicative of the presence of silicon atoms in their different oxidation states, cf. dashed line of figure 4 (a), which confirms again that PECVD film presents a structure more similar to the RBM. It is clear that the starting properties of the two materials determine the different evolution. EFTEM data suggest also that the thermal evolution of both PECVD and sputtered films does not really imply formation of new clusters, but simply the growth of those already existing in as-deposited films.

4. Conclusion

In conclusion, we have demonstrated that the structural properties of SRSO films strongly depend on the deposition method. In particular, while at lower annealing temperatures samples deposited by sputtering present an almost full agglomeration of the Si excess atoms, films grown by PECVD present agglomeration of only a small fraction of the Si excess. These data are explained on the basis of the different structures of the as deposited films.

References

[1] Iacona F, Franzò G and Spinella C 2000 *J. Appl. Phys.* **87**, 1295
[2] Pavesi L, Dal Negro L, Mazzoleni C, Franzò G and Priolo F 2000 *Nature (London)* **408**, 440
[3] Khriachtchev L, Rasanen M, Novikov S and Sinkkonen J 2001 *Appl. Phys. Lett.* **79**, 1249
[4] Min KS, Shcheglov KV, Yang CM, Atwater HA, Brongersma ML and Polman A 1996 *Appl. Phys. Lett.* **69**, 2033
[5] Zacharias M, Heitmann J, Scholz R, Kahler U, Schmidt M and Bläsing J 2002 *Appl. Phys. Lett.* **80**, 661
[6] Jambois O, Rinnert H, Devaux X and Vergnat M 2005 *J. Appl. Phys.* **98**, 046105
[7] Gourbilleau F, Portier X, Ternon C, Voivenel P, Madelon R and Rizk R 2001 *Appl. Phys. Lett.* **78**, 3058
[8] Iacona F, Bongiorno C, Spinella C, Boninelli S and Priolo F 2004 *J. Appl. Phys.* **95,** 3723
[9] Irrera A, Pacifici D, Miritello M, Franzò G, Priolo F, Iacona F, Sanfilippo D, Di Stefano G and Fallica PG 2002 *Appl. Phys. Lett.* **81**, 1866
[10] Miritello M, Lo Savio R, Iacona F, Franzò G, Irrera A, Piro AM, Bongiorno C and Priolo F 2007 *Adv. Mater.* **19**, 1582
[11] Spinella C, Bongiorno C, Nicotra G, Rimini E, Muscarà A and Coffa S 2005 *Appl. Phys. Lett.* **87**, 044102
[12] Barrera RG and Fuchs R 1995 *Phys. Rev. B* **52**, 3256
[13] La Magna A, Nicotra G, Bongiorno C, Spinella C, Grimaldi MG, Rimini E, Caristia L and Coffa S 2007 *Appl. Phys. Lett.* **90**, 183101
[14] Dal Negro L, Yi JH, Michel J, Kimerling LC, Chang TWF, Sukhovatkin V and Sargent EH 2006 *Appl. Phys. Lett.* **88**, 233109
[15] Temkin RJ 1975 *J. Non-Cryst. Solids* **17**, 215
[16] Philipp HR 1972 *J. Non-Cryst.* Solids **8-10**, 627
[17] Schlenz H, Kirfel A, Schulmeister K, Wartner N, Mader W, Raberg W, Wandelt K, Oligschleger C, Bender S, Franke R, Hormes J, Hoffbauer W, Lansmann V, Jansen M, Zotov N, Marian C, Putz H and Neuefeind J 2002 *J. Non-Cryst. Solids* **297**, 37
[18] Iacona F, Lombardo S and Campisano SU 1996 *J. Vac. Sci. Technol. B* **14**, 2693

Structural and compositional study of Erbium-doped silicon nanocrystals by HAADF, EELS and HRTEM techniques in an aberration corrected STEM

R J Kashtiban[1], U Bangert[1], I Crowe[2], M P Halsall[2], B Sherliker[2], A J Harvey[1], J Eccles[1], A P Knights[3], R Gwilliam[4] M Gass[5]

[1] Nanostructured Materials Research Group, School of Materials, The University of Manchester, P.O. Box 88, Manchester, M1 7HS, UK
[2] Microelectronics and Nanostructures Research Group, School of Electrical and Electronic Engineering, University of Manchester, M60 1QD, UK
[3] Department of Engineering Physics and the Centre for Emerging Device Technologies, McMaster University, 1280 Main Street West, Hamilton L8S 4L7, Ontario, Canada
[4] Surrey Ion Beam Centre, Advanced Technology Institute, University of Surrey, Guildford, GU2 5XH, United Kingdom
[5] SuperSTEM, STFC Daresbury Laboratory, Warrington WA4 4AD, UK

E-mail:Reza.jalilikashtiban@postgrad.manchester.ac.uk

Abstract: Er-doped SiO_2 and Si nano-crystals (NCs) embedded in a SiO_2 matrix were produced by ion beam implantation of Si (100) substrates. After annealing Er ions agglomerate in different positions with different compositional properties in samples with and without Si implants. HAADF and EELS show that in the sample with Si implants the Si and Er distribution is identical and within a band of ~110nm width ~75nm below theSiO_2 surface whereas in the sample with no excess Si, Er forms on average much larger, amorphous aggregates, presumably an Er-oxide, in the SiO_2 matrix with tendency to move towards the surface of the SiO_2 layer.

1. Introduction

SiO_2, doped with moderate concentrations of Er is currently used in Si-based optical amplifiers [1, 2]. As Er has a low absorption cross section highly doped structures could be a solution but Er clustering is a major problem with high Er concentrations. To overcome this problem, Er-doped silica containing Si-NCs have been introduced as a promising candidate [3-5].The broad continuous absorption band of the Si-NCs together with the possibility of using cheap LEDs instead of laser excitation, the larger excitation cross section of Er-sensitized Si-NCs compared to pure silica (~ 10^{-16} cm^2)[6] besides the improvement in the emission cross section of the sensitized Er ions (emission at 1.54 μm) [7] are advantages of this structure compared with Er-doped SiO_2 in absence of Si-NCs. In this structure Er ions absorb excitonic energy of optically excited Si-NCs and emit at 1.54 μm via intra-band transitions of the 4f shell,$^4I_{13/2}\rightarrow{}^4I_{15/2}$.

To improve the excitation rate at 1.54 μm an understanding of the position of Er and Si-NCs with respect to each other as well as of the Er content in the structure in order to reach an optimal

Journal of Physics: Conference Series **209** (2010) 012043 doi:10.1088/1742-6596/209/1/012043

concentration with the highest excitable fraction of Er is quite important. In previous work, implantation of Er ions into Si-NCs embedded in SiO_2 led to an amorphous Si structure evidenced by EFTEM, nevertheless these amorphous Si clusters were able to sensitize Er ions[8]. In photoluminescence measurements it has been shown that excitable Er ions were located in the surrounding SiO_2 with maximally 1-2 ions per nanocrystal [9]. In a recent EFTEM/ HAADF study involving Er concentrations larger than $1x10^{19}$ cm^{-3}, Er-rich clusters, 5-10 nm in size, were observed separate from the Si-NCs [10]. In the mentioned work it was claimed that Er segregation is independent of Si-NCs density.

In this report we have applied a variety of microscopic techniques to investigate the structure of Er-doped Si-NCs in a SiO_2 matrix. High-angle annular dark field (HAADF), electron energy loss (EEL) spectrum imaging, energy-filtered TEM (EFTEM), conventional dark field (DF) and bright field (BF) TEM and high-resolution TEM (HRTEM) were carried out to understand the compositional and structural properties in the most direct and accurate way available.

2. Experimental

HRTEM and DF diffraction contrast images and EFTEM maps (produced by passing electrons through the filter, which have lost 29 eV corresponding to the Er $O_{2,3}$ absorption edge) were acquired with a Tecnai FEGTEM microscope operating at 300 kV and the sample aligned with the <110> zone axes of the Si parallel to the electron beam. HRSTEM BF and HAADF images were acquired with an aberration corrected STEM at Daresbury Laboratories (SuperSTEM, the first UK *Cs* corrected cold field emission STEM, which employs a Nion Mark II quadrupole-octupole corrector[11]) operating at 100 keV. In this instrument Z-contrast lattice images with 1Å resolution can be obtained. The equipment includes a UHV Gatan Enfina system, with which EEL spectrum images (SI's) can be obtained.

In order to study Er-implant related microstructure and composition two kinds of samples grown on silicon substrates, with similar SiO_2 thickness and composition, were compared; these consisted of (1) solely Er-doped SiO_2, and (2) of Er-containing Si-NCs in SiO_2; in the latter case a beam of 85 keV Si ions to a dose of 8×10^{16} at/cm^2 was first targeted at a 700-nm thick SiO_2 layer grown on the top of a Si (1 0 0) substrate, and then Er-ion implantation was carried out with a beam of 320 keV and a dose of $3x10^{15}$ at/cm^2. Post implantation the samples were annealed at 1100 °C in an N_2 ambient for 1 hour in order to form the Si-NCs. The Er concentration peak in the samples was $3x10^{20}$ cm^{-3}, the silicon peak concentration $8x10^{21}$ cm^{-3}.

3. Results and Discussion

Figures 1 a)-c) show DF diffraction contrast images of sample (2) under various reflections. The NCs appear crystalline, of the same density and size as revealed in the HAADF image (acquired in the Daresbury SuperSTEM) in figure 1d). The lower magnification HAADF image in figure 1e) shows the entire 110 nm wide band of Er-doped Si-NCs at ~75 nm depth below the surface of the SiO_2. The HAADF image at higher magnification (figure 1d) shows 1-4 nm diameter structures with a density of $3.3x10^{17}$ cm^{-3}. The green rectangle in the image represents the area from which a spectrum image has been taken.

In order to evaluate the composition of the structure, EELS was performed. Figure 2 shows the intensity maps of the Er-$N_{4,5}$ and Si-$L_{2,3}$ absorption edges obtained from an EEL spectrum image [12] besides a simultaneously acquired HAADF image of the green-framed area shown in figure 1. The Er signal was extracted following a suitable power law background fitting very close to the onset of the Er-$N_{4,5}$ Edge at about 165 eV. The contrast of the maps is presented on the greyscale with white/bright signifying high and black/dark low concentration of the mapped materials. As can be seen from the maps there is an obvious correlation between the Si and Er distribution within the structure showing that Er is likely to accumulate within and around the Si-NCs after the annealing stage.

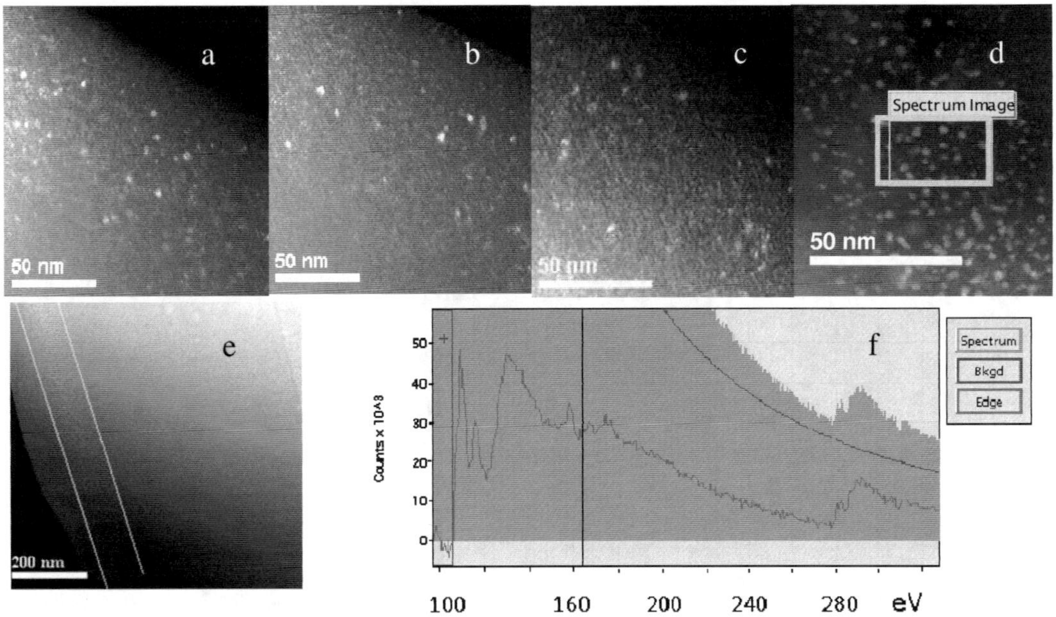

Figure 1. a)-c) DF images taken of the Er-doped Si NC region with <022>, <111> and <200> reflections, respectively. d) HAADF image of Er-doped Si-NCs within the band. The green rectangle represents the area in which an SI has been taken. e) HAADF image at lower magnification showing the entire ~110 nm wide band of Er-doped Si-NCs ~75 nm below the SiO_2 surface. f) EELS spectrum representing different edges. The strong peak at ~168eV representing the Er contribution can be seen clearly in the spectrum.

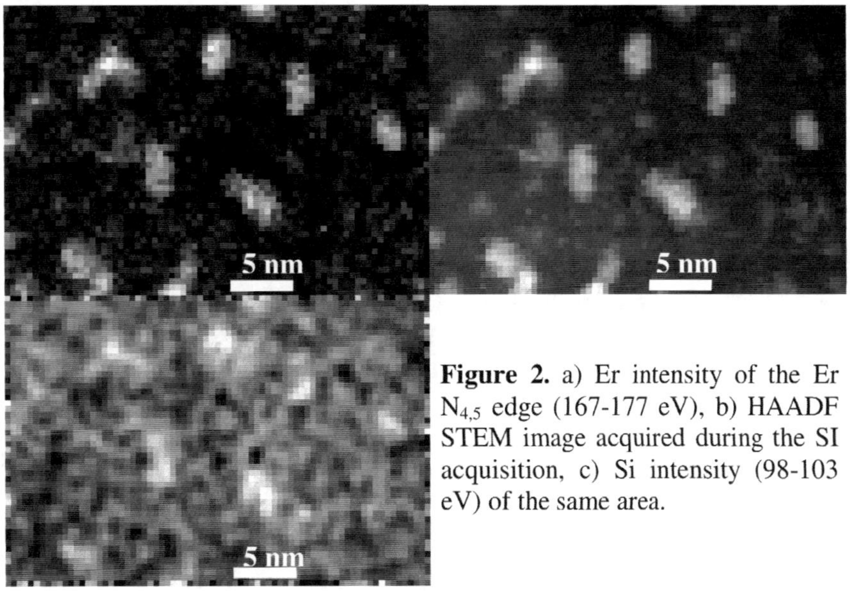

Figure 2. a) Er intensity of the Er $N_{4,5}$ edge (167-177 eV), b) HAADF STEM image acquired during the SI acquisition, c) Si intensity (98-103 eV) of the same area.

There is no significant Er-clustering within the SiO_2 away from the Si-NCs, suggesting there is no formation of separate Er_2O_3 particles in the presence of Si-NCs. Sample (1) with a similar concentration of Er but without implanted Si was investigated by conventional TEM and EFTEM. TEM revealed that implanted Er ions agglomerate in Er rich clusters with amorphous structure in a band near to and extending up to ~170 nm below the surface of SiO_2 as shown in figure 3.This is different from the band extending only to ~75nm into the SiO_2 as observed in the sample (2) with Si-NCs. This proves that Er segregation is not independent of the Si-NC concentration as claimed in recent research [10] and is strongly correlated with the Si concentration. It may suggest that silicon ions or NCs act as agglomeration centres for Er ions.

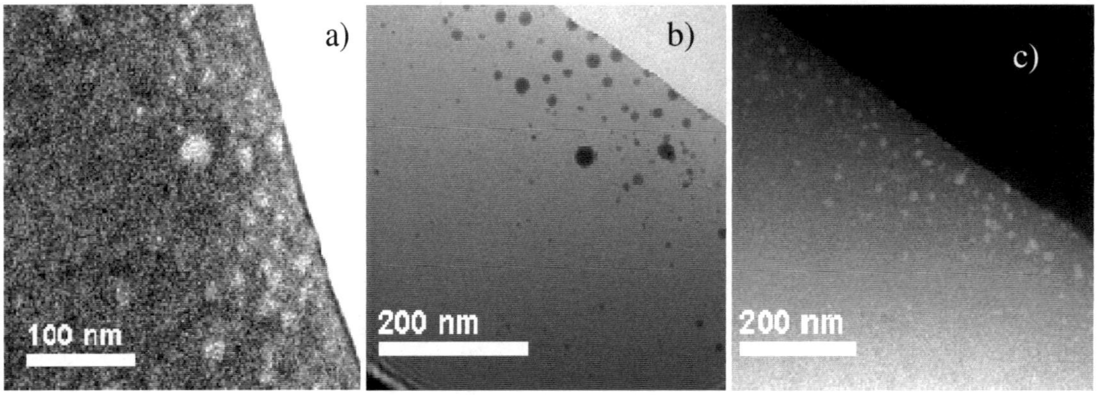

Figure 3. Er-doped SiO_2 sample without Si-NCs, showing Er rich clusters in a band close to the surface. a) EFTEM image of the Er $O_{2,3}$ edge at 29 eV; Er-rich clusters are bright over the dark SiO_2 background, b) BF image, c) HAADF image

4. Conclusion

When Er is co-implanted with Si and annealed, HAADF and EELS show that the Si and Er distribution is identical. The conventional DF images as well as HREM lattice images (not shown) strongly suggest that the aggregates are crystalline, thus constituting Er-rich Si-NCs. In the absence of Si, Er forms on average much larger, amorphous aggregates, presumably an Er-oxide, in the SiO_2 matrix.

References

[1] Miniscalco WJ 1991 *Journal of Lightwave Technology* **9**, 234
[2] Desurvire E, Simpson JR and Becker PC 1987 *Optics Letters* **12**, 888
[3] Kik PG, Brongersma ML and Polman A 2000 *Appl. Phys. Lett.* **76**, 2325
[4] Fujii M, Yoshida M, Kansawa Y, Hayashi S and Yamamoto K 1997 *Appl. Phys. Lett.* **71**, 1198
[5] Wojdak M, Klik M, Forcales M, Gusev OB, Gregorkiewicz T, Pacifici D, Franzò G, Priolo F and Iacona F 2004 *Phys. Rev. B* **69**, 233315
[6] Kenyon AJ, Chryssou CE, Pitt CW, Shimizu-Iwayama T, Hole DE, Sharma N and Humphreys CJ 2002 *J. Appl. Phys.* **91**, 367
[7] Kik PG and Polman A 2002 *J. Appl. Phys.* **91**, 534
[8] Pacifici D, Franzò G, Iacona F, Boninelli S, Irrera A, Miritello M and Priolo F 2003 *Materials Science and Engineering: B* **105**, 197
[9] Kik PG and Polman A 2000 *J. Appl. Phys.* **88**, 1992
[10] Pellegrino P, Garrido B, Arbiol J, Garcia C, Lebour Y and Morante JR 2006 *Appl. Phys. Lett.* **88**, 121915
[11] Krivanek, OL, Dellby N and Lupini AR 1999 *Ultramicroscopy* **78**, 1
[12] Hunt JA and Williams DB 1991 *Ultramicroscopy* **38**, 47

STEM EDX applications for arsenic dopant mapping in nanometer scale silicon devices

G Servanton[1], R Pantel[1], M Juhel[1] and F Bertin[2]

[1] STMicroelectronics, 850 rue Jean Monnet, F-38926 Crolles, France
[2] CEA-LETI, MINATEC, 17 rue des Martyrs, F-38054 Grenoble, France

E-mail: germain.servanton@st.com

Abstract. In this paper we evaluate sensitivity limits and applications of scanning transmission electron microscopy (STEM) energy-dispersive X-ray (EDX) spectroscopy technique for the mapping of arsenic dopant distribution in nanometer range silicon devices. First we show that lamella radiation damages, generated by the intense focused electron probe at 200 keV, are significantly reduced at 120 keV. This allows to use high electron doses and therefore to improve the EDX sensitivity. The analysis of 45nm nMOS transistor clearly shows the n doped areas and local segregation in gate and spacers.

1. Introduction

Highly doped shallow junctions are developed for short channel metal-oxide-semiconductor (MOS) transistors. The 2D mapping of dopants is critical to control their distribution and is also required for device improvements. In this paper, we evaluate the lamella radiation damages for primary beam energies ranging from 60 to 200 keV. Silicon knock-on damages is significantly reduced by lowering the primary energy to 120 keV. STEM EDX spectra can be integrated for longer acquisition times. The final work concentrates on STEM EDX 2D mapping arsenic distribution in a 45 nm complementary MOS (CMOS) device.

2. Experimental Set-Up

STEM EDX experiments are performed with a field-emission transmission electron microscope (FEG TEM) TECNAI F20. EDX spectra are acquired with an EDAX system. Samples are thinned using a focused ion beam (FIB) FEI HELIOS with a final cleaning at low energy (5 kV). Probe current is measured by the electron current on the phosphorous screen I(nA)=(2.15 x emulsion)/t(s), where t(s) is the exposure time in seconds. The lamella thickness is calculated using energy filtered TEM (EFTEM) imaging, with the formula: $t/\lambda = \ln\left(I_{Total}/I_{Elastic}\right)$, where t is the thickness and λ the inelastic mean free path.

3. Silicon radiation damage versus primary energy

FEG high brightness sources improve STEM EDX resolution and sensitivities but are susceptible to induce radiation damage in the silicon crystal. An example of STEM EDX line profile analysis with noticeable radiation damage at 200 keV is given in figure 1. These damages are undesirable because they can disturb the real arsenic distribution. Moreover structural defects inside silicon crystals (dislocations, grain boundaries, stacking faults) are susceptible to enhance dopant diffusion when radiation damage increases [1]. The first well known radiation phenomenon is the knock-on damage which is well documented [2, 3]. Nevertheless, this radiation effect is strongly energy dependant, in particular below a critical energy, the maximum electron momentum transferred to atoms is not high enough to displace silicon atoms. When the primary energy is below this threshold, the radiation damages become negligible.

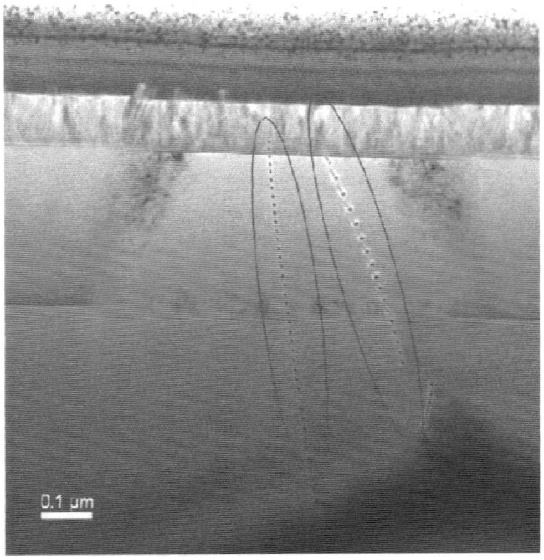

Figure 1. TEM image showing silicon sputtering along a STEM EDX line profile with (1 nA × 40 s) dose on the left and (1 nA × 200 s) dose on the right.

We experimentally evaluate the radiation damage versus the electron dose for an intense probe focused on pure silicon. The analysis area is a fixed point from where the Si-K EDX emission signal is monitored to track the radiation damage. Figure 2(a) presents the EDX signal plotted versus radiation time for a 200 keV primary energy beam, 2.5 nA probe current and for various crystal thicknesses of 70 nm, 100 nm and 135 nm. As the specimen is locally etched, the EDX signal decreases and finally a hole is observed (see inset in figure 2). A 70 nm-thick lamella is completely sputtered for a (2.5 nA × 100 s) electron dose at 200 keV. The 100 nm and 135 nm thick lamellae are also totally sputtered after few minutes.

Figure 2(b) presents similar experiments obtained for various primary energies of 200, 160, 120 and 80 keV, in order to determine the critical energy below which no radiation damage will be observed. In this graph, all curves can be easily compared in dose since the recorded primary current is the same for both energies (2.5 nA). At 160 keV, Si etching still occurs but its rate is about five times lower than at 200 keV. Finally at 120 keV (and 80 keV), the relative EDX signal remains constant during the acquisition time. Radiation damages below 120 keV are not detectable.

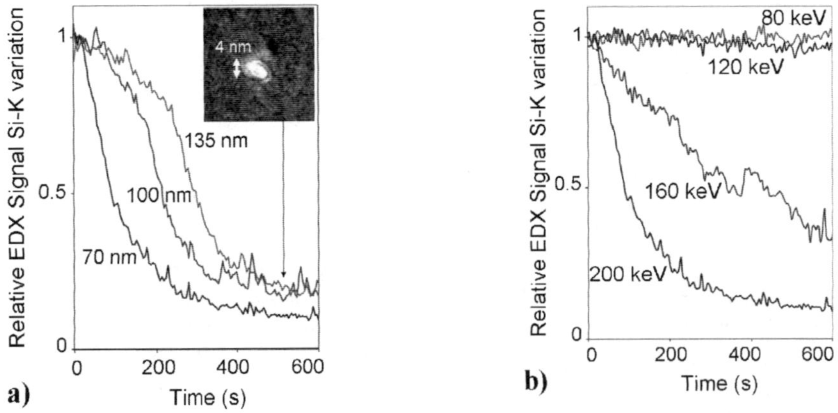

Figure 2. (a) EDX signal plotted versus radiation time for a 200 keV primary energy beam, 2.5 nA probe current and for crystal thicknesses 70 nm, 100 nm and 135 nm. (b) 200 keV, 160 keV, 120 keV and 80 keV primary energy beam radiation damage kinetics versus the exposure time.

As a consequence, all our EDX analyses were performed at 120 keV. Figures 3 (a, b and c) show the influence of electron dose on the arsenic detection sensitivity in an EDX spectrum of a 0.7% As doped Si sample. Figure 3 (a), low dose (0.9 nA x 3 s), shows a very noisy EDX spectrum where the As-K_α peak begins to be detectable but with a signal to noise ratio only about one. For higher doses (0.9 nA x 30 s in Fig. 3(b); 0.9 nA x 60 s in Fig. 3(c)), the As-K_α peak is very distinct from the background. From Figure 3(a) to Figure 3(c), the signal to noise ratio is improved by a theoretical factor around $\sqrt{20} \cong 4{,}5$. This points out that the key parameter in EDX analysis is the electron dose and this is the only way to lower the detection limits. This is possible at electron energy equal to or below 120 keV.

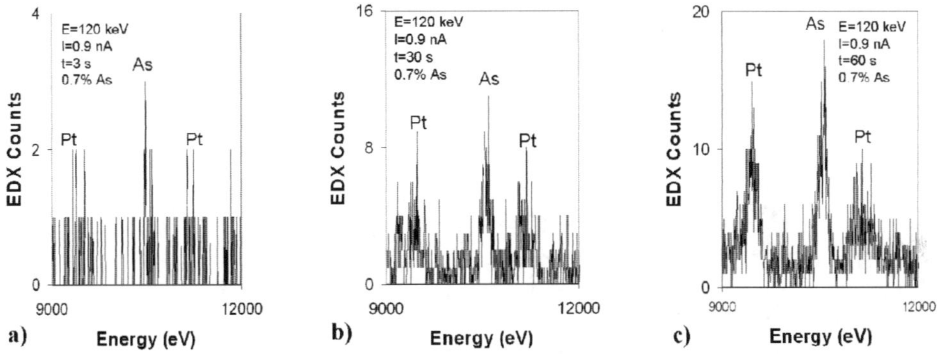

Figure 3. 120 keV STEM EDX spectrum from a 0.7% As doped Si acquired during 3 seconds (a), 30 seconds (b) and 60 seconds (c). The platinum peaks surrounding the arsenic one come from the FIB capping deposition.

4. STEM EDX quantitative arsenic mapping in a 45 nm SRAM device.

Dopant heterogeneities are undesirable since they can alter the transistor performances. The mapping of dopant segregation in nanometer scale devices is challenging regarding the low arsenic concentration to detect (less than 1%As in Si). Figure 4(a) shows a STEM image of a 45 nm nMOS transistor in SRAM. The regions of interest are indicated by white arrows and correspond to the arsenic low doped drain region (LDD). Applying optimization suggested in this paper, i.e. decreasing electron energy and increasing the dose, STEM EDX mapping at 120 keV (0.9 nA × 5 s) is performed in figure 5(b) on a 45 nm CMOS transistor.

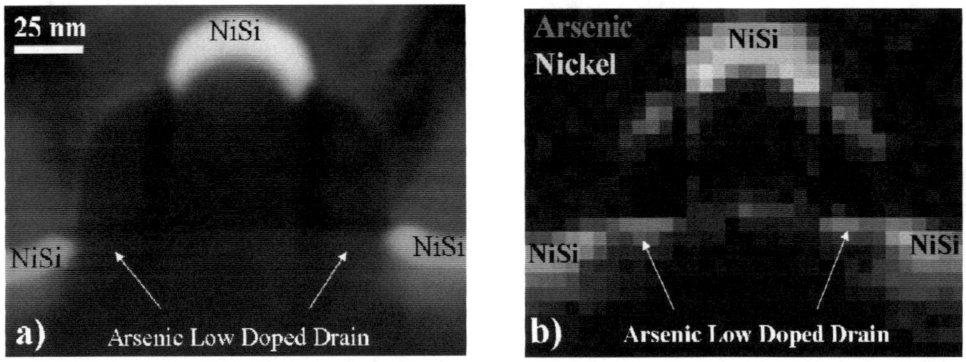

Figure 4. (a) STEM image of the 45 nm SRAM under analysis showing the region of interest (arsenic low doped drain). (b) STEM EDX mapping performed at 120 keV with (0.9 nA × 5 s) dose/pixel and for (37 × 25) pixels with a drift correction procedure for each 10 spectra.

The arsenic low doped drain regions (LDD), implanted at 2 keV energy and for 2.10^{15} $at.cm^{-2}$ concentration, are detected and indicated by the white arrows in figure 4(b). After the implantation process, the spacers (Si_3N_4) were built and another dose of arsenic was implanted to create the source and drain doped pockets. This is why arsenic is clearly detected at the top of the spacers. Arsenic segregation under the NiSi, probably at point defects, is also shown (see figure 5(b)). Also it seems that at the bottom of the gate, arsenic segregates just above the thin SiO_2 oxide. Arsenic segregation at SiO_2/Si interface is the subject of recent publications [4,5].

On figure 5(a), we present the result of several 120 keV EDX mappings that we recombined together as a global map to generate a more highly resolved zoom into the 45 nm SRAM. The different arsenic regions are detected and segregation at the oxide interface is also observed. Figure 5(b) presents a zoomed 120 keV EDX mapping of an n-doped silicided source/drain, showing the arsenic in the LDD but also arsenic segregation at the NiSi interface. This segregation occurs after the drain/source arsenic implantation and can be the place of NiSi encroachments [6].

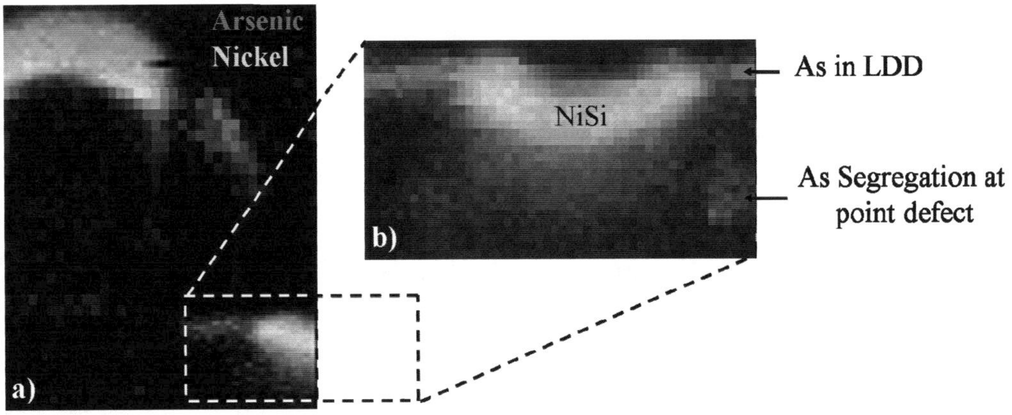

Figure 5. (a) STEM EDX mapping zoomed on the spacer region. (b) STEM EDX mapping revealing arsenic segregation at point defects.

5. Conclusion

This study shows that STEM EDX allow arsenic dopant detection with low concentration at the nanometer scale. The application of this technique is very useful in particular in the microelectronic field to map the dopant distribution and to evaluate possible heterogeneities. Arsenic mapping on a 45 nm scale CMOS shows arsenic presence in the LDD, at the top of the spacers, at the SiO_2/Si interface at the bottom of the grid and also unexpected dopant segregation at point defects.

References

[1] Hobbs LW 1979. Radiation effects in analysis of inorganic specimens by TEM, *Introduction to Analytical Electron Microscopy*, Plenum Press, New York, 1979, pp. 437-480.

[2] Egerton RF, Li P and Malac M 2004 *Micron* **35**, 399

[3] Williams DB and Carter CB 1996. *Transmission electron microscopy*, Plenum Press, New York

[4] Pei L, Duscher G, Steen C, Pichler P, Ryssel H, Napolitani E, De Salvador D, Piro A M, Terrasi A, Severac F, Cristiano F, Ravichandran K, Gupta N and Windl W 2008 *J. Appl. Phys.* **104**, 043507

[5] Parisini A, Morandi V, Solmi S, Merli PG, Giubertoni D, Bersani M and Van Der Berg JA 2008 *Appl. Phys. Lett.* **92**, 261907

[6] Imbert B. et al. Nickel silicide encroachment formation and characterization. Microelectronic Engineering (online June 2009).

Ion implantation enhanced formation of 3C-SiC grains at the SiO_2/Si interface after annealing in CO gas

B Pécz[1], J Stoemenos[2], M Voelskow[3], W Skorupa[3], L Dobos[1], A Pongrácz[1] and G Battistig[1]

[1] Research Institute for Technical Physics and Materials Science, Hungarian Academy of Sciences, P.O.Box 49, H-1525 Budapest, Hungary
[2] Aristotle University of Thessaloniki, Physics Department, 54006, Thessaloniki, Greece
[3] Institute of Ion Beam Physics and Materials Research, Forschungszentrum Dresden-Rossendorf, D-01314 Dresden, Germany

E-mail: pecz@mfa.kfki.hu

Abstract. SiC grains can be grown without voids at the SiO_2/Si interface using a simple method, i.e. annealing in CO gas. Present experiments aim to create nucleation centers for the SiC crystallite growth by carbon ion implantation. The formation of the nucleation clusters, as well as the morphology, the size and the density of the nanocrystals, were systematically studied by conventional and high resolution Transmission Electron Microscopy. The nanocrystallites were developed following two different modes of growth: The first develops facets along the <100> crystallographic direction giving tetragonal grains, and the second facets along the <110> direction resulting in elongated nanocrystallites. It was shown that combined low dose carbon implantation and subsequent high temperature annealing in CO leads to a substantial increase of the covering of the Si surface by high quality 3C-SiC nanocrystallites.

1. Introduction

Silicon carbide with its hexagonal and cubic polytypes is one of the wide band-gap semiconductors used for high temperature applications. Obviously the growth of cubic SiC on Si would be very advantageous, because very large, high quality substrates would then be available at relatively low cost. However, beside the problem of the large misfit, another general problem in that is the formation of voids inside Si.

Earlier experiments showed that SiC grains can be grown at the SiO_2/Si interface using a simple method, i.e. annealing in CO gas, and in this case no voids were formed [1]. Full SiC coverage of the silicon surface cannot be achieved due to the limits of that method. However, nanocrystals were prepared which were also separated later from the Si substrate [2].

The present experiments aim to create nucleation centers for the SiC crystallite growth by carbon ion implantation in order to increase the surface coverage of SiC. The formation of the nucleation clusters as well as the morphology, the size and the density of the nanocrystals were systematically studied by conventional and high resolution Transmission Electron Microscopy. The nanocrystallites were developed following two different modes of growth: The first develops facets along the <100>

crystallographic direction giving tetragonal grains, and the second facets along the <110> direction resulting in elongated nanocrystallites. It was shown that combined low dose carbon implantation and subsequent high temperature annealing in CO leads to a substantial increase of the coverage of the Si surface by high quality 3C-SiC nanocrystallites.

2. Experimental

Carbon implantation of SiO_2/Si wafers (150 nm thick oxide) was carried out at 40 keV with doses of $8\times10^{14}cm^{-2}$, $2\times10^{15}cm^{-2}$ and $4\times10^{15}cm^{-2}$ both at RT and at 550°C. The implanted wafers were annealed in CO atmosphere at 1080°C (and some of them, as reference samples, in nitrogen atmosphere) for 130 minutes. The specimens were characterized by conventional and high resolution transmission electron microscopy (TEM) in cross-section and in plan view as well.

Transparent TEM specimens were prepared by Ar^+ ion milling at 10 keV and finished at 3 keV. In the case of plan view samples the SiO_2 layer was removed from the top surface by HF etching prior to ion milling from the backside. Conventional TEM images were taken on a Philips CM20 transmission electron microscope working at 200 kV as well as on a JEM 120 working at 100 kV, while high resolution images were taken on a JEOL 3010 operating at 300 kV.

3. Results

Despite careful analysis carried out on samples implanted to the lowest dose ($8\times10^{14}cm^{-2}$) and annealed in nitrogen, SiC precipitates were not found in the sample. However, a few grains of cubic SiC were detected in a sample implanted by carbon ions to the highest dose ($4\times10^{15}cm^{-2}$) and subsequently annealed in nitrogen (Fig. 1). Another important observation was the high number of bright features appearing on a dark field image taken using the silicon (220) reflection (Fig. 2). They appeared as black and white contrast in the high resolution images [3] on the scale of about 5 nm and that contrast was attributed to Si-C complexes [3], which were considered as possible nucleation sites for 3C-SiC formation during subsequent annealing in CO.

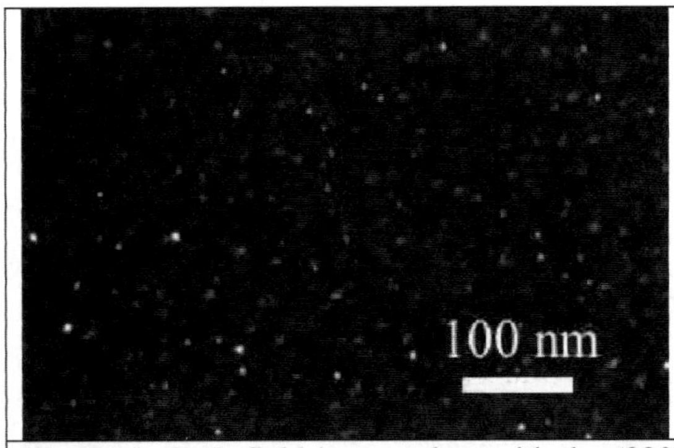

Figure 1. Dark field image taken with the (220) reflection of SiC. Dose: $4\times10^{15}cm^{-2}$, annealed in nitrogen

Figure 2. Dark field image taken with the (220) reflection of silicon.

Another reference sample was annealed in CO, but not implanted. That specimen showed well developed 3C-SiC grains formed exactly at the SiO_2/Si interface with a mean size of 18 nm and a characteristic thickness of about 10 nm. All of those crystallites are epitaxial to the silicon and are tetragonal with sides along the [100] and [010] directions. About 24% of the silicon surface is covered by the 3C-SiC crystallites.

In all of the samples implanted by carbon and subsequently annealed in CO, a high number of 3C-SiC grains were formed, all of them epitaxially aligned to the single crystalline silicon matrix. Moiré patterns show those grains clearly. The relatively large SiC grains (shown in bright field in Fig.3) are situated down to the depth of 50 nm. This is a remarkable difference from the former results, in which implantation was not carried out. Carbon implantation enhanced the nucleation of SiC successfully, as it was planned.

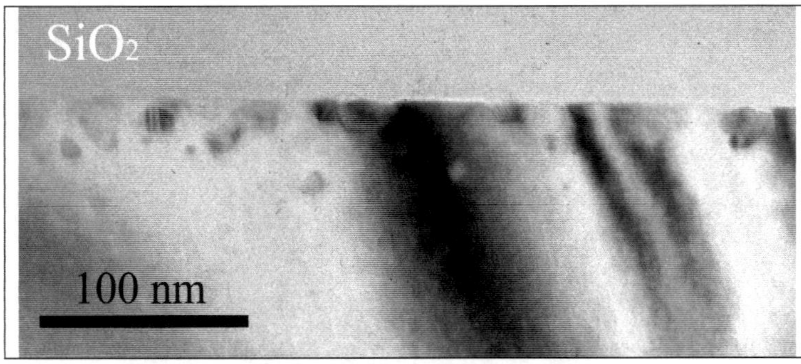

Figure 3. Cross section of a sample implanted to the dose of $2\times10^{15}/cm^2$, at room temperature, subsequently annealed in CO.

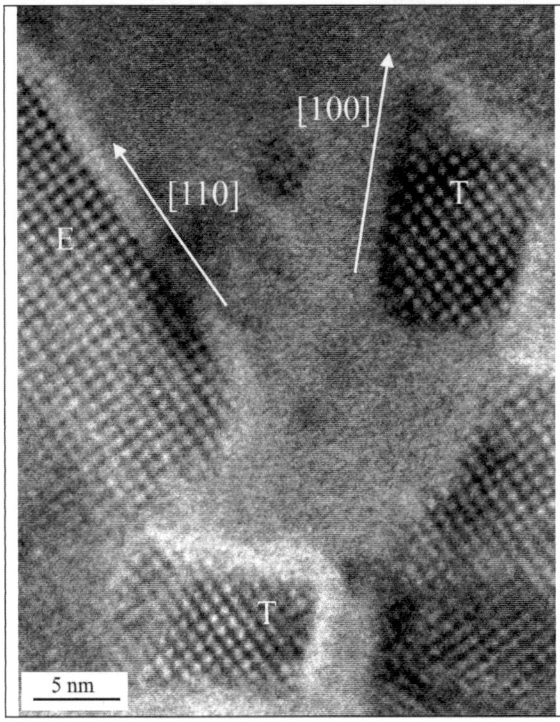

Figure 4. High resolution image taken in plan view on a sample implanted to a dose of $4\times10^{15}cm^{-2}$ at 550°C and annealed in CO at 1080°C for 130 min. Tetragonal grains are marked by letter T, while an elongated rectangular one is marked by E.

However, two types of the cubic SiC grains were formed (in all cases, when carbon implantation was done prior to annealing in CO), the first is tetragonal with sidewalls parallel to the <100> type directions (exactly like in the case of the former results achieved without ion implantation), the second is elongated, rectangular grains with sides along the <110> type directions (Fig. 4). The longer sidewall of the latter grains is 3-4 times longer than the shorter one.

When the sample was implanted to the lowest dose and annealed subsequently (see Fig. 5) the number of 3C grains formed was exactly the same as in the case of the annealed, non-implanted sample. It seems, that there is a threshold for this process below which new nucleation sites for the later growing SiC are not formed.

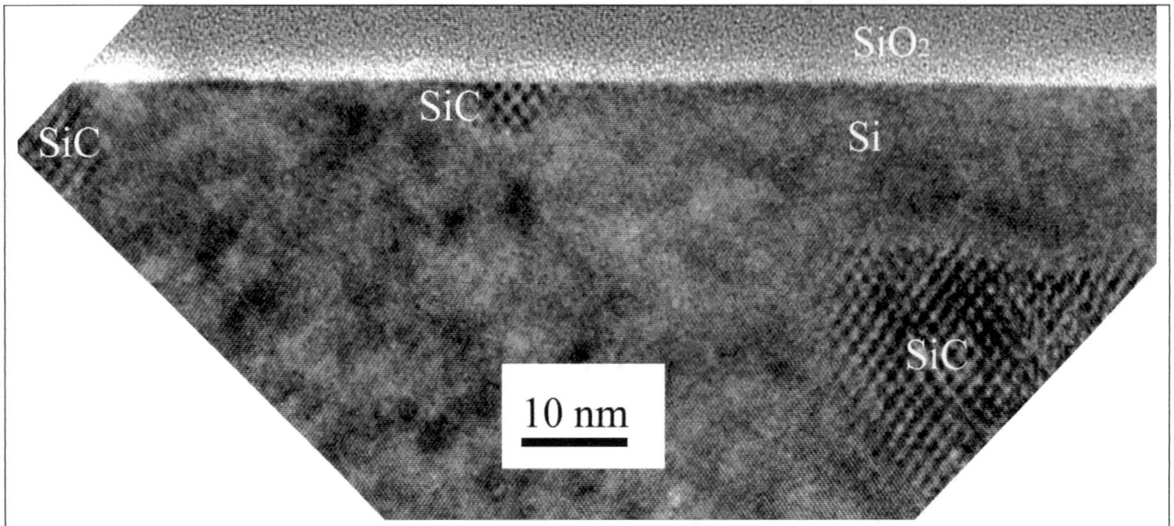

Figure 5. High resolution image a sample implanted to a low dose of carbon and annealed in CO.

Despite the implantation processing and the high density of the nanocrystallites which were formed during annealing in CO, no extended defects were observed in the Si matrix. The formation of the 3C-SiC nanocrystallites is not a conventional epitaxy where the growth starts on the Si surface and the grains grow into the free space above the surface. In the present case the 3C-SiC nanocrystals start growing from the surface towards the Si substrate. It is easy to show that the excess volume in the Si lattice due to formation of 3C-SiC crystallites is only 3.25%. Due to this very small increase in the volume the stress imposed in the Si matrix by the nanocrystallites with the observed size can not reach the critical value to create extended defects [4]. It is worth noticing that the critical size of the SiC precipitates required to generate defects in the Si matrix is above 150 nm [5], almost one order of magnitude larger than in the present case.

4. Conclusions

Thanks to the creation of nucleation sites the density of SiC epitaxial grains formed during annealing in CO gas was increased. Parallel to that the surface coverage by 3C-SiC was also increased to the range of 30-32%. The mean size of the cubic SiC crystallites was decreased down to 10-11 nm. Further experiments are planned in which that crystallite size will be decreased further in order to use their quantum dot properties.

Acknowledgements
This work was financially supported by the OTKA project (Hungary) No. K75735. The access to the large scale facility (ion implantation) of AIM Rossendorf (Germany) was provided via RITA - Research Infrastructures Transnational Access Contract Number 025646.

References

[1] Krafcsik OH, Josepovits KV, Toth L, Pecz B and Deak P 2002 *J. Electrochem. Soc.* **149,** G297
[2] Makkai Z, Pécz B and Bársony I 2005 *Appl. Phys. Lett.* **86,** 253109
[3] Pécz B, Stoemenos J, Voelskow M, Skorupa W, Dobos L, Pongrácz A and Battistig G 2009 *J. Appl. Phys.*, in press
[4] Ashby MF and Johnson L 1969 *Phil. Mag.* **20,** 1009
[5] Nejim A, Hemment PL and Stoemenos J 1995 *Appl. Phys. Lett.* **66**, 2646

Electron Tomography of Gate-All-Around Nanowire Transistors

PD Cherns[1], F Lorut[2], C Dupré[1], K Tachi[1], D Cooper[1], A Chabli[1] and T Ernst[1]

[1] CEA, LETI, MINATEC, F38054 Grenoble, France
[2] ST Microelectronics, 850 rue Jean Monnet, 38926 Crolles, France.

E-mail: peter.cherns@cea.fr

Abstract. We present a study of gate-all-around (GAA) Si nanowire transistor structures using high angle annular dark field (HAADF) STEM tomography. Device structures have been prepared in needle shaped samples using a focused ion beam (FIB), in order to allow sample rotation to +/- 80°. Tomograms are presented, both from a full three channel device structure and also from a single wire test structure, without the hydrogen annealing step. It is observed that hydrogen annealing alters the rectangular cross section of the nanowires, narrowing them and smoothing the corners.

1. Introduction

Developments in transistor design have until recently involved scaling of the complementary metal oxide semiconductor (CMOS) architecture. However, as gate lengths are reduced below 50nm, short channel effects become increasingly significant, and high leakage currents become a problem. One promising solution to this is the gate-all-around (GAA) architecture [1]. The fabrication of these devices involves gate stack integration around semiconductor nanowires, as seen in figures 1 and 2. This design offers excellent electrostatic control combined with high I_{ON} current and immunity from short channel effects.

The characterization of this new generation of devices presents fresh challenges in transmission electron microscopy (TEM). As the length scales of devices are reduced beyond TEM sample dimensions, 2D projections through devices are no longer sufficient. For this reason we present characterization of GAA device structures using electron tomography. This technique allows us to study our samples with a spatial resolution of ~2nm in all three dimensions [2]. The morphology of the nanowires in these devices has a significant effect on their electrical characteristics. With electron tomography we can study changes in morphology along the length of each channel, something which is not available from conventional TEM. To demonstrate this we have examined the effect of the hydrogen annealing step in device fabrication. It has been previously shown that this step rounds sharp corners and reduces roughness [3], which are both important steps towards improved device performance.

The overall aim of this work is to establish methods of sample preparation, data acquisition and treatment for successful and reliable 3D characterization of these samples.

Journal of Physics: Conference Series **209** (2010) 012046 doi:10.1088/1742-6596/209/1/012046

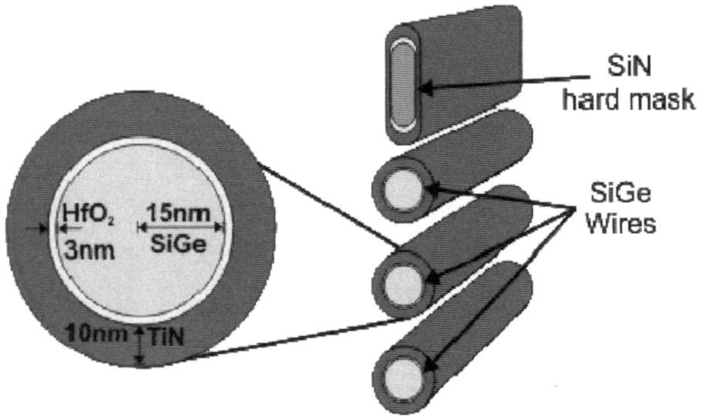

Figure 1. Cross section and 3D schematic of a three channel GAA transistor.

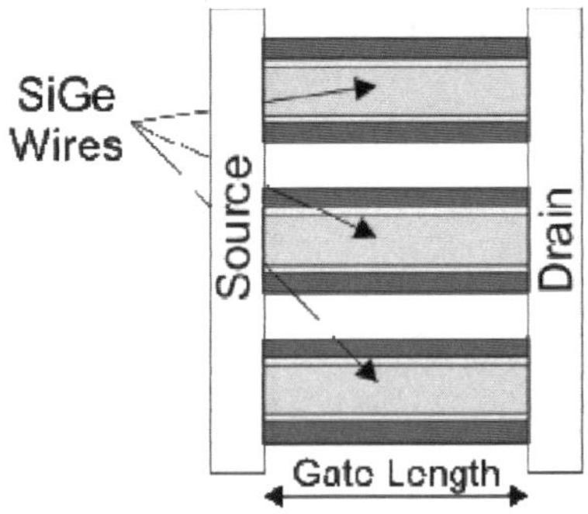

Figure 2. Schematic of GAA transistor showing three channels running from source to drain.

2. Samples

The structures that have been studied in this project include a GAA three channel device and also single wire test structures. The GAA device is processed from a ($Si.Si_{0.8}Ge_{0.2}$) superlattice, which is epitaxially grown on top of a silicon-on-insulator (SOI) substrate. Si/SiGe fins are formed by anisotropic plasma etching. The Si between the SiGe wires is then removed, the wires are annealed in H_2 at 700°C and gate stack integration of 3nm HfO_2 and 10nm TiN takes place. The resultant wire cross section is seen in figure 1. The single wire structures have the same gate stack integration, but have been prepared both with and without H_2 annealing at 750°C.

3. Experimental Work

3.1. Sample Preparation

During initial studies these device structures were prepared for TEM examination inside parallel-sided lamellae, using standard lift-out techniques in the focused ion beam (FIB). It was found that although this type of sample preparation was sufficient to obtain high quality TEM images at zero tilt, several problems arose when using the same sample geometry for electron tomography. At high tilt angles, sample thicknesses became problematic, causing a degradation in image quality. Also, at these high angles shadowing from heavy element (Pt or W) capping layers caused images to be unusable. Both these problems limited the available tilt range and therefore increased distortions caused by the missing wedge. In this work, we have used needle shaped specimens [4], prepared using annular milling in the FIB. The end result is that the device is isolated inside a 350nm diameter needle (see figure 3). The preparation in this case is more specialized and therefore more time-consuming, but the the final sample can be tilted over the full tilt range of the holder without issues of shadowing or thickness changes, as demonstrated in figure 4.

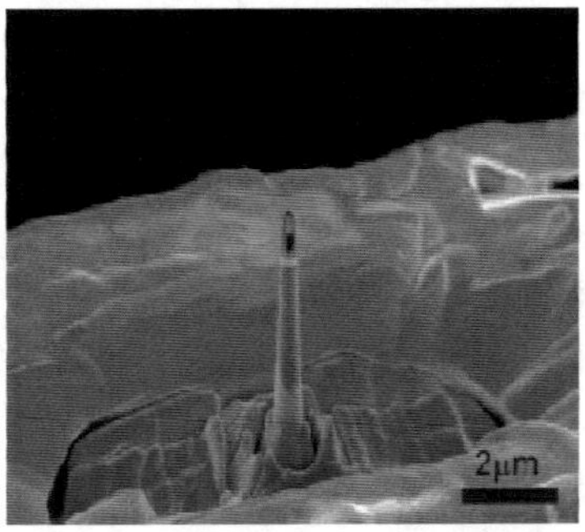

Figure 3. SEM image of needle shaped specimen, containing a GAA device, prepared in FIB.

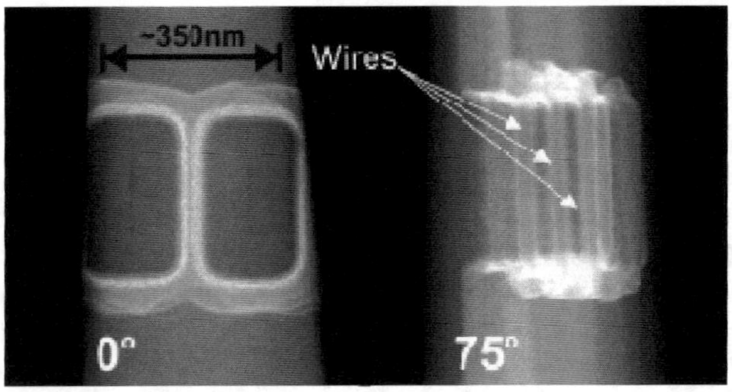

Figure 4. Two HAADF images of a 350nm diameter needle shaped specimen, containing a GAA device, with nanowires indicated.

3.2. Electron Tomography

All the TEM characterisation on these samples has been carried out on the FEI Titan 300kV fitted with a probe Cs corrector, located at Minatec, CEA-Grenoble. To allow high tilt for tomography, a Fischione 2020 sample holder is used. All tilt series have been acquired using high angle annular dark field (HAADF) STEM imaging. Each tilt series consists of 151 images, acquired with 1° steps, between +/- 75°. The post-acquisition tilt series alignment and tomographic reconstruction was carried out using the FEI Inspect 3D software. The simultaneous iterative reconstruction technique (SIRT) algorithm, with 25 iterations, was used for the reconstruction. Visualisation and analysis has been carried out using the visualisation software suites Amira and Chimera.

4. Results

4.1. Three channel GAA device

Figures 5 and 6 demonstrate that the reconstruction from the GAA device has been successful. An iso-surface representation of the tomogram is presented in figure 5. The threshold in this image is set at the intensity level of the TiN encapsulating layer. In this case, the SiGe wire is not directly represented, but the location of its boundary is inferred from the shape of the 3nm HfO_2 surround. In figure 6, two slices through the raw tomogram are shown. From the slice on the left, through the diameters of the wires, the anisotropic loss of resolution and associated distortion due to the missing wedge can be observed, as indicated. This presents an obstacle to a full characterization of the morphology of the wire. Ultimately the solution lies in the acquisition of data over the full 180° tilt range, using an appropriate sample holder. This is not a problem in the plane of the second slice, on the right of the figure. It can be seen that the 3 different layers present can be resolved easily, allowing analysis of changes in thickness and morphology.

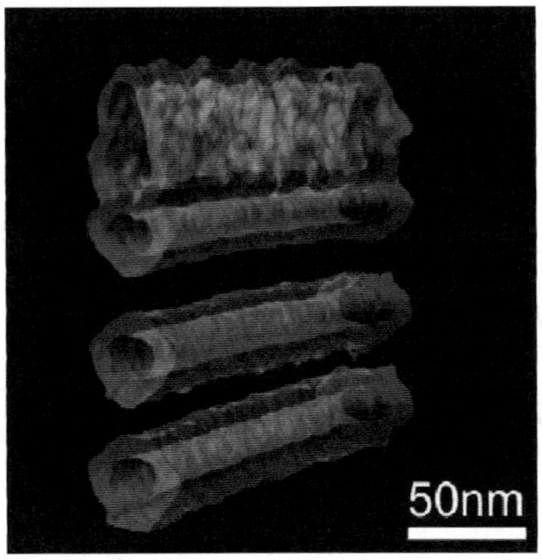

Figure 5. Isosurface representation of tomogram obtained from 3-channel GAA transistor.

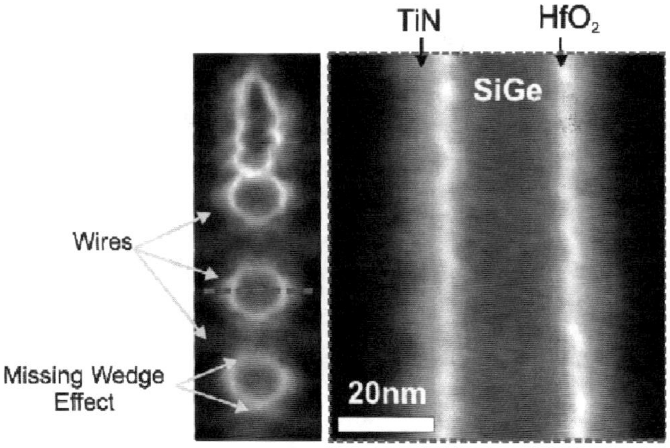

Figure 6. Two slices through tomogram from 3-channel GAA transistor. Orientation of second slice is indicated by red dashed line.

4.2. Effect of H_2 annealing

Figure 7 shows a zero degree HAADF image, and two volume representations of the reconstruction from the single wire sample without H_2 annealing. The location of the encapsulated Si nanowire is indicated in all three images. The HAADF image indicates that the shape of the wire is significantly different from that seen in the finished device structure, though it contains no information about the wire cross section. Figure 7 (c) has the tomogram tilted in order to show the Si wire in cross section. It can be seen that before the annealing stage of processing the nanowires have cross sections that are approximately rectangular. This can be compared to the cross sections we observe in figures 5 and 6

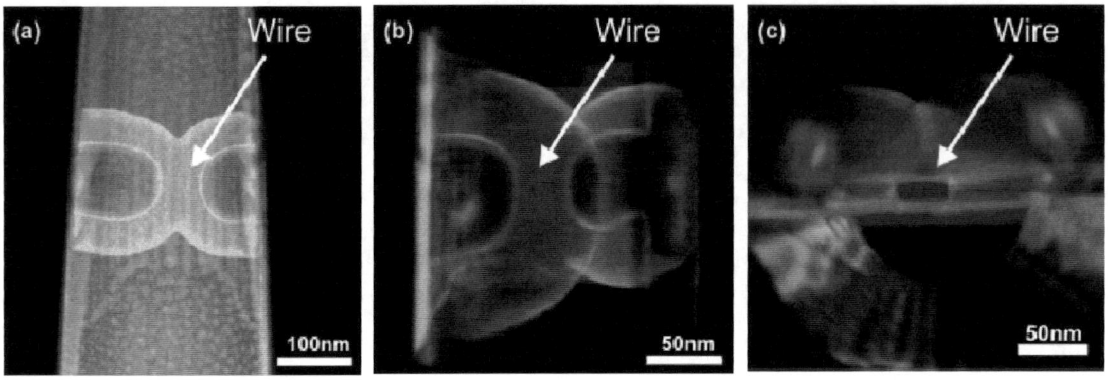

Figure 7. (a) HAADF image of needle shaped sample, containing a single Si nanowire, encapsulated as described in section 2. Si nanowire has not been H_2 annealed. (b) Volume representation of tomogram from the same sample, with location of nanowire indicated. (c) Volume representation taken end on to the wire, showing the rectangular cross section of the wire.

from wires in the full device structure. It is seen that during the H_2 annealing step, the wires are significantly narrowed and the edges are rounded.

5. Conclusion

It has been demonstrated that electron tomography, combined with advanced FIB sample preparation offers a unique 3D characterization of novel GAA transistors. From our tomograms we are able to resolve the SiGe, HfO_2 and TiN layers in these samples. Due to the missing wedge, we currently do not have sufficient resolution to carry out this analysis in all three directions. This can be solved in future by combining the same method of sample preparation with a holder capable of 360° rotation.

By preparing a single wire test structure we have demonstrated the change in wire cross section induced by H_2 annealing during processing.

References

[1] Dupré C, Ernst T, Maffini-Alvaro V, Delaye V, Hartmann J-M, Borel S, Vizioz C, Faynot O, Ghibaudo G and Deleonibus S 2008 *Solid State Electron.* **52**, 519

[2] Midgley PA and Weyland M 2003 *Ultramicroscopy* **96**, 413

[3] Dornel E, Ernst T, Barbe J C, Hartmann J-M, Delaye V, Aussenac F, Vizioz C, Borel S, Maffini-Alvaro V, Isheden C, and Foucher J 2007 *Appl. Phys. Lett.* **91**, 233502

[4] Kawase N, Kato M, Nishioka H and Jinnai H 2007 *Ultramicroscopy* **107**, 8

Acknowledgements

This work was performed within the IBM-STMicroelectronics-CEA/LETI MINATEC collaboration.

Journal of Physics: Conference Series **209** (2010) 012047 doi:10.1088/1742-6596/209/1/012047

Characterization of a FinFET 6T-SRAM cell by tomography

O Richard, S Demuynck, A Veloso, P Van Marcke and H Bender

IMEC, Kapeldreef 75, B-3001 Leuven, Belgium

E-mail: olivier.richard@imec.be

Abstract. A 0.186μm^2 6T-SRAM cell composed of FinFETs is analysed by high-angle annular dark field scanning transmission electron microscopy (HAADF-STEM) tomography. It allows to reveal features that are not observed in the conventional (S)TEM images due to the overlap of different materials over the thickness of the TEM specimen. The influence of the specimen thickness on the quality of the obtained tomograms is studied.

1. Introduction

Due to its good short channel characteristics and the possibility to control the channel potential without heavy doping the fin field effect transistor (FinFET) is considered one of the most promising structures in order to continue the scaling of the 6 transistors static random access memory (6T-SRAM) device beyond the 32nm node [1,2].

In this study, a 6T-SRAM cell with an area of 0.186μm^2 with ~20nm wide fins (pitch 124nm) and ~60nm wide metal/polysilicon gate lines (pitch 150nm) crossing in the orthogonal direction is characterized. Contact vias with a diameter of ~60nm and boomerang-shaped vias are patterned with full-field extreme ultra-violet (EUV) lithography. The process flow is stopped before the tungsten metallization and after the deposition of a Ti/TiN contrast layers stack in the contact and boomerang-shaped vias. Due to the complex geometry of this device (figure 1) and the small dimensions of the different structures (less than the thickness of the TEM specimen), it is not possible to avoid the overlap of different materials (metal/polysilicon gate, silicided fin, Si fin, high-k dielectric, nitride spacers, contact and boomerang-shaped vias, Ti/TiN contrast layers, residues…) over the thickness of the TEM specimen in conventional (S)TEM images. Moreover, the preparation of such specimens at the exact position of interest is difficult and one loses the 3D overview of the full cell in thin specimens. In order to overcome these problems, this device is analysed by STEM tomography.

2. Experimental

Different specimens with different thicknesses and at different positions from similar 6T-SRAM cells are prepared by FIB milling. In order to protect the specimen during the FIB milling a CVD glass layer and a sputtered Pt layer are deposited beforehand on the area of interest. The CVD glass layer is ~85nm thick for the thin specimens and ~350nm thick for the thicker TEM specimens in order to avoid the overlap of the device with the Pt layer during the image acquisition at high tilt angle.

The TEM specimens are mounted on a single tilt tomographic holder (Fischione) with the tilt axis parallel to the oxide/silicon interface (figure 2).

The device is composed of different (poly)crystalline materials (silicide/polysilicon gate, silicided fin, Si fin, Si substrate, Ti/TiN contrast layers) inducing diffraction contrasts in the conventional TEM image series and breaking the projection criterion needed for tomography. The HAADF-STEM mode where the contrast is proportional to the TEM specimen thickness and to $\sim\langle Z\rangle^2$ where Z is the atomic number is used in order to overcome this problem.

An FEI Tecnai (F30) FEG transmission microscope operating at 300kV equipped with the Xplore3D software (FEI) is used in order to acquire the tilt image series. The tilt range is taken for each series as large as possible till the structure is shadowed by the silicon substrate or/and by the protective Pt layer, with 1° tilt increment. The image series alignment is performed using 10nm gold beads as fiducial markers and the 3D reconstructions are carried out with the Inspect3D (FEI) software. Two 3D reconstruction schemes, the weighted back projection (WBP) and the simultaneous iterative reconstruction technique with 20 iterations are applied [3,4]. The obtained volumes (3D reconstruction) are then sliced using the Amira software.

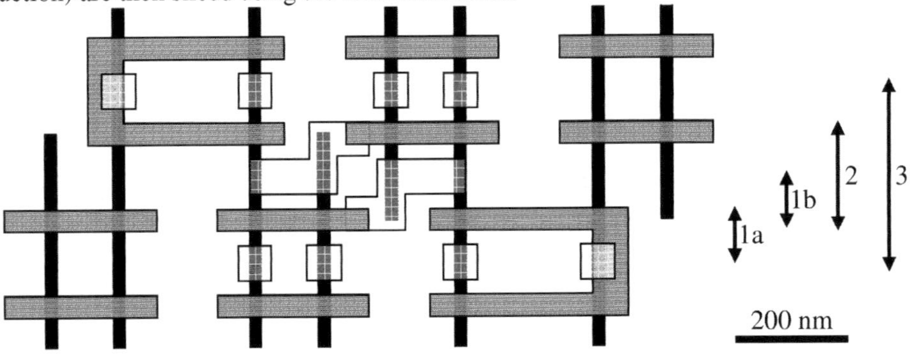

Figure 1. Layout of the 6T-SRAM device. The vertical black lines are the fins, the horizontal grey lines are the high-k/TiN/polysilicon gate lines running over the fins, the white squares and boomerang shapes are the contacts. The arrows mark the position of the different TEM specimens.

3. Results and discussion

3.1 Thin specimens (specimens # 1a and 1b)

Even with thin specimens (~100nm thick), the correct interpretation of conventional (S)TEM images is not possible due to projection overlaps. Figures 2a and b show two examples for specimens prepared at different positions through the cell. HAADF-STEM tomography is performed with both specimens. Different slices of the 3D reconstruction (tilt range: -69°, +70°) obtained with the specimen of figure 2a are shown in figure 3. As can be seen on the slices taken along the fins (figure 3b-d); the contact vias are not in the middle of the TEM specimen thickness.

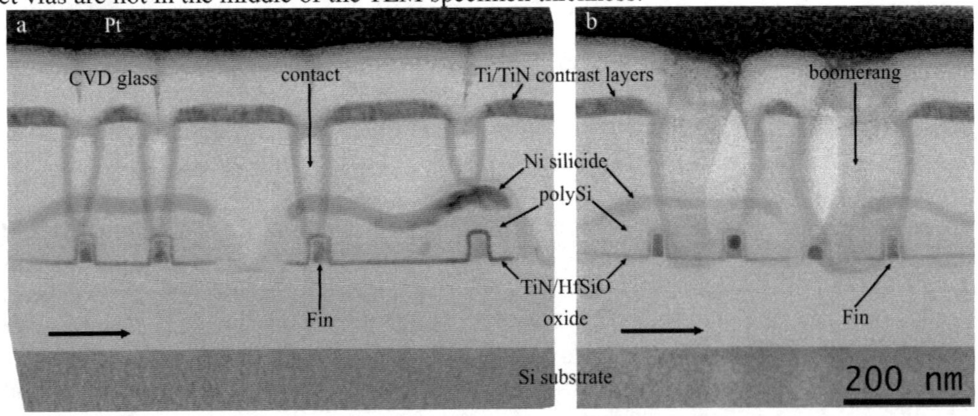

Figure 2. Conventional TEM images obtained with two thin specimens prepared at the level of the contact vias (a) and of the boomerang-shaped vias (b). The tomographic tilt axis is represented by the black arrow parallel to the oxide/Si substrate interface.

The process flow used for the device fabrication was not fully optimized and therefore the following process problems are clearly revealed on the tomographic slices whereas they are obscured in the TEM images.

The contact vias are slightly shifted to the left with respect to the fins (figure 3a) and, more importantly, not in contact with the fully silicided fins (figure 3a and white arrows b-d), except for the right contact via landing on the polysilicon/silicide gate which is locally in contact with the silicide.

The parts of the fins uncovered by the polysilicon gate exhibit a bright contrast over their whole height indicating that they are fully silicided and silicide is for some of them also observed slightly under the HfSiO/TiN layer stack (white triangle figure 3b).

Metallic residues exhibiting bright contrast are observed on the nitride spacers at the side of the polysilicon gate lines (white arrow figure 3a and other slices (not shown here) parallel with the slice of the figure 3a).

The cross shaped features due to the missing wedge are more visible on the slices coming from the WBP than from the SIRT 20 reconstruction but the contrast exhibited by the different materials is better for the WBP slices (compare figure 3d versus 3c).

The bright points observed in the bottom of the contact vias (e.g. white open triangle in figure 3a) are due to platinum coming from the protective layer whereas the bright features observed in the Si substrate and in the oxide layer (white solid triangles in figure 3a) are due to gold beads indicating that this slice is close to the edge of the TEM specimen.

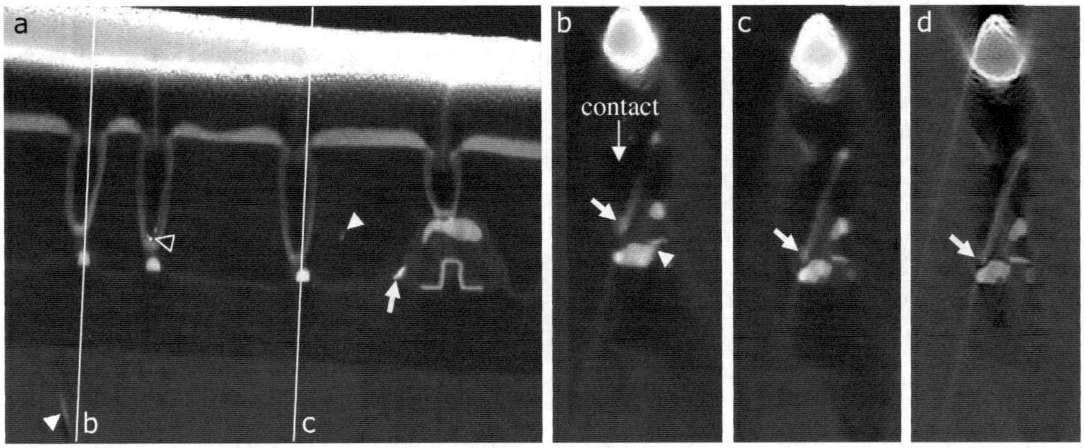

Figure 3. Slices from the SIRT 20 3D reconstruction in the middle of the contacts (a), along fins (b) and (c). For comparison the slice (d) is taken at the same position as the slice (c) but is taken from the WBP 3D reconstruction. The bottom of the vias is marked with thick white arrows (b-d).

3.2 Thick specimen (specimen #2)

HAADF-STEM tomography (tilt range: -62°, +67°) is performed with a thick specimen (~200nm) with the central wing of the boomerangs centered in the middle of the TEM specimen and with two polysilicon gate lines in the thickness of the TEM specimen. As shown on figure 4, despite the thickness of the TEM specimen it is possible to reconstruct and observe the fins, boomerang vias and polysilicon gate lines. It is nevertheless worth noting that one of the polysilicon gate lines (figure 4a) is in focus whereas the other polysilicon gate line (figure 4b) is slightly out of focus. During the tilt acquisition series and especially for the images at high angle, it is not possible to focus the whole HAADF-STEM images. At high angle (65°), the effective thickness is ~470nm.

Spherical voids are observed in the oxide (white arrows figure 4b). These voids are also observed on the slices coming from the WBP 3D reconstruction and are therefore most likely real features of the structure and not reconstruction artifacts.

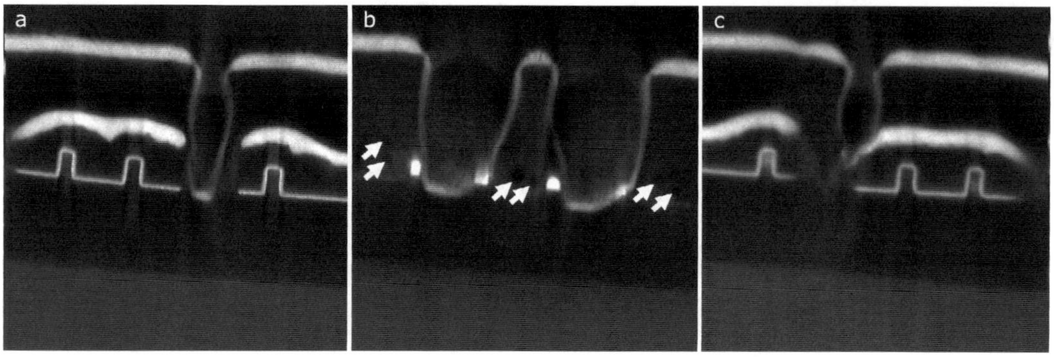

Figure 4. Slices from the SIRT 20 3D reconstruction in the middle of the polySi gate line (a), in the middle of the central wing of the boomerangs (b) and in the middle of the second polySi gate line (c).

3.3 Very thick specimen (specimen #3)

HAADF-STEM tomography (tilt range: -58°, +58°) is performed with a very thick specimen (~340nm) with the central wings of the boomerangs centered in the middle of the TEM specimen and with two rows of contact vias in the thickness of the TEM specimen. The contact vias are clearly revealed at one side of the 3D reconstruction (figure 5a) whereas at the other side the slice is not well in focus but still the main properties of the structure are revealed (figure 5b). Notice that at high angle (58°), the effective thickness is ~640nm.

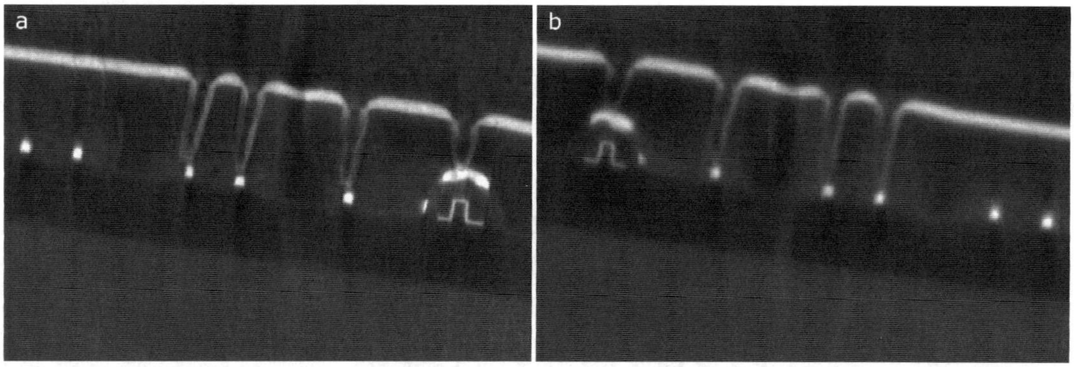

Figure 5. Slices from the SIRT 20 3D reconstruction in the middle of the first row of contact vias (a) and in the middle of the second row of contact vias (b).

4. Conclusions

HAADF-STEM tomography analysis of the FinFET 6T-SRAM cell allows to reveal features which are not observed in conventional (S)TEM images due to the overlap of different materials over the TEM specimen thickness; e.g. presence of residues on the nitride spacers, silicide distribution in the fins, shape and etch depth of the contacts. Moreover in order to obtain interpretable (S)TEM images, very thin TEM specimens at accurate positions would be needed probing only a limited volume of the cell. Although the focusing problem arises for specimens thicker than ~200nm, valuable information on the structure of the device can still be obtained.

References

[1] Collaert N *et al.* 2008 *Tech. Dig. Int. Conf. on IC Design and Technology* p 59
[2] Veloso A *et al.* 2008 *Tech. Dig. Int. Electron Devices Meeting* p 861
[3] Midgley P A and Weyland M 2003 *Ultramicroscopy* **96**, 413
[4] Kübel C *et al.* 2005 *Microscopy and Microanalysis* **11**, 378

Journal of Physics: Conference Series **209** (2010) 012048 doi:10.1088/1742-6596/209/1/012048

InAs/GaAs(001) molecular beam epitaxial growth in a scanning tunnelling microscope

F Bastiman, AG Cullis and M Hopkinson

The University of Sheffield, Department of Electronic and Electrical Engineering, Mappin Street, S1 3JD, UK

E-mail: f.bastiman@shef.ac.uk

Abstract. The growth on InAs on GaAs(001) has attracted great interest and investigation over the past few decades primarily due to the opto-electronic properties of the self-assembled quantum dot (QD) arrays formed. Scanning tunnelling microscopy (STM) has been extensively employed to investigate the complicated and spontaneous mechanism of QD growth via molecular beam epitaxy (MBE). Classically, combined MBE-STM requires quenching the sample after growth and transferring it to an arsenic-free high vacuum chamber which houses the STM system. However, without access to the phenomenon as a dynamic process a basic understanding remains elusive. In order to access surface dynamics, MBE and STM must be combined into a single element. The system herein discussed allows the operation of MBE sources in an STM system relating to InAs/GaAs(001) surfaces.

1. Introduction

GaAs(001) exhibits a myriad of reconstructions depending on As overpressure, Ga:As flux ratio and sample temperature [1], however the (2×4) and c(4×4) reconstructions have received that greatest attention [2,3] due to their existence within the temperature and flux range typically used for high quality GaAs and InGaAs growth. An improved fundamental understanding of these surfaces could play a significant role in future device fabrication, especially for self-assembled InGaAs quantum dots, since here a partial transition from (2×4) to c(4×4) can be observed in the early stages of growth just before an InAs wetting layer (WL) forms.

It is well established that the 2D WL growth is thinner than would be accounted for by the actual deposited In and that at the point when quantum dots (QDs) form the material constituting the dot accumulates faster than deposition rates supply [4-6]. The In segregation mechanism explains the mobile In population as due to segregation induced surface strain, which inhibits In incorporation on lattice sites [7-9]. This principle explains why the In is so mobile on the surface, however, consider the mobile In: it does not form large metallic clusters during epitaxy, as this liquid metal would be detrimental to growth. It can form small clusters, trimers, for instance, where In(As) occupation has occurred. Otherwise, In terminated As-rich, or As-terminated In-rich, reconstructions could accommodate excess In locally, thus providing temporary storage.

Scanning tunneling microscopy (STM) has been employed to study GaAs surfaces since the 1980s. More recently, *in vacuo* molecular beam epitaxy (MBE) combined with STM has allowed preservation of the growth surface affording a wealth of understanding of the fundamental properties of MBE [10-12]. In such systems the MBE and STM chambers are isolated, and growth must be terminated and the sample cooled prior to imaging. This so called quenching process offers snapshots of the growth surface, however the unknown effects of the quenching process may ultimately alter the "as grown" morphology [11]. Combining MBE and STM into a concurrent system allows access to growth dynamics and transient phenomena without the negative influence of sample quenching.

2. MBSTM System

In general, STM can be applied to investigate either atomic binding/reconstruction phenomenon at high resolution (~0.1nm) or islanding/step edge profiles at low resolution (100nm). The ultimate resolution capabilities are usually dominated by tip quality and various noise effects. MBE is a dynamic process, where single ML deposition times can range from 10 to 1000 s and individual adatom site hopping is significantly faster. Similarly, STM is governed by image capture cycles of ~200-400s.

In our molecular beam scanning tunnelling microscope (MBSTM) system, atomic resolution has been observed at up to 450°C on GaAs(001). Thermal drift is substantial during sample temperature ramping (~0.5nm/s), however this settles to a constant 0.01-0.05nm/s after several hours and hence can be easily compensated. Images are assembled from rastered line scans of the surface, and capture times are typically 400s for high resolution images.

Heating is supplied through a custom heating plate, where a piece of pyrolytic boron nitride (PBN) is held in close proximity to the sample (figure 1). Current contacts available in both MBE and MBSTM chambers provide heating current for the PBN element. The plate provides stable heating up to 750°C and allows near 360° reflection high energy electron diffraction (RHEED) access in the MBE chamber.

E-beam sources are utilized for group III and V deposition; after careful alignment the small capacity crucibles (0.7cc) can supply very low fluxes, suitable for slow MBE growth. Each cell is equipped with an integral flux monitor which is calibrated to known beam equivalent pressure (BEP) (figure 2). The sources are more suited to group III evaporation, where the radiant heating effects of the filament (~300°C) do not effect the evaporation rate, however stable group V fluxes from 100nA-10μA can be obtained. The main advantage of e-beam sources for STM is that the low cell temperatures do not induce additional sample drift during shutter operation.

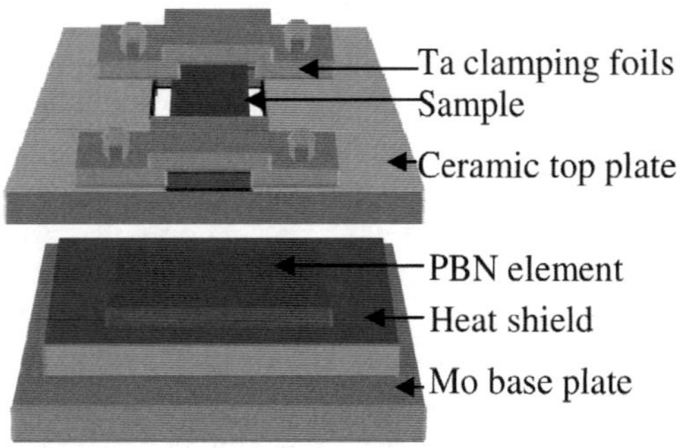

Figure 1. Construction diagram of the PBN heating plate.

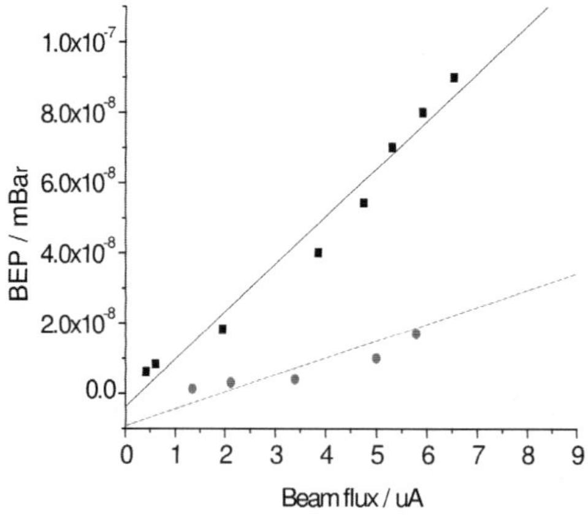

Figure 2. BEP vs. flux for As e-beam cell. Squares and black line are As data and linear fit respectively. Circles and red line are corresponding In cell data.

Tip-flux interaction creates a small but non-negligible shadowing effect on the sample surface. In order to minimize shadowing, the tip is rastered toward the sources (into the beam), hence the surface is imaged prior to shadowing effects. Ultimately this creates a deposition gradient across the scan area, which is partially compensated by adatom diffusion into the partially-shadowed region. The upper section of the image, therefore, receives a larger flux than the lower section.

A similar system has been outlined by Tsukamoto [13], where an STM module was integrated into an existing MBE chamber, so called STMBE. The challenges presented in such a system are biased toward STM operation and dampening. Here results are presented from a series of high temperature images captured in the presences of none, one and all of the In and As fluxes. The intention has been to observe GaAs(001) surface stability, dynamics and InAs/GaAs(001) wetting layer formation.

3. Experimental Procedure

The molecular beam scanning tunnelling microscope (MBSTM) is a commercially available Omicron system combining two e-beam MBE sources with an atomic resolution variable temperature scanning tunnelling microscope (VT-STM) head. For high quality InAs MBE an As:In flux ratio of 10:1 was utilised where applicable. The sources are tailored to low deposition rates and can supply stable fluxes at <0.001ML/s growth rates. Hence a typical ~1.5ML Stranski-Krastanow (S-K) transition for InAs/GaAs at 400°C can be imaged over several hours.

A Si-doped vicinal GaAs(001) epi-ready wafer was cleaved to 10 x 3.9mm^2 in order to accommodate the sample mounting. This plate has been shown to provide uniform heating with 2°C variation across the entire sample when observed by thermography camera. With no additional *ex situ* cleaning, standard oxide removal was performed in the MBE chamber under a 3.0E-6 mbar As_4 BEP at 580°C and then a 0.5μm buffer layer was grown at this temperature with As:Ga flux ratio of 30:1. This resulted in a strong (2x4) RHEED pattern and STM observations showed a smooth surface comprised of single bilayer steps.

The sample was then quenched to ~420°C in a similar manner to that outlined by Yang et al. [14]. The RHEED pattern was monitored throughout the quenching, with particular attention given to the intensity of the 2nd order diffraction rods. With the sample held at 420°C all additional As was pumped from the growth chamber before the sample heating was removed and the sample was swiftly transferred into the MBSTM chamber.

STM tips were prepared with a 2 step process. The first involved electro-chemically etching high purity <111> 0.4mm W wires in 3M NaOH solution. Tips that showed short shanks and sharp apices under light optical microscopy were then placed in a custom designed heating plate and inserted into the vacuum system. Six 10s heating cycles were then applied to the apex at 900°C, subliming the oxide.

Initial STM scanning at room temperature revealed a highly ordered β2(2×4) reconstruction across the entire surface. The sample was then ramped to 250-450°C within the STM stage and left for 5 hours to stabilise. Scanning commenced and continued until the *z*-drift was negligible and the *xy*-drift was in the order of 0.05 nm/s and could be compensated.

Whilst this range is lower than the temperatures normally used for GaAs/GaAs (570-590°C) or InAs/GaAs (480-510°C) it is necessary in this instance to choose a lower temperature. We believe that As-rich (2×4) reconstruction is only stable in the absence of an As flux from 400-430°C. Since the sample temperature must be allowed to stabilize for several hours prior to imaging in the absence of an As flux, this temperature was chosen to maintain a highly ordered (2×4) surface. Since this represents the lowest temperature at which (2×4) is stable in the absence of As, any external flux will instigate a transformation to c(4×4). This allows a very low flux to be applied, hence prolonging the transition to accommodate multiple STM images. In the case of pure As deposition the flux was 200nA, equivalent to 2.0E-8mBar, for InAs growth the fluxes were In: 100nA=1.0E-8mBar, As: 1μA=1.0E-7mBar.

4. Results and Discussion

4.1. Variable Temperature STM (VT-STM)

This section demonstrates the high temperature atomic resolution of the MBSTM. Atomic resolution is essential for identifying reconstruction domains during high temperature MBE growth. The minimum required resolution is 0.4nm, corresponding to a 1× pattern in the (1×3) and (3×1) reconstructions.

High temperature STM on GaAs(001)-(2×4) surfaces up to 320°C is arguably similar to RT-STM. In the absence of impinging flux, the surface remains static, with the low thermal energy insufficient to enhance adatom mobility. After compensating thermal drift, the (2×4) surface appears identical to RT surfaces. This range represents the temperature at which all surfaces are stable.

Upon heating to 320-350°C the dimerized As surface atoms become thermally mobile. The surface begins to transform into a c(4×4) reconstruction, which is energetically favourable at this temperature. The limited mobility prevents a surface-wide reconstruction change. STM of this surface can detect both 0.8nm (2×) and 1.6nm (4×) atomic spacings in the [110] direction, and hence distinguish between (2×4) and c(4×4) domains (figure 3a).

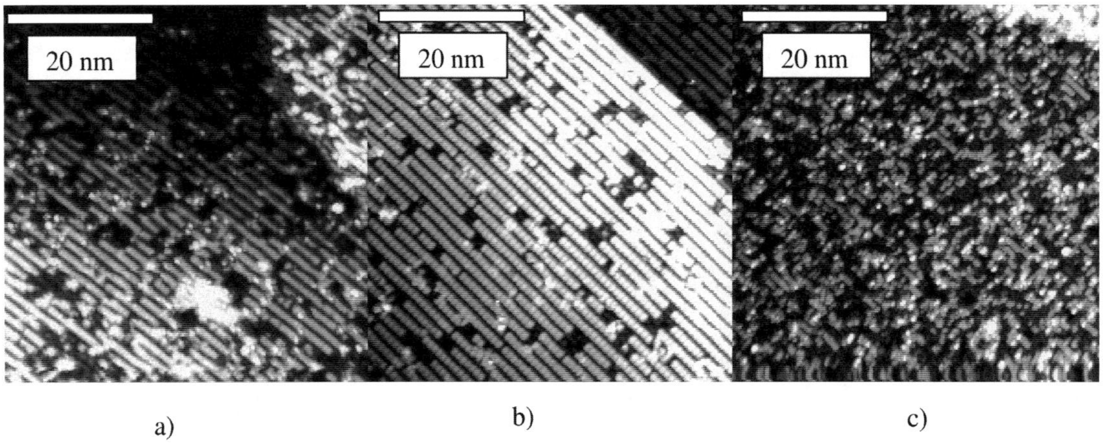

Figure 3. High temperature STM images of GaAs(001). a) $50x50nm^2$ image at 350°C: (2×4) dimer rows can be seen running from top-left to bottom-right, however c(4×4) is clearly visible on the step edge in the upper-right, in the dark pits and on the white central island that start as an isometric (2×4) island. b) $50x50nm^2$ image at 400°C: (2×4) dimer rows are dominant running from top-left to bottom right, some c(4×4) pits are visible after 24 hours of heating. c) $75x75nm^2$ image at 415°C: surface roughening. Some (2×4) dimer rows are visible, but mainly the surface is littered with various 2× and 3× domains.

Heating directly from RT to 350-410°C has a different effect. Whilst the As surface atoms are thermally mobile, arguably more thermally mobile than between 320-350°C, the surface (2×4) reconstruction remains stable (figure 3b). The majority of the surface remains (2×4) even after 24 hours of heating, with small c(4×4) pitting.

Heating to 410°C results in large scale As desorption into the vacuum. The surface is now unstable, due to the volatility of As at this temperature, and hence becomes increasingly Ga rich. The original (2×4) can be identified with 1.6nm (4x) spacing in the [110] direction, as previously, however, 0.8nm and 1.2nm spacings are evident, implying the excess Ga has caused localized, transient reconstruction domains before the surface stabilizes at (3×1) (figure 3c).

4.2. Molecular Beam STM (MBSTM)

Operation of the STM during flux irradiation adds a further complication, especially for III-V MBE where the group V flux typically exceeds the group III flux by 30 times. This section discusses the effects of e-beam sources on STM imaging.

4.2.1. As flux.

Typically the STM is operated in UHV with a background pressure of 1.0E-11 mBar. The operation of the e-beam cell in the STM chamber, without liquid nitrogen cooling, results in a high background pressure, which is ~1/4 of the beam flux. Hence, during initial heating of the cell to achieve a stable flux, the effect of the background As on the surface reconstruction is noticeable. Ultimately, at a sample temperature of 400 °C, the As background pressure is sufficient to induce a partial transformation to c(4×4). This is shown in figure 4a, where As dimers have appeared on top of the (2x4) and similar sized pits decorate the surface. Nevertheless, the β2(2×4) (figure 4c) is still dominant and the induced roughness is similar to that of the γ(2×4) surface (figure 4d).

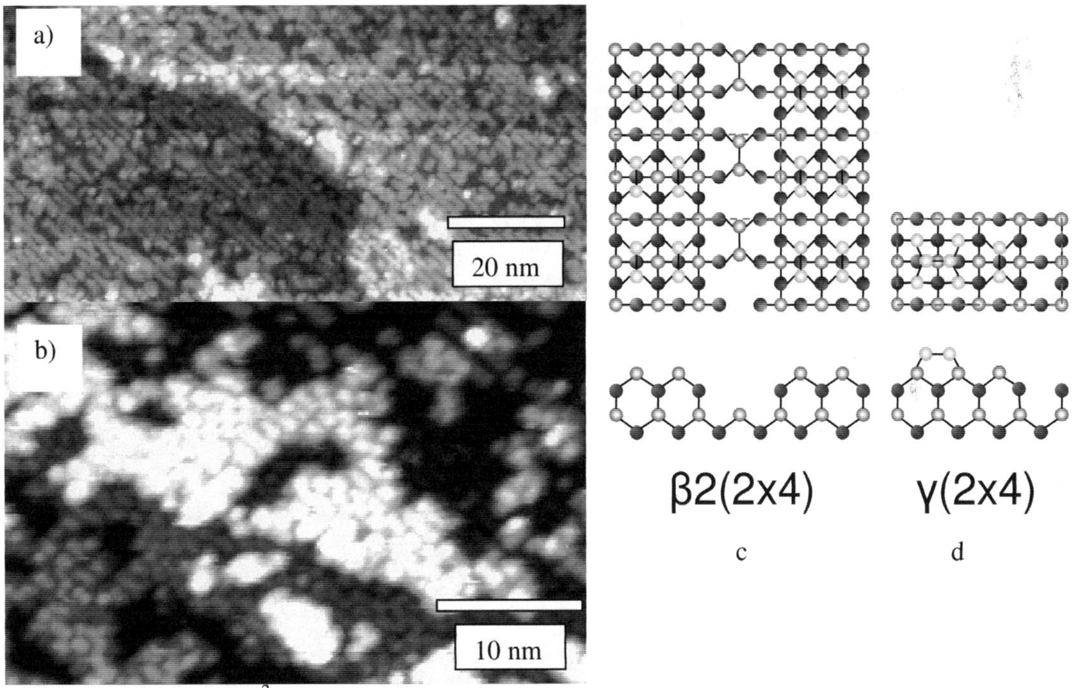

Figure 4. a) 100x50nm^2 image at 400°C, (2×4) dimer rows run from upper-left to lower-right: bright spots indicate adsorbed As from the background flux. b) 40x24nm^2 image at 400°C, c(4×4) is now dominant, only isolated (2×4) domains remain (upper right of image). c) β2(2×4); dark circles: Ga, light circles: As. d) γ(2×4): dark circles: Ga, light circles: As.

The first shutter operation results in a brief loss in imaging capabilities, due to the initial interaction of the flux with the tip, this typically lasts <1s. From this point on the resolution remains remarkably stable throughout the subsequent images, with a few scan lines and contrast changes due to adatom-tip interaction. The surface slowly undergoes a transformation from (2×4) to c(4×4), with c(4×4) islands forming 1 atomic layer above the (2×4) and pits forming 1 atomic layer below [15]. The resolution of the 150x150nm^2 images is high enough to identify the 2 domains and the 0.8nm spacing of the c(4×4) reconstruction is clearly visible. Reconstruction change preferentially takes place at step-edges, which are known to act as catalysts for growth phenomena [16]. The reconstruction change continues from image to image, with step edges spreading and small islands coalescing until the c(4×4) reconstruction is dominant and only isolated dimer rows remain of the original (2x4) (figure 4b).

4.2.2. In and As flux.

Imaging of InAs/GaAs wetting layer formation has been performed at low magnification (500x500nm^2) in order to image step edge profiles, 2D roughening and 3D island formation. Growth of InAs on GaAs(001) comprises of a number of phases. Initially the InAs adopts the lattice spacing of GaAs(001)-c(4×4) and 2D growth ensues under an alloyed (1×3) reconstruction (figure 5a). Both c(4×4) (figure 5c) and (1×3) (figure 5d) occupy the same atomic layer, hence the surface undergoes planarization. This continues up to around 1.1ML where the surface begins to roughen and enter into a transition phase (figure 5b). Near the segregation triggered S-K transition [7, 8] 2D growth slows, due to increasing In segregation toward the surface and, as a result of the increasing strain, limited sites remain for In incorporation.

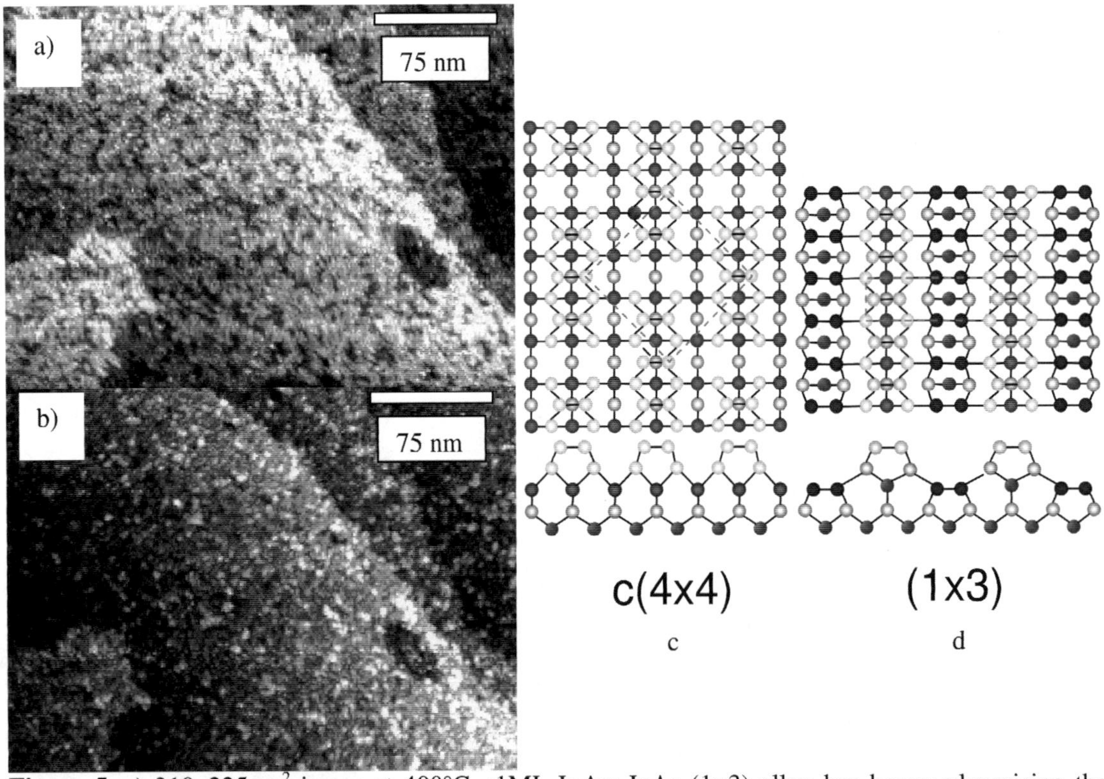

Figure 5. a) 310x225nm^2 image at 400°C <1ML InAs: InAs (1×3) alloy has begun planarising the 2ML high c(4×4) starting surface. b) 310x225nm^2 image at 400°C ~1.1ML InAs: surface has roughened due to lack on In bonding sites in upper layer. c) c(4×4); dark circles: Ga, light circles: As. d) alloyed (1×3); dark circles: Ga; light circles: As; blue circles: In.

The images frequently exhibit scan lines as a consequence of In and As atom contamination of the tip interfering with tip-sample interaction. The scan lines are transient, meaning any material adsorbing onto the tip quickly desorbs. The tip, held in close proximity to the surface, reaches an equilibrium temperature of an order with the substrate. Hence impinging flux does not build up on the apex, provided low fluxes are applied.

Shortly before QDs form, a number of small pre-QD structures appear with a diameter of ~10-15nm and a height of 1.5nm (~5ML) (figure 6). These islands are quasi-stable, appearing and disappearing through a series of 10 images, similar to previous observations [17]. The tip interaction at nucleation sites is witnessed in several pre-QDs where the formation is "clipped" giving the dot a flat edge. This clipping does not seem to effect nucleation, as clipped pre-QD are present and whole in the following image.

The pre-QDs appear randomly across the surface, though are limited to the upper ¼ of the image, where tip shadowing is significantly reduced. This shows the sensitivity of the S-K transition to sub-ML growth profiles. The nucleation sites are not limited to step edges, nor do the nucleation sites appear distinct from other areas of the upper ¼ of the image where pre-QDs do not form. There the surface is covered in 2D structures, some displaying bright spots where local reconstruction has added an additional atomic layer. Some ripen into pre-QDs, though the ripening process appears random. This re-entrant behaviour continues until WL formation is complete at 1.4ML and full 3D structures with 60nm diameter and 7nm height form.

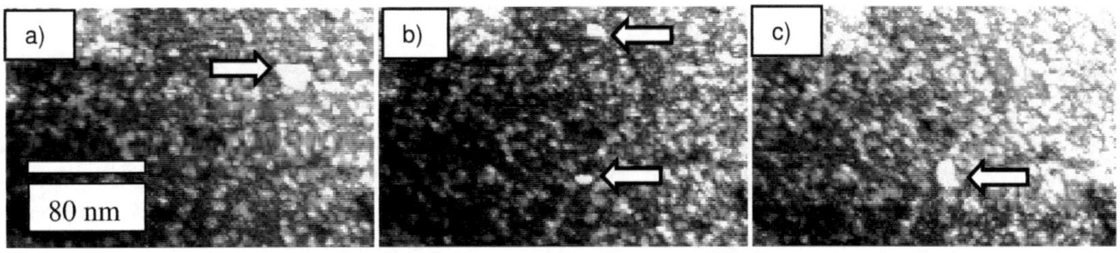

Figure 6. Series of sequential STM images showing pre-QD structures in the same sample area. Height 1.5nm, diameter 10-15nm.

5. Conclusion

The operation of e-beam sources to achieve MBSTM has been discussed for the III-V system, concentrating specifically on InAs/GaAs(001) epitaxial growth. Atomic resolution at high temperature and in the presence of an As flux has been used to observe reconstruction change from (2×4) to c(4×4). Low resolution imaging in the presence of In and As fluxes has further been shown to allow WL growth to be observed and provide dynamic results on pre-QD structures. Tip-flux-sample interactions in the form of shadowing, contamination and QD growth alteration have been shown to require consideration when analyzing observations.

Acknowledgements

The authors would like to thank Richard Frith for encouragement and technical support during the duration of the project.

References

[1] Daweritz L and Hey R 1990 *Surf. Sci.* **236**, 15
[2] Biegelsen DK et al. 1990 *Phys. Rev. B* **41**, 5701
[3] Hashizume T et al. 1994 *Phys. Rev. Lett.* **73**, 2208
[4] Joyce PB et al. 1998 *Phys. Rev. B* **58**, R15981.
[5] Joyce BA and Vvedensky D D 2004 *Mat. Sci. & Eng. R-Rep.* **46**(6), 127
[6] Joyce BA et al. 1997 *Jpn. J. Appl. Phys.* **36**(6B), 4111
[7] Cullis AG et al. 2002 *Phys. Rev. B* **66** R081305.
[8] Walther T et al. 2001 *Phys. Rev. Lett.* **86**, 2381.
[9] Cullis AG et al. 2005 *Appl. Surf. Sci.* **244**, 65
[10] Orr BG, Snyder CW and Johnson M 1991 *Rev. Sci. Inst.* **62**, 1400

[11] Nosho BZ et al. 1999 *Appl. Phys. Lett.* **74**, 1704
[12] Itoh M et al. 1998 *Phys. Rev. Lett* **81**, 633
[13] Tsukamoto S, Bell GR and Arakawa Y 2006 *Micro. J.* **37**, 1498
[14] Yang H et al. 1999 *J. Vac. Sci. Tech. B* **17**, 1778
[15] Bell GR et al. 1999 *Phys. Rev. B* **59**, 2947
[16] Ouerghi A et al. 2008 *Surf. Sci* **602**, 1631
[17] Ramachandran TR et al. 1997 *J. Crys. Growth* **175**, 216

Scanning capacitance microscopy studies of GaN grown by epitaxial layer overgrowth

RA Oliver, SE Bennett, J Sumner, MJ Kappers and CJ Humphreys

Department of Materials Science and Metallurgy, University of Cambridge, Pembroke St., Cambridge, CB2 3QZ, UK.

E-mail: rao28@cam.ac.uk

Abstract. Epitaxial layer overgrowth (ELOG) is a common technique for dislocation density reduction in GaN heteroepitaxy. Here, scanning capacitance microscopy is used to study the variations in unintentional doping arising from the ELOG process and reveals facet-dependent incorporation of n-type dopants during the initial regrowth of GaN, and then p-type doping arising from the use of bis(cyclopentadienyl)magnesium to enhance lateral growth during the coalescence stage.

1. Introduction

GaN is currently in widespread use in optoelectronic devices – particularly in blue light emitting diodes (LEDs) and laser diodes (LDs) – and in this context various routes for the reduction of threading dislocation (TD) densities in GaN-based heteroepitaxial structures have been developed. The applications of the nitride semiconductors are now expanding into the electronics field – for example in heterostructure field effect transistors (HFETs) [1] – and here again TD density reduction is a priority since TDs result in increased leakage currents and decreased carrier mobilities [2]. To optimise electronic devices, however, it is necessary to control not only the TD density but also any unintentional doping in the structure [1].

Epitaxial layer overgrowth (ELOG) is a route to TD density reduction which involves patterning a GaN seed layer with stripes of an amorphous masking material. GaN is then grown up through the gaps in the mask and overgrows the masked region forming wings of material with a low dislocation density. In previous studies, we have used scanning capacitance microscopy (SCM) to show that other dislocation density reduction routes may lead to the introduction of unintentionally-doped (uid) layers in the structure [3]. Here, we assess the impact of ELOG on the location and dopant concentration of uid regions. We will compare the cross-sectional SCM data with transmission electron microscopy (TEM) data showing the defect structure and to help provide some insight into how the variations in unintentional doping arise.

2. Methods

The sample for SCM characterisation was grown on c-plane sapphire substrates using a 6×2” Thomas Swan close-coupled showerhead metalorganic vapour phase epitaxy (MOVPE) reactor.

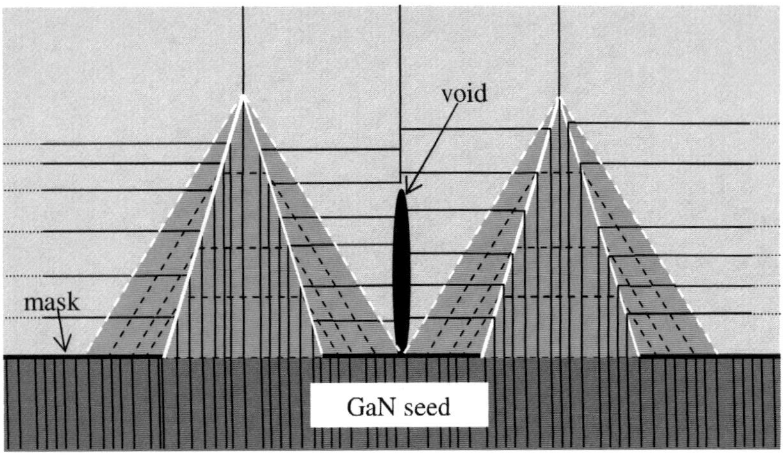

Figure 1. Schematic of the expected structure of the ELOG sample (based on [4]). Fine black lines represent dislocations. Dashed lines represent the shapes of the stripes of GaN as it grows through the windows in the mask. Solid white lines mark the position of the edges of the stripes of GaN as the shape changes during low V:III ratio growth.

Trimethylgallium (TMG), silane (SiH_4), bis(cyclopentadienyl)magnesium (Cp_2Mg) and ammonia (NH_3) were used as precursors. Hydrogen (H_2) was the carrier gas. Initially, a ca. 30 nm GaN nucleation layer was deposited at 540 °C, and thereafter a 500 nm GaN seed layer was grown at 1005 °C. Next, a SiN_x masking was deposited *in situ* for one hour at 860 °C. The mask was patterned *ex situ* using optical lithography to form 5 μm stripes and 5 μm windows parallel to the *m*-direction ($[1\bar{1}00]$ direction). Initial uid-GaN regrowth through the windows occurred at 980 °C, 400 Torr and a V:III ratio of 200 to form stripes of material with a triangular cross-section. Following initial stripe growth, coalescence was carried out, aided by an additional Cp_2Mg flux, at 1020 °C, 100 Torr and a V:III ratio of 1300.

Based on similar approaches employed by other researchers [4], we anticipate a sample structure similar to that seen in Figure 1: the first GaN grown through the gaps in the mask should initially form stripes of trapezoidal cross-section, with $\{11\bar{2}2\}$ side facets and a (0001) top facet. As growth at the lower V:III ratio is continued, we expect the (0001) facet to shrink, eventually leaving a stripe of triangular cross section. TDs which intersect the inclined side facets are expected to bend over so that their line is in the (0001) plane. Thereafter, the growth conditions are adjusted to encourage lateral growth. Where the two wings coalesce, a void may form. TDs may terminate at this void, or may meet and annihilate at the coalescence boundary. To check whether the expected growth modes occur we have used TEM to inspect an ELOG sample, which was grown as described above but without a Cp_2Mg flux. This sample was then examined using a Philips CM30 300 kV analytical TEM with a LaB_6 source. Weak-beam dark field (WBDF) images were taken using a ($\boldsymbol{g}$,$-3\boldsymbol{g}$) diffraction condition.

SCM was performed on sample cross sections prepared by cleaving using an optimised method described in Sumner *et al.* [5]. A Veeco Dimension 3100 atomic force microscope (AFM) equipped with a commercial SCM module was employed. SCM is a technique based on contact mode AFM in which an alternating bias is applied to a metal-coated tip. (Platinum/iridium-coated tips were used here). The amplitude and phase of the rate of change of tip-sample capacitance with voltage (dC/dV) are recorded and provide information about the amount and sign of the carriers in the sample, respectively. In the open loop mode, which has been used here, the dC/dV amplitude signal increases as the carrier density decreases (although no dC/dV amplitude is recorded if the carrier density is below the detection limit of the system). For our setup, the dC/dV phase is negative for n-type material, positive for p-type material [5], and around zero for insulating or metallic regions. More details concerning the SCM technique are available elsewhere [6]. To assess dopant densities quantitatively using SCM data requires either extensive modeling or the use of calibration samples [6]. Here, we have chosen the latter approach: a sequence of layers with varying dopant concentrations was grown on top of the ELOG sample, and the doping density in these layers was assessed using secondary ion mass spectrometry (SIMS) on a separate set of layers grown under the same conditions.

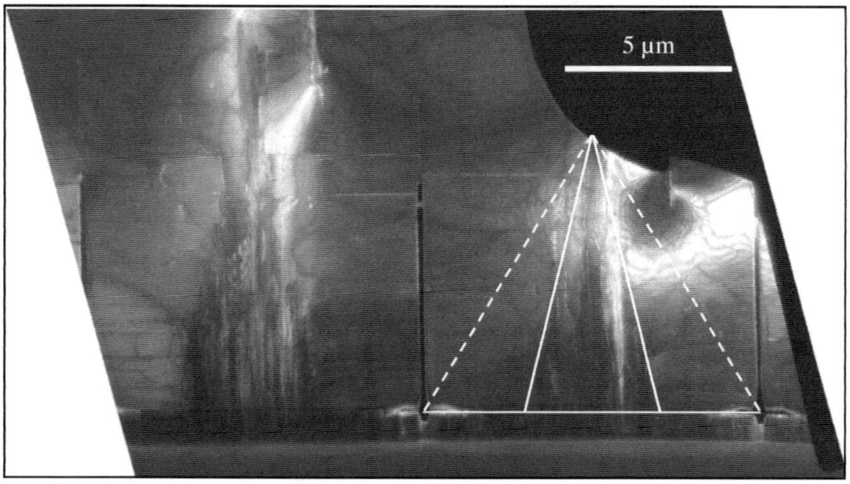

Figure 2. (0002) WBDF TEM image of an ELOG GaN sample. The white lines on the right hand half of the image correspond to the white lines indicating the growth of the GaN stripe through the window region in Figure 1.

The multilayer structure was then imaged simultaneously with the unintentionally doped regions in the ELOG sample, and a calibration curve was drawn relating the SCM d*C*/d*V* amplitude signal to the known doping concentrations. Using this curve, doping concentrations for uid regions were then calculated. This method has previously been shown to provide dopant concentrations for uid regions similar to those found by SIMS in subsequent experiments [3].

3. Results and Discussion

Figure 2 shows a (0002) WBDF TEM image of an ELOG GaN sample. Many of the features shown in the schematic in Figure 1 are evident, including horizontal dislocations and voids at the coalescence boundary between the two 2 wings. The figure shows two periods of the masking pattern. As a guide to the eye, the white lines from the schematic in Figure 1 which indicate the low V:III ratio growth of the initial GaN stripes are superimposed on the right hand half of the TEM image. Regions with a high density of dislocations occur over the window region between the mask stripes, but the dislocations bend over where they encounter a {1122} facet, as expected. Hence, these data suggest that our schematic provides a good indication of the way in which the growth proceeds.

Figure 3 provides an overview of the SCM data. The topography of the cleaved surface is shown in Figure 3(a). Note that the void seen in the TEM data is not clearly seen. A large step in the topography is evident at the same position, but the tip does not penetrate far into the void. The SCM d*C*/d*V* phase data reveals both n- and p-type regions. Within the triangle of material which originally grew through the window, the material is n-type, but in the wings, where Cp_2Mg was used to aid coalescence, the material is p-type, despite the fact that p-doping of GaN normally requires an anneal step to activate the p-dopants, which was not carried out here. Unfortunately, the sensitivity of SCM to p-type doping in GaN is not yet well known, so we cannot estimate the minimum hole density

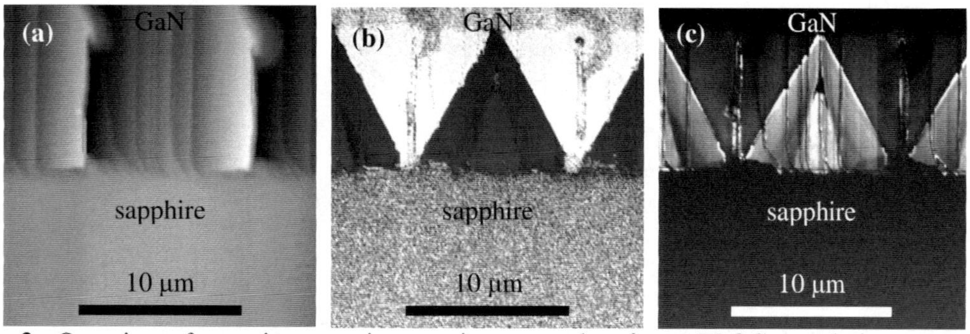

Figure 3. Overview of scanning capacitance microscopy data for an ELOG sample in cross section: (a) topography, (b) d*C*/d*V* phase and (c) d*C*/d*V* amplitude.

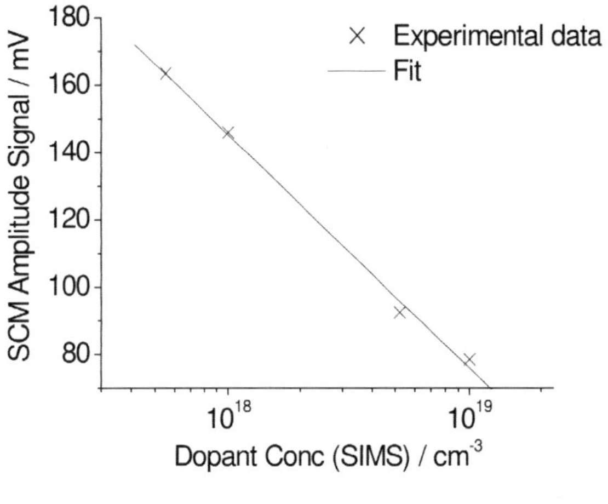

Figure 4. Graph showing the relationship between SCM d*C*/d*V* amplitude signal and the dopant concentration as measured by SIMS for the multilayer structure which was grown on top of the ELOG structure and imaged simultaneously with the n-type uid regions in the ELOG material.

which could give rise to the observed contrast. Lastly, the d*C*/d*V* amplitude data shows that material which grew on a (0001) facet exhibits a higher signal – and thus a lower electron density – than the material which grew on the $\{11\bar{2}2\}$ side facets. This suggests that incorporation of the n-type dopant – which may be silicon arising from the mask material – is facet dependent in this system.

Similar data to those shown in Figure 3(c) were recorded including the additional multilayer grown on top of the ELOG sample which contained material with known dopant concentrations. Hence a calibration curve was drawn (see Figure 4) relating the dopant concentration to the SCM d*C*/d*V* amplitude signal. As in our previous studies, when the SCM data are plotted against the logarithm of the dopant concentration a roughly linear relationship is seen. Hence using the fitted curve shown in the Figure, the SCM signal levels in the n-type uid regions of the ELOG sample may be converted to dopant concentrations. This analysis suggests that the dopant concentration in the material grown on (0001) facets is ca. 1×10^{18} cm^{-3}, whilst in the material which grew on the {1122} facets the dopant concentration varies from about 2×10^{18} cm^{-3} to about 9×10^{18} cm^{-3}, with the dopant concentration being highest closest to the mask – which might act as a source of dopants.

4. Conclusions

Epitaxial layer overgrowth of GaN results in unintentional doping of the material which grows between the mask stripes. Dopant incorporation appears to be faster on the {1122} facets than on the (0001) facets. During coalescence, if Cp_2Mg is used to enhance the coalescence rate, p-type conductivity may result despite the absence of an anneal step to activate the acceptors. These variations in unintentional doping might affect electronic devices grown on ELOG GaN pseudo-substrates.

References

[1] Lee W, Ryou J-H, Yoo D, Limb J, Dupuis RD, Hanser D, Preble E, Williams NM, and Evans K 2007 *Appl. Phys. Lett.* **90**, 093509-1-3
[2] Freitas Jr. JA, Tischler JG, Kim J-H, Kumagai Y, and Koukitu A 2007 *J. Cryst. Growth* **305**, 403
[3] Das Bakshi S, Sumner J, Kappers MJ, Oliver RA 2009 *J. Cryst. Growth.* **311**, 232
[4] Gibart P 2004 *Reports on Progress in Physics* **67**, 667
[5] Sumner J, Oliver RA, Kappers MJ, and Humphreys CJ, 2007 *phys. stat. sol.c* **4**, 2576
[6] Oliver RA 2008 *Reports on Progress in Physics* **71**, 076501-1-37

Observation of dopant profile of transistors using scanning nonlinear dielectric microscopy

K Honda[1], K Ishikawa[2] and Y Cho[2]

[1] Device and Materials Laboratory, Fujitsu Laboratories, Atsugi 243-0197, Japan
[2] Research Institute of Electrical Communications, Tohoku University, Sendai 980-8577, Japan

E-mail: honda.koichiro@jp.fujitsu.com

Abstract. We have demonstrated that scanning nonlinear dielectric microscopy (SNDM) exhibits high performance and high resolution in observing the dopant concentration profile of transistors. In this study, we have measured standard Si samples, which are known to have one-dimensional dopant concentration values, calibrated by using conventional secondary ion mass spectrometry (SIMS). Good quantitative agreement between the SNDM signals and dopant density values was obtained by SIMS. We succeeded in visualizing high-resolution dopant profiles in n- and p-type channel MOSFET with 40 nm gate channels. It is considered that SNDM would be an effective method in measuring the quantitative two-dimensional dopant profiles of semiconductor devices. Finally, we have observed the dopant depth profiles of an SRAM memory cell by using SNDM, and succeeded in detecting the insufficient extension ion implantation in the PMOS transistor area.

1. Introduction

It is very important to design and analyze Large Scale Integrated Circuits (LSIs) to measure two dimensional density profiles of the dopant injected into the source/drain and channel area of transistors with high accuracy. In this regard, several analytical methods, which include Scanning Probe Microscopic methods such as Scanning Capacitance Microscopy (SCM) [1] and Scanning Spreading Resistance Microscopy (SSRM) [2], and Electron Holography with a Transmission Electron Microscope [3], have been developed. In those methods, the Scanning Probe Microscopy (SPM) technology is thought to be the most advantageous mainly due to the ease of the sample preparation. The SPM method uses an Atomic Force Microscope (AFM) especially equipped with an electrically conductive cantilever. SCM is a method to measure the distribution of the capacitance change due to an external impressed voltage by using conductive AFM, and SSRM is a method to measure the resistivity distribution.

From our previous studies we have shown the effectiveness of Scanning Nonlinear Dielectric Microscopy (SNDM) of the SPM category [4-7]. This technique has been mainly applied to observe the distribution of the polarization on the surface of ferroelectric materials, for example the polarization distribution of $LiTaO_3$, the thickness observation of the polarization wall of $Pb_xZ_{1-x}T$, and the application of the ferroelectric memory that uses nano-size domains, and so on.

In the case of semiconductor applications, the 7×7 surface reconstruction of clean Si surfaces was detected by the high-resolution observation in a high vacuum [8]. We have also succeeded in visualizing the charges in nonvolatile flash memory by using SNDM [9, 10]. In order to apply SNDM to LSI analysis here, we report on the measurement of the dopant concentration profile of the source-channel-drain area in transistors.

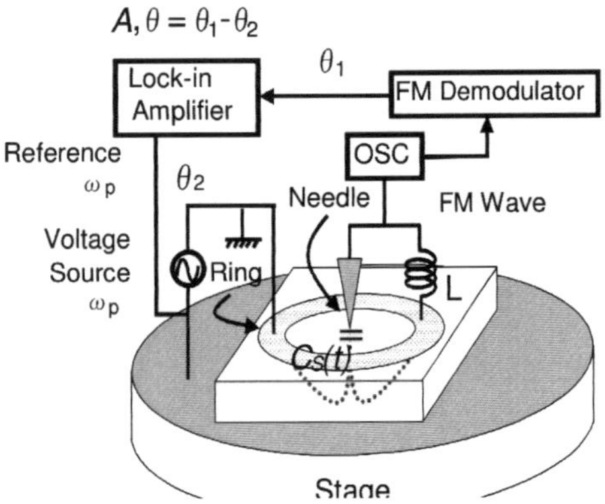

Figure 1. Schematic diagram of SNDM system.

2. SNDM

A schematic diagram of the SNDM system [4-7] is shown in Figure 1. SNDM is one of the microwave microscopy measurement techniques that use an AFM where a ring electrode is used together with a cantilever. In this method, an alternating electric field $E\cos\omega_p t$ ($E = V/d$, where V is the amplitude of the applied voltage, d is sample thickness, and ω_p is angular frequency) is applied between the electrode and the sample, and the capacitance ($C_s(t)$ of Figure 1) variation of the sample surface under the cantilever tip is detected as a change in the resonance frequency of the LC resonant circuit: due to the change in capacitance, the SNDM probe generates a frequency shift of the LC circuit as a frequency modulated (FM) signal. By detecting the FM signal with an FM demodulator and a lock-in amplifier, we can obtain a voltage signal proportional to the capacitance variation. In Figure 1, the distance between the ring electrode and the tip is much shorter than the wavelength of the oscillating microwave, so that the stray impedance in the electrical circuit is sufficiently small, and consequently quite a small variation in capacitance is able to be detected. The sensitivity of the capacitance variation of SNDM is about 10^{-22}F [6], and this is highly sensitive compared with that of the SCM in which the sensitivity is only 10^{-18}F. The relative capacitance variation $\Delta C_s(t)=C_s(t) - C_{s0}$ is

$$\Delta C_s(t)/C_{s0} = (\varepsilon_{333}/\varepsilon_{33})\, V + (\varepsilon_{3333}/2\varepsilon_{33})V^2 + \text{higher orders} \quad (1),$$

$$\Delta C_s(t)/C_{s0} = (\varepsilon_{333}/\varepsilon_{33})\, E\cos\omega_p t + (\varepsilon_{3333}/4\varepsilon_{33})E^2\cos 2\omega_p t + \text{higher orders} \quad (2)$$

In these equations, ε_{33} is a linear permittivity, ε_{333} is the lowest-order and ε_{3333} is the 2nd order nonlinear permittivity. The static capacitance, C_{s0}, is proportional to ε_{33}. The ratios $\varepsilon_{333}/\varepsilon_{33}$ and $\varepsilon_{3333}/\varepsilon_{33}$ can be obtained by detecting the $\cos\omega_p t$ (ω-component) and the $\cos 2\omega_p t$ (2ω-component) of SNDM signals, respectively.

When the sample surface has the MOS type structure, the alternating electric field ($E\cos\omega_p t$) which is applied between the electrode and the MOS sample would change charge density at the interface of the SiO_2-Si substrate. Therefore, we detect the capacitance deviation underneath the SiO_2 film of the MOS system. From the equation (1), the 1st order and the 2nd order differentials are $dC/dV \sim \varepsilon_{333}/\varepsilon_{33}$, and $d^2C/dV^2 \sim \varepsilon_{3333}/\varepsilon_{33}$. Therefore, dC/dV and d^2C/dV^2 can be obtained by detecting the ω- and 2ω-components of the SNDM signals, respectively. The average capacitance change of the MOS interface according to the difference of the dopant density can be measured by SNDM signals similar to SCM [1]. In other words, the dopant density can be

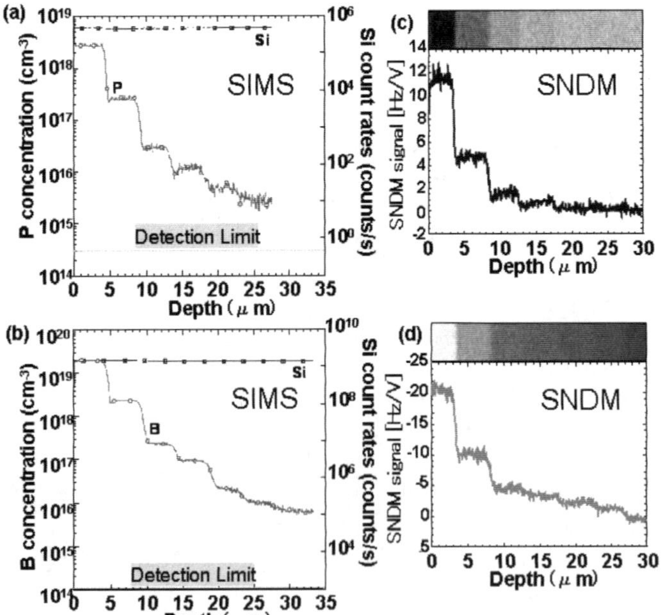

Figure 2. Dopant concentrations of n- type and p-type samples. (a),(b): measured by SIMS. (c),(d): SNDM signal.

obtained from the capacitance change.

3. Results and Discussion

3.1 Standard Sample

As the first stage, the measurement of the distribution of the dopant profile of standard samples with control dopant density was made by SNDM. The standard samples used in this study were epitaxial multilayer Si on n- and p-type wafers. The dopant densities were different from layer to layer. For SNDM studies, the epitaxial Si wafers were cut, and their cross sections were mirror polished.

The dopant concentrations were measured by SIMS, and the SNDM signals and the dopant concentrations obtained from SIMS were correlated. Figure 2 shows an example of the SNDM measurement result of the epitaxial multilayer film of the standard specimen. Figures 2(a) and 2(b) are SIMS images of n-type and p-type samples, respectively. Figures 2(c) and 2(d) are the SNDM signal strength distributions obtained by averaging the scanning data of each image of 32 SNDM lines. In the SNDM images of 2(c) and 2(d), the highly doped layer exhibits a black contrast in the n-type sample whereas a bright contrast in the p-type sample. The existence of 5 layers and 7 layers with different dopant concentrations were confirmed in n-type and p-type samples, respectively. The width of the plateau area corresponding to each layer was found to be about 4-5μm, and this was found to be in good agreement with the SIMS results. Figure 3(a) gives the calibration curves of the SNDM signal vs. density distribution (obtained from SIMS) and Figure 3(b) provides calibrated results. SIMS and SNDM agree well in the density range from 1×10^{17} cm^{-3} to 1×10^{19}cm^{-3} for p-type and from 1×10^{16} cm^{-3} to 1×10^{18} cm^{-3} for n-type. The higher concentration ranges of these calibration curves can be approximated by a simple power law. As mentioned above, SNDM signals corresponding to the dopant density of each level of the epitaxial multilayer film were obtained.

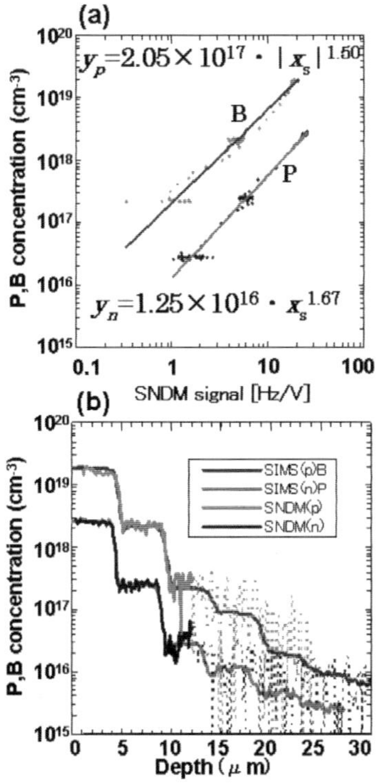

Figure 3. (a) Calibration curve in standard sample. (b) Calibrated SNDM signal.

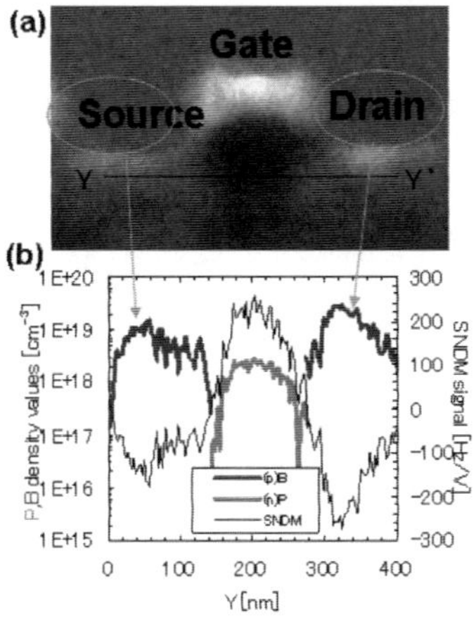

Figure 4. (a) SNDM image and (b) y-y' line signal strength of 90 nm node PMOS cross section.

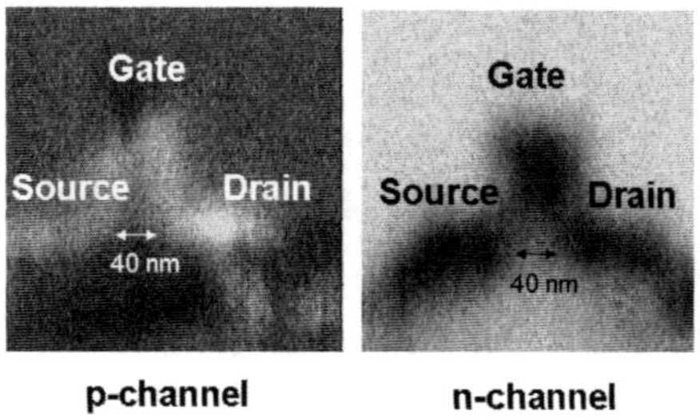

Figure 5. SNDM image of 65 nm node N- and PMOS cross section.

3.2 Transistor Observation

Next, the dopant profile of a transistor was observed as a true application of the above mentioned experiments. We succeeded in obtaining density profiles of the cross sections of both n- and p-channel transistors by measuring SNDM and using the calibration curves of the correlation samples. Figure 4 gives SNDM data of the cross section of a 90 nm node p-channel transistor. This transistor sample was made using the same conditions as for the standard specimen. Figure 4(a) illustrates SNDM image of the transistor of the cross section and Figure 4(b) gives the *y-y'* line signal strength distribution across the source and the drain. We used the calibration curve of Figure 3 to calculate the SNDM signal of the density distribution. As a result, two dimensional concentration profiles of the transistor sections were obtained. Therefore, a quantitative measurement of the two-dimensional distribution of the dopant concentration in the semiconductor device is thought to become possible by using the method mentioned above. However, it is necessary to compare this data with SSRM to find out whether the dopant profiles by the SNDM measurement are correct. Figure 5 shows the SNDM image of 65 nm node N- and PMOS transistors. Both NMOS and PMOS, 40 nm length channel areas were clearly resolved by this method [11].

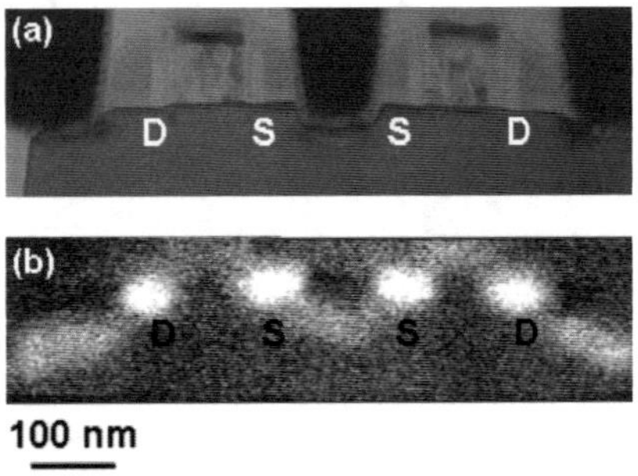

Figure 6. (a) TEM image and (b) SNDM image of a SRAM cell

3.3 Application to failure analysis

We will explain an example of analyzing a single defective bit of SRAM of 90 nm design as an application of this method to defect analysis. Figure 6 shows a section of a TEM image of p-channel of an SRAM cell transistor together with the SNDM image of the same part. As can be seen from this figure, the source/drain and the channel of a pair of transistors are clearly visible.

Next we will explain the results of a failed SRAM transistor. From the results of the failed PMOS transistor, it can be said that the main causes of the transistor defects would be the high resistance due to a lack of injection of ions into the diffusion area of the transistor, the abnormal high resistivity of the contact, and so on. Figure 7 shows an electron beam induced current (EBIC) image of a failed bit cell transistor and the SNDM image of the same part. In this figure, the diffusion area is marked with a dotted circle. The response is weak in this area in the EBIC image. On the other hand, the extension part of one side of the diffusion zone seems to be narrower than the other side of the SNDM image, and the channel seems to be longer. This shows that the amount of the ion implantation into this area is insufficient. In this study, we have succeeded in making visible the lack of extension of ion implantation into the region underneath the spacer of a PMOS transistor in an actual device of a 90nm SRAM cell.

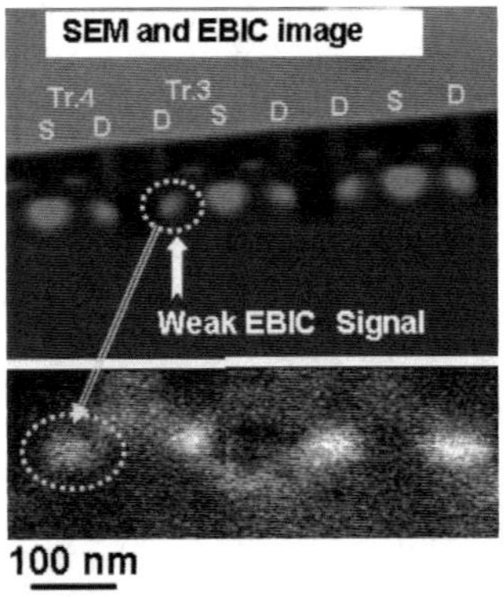

Figure 7. EBIC image and SNDM image of a cell transistors.

4. Conclusion

With the present study we have successfully paved the way to determining dopant profiles and analyzing the defects of transistors through SNDM measurements.

References

[1] Zavyalov VV, McMurray JS and Williams CC 1999 *J. Appl.Phys.* **85,** 7774
[2] De Wolf P, Clarysse T, Vandervorst W, Hellemans L, Niedermann Ph and Haenni W 1998 *J. Vac. Sci. Technol. B* **16,** 401
[3] Volkl E and Lehmann M, 1999 *Introduction to Electron Holography*, eds. Volkl E, Joy D and Allard L A (New York: Plenum)
[4] Cho Y, Kirihara A, and Saeki T 1996 *Rev. Sci. Instr.* **67,** 2297
[5] Cho Y, Kazuta S, and Matsuura K 1999 *Appl. Phys. Lett.* **75,** 2833
[6] Odagawa H, and Cho Y, Funakubo H, and Nagashima K 2000 *Japan. J. Appl. Phys.* **39**, 3808
[7] Cho Y, Fujimoto K, Hiranaga Y, Wagatsuma Y, Onoe A, Terabe K, and Kitamura K 2002 *Appl. Phys. Lett.* **81,** 4401
[8] Hirose R, Ohara K and Cho Y 2007 *Nanotechnology* **18,** S185
[9] Honda K and Cho Y 2005 *Appl. Phys. Lett.* **86,** 013501
[10] Honda K, Hashimoto S and Cho Y 2006 *Nanotechnology* **17,** 084014
[11] Ishikawa K, Honda K and Cho Y 2007 *Nanotechnology* **18,** 084015

Specimen preparation for off-axis electron holography using focused ions, energy filters and laser beams.

D Cooper[1], B Aventurier[1], F Templier[1], A Chabli[1], Ph Salles[2] and G Benassayag[2]

[1] CEA-LETI, Minatec, 17 rue des Martyrs, F38054 Grenoble, France
[2] CEMES, 29 rue Jeanne Marvig, BP94347, 31055 Toulouse, Cedex 4, France

E-mail: david.cooper@cea.fr

Abstract. Focused ion beam milling is routinely used to prepare specimens with nm-scale site specificity for examination by transmission electron microscopy. Although low-energy milling techniques can be used to prepare excellent specimens for many TEM-based techniques, the case is more complicated for dopant profiling. Off-axis electron holography can in principle be used to provide 2D dopant maps of semiconductor devices. However, artefacts such as Ga implantation and the introduction of defects deep in the crystalline regions of the specimens significantly change the properties of the doped semiconductors. In this paper we discuss methods that can be used to remove the artefacts that are present in FIB-prepared semiconductors.

1. Focused ion beam milling for dopant profiling.

Focused ion beam (FIB) milling is the only technique that can be used to prepare specimens with nm-scale site specificity for examination by transmission electron microscopy. The FIB typically uses a beam of gallium ions that are accelerated to 30 kV to remove the material from around the region of interest to provide a thin lamella of between 50 and 500 nm, depending on the TEM technique and material. It is well known that the FIB introduces artefacts into the specimens in the form of amorphous surface layers (25 nm in Si for 30 kV Ga ions) and Ga implantation. Another artefact which is often not discussed, is the creation of defects deep in the crystalline regions of the specimen. These defects can cause a significant problem when measuring the active dopant concentrations using off-axis electron holography, by scanning electron microscopy and scanning probe microscopy as they can trap the dopants in the specimens. In this paper, the focus is on the development of specimen preparation for dopant profiling using off-axis electron holography, however, the ideas that are presented here are transferable to other techniques such as scanning electron microscopy and scanning spreading resistance microscopy.

Off-axis electron holography uses an electron biprism to interfere an object wave that has passed through a specimen with a reference wave that has passed through only vacuum. From the resulting interference pattern, or hologram, phase and amplitude images of the specimen can be reconstructed. As the phase of an electron is very sensitive to changes in electrostatic potential, such as from the presence of dopants, electron holography can in principle fulfil the requirements of the semiconductor industry for a technique that can provide 2D dopant maps with nm-scale resolution. As the phase of an electron is very sensitive to changes in specimen thickness, composition and strain, as well as from active dopants, specimens of constant thickness over an area of up to 1.0 μm^2 are required with nm-scale site specificity. This is only possible with FIB milling, however the defects that are introduced deep in the specimen can trap the dopants which is one of the principle causes of what is known as the electrically inactive thickness. In addition, the electrically inactive thickness is strongly dependent on the FIB operating voltage and also the dopant concentration in the specimens [1]. Therefore in order to be able to extract quantitative information from phase images of semiconductors, this inactive layer must be better understood or removed.

2. The electrically inactive thickness and *in situ* annealing

In order to improve the understanding of the effects of FIB-milling on semiconductors, simple Si *p-n* junctions were grown using reduced-pressure vapour chemical deposition. The junctions examined comprised a 1.0-μm-thick, $2x10^{18}$ cm^{-3} phosphorus (*n*-doped) layer on a 1.0-μm-thick, $2x10^{18}$ cm^{-3} boron (*p*-doped) layer, on a lightly *p*-doped Si substrate. Two different specimens were prepared using a FIB operated at 30 kV using conventional milling procedures. Each specimen comprised lamellas of thicknesses in the range 300 to 600 nm. Electron holograms of the junctions were acquired using a FEI Titan TEM operated at 200 kV. In order to extend the field of view a Lorentz lens was used with the objective lens switched off. Figure 1(a) shows a reconstructed phase image of a 400-nm-thick specimen containing a *p-n* junction. The *n*-doped region appears brighter than the *p*-doped region as it is at a higher potential. It is also clear that the *p-n* junction does not extend to the edge of the specimen as the dopants in this region are not active. This is evidence of the damage introduced into the specimen during FIB milling. Figure 1(b) shows the step in phase extracted across the region indicated in (a). Figure 1(c) shows the step in phase measured across a series of junctions as a function of crystalline specimen thickness measured by convergent beam electron diffraction (CBED) for the two different specimens. From the *x*-intercept, the electrically inactive thickness can be determined. For both of the specimens it is approximately 140 nm, corresponding to 70 nm on each face of the specimen in addition to the 25 nm thick amorphous layer.

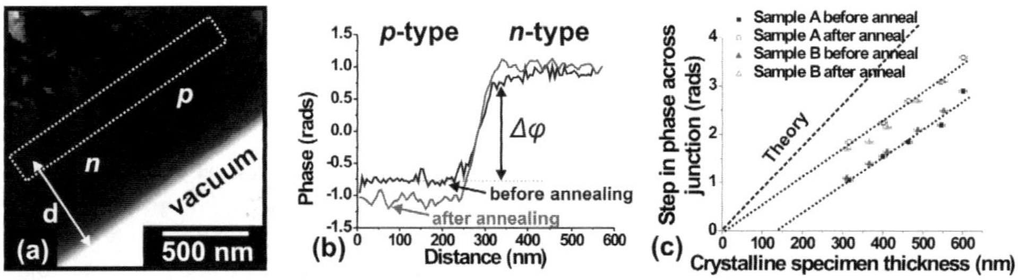

Figure 1. (a) Shows a phase image of a *p-n* junction. (b) Shows profiles extracted from a 400nm thick specimen before and after a 350 °C anneal. (c) Shows the step in phase as a function of the crystalline specimen thickness for two specimens with a range of different thickness lamellas both before and after annealing.

It is known that by annealing specimens in the TEM using an *in situ* heating holder at only 300 °C the damage in the specimens can be removed leading to an increase in the step in phase measured across the junctions [1]. Figure 1(b) shows the phase measured across a 400 nm thick junction before and after annealing. In Figure 1(c) it can be seen that the electrically inactive thickness is almost completely removed for both of the specimens, demonstrating that the technique is reproducible.

In situ annealing removes the defects that are present in the crystalline regions in the specimen that are introduced during specimen preparation. The gradient of the graph gives the built-in potential in the specimen; even after annealing it is clear that the measured potential is less than predicted by theory. *In situ* annealing has been successful here, as conventional FIB-milling has been used to prepare the specimens. In the case of specimens that have been prepared using lift out where a small lamella is stuck to a TEM grid, annealing is more problematic as the heat from the *in situ* holder cannot be transferred to the specimen without destroying the Pt/W that is used to attach the specimen to the grid. In order to apply the heat directly to the specimen, laser annealing has been developed.

3. Low energy excimer laser annealing

Lasers have been used to anneal ion implanted semiconductor devices for many years in order to re-crystallise the amorphous specimens and reactivate the implanted dopants. Lasers are used as their rapid pulse lengths can lead to very low distances of dopant diffusion. High laser powers (>800 mJcm^{-2}) are typically used in order to melt and re-crystallise bulk specimens, however, in order to transfer only a small amount of energy to the specimen the laser has been operated at between 145 and 450 mJcm^{-2} [2]. Here, two different specimens have been annealed using 20 ns pulses with the laser operated at two different powers. Figure 2(a) shows the step in phase measured across a series of different lamella as a function of crystalline specimen thickness measured by CBED. Directly after FIB preparation an electrically inactive thickness of around 140 nm is revealed from the intercept. However, after the specimen has been annealed using a laser power of 235 mJcm^{-2}, the inactive thickness is reduced to around 5 nm. In this case, the laser power seems to be too high as two of the thinner specimens melted during annealing and were not suitable for subsequent examination. Although these results show that the Si *p-n* junctions can be repaired using either an *in situ* biasing holder or laser irradiation, the theoretical built in potential has not been recovered in either case. This is due to the build-up of charge in the specimens during examination.

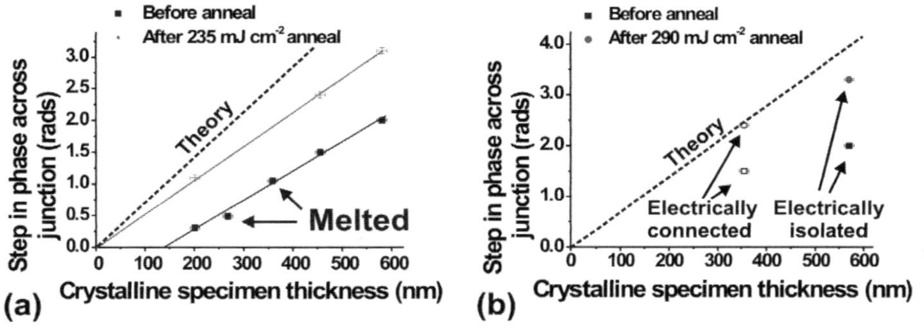

Figure 2. (a) Shows the step in phase as a function of the crystalline specimen thickness for a specimen both before and after annealing using a XeCl laser with a power of 235 mJ cm^{-2}. (b) Shows two specimens, with and without electrical connections before and after annealing.

Figure 2(b) shows two different specimens before and after laser annealing using a power of 290 mJcm^{-2}. One of the specimens has one of the sides of the *p-n* junction electrically isolated and the other specimen has both sides of the junction connected to earth by using a thick layer of sputtered platinum. For the electrically isolated specimen, even after the damage has been removed from the specimen by the laser treatment, the phase measured across the *p-n* junction is much less than expected from theory. However, after annealing the specimen with both sides of the junction connected to earth, the theoretical phase across the junction is recovered. Therefore, in order to recover the theoretical potential from a Si *p-n* junction, both the FIB-related damage and the build-up of charge during examination must be removed from the region of interest.

4. Specimen preparation with silicon, gallium and gold

It has been shown that the electrically inactive thickness can be removed from Si semiconductor specimens that have been prepared by FIB-milling. However, it is known that such specimens will be heavily implanted with Ga. As Ga is a dopant in Si, it would be preferable to clean the surfaces of the specimens before any annealing treatment.

Preliminary experiments have been performed using an energy filtered Orsay Physics EvB FIB using an AuSi source [3]. Specimens were prepared to a thickness of 5 μm using conventional Ga milling and then finished using the energy filter to select either 30 kV Au+ or Si++ ions for cleaning. Figure 3(a) shows the step in phase measured across the junction as a function of electrically inactive thickness measured using CBED for specimens prepared using Au+, S++ and Ga+ ions. The electrically inactive thickness which can be measured from the *x*-intercept is strongly dependent on the choice of ion and is 80, 140 and 410 nm for the specimens prepared using the Au+, Ga+ and Si++ ions respectively. Figure 3(b) shows both the inactive thickness plotted against the simulated range of the ion in Si calculated using SRIM. The graph shows that the electrically inactive thickness in the specimen is directly proportional to the simulated ion range. To test this relationship, a further specimen was prepared using an 8kV Ga ion beam and the inactive thickness was found to be 60 nm (indicated using the open symbol) which fits the general trend shown in Figure 3(b).

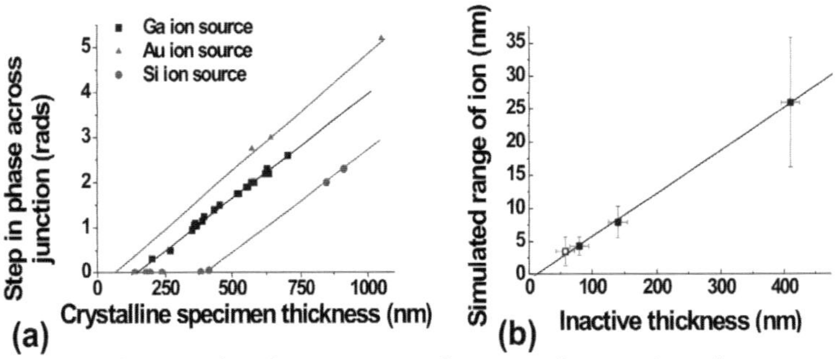

Figure 3. (a) Shows the step in phase measured across the *p-n* junctions as a function of the crystalline specimen thickness measured using CBED for specimens prepared using different ions. (b) Shows the inactive thickness for each of the different ions used as a function of the simulated range of the ions.

5. Discussion

The presence of an electrically inactive thickness has been shown to arise principally from the introduction of physical damage during FIB preparation. The inactive thickness can be almost entirely removed by annealing the specimens either *in situ* in the TEM or by using an excimer laser operated at low power. Even after the inactive thickness has been removed, charging of the specimen during examination will result in a measurement of the built-in potential that is less than predicted by theory. By using improved electrical connections to earth, the build-up of charge can be removed and the correct potential recovered from the junctions.

It has been shown that the Ga can be cleaned from the FIB-prepared specimens by using a Si ion beam. However, this beam introduces a large amount of damage into the specimens. In order to use Si ions for routine cleaning, it will be necessary to reduce the operating voltage of the FIB. In summary, a range of techniques have been used to improve semiconductor specimens prepared in the FIB which have improved our understanding of the artifacts in the specimens and have allowed the theoretical potential to be recovered from a Si *p-n* junction specimen.

References

[1] Cooper D, Twitchett AC, Somodi PK, Midgley PA, Dunin-Borkowski RE, Farrer I and Ritchie DA 2006 *Appl. Phys. Lett.* **88**, 063510
[2] Cooper D, Hartmann JM, Aventurier B, Templier F and Chabli A 2008 *Appl. Phys. Lett.* **93**, 183509
[3] Cooper D, Bertin F, Salles P and Benassayag G 2008 *Appl. Phys. Lett.* **93**, 043510

Focused ion beam patterning to dielectrophoretically assemble single nanowire based devices

V La Ferrara[1], B Alfano[2], E Massera[1] and G Di Francia[1]

[1] ENEA Research Center, P.le E. Fermi 1, 80055 Portici (NA), Italy
[2] University of Naples "Federico II", P.le Tecchio 80, 80125 Naples, Italy

E-mail: vera.laferrara@portici.enea.it

Abstract. Direct-write processing is increasingly taking place in nanodevice fabrication. In this work, Focused Ion Beam (FIB), a powerful tool in maskless micromachining, is used for electrode patterning onto a silicon/silicon nitride substrate. Then a single palladium nanowire is assembled between electrodes by means of dielectrophoresis (DEP). The nanowire morphology depends on the electrode pattern when DEP conditions are fixed. FIB/DEP combination overcomes the problem of nanowire electrical contamination due to gallium ion bombardment and the as-grown nanowire retains its basic electrical properties. Single nanowire based devices have been fabricated with this novel approach and have been tested as hydrogen sensors, confirming the reliability of this technology.

1. Introduction

Direct patterning of electrodes is a promising route for the fabrication of complex nanoelectronic circuits. Generally, device electrodes are realized using electron beam or optical lithography, metallisation and lift-off. Direct-write processes avoid these multi-step procedures and fabricate devices with electrical pads defined directly onto the substrate. Focused Ion Beam (FIB), a well known direct-write tool in nanodevice processing, is often used for electrode patterning in nanodevices. Its operating mechanism relies on a highly focused and energetic gallium ion beam that interacts with a platinum organometallic precursor to deposit platinum nanocontacts. In particular, FIB is often employed to make electrical contacts at the nanowire end after its positioning onto the substrate [1]. Recently we have demonstrated the feasibility of a process to realize a single Pd nanowire hydrogen sensor grown by dielectrophoresis (DEP) using electrodes previously deposited by FIB [2]. This procedure overcomes gallium ion contamination, which creates defects and gallium impurities along the length of the nanowire, and modifies the actual resistivity.

In this paper, we show how it is possible to vary the single palladium nanowire morphology by changing the FIB electrode pattern at fixed DEP working conditions. Finally the electrical characterization of the sensors based on a single Pd nanowire is discussed.

2. Experimental

A crystalline silicon wafer coated with 100 nm of Si_3N_4 was used as starting substrate. A FIB column, integrated in a FEI Quanta 200 3D dual beam system, was used for platinum nanoelectrode patterning.

The highly focused gallium ion beam interacts with the platinum organometallic precursor gas. Metal-organic compounds are dissociated and the non-volatile part of the compound is deposited onto the sample surface to form electrodes. Using 30 kV as accelerating voltage and 10 pA as emission current for ion beam emission, platinum nanoelectrodes are deposited onto the substrate. Each platinum electrode, 300 nm in width and 500 nm in height, has a length of about 100 μm, ending in 0.25 cm^2 Cr/Au pads, previously deposited by electron beam-assisted evaporation.

The feed solution for DEP processing was prepared by dissolving crystalline Pd(acetate)$_2$ (($C_2H_3O_2$)$_2$Pd, Sigma Aldrich) in 10 mM HEPES ($C_8H_{18}N_2O_4S$, Sigma Aldrich) buffer solution, with a pH of 6.5. To ensure complete and homogeneous dissolution, the solution was maintained hermetically closed and in the dark, on a magnetic stirrer for three hours and then centrifuged at 13000 rpm to separate the precipitate. A function generator (AGILENT 33220A) and an oscilloscope (LECROY LC684D) were used as the alternating current source and for monitoring the applied electrical signal, respectively. 2 μl of saturated solution were deposited by casting onto the substrate, previously washed in isopropyl alcohol, deionised water and dried in a nitrogen flow.

Palladium nanowires were grown between the electrodes by applying a 10 V_{pp} ($V_{peak\text{-}peak}$) sinusoidal electric field at 300 kHz. The process stops when the opposite electrodes are short-circuited by the forming nanowire. The single palladium nanowire based devices work as a hydrogen sensor.

Devices were first characterized in direct current (DC) condition at room temperature and in ambient air. A volt-amperometric technique, at constant bias, was then employed for sensor DC electrical characterization in a controlled gas-flow environment, pre-mixed with dry carrier gas in the desired percentage by mass flow meters and continuously controlled by means of an on-line Fourier transform infrared spectrometer [3]. All the tested devices were biased at 5 mV. The total gas flow was set at 500 sccm.

3. Results and discussion

Platinum patterns with different lengths of electrode overlap were fabricated keeping a fixed gap size of about 6 μm and fixed DEP working conditions. Electrode geometry affects the electric field gradient and consequently the shape and the number of nanowires and the growth sites.

In figure 1 we report three different electrode patterns which show how it is possible to obtain either one or two palladium nanowires if the electrodes are more or less overlapped. In figure 1a the two electrodes are not overlapped and a single main (M) palladium nanowire is formed. It is branched, thin, and maintains the same morphology along the wire. Its diameter is about 60 nm. In figure 1b the electrodes partly overlap each other and a single main (M) nanowire is still formed, although the growth of a secondary shorter nanowire (S) with a different nucleation site is observed. This wire cannot complete its growth by connecting the opposite electrode, because the growth of the main nanowire is faster and the process stops when it reaches the opposite electrode. The main nanowire presents an unusual morphology composed of two shapes: the topside consists of aggregated grains ranging from 100 to 200 nm while the underside shows a central trunk with many branches, which is thin and has a width of 60 nm. The electric field generated by this electrode pattern favours the growth of aggregated grains. As the nanowire grows close to the other electrode, more intense electric fields are produced at the edge of the palladium growth [4] and the trunk-structure growth appears dominant. For increased electrode overlap, about 80 μm, the growth of two main and secondary palladium nanowires is evident (figure 1c). The presence of the nanowires, described above, are also predicted by software simulation, modelling the electric field streamlines when pattern and working conditions are defined [2].

Any of the fabricated devices present gallium contamination in agreement with what has been already reported [2]. A Pt halo around each electrode is clearly visible. It is generated by the large spread in the interaction of the gallium ion beam with the organometallic platinum precursor gas [5]. Halos do not overlap, preserving electrical insulation of the transducer, but they reduce the gap size modifying the electric field gradient and consequently the single nanowire growth.

Journal of Physics: Conference Series **209** (2010) 012052 doi:10.1088/1742-6596/209/1/012052

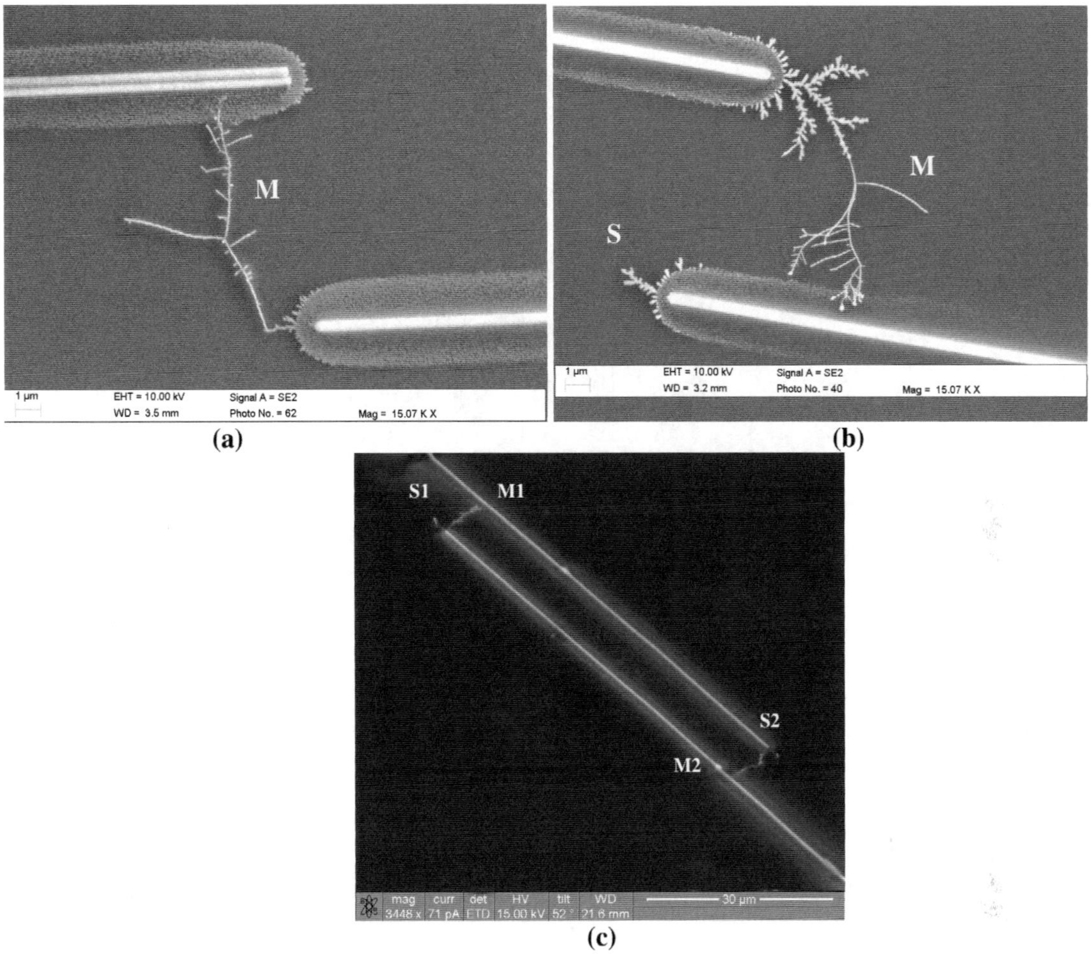

Figure 1. SEM images of palladium nanowires are shown: (a) single Pd nanowire has grown using an electrode pattern with no overlap; (b) a main nanowire (M) and a secondary nanowire (S) are formed when the electrode pattern is partly overlapped; (c) two main (M1 and M2) palladium nanowires originate between each electrode tip and the opposite one. Also in this case, secondary (S1 and S2) shorter, non short-circuiting Pd nanowires are formed. The electrode overlap is about 80 µm.

Devices based on single palladium nanowires are conductive showing an ohmic current-voltage (*I-V*) characteristic between –1V and +1V, with electrical resistance ranging from 40 kΩ to 120 kΩ. A typical electrical response, when the device is under hydrogen gas, is shown in figure 2. The device was maintained, for 30 minutes, in a dry air flow for zero current (baseline) monitoring. Hydrogen concentrations ranging from 1% to 0.1% were then introduced in the test chamber and the electrical response of the device was recorded.

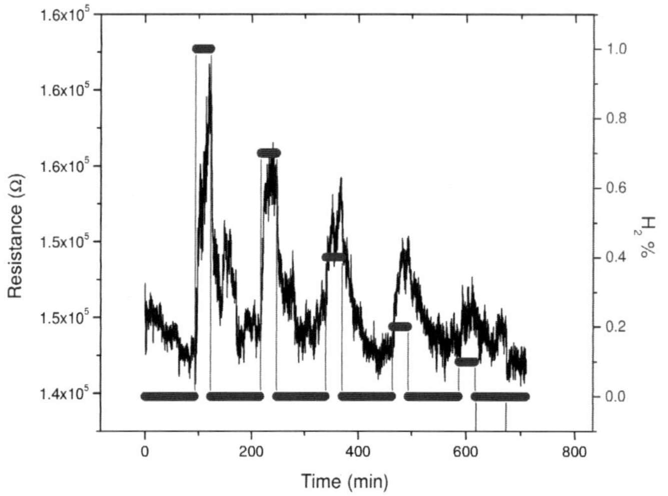

Figure 2. Sensor resistance (left axis) vs time, recorded at room temperature ranging from 1% to 0.1% H_2 concentration in dry air (right axis).

Upon exposure to hydrogen, palladium reacts to form the more resistive palladium hydride (PdH_x where x is the atomic ratio H/Pd); as a consequence the electrical resistance increases. The permeation of hydrogen through the palladium nanowire is in fact a reversible dissociative chemisorption process, and when dry air is introduced into the test chamber, the device presents a quite reversible behaviour [6].

4. Conclusions

We have shown how single Pd nanowire based hydrogen sensor devices can be realized without any manipulation of the wire itself, using a combination of DEP and FIB for the fabrication process. Nanowire morphology can be tuned by changing the electrode pattern. Moreover, the deposition of FIB platinum electrodes, before nanowire fabrication, avoids gallium contamination, preserving the electrical properties of the nanowires. The device operates as a hydrogen sensor at room temperature. Within the testing range, the sensor is sensitive and quite reversible.

References

[1] Romano-Rodriguez A and Hernandez-Ramirez F 2007 *Microel. Engn.* **84**, 789
[2] La Ferrara V, Alfano B, Massera E and Di Francia G 2009 *J. Nanosci. Nanotechnol.* **9**, 1
[3] Quercia L, Cerullo F, La Ferrara V, Di Francia G, Baratto C and Faglia G 2000 *Phys. Stat. Sol.* **182**, 473
[4] Hu Y, Perello D, Mushtaq U and Yun M 2008 *IEEE Trans. on Nanotechnology* **7**, 693
[5] Cronin SB, Lin Y-M , Rabin O, Black MR, Ying JY, Dresselhaus MS, Gai PL, Minet JP and Issi JP 2002 *Nanotechnology* **13**, 653
[6] Cabrera AL and Aguayo-Soto R 1997 *Catalysis Letters* **45**, 79

Energy filtered scanning electron microscopy: applications to dopant contrast

C Rodenburg[1], MAE Jepson[1], BJ Inkson[1], E Bosch[2], AKW Chee[3] and CJ Humphreys[3]

[1] Department of Engineering Materials, The University of Sheffield, Mappin Street, Sheffield, S1 3JD, United Kingdom
[2] FEI Electron Optics, Achtseweg Noord 5, Acht, 5600 KA Eindhoven, The Netherlands
[3] Department of Materials Science and Metallurgy, University of Cambridge, Pembroke Street, Cambridge, CB2 3QZ, United Kingdom

E-mail: C.Rodenburg@shef.ac.uk

Abstract. In the past we have demonstrated that the use of an energy filtered scanning electron microscope (SEM) can enhance SEM dopant contrast. Here we have concentrated on a technique based on energy filtering in the SEM that allows the delineation of p-n junctions.

1. The Energy Filtering (EF) principle used

The work described here was carried out in an XL 30 field-emission SEM, equipped with a through-the-lens-detection (TTL) option. The TTL system consists of an extractor tube that can be positively biased (tube bias, *T*) to direct secondary electrons (SEs) emerging from the specimen towards three layers of electrodes of opposite potential. Each electrode pair generates an electrostatic deflection field perpendicular to the optic axis. The upper two layers are required to minimise aberrations that could be caused by the lowest layer which is the electrode that deflects SEs towards the scintillation detector (Fig.1a). The strength of the deflection field can be adjusted by changing the deflector voltage, *D*. For standard SEM operation where *D*=60 V and *T*=250 V, the resulting field strength is sufficient to deflect SEs over the entire SE energy range (0-50 eV) towards the detector, although only SEs with energies $E_{SE} < 10$ eV are detected for all emergence angles, θ (Fig 1b). A reduction in *D* results in reduced deflection such that only low energy SEs are sufficiently deflected in the direction of the detector, as shown in Fig. 1c for *D*=10 V. Here the majority of SEs detected have energies below 2 eV for all angles of emergence. Hence the TTL can act as a low pass filter capable of detecting only those SEs with an energy less than a user defined maximum. However, Fig. 1d demonstrates that at *D*=20 V, the maximum energy of detected SEs differs for different emergence angles. For larger deflector voltages, this effect becomes even more pronounced and hence, for these deflector voltages, the filtering effect is not well defined.

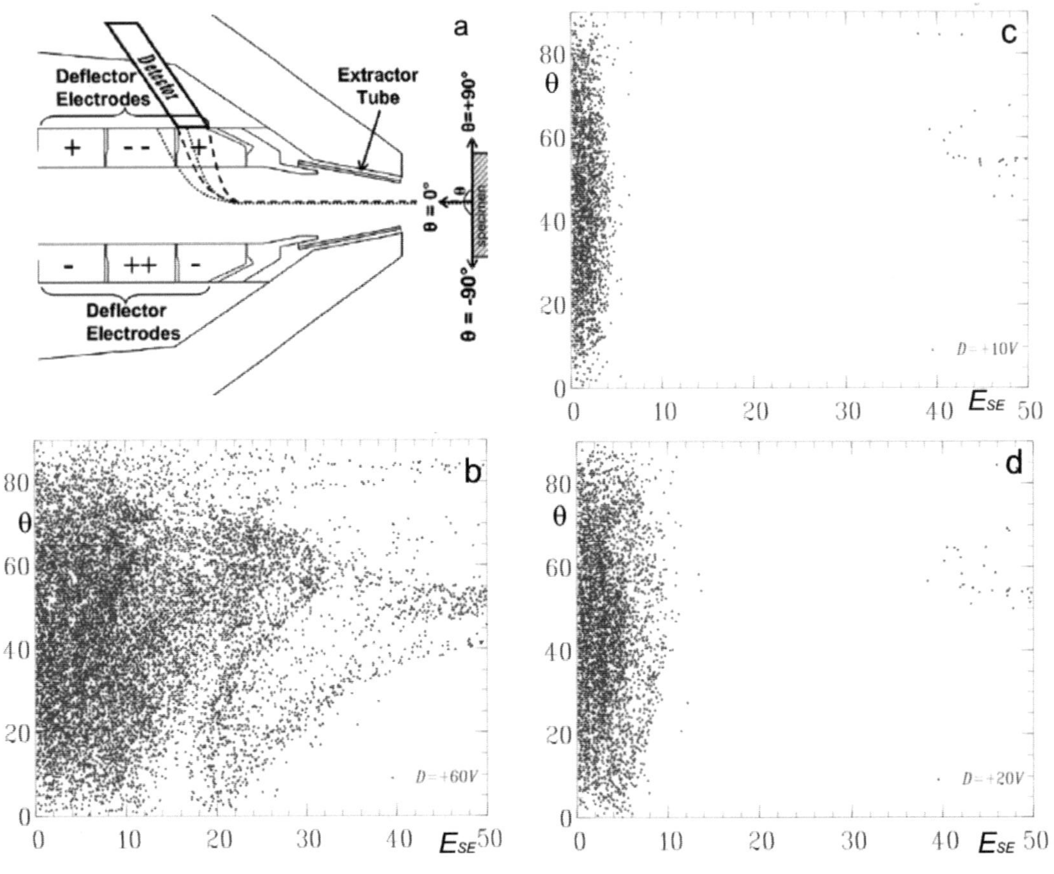

Figure 1. a) Schematic *x*-section of TTL, b-d) calculated acceptance diagrams for SEs generated by 1keV incident beam, at working distance 3mm, *T*=250V: b) *D*=60V, c) *D*=10V, d) *D*=20V

2. Junction delineation

A particular application of energy filtering to dopant mapping is the ability to delineate p-n junctions, on the basis that many of the suggested contrast mechanisms predict a strong deviation of the contrast, at the location of the junction, from that present some distance from the junction. Patch fields [1], strain [2] and charging [3] were all suggested as basic mechanisms of dopant contrast and a strong inhomogeneity in contrast is expected near the junction for all of these mechanisms.

For our experiment a test specimen consisting of a thick n-doped layer (As $5x10^{18}cm^{-3}$) on a p-type (B $1x10^{19}cm^{-3}$) substrate was used. Embedded in the thick n-doped layer are thin p-doped regions with the following doping levels: $6x10^{17}cm^{-3}$, $1x10^{18}cm^{-3}$, $4x10^{18}cm^{-3}$, $8x10^{18}$ cm^{-3} (starting at the left). The structure of this specimen is reflected in the unfiltered SEM image presented in Fig. 2a, in which n-type regions appear darker than p-doped areas. Fig. 2a also contains the definition of coordinates used later. In order to assess whether an inhomogeneity in contrast at the junction exists, we have adjusted *D* to produce ~0 % contrast between n and p doped regions far away from the junction. Using *T*=20V this was achieved with *D*=3V.The resulting image is presented in Fig. 2b, in which a double layer is clearly visible at the location of the junction between n-doped layer and p substrate. That the well defined positive contrast peak in Fig. 2b indeed marks the position of this junction is evident from the contrast profile in Fig. 3a, which compares the obtained contrast to the calculated strength of the electric field E_x 10 nm above the surface (*x* direction as in Fig. 2a). The maximum in the contrast coincides precisely with the maximum of E_x, thus allowing accurate junction delineation. For the thin p-type layers in this structure, the situation is more complex because the EF settings for accurate junction delineation are not fulfilled. In spite of this, there seems to be some correlation between the calculated E_x and the contrast profile although not all contrast features can be explained in this way, especially the case for the negative contrast at a depth of 3.5 μm. The contrast minimum at 3.5 μm does correlate to the location and extent of E_z as can be seen by comparing Fig. 3a and b. In the p-type region E_z has a focusing effect bringing SEs emitted at large θ closer to the optic axis. From Fig. 1c it is evident that SEs entering the tube almost parallel to the optic axis are less likely to be detected, hence causing reduced intensity and negative contrast. In addition, the energy of these SEs is increased such that some are not detected as they exceed the maximum detectable energy for a certain EF setting, again causing negative contrast. These two effects will only have their full impact at locations where E_x=0, as E_x counteracts the focusing action of E_z. In Fig. 3 it can be seen that E_z extends much further into the p substrate than E_x, hence a double layer is formed. Positive contrast occurs in the region $E_z \neq 0$, $E_x \neq 0$ as E_x is stronger than E_z (see Fig. 3) and negative contrast in the region of $E_z \neq 0$, E_x=0.

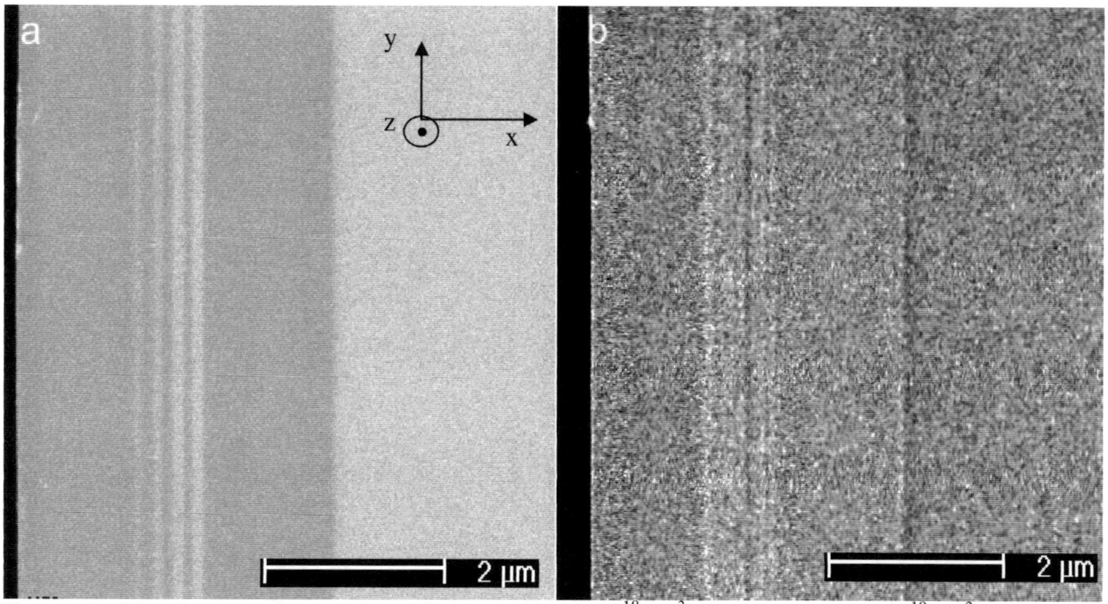

Figure 2. SEM images of x-section of n-type (As $5\times10^{18}cm^{-3}$) on p-type (B $1\times10^{19}cm^{-3}$) substrate, embedded in n-type are four p-type layers: $6\times10^{17}cm^{-3}$, $1\times10^{18}cm^{-3}$, $4\times10^{18}cm^{-3}$, 8×10^{18} cm^{-3}; a) standard SEM b) energy filtered T=20V, D=3 V

In Fig 3b E_z is plotted, together with the contrast profile obtained from Fig 2a. It is very noticeable that for the four thin p doped layers the calculated field distribution in z direction is very similar to the contrast profile of the standard SEM image. However, this is not true for the junction between n-type layer and p-type substrate, as we would expect to find increased intensity near the junction in the p-type region relative to that observed in the p-type region further away from the junction. However, the opposite is the case: the intensity near the junction is rising slowly to reach that of the bulk p-type region, making accurate junction delineation much more difficult than in EF images.

3. Conclusion

Our results indicate that EF SEM can be used for accurate junction delineation. We found that the observed contrast (double layer) at the junction may be explained as a result of patch fields and their influence on angular and energy distribution of SE. For a more detailed description of the exact mechanism for junction delineation, further calculations are required.

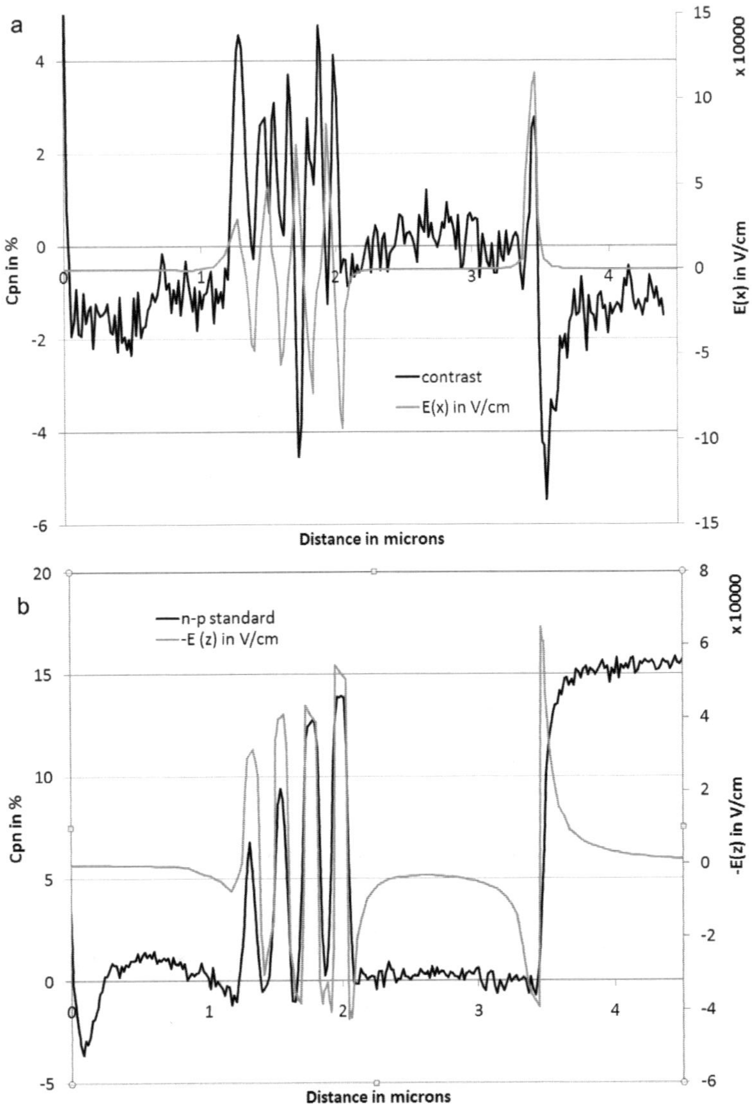

Figure 3. (a) Contrast profile EF: T=20V, D=3V and calculated field component parallel to specimen surface (E_x) present at 10nm above surface using ATLAS and model described in [4], (b) Contrast profile standard SEM and calculated field component perpendicular to specimen surface (E_z) present at 10nm above surface using ATLAS and model described in [4].

Acknowledgments
C. Rodenburg would like to thank the Royal Society for their support in the form of a Dorothy Hodgkin Fellowship. The authors would also like to thank the EPSRC for supporting this project (EP/E030602/1).

References

[1] Sealy CP, Castell MR and Wilshaw PR 2000 *J. Electr. Microsc.* **49**, 311
[2] Hsiao W-C and Liu C-P 2006 *Appl. Phys. Lett.* **88**, 212104
[3] Chakk Y and Horvitz D 2006 *J. Mater. Sci.* **41**, 4554
[4] Chee AKW, Rodenburg C and Humphreys CJ 2008 *Mater. Res. Soc. Symp. Proc.* **1026**, C04-02

Journal of Physics: Conference Series **209** (2010) 012054 doi:10.1088/1742-6596/209/1/012054

Progress towards quantitative dopant profiling in the SEM

KWA Chee[1,2], EGT Bosch[3] and CJ Humphreys[1]

[1] University of Cambridge, Department of Materials Science and Metallurgy, Pembroke Street, Cambridge, CB2 3QZ, UK
[2] University of Cambridge, Department of Engineering, Electrical Engineering Division, Centre for Advanced Photonics and Electronics, 9 J J Thompson Avenue, Cambridge, CB3 0FA, UK
[3] FEI Company, Building AAE, Achtseweg Noord 5, Eindhoven, The Netherlands

E-mail: kwac2@eng.cam.ac.uk

Abstract. Dopant contrast from secondary electron (SE) imaging is essentially determined by the local electronic band structure at the surface of the solid specimen being analysed in the scanning electron microscope (SEM). To study dopant contrast mechanisms on inhomogeneously doped semiconductors, we have developed a quantitative model by using a numerical method based on a combination of finite-element carrier and potential modelling and Monte Carlo simulation of SE emission and contrast detection. This combined model enables distinction between the effects of specimen internal fields (surface band-bending), microscopic external fields (patch fields) and any macroscopic external fields applied in the SEM.

1. Introduction

The quantification of dopant distributions directly from two-dimensional secondary electron (SE) contrast imaging is in principle a rapid and non-destructive *in-situ* characterisation procedure. Scanning electron microscopy therefore provides a promising contactless technique to study dopant distributions in semiconductor samples with sensitivity over a wide dynamic range (~ 10^{16} - 10^{20} dopants cm^{-3}) and at sub-10 nanometre spatial resolution [1]. However, its general implementation in industry has thus far been inhibited by the lack of an accurate and robust quantification procedure. Although it is known that the dopant contrast observed across a p-n junction is a function of the built-in voltage across the junction, surface band-bending and external local fields (patch fields) above the specimen, the lack of understanding of the respective contributions from all of these factors have hampered accurate quantification. To analyse the problem, we have used detailed computer modelling to investigate the effects of surface states and doping concentrations on the surface band-bending and external patch fields from the specimen, as well as their effects on SE yields and detected contrast in the SEM. The aim of our work presented here is to describe from first principles, a solid quantitative model for dopant profiling in the SEM, and compare it with experimental observations.

Journal of Physics: Conference Series **209** (2010) 012054 doi:10.1088/1742-6596/209/1/012054

2. Problem and solution approach

One obvious issue in SEM doping profiling quantification is that there is no simple analytical relation between the band-bending and trap states at the surface, which constitute a significant effect on dopant contrast [2]. To deal with this problem, numerical modelling was carried out using the finite-element method to calculate the surface charges as well as their effects on surface band-bending and external patch fields. Computations of the electrostatic fields were incorporated into a state-of-the-art Monte Carlo model which takes into account the physics of low energy electron scattering [3]. The Mott or acoustic phonon scattering cross section was used to describe (quasi)elastic interactions in the solid, and a dielectric formulation approach was adopted to determine inelastic scattering and SE generation. SE emission and contrast detection calculations were initially performed for p-n junction samples without surface band-bending (*i.e.* no surface states). In this case, there are no internal surface fields arising from surface states but only patch fields above the surface. Surface states were next included in the model to simulate their effects on dopant contrast.

3. Results and discussion

Iteratively converged solutions providing the equilibrium carrier and electrostatic field distributions were self-consistently obtained using the finite-element method; the semiconductor Poisson equation as well as the Laplace equation for a semiconductor-vacuum system was solved, including the coupled carrier continuity and drift-diffusion transport equations where applicable. A modified Shockley-Read-Hall (SRH) recombination model was applied to describe recombination effects at the specimen surface [4]. Fermi-Dirac statistics was used and bandgap narrowing effects were accounted for at high doping concentrations. When simulating typical surface band-bending effects, an amphoteric model for surface states was used, with a density of 4×10^{12} cm^{-2}. The value of the surface state density was determined from finite-element analysis in combination with systematic measurements of surface band-bending effects [5], and was found to agree with typical experimental values for a Si-SiO_2 system [6]. The donor- and acceptor-like surface states were specified localised in the bandgap at 0.35 and 0.73 eV above the top of the valence band respectively, and the electron- and hole-capture cross sections for the surface traps are $\varsigma_n = 10^{-15}$ and $\varsigma_p = 10^{-16}$ cm^2, respectively. In the foregoing calculations, it was assumed that the surface state density and energy are an intrinsic property of the semiconductor, and they do not depend on the local doping concentration [7]. Moreover, the spatial distribution of the surface states was assumed to be uniform in the semiconductor plane. Where ε_{Si} and ε_{SiO_2} are the dielectric permittivities of the semiconductor and the native oxide respectively, surface band-bending was computed according to a boundary condition at the semiconductor-oxide interface as follows:

$$\hat{n}\varepsilon_{Si}\vec{E}_{Si} - \hat{n}\varepsilon_{SiO_2}\vec{E}_{SiO_2} = \rho_s \tag{1}$$

$\hat{n}$ is the unit vector normal to the surface. The difference between the normal components of the electric fields at the respective regions adjacent to the surface boundary must be equal to the sheet charge density ρ_s (Gauss' Law). This charge term includes both the mobile (carriers) and fixed (ionised dopants and ionised surface states) charges.

To simulate electron beam-specimen interactions, a focussed probe consisting of 10^5 particles per pixel incident on the sample was defined with an accelerating voltage of 1 kV. The number of primary electrons per pixel used in the simulations provides reasonable statistical accuracy to within about ± 0.3 % of the calculated SE yield. The simulated primary electron beam energy is within the optimal range for dopant contrast imaging, allowing direct comparison with experimental data. The incident electrons are tracked inside the crystal, which generate further cascades of SEs. All particles are traced through any local electric fields, both inside and outside the solid. When computing the ratio and angular distribution of the emitted SEs, a quantum-mechanical model for transmission and refraction or reflection at a potential transition was employed based on the formulation referred to in

[8] to describe electron scattering at the surface boundary. A planar detector was defined directly on the surface plane, which virtually detects all particles that are emitted from the specimen. Note that this detector effectively has a potential that is equal to the locally existing surface potential of the specimen. A second virtual planar detector parallel to the surface plane was defined, centred on the metallurgical junction at a macroscopic distance above the sample (z = 0.5 mm) - which was outside the effective range of the specimen's microscopic patch fields. This virtual detector was specified with dimensions of 1× 1 mm^2 and with zero potential relative to the specimen. Therefore in general, particles having emission angles less than approximately 45° measured from the optical axis are collected, and those with larger emission angles are considered to be undetected. The theoretical absolute values for contrast were calculated using the following formula:

$$C_{cal} = \frac{\Delta N}{N_p} = \frac{N_p - N_n}{N_p} \tag{2}$$

where N_p and N_n are the number of calculated SEs from the p- and the n-type regions respectively.

The computed contrast values as at the surface over a wide range of doping concentrations are shown in Fig. 1. When no surface band-bending in the specimen is present, no systematic variation could be observed of the calculated number of SEs emitted from the p- and n-type regions. This is an expected result *a priori* because of the following reasons. When deriving the differential inelastic cross sections, the optical data for the semiconductor were equivalently applied to the p- and the n-type material, and therefore the calculated generation rate of SEs in the p- and the n-type region is expected to be similar. The inner potential parameters for the p- and the n-type regions were also treated as equal, assuming that these values are not significantly dependent on doping. Hence, the surface boundary transmission probabilities at the respective doped regions are equivalent. The computed contrast varies systematically with doping concentration on the output of the virtual detector at 0.5 mm above the surface (see Fig. 1).

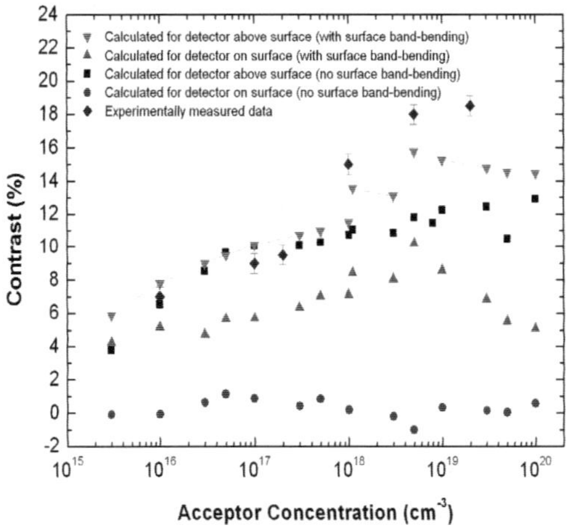

Figure 1. Theoretical calculations of dopant contrast (in %) for Si specimens containing a p-n junction with and without surface band-bending. Experimental values of contrast are included for comparison.

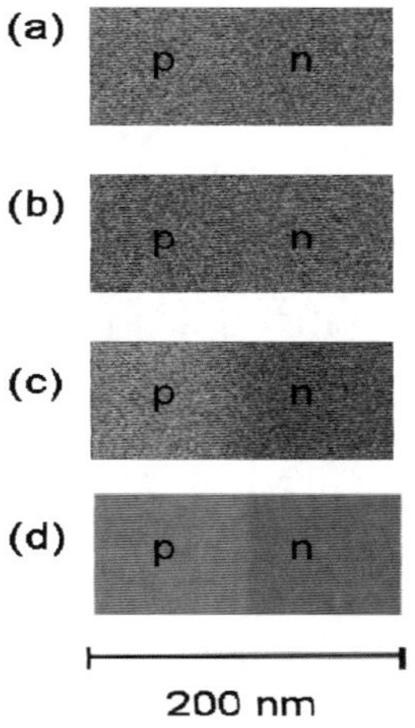

Figure 2. (a) A simulated SE image as on the surface, a simulated (b) BSE and (c) SE image as on the output of the virtual detector at 0.5 mm above the surface, and (d) a corresponding experimental SE image.

When surface band-bending effects are accounted for, there is a monotonic rise of the calculated contrast with acceptor concentration of up to about 10^{18} cm^{-3}. For higher doping concentrations, the calculated surface depletion width reduces and becomes comparable or even smaller than the inelastic mean free path for the SEs [9]. Experimental contrast data were also included in Fig. 1 for comparison. These experimental values were obtained on the FEI XL30 sFEG SEM using a through-the-lens detector (TLD) with an extractor voltage of 20 V. The collection angle of this TLD for the operating conditions used is around 45° for a working distance of ~ 6 mm, and the results are therefore directly comparable with the computed output of the virtual detector at z = 0.5 mm. It is clear from this comparison that, for the examples chosen, there is quite a convincing agreement between the experimental data and the theoretical plots including surface band-bending effects. BSE and SE images were numerically generated using 10^3 primary electrons per pixel for an electron probe incident normally on the surface with a landing energy of 1 keV. The plural scattering approach was adopted here in order to accelerate the computation speed. A numerical scan on the sample was performed over an area of 200×100 pixels^2. Comparison with a corresponding experimental SE image is provided as shown in Fig. 2. These images were obtained from a Si specimen consisting of a symmetric p-n junction, having a net acceptor and donor concentration of ~ 5 × 10^{18} cm^{-3} in the p- and the n-type regions respectively. Typical surface band-bending characteristics were assumed on this specimen. No significant dopant contrast is observable from the numerical BSE image as expected. Dopant contrast from the simulated SE image as at the surface was found to be too low, but that from the output of the virtual detector above the surface was comparable to the experimental SE image. Numerical accuracy for the simulated image is expected to improve further when using a primary electron count of a few orders of magnitude higher, but such simulations will take much longer to complete.

4. Conclusion

A rigorous quantitative model for SEM dopant profiling has been developed from basic principles and compared with experimental observations. To calculate dopant contrast, the combined numerical model has to account for both the effects of internal surface band-bending and external patch fields from inhomogenously doped specimens. Simulation results show close agreement with experimental data for the chosen modelling parameters, at least within the limits of experimental uncertainty. The results of this work lead to a deeper understanding of the physical mechanisms underpinning dopant contrast, thereby enabling the future development of a quantitative procedure for dopant profiling of semiconductors in the SEM.

Acknowledgements
KWA Chee acknowledges the Overseas Research Studentship Award Scheme, the Cambridge Commonwealth Trust, the Cambridge Brockhouse Engineering Studentship and an FEI company industrial bursary for doctoral financial sponsorship. The authors are grateful to Silvaco company for access to their program source code and CERN for their open-source simulation toolkit, and Gerard Van Veen of FEI company for initiating the European consortium project to support this work.

References

[1] Elliott SL, Broom RF and Humphreys CJ 2002 *J. Appl. Phys.* **91**, 9116
[2] Chee KWA, Rodenburg C and Humphreys CJ 2008 *MRS Symp. Proc.* **C04-02**, 1026
[3] Kieft E and Bosch EGT 2008 *J. Phys. D: Appl. Phys.* **41**, 215310
[4] Roulston DJ, Arora ND and Chamberlain SG 1982 *IEEE Trans. Elec. Dev.* **29**, 284
[5] Chee KWA, Rodenburg C and Humphreys CJ 2007 *J. Phys. Conf. Ser.* **120**, 407
[6] Stesmans A and Van Gorp G 1990 *Phys. Rev. B.* **42**, 3765
[7] Chung MF 1971 *J. Phys. Chem. Solids* **32**, 475
[8] Cohen-Tannoudji C, Diu B and Laloe F 1977 *Quantum Mechanics* (Hermann, Paris)
[9] Chee KWA, Bosch EGT and Humphreys CJ *to be published*

The chemisorption of oxygen and its effect on the secondary electron emission from doped semiconductors

FN Zaggout, CGH Walker and MM El Gomati

Department of Electronics, University of York, Heslington, York, YO10 5DD

E-mail: mmg@ohm.york.ac.uk

Abstract. The determination of the dopant concentration in semiconductors at the nanometre scale is an important technological goal. One possible approach is to use the dopant contrast (DC) effect in low voltage scanning electron microscopy (LVSEM). The DC effect normally causes p-doped semiconducting material to appear brighter than n-doped in a LVSEM and the contrast is dependent on the dopant concentration. In this report we have studied highly doped p- and n-type silicon with thick oxide layers and after the oxide layer was removed by dipping in diluted HF. DC was measured as a function of primary beam voltage. It was found that the DC could be reversed (i.e. n-type brighter than p-type) at higher primary beam voltages for the thicker oxide, but such reversals were not apparent for the thinner oxide. By comparing with results from different electron emission techniques such as photoemission and field emission it can be concluded that oxygen plays an important role in DC not just for monolayer thicknesses as previously recognised, but also for much thicker oxides.

1. Introduction

The discovery that differently doped regions of a semiconductor can produce differing levels of secondary electron (SE) emission in a scanning electron microscope (SEM) was first made by Chang and Nixon [1]. The contrast between the two regions is known to be dependent on the n-type dopant concentration [2]. The dopant contrast (DC) effect could be a useful tool for the semiconductor manufacture industry if it could be fully understood and quantified [3]. However, the root cause of the effect has been in dispute for a number of years. The most often quoted theory is the "Patch field" theory [4] which explains DC as being due to electrostatic fields external to the sample. Another theory attributes DC to the band bending near the surface caused by the native oxide [5] and another that there are differing inelastic mean free paths (IMFP) for the SEs in the differently doped semiconductors [6]. Chakk and Horvitz [7] suggested a dynamic charging mechanism. The York group have promoted yet another alternative theory suggesting that surface contamination – principally carbon – causes a surface energy barrier similar to that of a Schottky barrier [8-10]. Duraud et al. [11] proposed that oxygen played a role in the DC which was mainly due to varying thicknesses of the oxide layer but a deeper explanation was lacking.

In this paper we present data on the effect of the thickness of the oxide layer on the contrast as a function of the primary beam. We then discuss these results in the context of other reports using different electron emission techniques (such as field emission and photoemission).

2. Experiment

Boron patterns were diffused into a (111) silicon wafer (phosphorus-doped n-type silicon substrate) at a temperature of ~1000 K. The n-type substrate was doped to a level of ~ $10^{15}cm^{-3}$ while the p^+ regions were doped at ~$10^{19}cm^{-3}$ to a depth of ~3 μm.

The native oxide was thinned by ultrasonic cleaning in IPA followed by dipping in 10:1 H_2O:HF. An SEM (type Tescan, TS5136MM) was used to image the sample at a variety of beam energies. The SEM was equipped with an Everhart-Thornley (ET) detector for the acquistion of the SE signal. The equation to determine the DC, $C_{A/B}$, between two differently doped areas A and B is given by:

$$C_{A/B} = I_A - I_B \quad (1)$$

where I_A and I_B are determined from the greyscale level of the image (white equals 1 and black equals 0). It was ensured that the sensitivity of the ET detector did not vary between the acquired images. The highly doped p+ type region is assigned to region A and the lightly doped n type areas is assigned to region B. In order to allow a substantial oxide layer to grow, the samples were stored in air at atmospheric pressure for several months prior to the experiments. The first set of SEM images were carried out on the oxidised samples. The samples were then subjected to the procedure described above to remove the oxide layer and then re-imaged in the SEM. The reader is directed to [12] for further details.

3. Results

Upon imaging the oxidised sample at a primary energy (PE) of 1 keV it was found that the p^+ regions have a greater SE emission than the n regions as has been found many times before. However, a DC inversion was found at higher PE (see Figure 1).

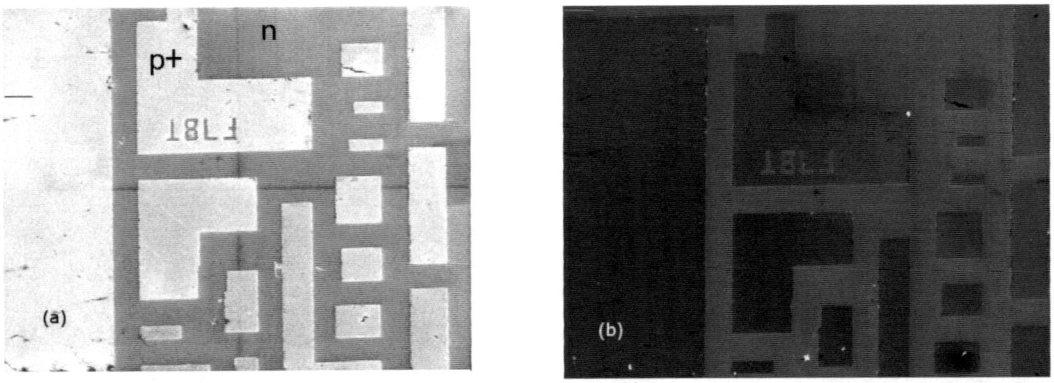

Figure 1. SE images of an oxidised Si (111) sample (a) 1keV and (b) 7keV. Image size: 1.04mm x 0.87mm.

Upon removing the oxide, it was found that the DC continued to vary with PE, although no contrast reversal was found. Using equation (1), the DC was determined for a range of PEs for samples with and without the thick oxide layer. The results are shown in Figure 2. Sealy et al. [4] also studied DC as a function of PE, but they did not find DC reversal.

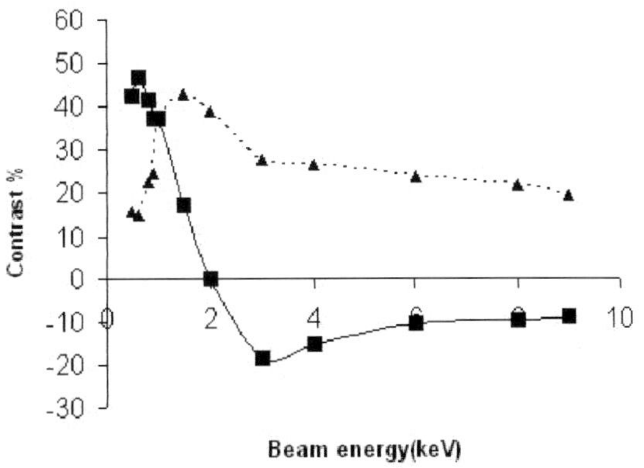

Figure 2. DC between the differently doped regions vesus PE. = thick oxide, ▲ = thinned oxide.

4. Discussion

The results shown above clearly indicate that thick oxide layers have a major influence on the DC effect. The cause for this could be due to charging within the surface oxide layer, although why the charging should influence the SE emission from one doped region more than another is unclear. In addition, the results shed little light on the current competing theories which attempt to explain DC.

4.1. Electron emission from semiconductors – a look at the literature

More insight into the DC mechanism can be obtained from papers using other electron emission techniques. These techniques include photoemission (PE) and field emission (FE).

4.1.1. Photoemission.

It is well known that the native oxide on Si can induce considerable surface band bending. Indeed, in a PE study undertaken by Wagner and Spicer [13] using in-situ cleaved doped Si surfaces it was revealed that the valence band maximum of n-type Si shifts to a ~0.4eV higher binding energy upon exposing the surface to a single monolayer of oxygen. This fact has consequences for the foundations of the "patch field" theory [4] as this theory assumes that the energy required to remove an electron from the semiconductor into the vacuum is independent of whether it is n-type or p-type (see Fig 8 of ref [4]). The results of Wagner and Spicer [13] call this assumption into question as it is not clear whether the argument proposed by Sealy et al. [4] is for the case where no oxygen is on the Si surface or when a layer of oxygen is deposited on the surface. Ballarotto et al. [14] used Photoelectron Emission Microscopy (PEEM) to study doped semiconductors and also used the presence of oxygen to explain their results. Their theory has very close similarities to that of Perovic et al. [5]. If significant external electrostatic fields were present, then one would expect the PEEM results of Ballarotto et al. [14] to appear distorted – they are not.

4.1.2. Field emission

Johnson et al. [15] studied FE from p- and n- doped Si whiskers as the extraction voltage was varied. It was found that there was a delayed onset to the FE from n-type Si. This suggests that there was an energy barrier that the electrons emitting from the n-type semiconductor had to overcome. It is difficult to imagine how an electrostatic field external to the sample could induce such an effect. In addition, such a delayed onset of the FE is not consistent with the hot electron IMFP theory [6].

4.2. Electronic effects due to chemisorption

Chemisorption on semiconductors is known to behave differently than on other materials, e.g. metals [16] and the theory describing how gases adsorb onto semiconductors underpins much of the gas sensor technology available today. The adsorbtivity is controlled according to whether the adsorbing gas is an "acceptor" (i.e. accepts an electron from the substrate) or a "donor"(i.e. donates an electron to the substrate) and the position of the Fermi level within the band gap (i.e. whether the substrate is n-type or p-type). From the predictions on absorptivity [16], one might expect the native oxide to be thicker on n-type than on p-type semiconductor. This seems to be borne out by experiment [17]. The deposition of different materials on the surface will have different effects on the band bending which has been discussed within the context of DC before [8-10].

4.3. The effect of a thicker oxide on the dopant contrast

Since the native oxide is an insulator, this could well influence the SE emission (especially for thicker oxides), as has been extensively discussed by Cazaux [18]. Contrast reversals due to variations in primary beam voltage, charging effects and scanning frequency were discussed. The possibility that the oxides had different thicknesses on the n- and p- type substrates may explain their different behaviours under the primary electron beam.

5. Conclusions

From the results presented here, the thickness of an oxide layer clearly influences the DC of heavily doped semiconductors. This may be due to charging effects on thick oxides. In addition, by considering results from the literature it is clear that oxygen plays a role in DC even for very thin oxide layers. Well understood chemisorption on semiconductor theory can explain the literature results for photoemission and field emission but cannot be used to explain the contrast reversal as a function of primary beam energy for thick oxides found here.

Nevertheless, it is important that, in future, all interpretations of DC results consider the presence of oxygen (and other contaminants) whether sub-monolayer or several microns in thickness.

References

[1] Chang THP and Nixon WC 1967 *Solid State Electron.* **10,** 701
[2] Venables D and Maher DM 1996 *J. Vac. Sci. Technol. B* **16**, 362
[3] National Technology Roadmap for Semiconductors (1997) Semiconductor Industry Association, San Jose, CA
[4] Sealy CP, Castell MR and Wilshaw PR 2000 *J. Electron Microsc.* **49**, 311
[5] Perovic DD, Castell MR, Howie A, Lavoie C, Tiedje T and Cole JSW 1995 *Ultramicr.* **58**, 104
[6] Frank L, Mika F, Hovorka M, Valdaitsev D, Schönhense G and Müllerová I 2007 *Mat. Trans.* **48**, 936
[7] Chakk Y and Horvitz D 2007 *J. Mater. Sci.* **41**, 4554
[8] El-Gomati MM, Wells TCR, Müllerová I, Frank L and Jayakody H 2004 *IEEE Trans. Electron Dev.* **51**, 288
[9] El-Gomati MM and Wells TCR 2001 *Appl. Phys. Lett.* **79**, 2931
[10] El-Gomati MM, Zaggout F, Jayacody H, Tear S and Wilson K 2005 *Surf. Interf. Anal.* **37**, 901
[11] Duraud JP, Le Moël A, Le Gressus C, Pantel R and Chornik B 1984 *Scanning Microsc.* **1**, 49
[12] Zaggout FN, "Quantification of SE dopant contrast in low voltage scanning electron microscope", D.Phil. Thesis University of York, (2007).
[13] Wagner LF and Spicer WE 1974 *Phys. Rev. B* **9**, 1512
[14] Ballarotto VW, Siegrist K, Phaneuf RJ and Williams ED 2002 *J. App. Phys.* **91**, 469
[15] Johnson S, Markwitz A, Rudolphi M, Baumann H, Oei SP, Teo KBK and Milne WI 2006 *Curr. Appl. Phys.* **6**, 503
[16] Wolkenstein T "Electronic Processes on Semiconductor Surfaces during Chemisorption" (1991) Consultants Bureau.
[17] Morita M, Ohmi T, Hasegawa E, Kawakami M and Suma K 1989 *Appl. Phys. Lett.* **55**, 562
[18] Cazaux J 1999 *J. Appl. Phys.* **85**, 1137

Journal of Physics: Conference Series **209** (2010) 012056 doi:10.1088/1742-6596/209/1/012056

Characterization of ytterbium silicide formed in ultra high vacuum

A Łaszcz[1], J Ratajczak[1], A Czerwinski[1], J Kątcki[1], V Srot[2], F Phillipp[2], P A van Aken[2], D Yarekha[3], N Reckinger[4], G Larrieu[3] and E Dubois[3]

[1] Institute of Electron Technology, Al. Lotników 32/46, 02-668 Warsaw, Poland
[2] Max-Planck-Institut für Metallforschung, Heisenbergstr. 3, D-70569 Stuttgart, Germany
[3] IEMN/ISEN, UMRS CNRS 8520, Avenue Poincare, Cite Scientifique, BP 69, 59652 Villeneuve d'Ascq Cedex, France
[4] Université catholique de Louvain, Place du Levant 3, B-1348 Louvain-la-Neuve, Belgium

E-mail: laszcz@ite.waw.pl

Abstract. The formation of ytterbium silicide fabricated by annealing at 480 °C for one hour has been studied by means of high resolution transmission electron microscopy (HRTEM) and energy dispersive X-ray spectroscopy (EDS). The annealing process has been performed under ultra high vacuum (UHV) conditions. The formation of an amorphous silicide layer was observed between the Yb-layer and the silicon substrate in the as-deposited sample. Ytterbium silicide observed after annealing consists of two different layers: crystalline and amorphous ones. The studies confirmed that the formed crystalline layer is of the $YbSi_{2-x}$ phase, however, the structure is different from the hexagonal AlB_2 type.

1. Introduction

Ytterbium silicide has the lowest electric work function [1] amongst the rare-earth (RE) metals (Gd, Dy, Ho, Er) already used for low Schottky barrier [2] applications. Therefore, ytterbium silicide is a very promising material as a silicide source/drain in n-channel Schottky Barrier (SB) MOSFETs [3]. The most commonly used method of the ytterbium silicide fabrication is a thermal annealing of the Yb-film deposited on silicon. Interdiffusion of ytterbium and silicon atoms during annealing at the metal/silicon interface makes the solid-state reaction possible. There are some reports on fabrication of ytterbium silicide by furnace or rapid thermal annealing (RTA) at different temperatures e.g.: 300-360 °C [1], 450 °C [4], 530 °C [5], 600 °C [3], and 200-800 °C [6].

Among many ytterbium silicides (e.g. Yb_2Si, Yb_5Si_3, Yb_5Si_4, YbSi, Yb_3Si_4 and Yb_3Si_5) the most known is the $YbSi_{2-x}$ phase [7-8]. Its structural properties are very similar to the well known $ErSi_{2-x}$ phase [9]. The maximum silicon content in $YbSi_{2-x}$ was established to be equal to 63.5 atomic % of Si [8]. The value of 63.5 atomic % Si is close to the $YbSi_{1.8}$ formula and it is a silicon-rich ytterbium silicide phase, which crystallizes with the hexagonal AlB_2 type. Ytterbium silicide and other RE silicides with the hexagonal AlB_2 crystal structure have an excellent lattice match with silicon, so that high quality silicide thin films can be grown on silicon [9].

2. Experimental

The 20 nm thick ytterbium film was deposited by e-gun evaporation on the silicon substrate. One hour annealing of the Yb/Si structure at 480 °C has been applied to form ytterbium silicide. RE metals are known to react with oxygen. In order to omit applying a capping layer, the formation of ytterbium silicide was performed under special conditions, which also protect well against oxidation. For this reason, the annealing of the studied Yb/Si structure was carried out under ultra-high vacuum (UHV) conditions, without breaking the vacuum between deposition and annealing. For the investigation of the mechanisms of the ytterbium silicide formation in the Yb-Si/Si structure, transmission electron microscopy (TEM) specimens were prepared by the conventional cross-section method using low-temperature argon ion milling. TEM studies were focused on microstructural observations of silicide layers and silicide/Si interfaces at an atomic scale by HRTEM (JEOL-4000FX), as well as on the identification of formed silicide phase using EDS in scanning TEM (VG HB501UX, with the electron beam size diameter of 0.7 nm, as determined by the full width at half maximum (FWHM)). A non-annealed sample was also investigated for comparison.

3. Results and discussion

The HRTEM image presented in Fig. 1a clearly shows that during the deposition of the Yb-layer on the silicon substrate, the reaction of ytterbium with silicon took place. As a result a 5 nm thick amorphous silicide layer was formed at the Yb/Si interface. This layer has very flat interfaces with the Yb-layer and the silicon substrate, respectively. The thickness of the Yb-layer after deposition is 15 nm. It means that about 25% of the Yb-layer reacted with silicon during the deposition process. The Yb-layer consists of columnar grains. The formation of an amorphous interlayer between the metal and the substrate just after metal deposition on silicon was already proved in many metal/silicon systems [10]. The growth of the amorphous phase stops when the critical thickness, usually several nanometers, is reached.

Figure 1b presents an image of the Yb/Si structure annealed at 480 °C for one hour. During annealing at this temperature, the entire Yb-layer reacted with silicon, forming a silicide layer that consists of two phases, in the form of continuous layers. The bottom layer is crystalline and the upper one is amorphous. The total thickness of both layers is about 18-20 nm and the thicknesses of the crystalline and amorphous parts constitute 2/3 and 1/3 of the whole thickness, respectively. The top surface and all interfaces in the annealed sample are rough. The phase at the bottom, i.e the crystalline layer, was identified by EDS as the $YbSi_2$ phase (Fig. 2a). Although the EDS measurements showed that the ratio of atomic concentration of ytterbium and silicon in crystalline silicide is about 1:2 or more, however the crystalline silicide phase probably corresponds to the Si-rich $YbSi_{(\sim 1.8)}$ phase [7-8]. The small differences in EDS measurements result from the sputtering of the material from the Si substrate caused by the ion milling process.

According to the equilibrium phase diagram for Si-Yb, obtained experimentally using furnace annealing, the Si-rich $YbSi_{2-x}$ phase is not expected to form below 770 °C [7-8]. However, many other detailed studies (for both, furnace and RTA annealing) showed that this phase is formed already at significantly lower temperatures [1,3-6,11], sometimes even at 300 °C. It brings the conclusion that although the Yb-Si systems have already been studied by some research groups, the kinetics involved in the interface formation of ytterbium silicides or others RE silicides is still poorly understood [7].

It was reported that RE disilicides can crystallize with three closely related types of crystallographic structure: $ThSi_2$, $GdSi_2$ and AlB_2 [8]. $GdSi_2$ can be considered as an orthorhombic deformation of the tetragonal $ThSi_2$ type and the most frequent hexagonal AlB_2 type is a representative of a transposition structure of $ThSi_2$.

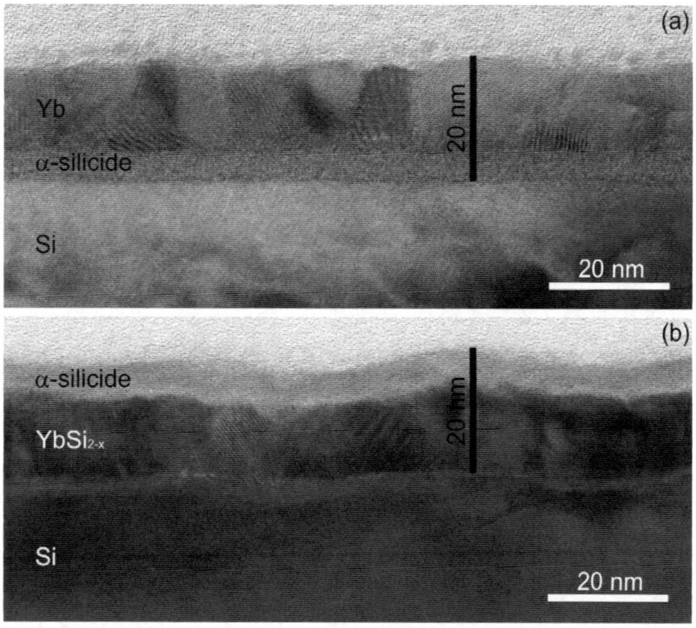

Figure 1. HRTEM images of (a) an as-deposited Yb/Si structure and (b) structure annealed at 480 °C for one hour.

The HRTEM image presented in Fig. 2b reveals a very distinct silicide/silicon interface, as well as differences in lattice parameters between silicide and silicon. An arrangement of ytterbium silicide atoms visible in the HRTEM image does not correspond to the $YbSi_{2-x}$ phase with the hexagonal AlB_2 type structure. Probably, the $YbSi_{2-x}$ phase formed in our sample after annealing at 480 °C corresponds to either $ThSi_2$ or $GdSi_2$ types of structures.

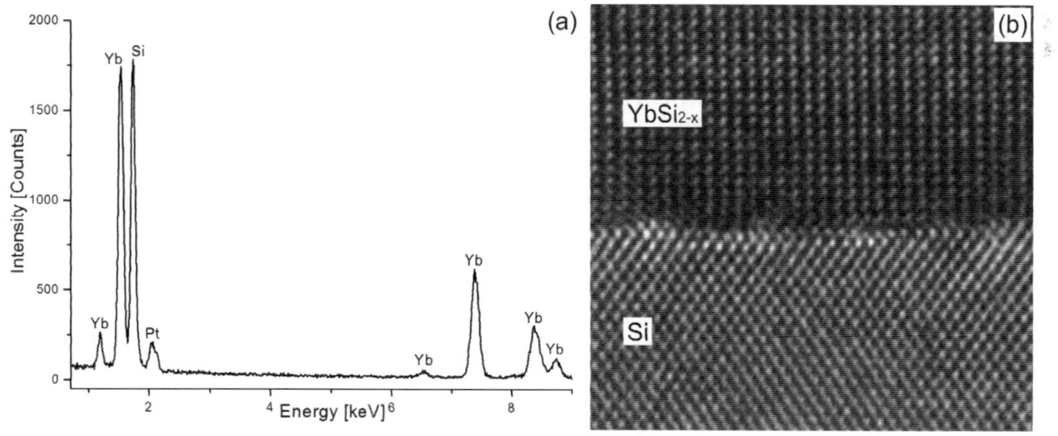

Figure 2. The Yb/Si structure annealed at 480 °C for one hour: (a) EDS spectrum obtained from the crystalline silicide layer and corresponding to the $YbSi_{2-x}$ phase, and (b) HRTEM image of the silicide/silicon interface.

EDS measurements showed that atomic concentrations of ytterbium and silicon in the amorphous silicide layer, which was formed above the crystalline layer, are similar to results obtained for the crystalline $YbSi_{2-x}$ layer. The formation of the amorphous phase above the crystalline phase in our annealed sample can be explained by the following hypothesis. In the first step of the solid-state reaction, a thin amorphous silicide layer forms at the Yb/Si interface, and then the amorphous phase moves up to the surface, consuming the Yb-layer. Then a crystalline phase nucleates at the α-silicide/Si interface and consumes the amorphous phase moving up to the surface. However, in our case the crystalline phase grew up to 2/3 of the silicide layer thickness. In this case the conditions of the process were insufficient to form a crystalline structure in the entire thickness of the silicide layer. This effect should be taken into account during optimization of annealing conditions.

4. Conclusions

The formation of ytterbium silicide fabricated by annealing at 480 °C has been studied by means of advanced TEM techniques. The formation of an amorphous silicide layer between the Yb-layer and the silicon substrate was already observed in the as-deposited sample. Annealing of the Yb/Si structure at 480 °C forms ytterbium silicide. However, ytterbium silicide consists of two different layers: a thicker crystalline layer and a thinner amorphous one. The phase formed in the crystalline layer corresponds to $YbSi_{2-x}$. However the observed crystal structure is different from the hexagonal AlB_2 type that is the most frequently observed for RE disilicides.

Acknowledgements

This publication is based on the research partly supported by: European Commission under the project "Metallic Source/Drain Architecture for Advanced MOS Technology" (IST-016677 METAMOS) and Polish Ministry of Science and Higher Education under the grant N N515062833.

References

[1] Lee RTP, Lim AE-J, Tan K-M, Liow T-Y, Lo G-Q, Samudra GS, Chi DZ and Yeo Y-C 2007 *IEEE Electron Device Letters* **28**, 164
[2] Tu KN, Thomson RD and Tsaur BY 1981 *Appl. Phys. Letters* **38**, 626
[3] Zhu S, Chen J, Li MF, Lee SJ, Singh J, Zhu CX, Du A, Tung CH, Chin A and Kwong DL 2004 *IEEE Electron Device Letters* **25,** 565
[4] Nowak JD, Song SH, Campbell SA and Carter CB 2007 in *Microscopy of Semiconducting Materials*, A.G. Cullis and P.A. Midgley Editors, (Springer, Berlin) p. 333
[5] Kuzmin M, Perälä RE, Vaara RL, Laukkanen P and Väyrynen IJ 2004 *J. Cryst. Growth* **262**, 231
[6] Chi KS and Chen LJ 2001 *Materials Science in Semiconductor Processing* **4**, 269
[7] Grytsiv A, Kaczorowski D, Leithe-Jasper A, Tran VH, Pikul A, Rogl P, Potel M, Noël H, Bohn M and Velikanova T 2002 *J. Solid State Chemistry* **163**, 178
[8] Palenzona A, Manfrinetti P, Brutti S and Balducci G 2003 *J. Alloys and Compounds* **348**, 100
[9] Knapp JA and Picraux ST 1986 *Appl. Phys. Letters* **48,** 466
[10] Chen LJ 2000 *Material Science and Engineering* **R29**, 115
[11] Chi KS, Tsai WC and Chen LJ 2003 *J. Appl. Physics* **93**, 153

16th International Conference on Microscopy of Semiconducting Materials IOP Publishing
Journal of Physics: Conference Series **209** (2010) 012057 doi:10.1088/1742-6596/209/1/012057

Chemical analysis of nickel silicides with high spatial resolution by combined EDS, EELS and ELNES

E Verleysen[1,2], H Bender[1], D Schryvers[3] and W Vandervorst[1,2]

[1] IMEC, Kapeldreef 75, 3001 Leuven, Belgium
[2] Instituut voor Kern- en Stralingsfysica, K. U. Leuven, Celestijnenlaan 200D, 3001 Leuven, Belgium
[3] Universiteit Antwerpen, EMAT, Groenenborgerlaan 171, 2020 Antwerpen, Belgium

E-mail: eveline.verleysen@imec.be

Abstract. The purpose of this study is to develop methodologies for the characterization of Ni-silicides by analytical TEM techniques. The measurements are performed in STEM mode to allow a good spatial resolution. Initial examination of the Ni-silicide phases is done by energy-dispersive X-ray spectroscopy (EDS). It is shown that quantification results with high accuracy and precision are obtained when a Ni-silicide reference layer is used as a standard, and an absorption correction is applied to the data. The EDS quantification results are then used as a reference for the quantitative electron energy-loss spectroscopy (EELS) analysis. It is shown that EELS analysis yields quantification results with high accuracy and precision, when the EELS data are treated with a model based approach. The EELS analysis is combined with a study of the Ni-$L_{2,3}$ near-edge structure (ELNES). The ELNES of the phases can be distinguished either visually or by calculating the branching ratio from the spectra with the model based approach. This paper shows that the different Ni-silicide phases can be distinguished with these EELS based methods. The accuracy and precision of the EDS and EELS results are compared.

1. Introduction

The continuous scaling in semiconductor technology has made accurate characterization of transistor components more challenging. Often, the main difficulty is to obtain both high spatial resolution and information on the chemical composition of the material. In this context, high resolution spectroscopic methods, such as EDS and EELS, play a crucial role in facilitating new materials integration and in the development of non-classical devices.

In this study, Ni-silicides are selected as a test-vehicle to develop quantitative methodologies for the accurate and precise characterization of transistor components. The Ni-silicide system consists of a wide range of different phases [1]. These phases are formed in polycrystalline layers with generally small grain sizes. Since this makes it difficult to distinguish the phases based on their lattice spacings, it is essential to combine HREM studies with chemical analysis. In this paper the results of the characterization of the full range of Ni-silicide phases with EDS, EELS and ELNES are presented.

2. Sample preparation and experimental setup

Ni-silicide samples are prepared by rapid thermal annealing of a Ni-layer and a thin polycrystalline Si film, deposited on top of a SiO_2/Si substrate. The temperature and thickness ratio of the Ni and Si-layer are varied to obtain different phases. The samples are prepared for TEM by FIB lift-out. Wedge samples are used to study the effect of specimen thickness on the quantification results. EDS, EELS and ELNES measurements are carried out on a TECNAI F30 operating at 300keV, equipped with an EDAX detector and a Tridiem GIF energy filter. The measurements are performed in STEM mode with a spot size of 2 nm. To avoid beam convergence effects, a condenser aperture of 0.5 μm is used for the EELS measurements. To avoid saturation of the GIF, EELS spectra are recorded with an acquisition time that is five times shorter than the acquisition time of the EDS spectra. Additional ELNES measurements are performed at TU Delft on a TECNAI operating at 200 keV, equipped with a monochromator allowing an energy resolution of 0.2eV. These spectra are recorded in TEM mode.

3. Results and discussion

3.1. Quantitative EDS analysis

After initial qualitative examination, the EDS spectra are processed by removing the background, integrating the Si-Kα and Ni-Kα peaks and converting the integrated peak intensities to concentrations with the Cliff-Lorimer equation [2]. The k-factor in the Cliff-Lorimer equation is determined by a Ni-silicide standard: $Ni_{31}Si_{12}$, measured and processed under identical conditions as the unknown Ni-silicide samples. Since EDS recordings from various thicknesses of the samples indicate strong absorption of the Si-K peak, an absorption correction is applied to the data. The results of the quantitative EDS analysis are shown in table 1a. The estimated concentration ratios are presented, together with the expected concentration ratios [1] between brackets. The standard deviations from the N repetitive measurements at different thickness positions are shown. For the Ni_3Si phase, a possible variation in the Si/Ni concentration ratio between 0.295 and 0.325 is reported in the literature [1]. The accuracy is then expressed as a range calculated relative to these values.

3.2. Quantitative EELS analysis

The Si-K edge and the Ni-L edge are used for the quantifications. EELS spectra are processed using a model-based approach with the EELSMODEL program [3]. This means that a parametric model describing the expectations of the EELS observations is derived. The unknown parameters are estimated by fitting this model to the experimental spectrum, using a criterion of goodness of fit (maximum likelihood). The model is constructed from a background power law, a hydrogenic cross-section for the Si-K edge, a hydrogenic cross-section for the Ni-L edge with Gaussian peaks to model the white lines, and a convolution with the experimental low loss spectrum to simulate multiple scattering. Since the difference in energy loss is too large, it is not possible to record the Si and Ni core edges simultaneously. The results of the quantitative EELS analysis are reported in table 1b. It is shown that quantification results with a good accuracy and precision are obtained, except for the Ni_3Si phase. The reason for this is that the Si-K edge becomes very small and noisy for this phase. The

Table 1. Results of the quantitative analysis.

a) Results of the quantitative EDS analysis

Phase	N	Ratio Si/Ni (expected)	St. dev.	Rel. prec. (%)*	Rel. acc. (%)*
Ni_3Si	15	0.299 (0.295;0.325)	0.022	7.4	1.4;-8.0
$Ni_{31}Si_{12}$	15	0.389 (0.387)	0.027	6.9	0.5
Ni_2Si	15	0.501 (0.500)	0.028	5.6	0.2
NiSi	15	1.012 (1.000)	0.066	6.5	1.2
$NiSi_2$	15	2.023 (2.000)	0.189	9.3	1.2

b) Results of the quantitative EELS analysis

Phase	N	Ratio Si/Ni (expected)	St. dev.	Rel. prec. (%)*	Rel. acc. (%)*
Ni_3Si	20	0.264 (0.295;0.325)	0.052	19.7	10.5;18.8
$Ni_{31}Si_{12}$	20	0.395 (0.387)	0.022	5.6	2.1
Ni_2Si	20	0.494 (0.500)	0.058	11.7	1.2
NiSi	20	1.018 (1.000)	0.120	11.8	1.8
$NiSi_2$	20	2.009 (2.000)	0.159	7.9	0.5

Relative accuracy is defined as 100(estimated value - expected value)/expected value, relative precision as 100*(St.dev.)/estimated value.

EELS results are compared to the EDS results in figure 1. It is shown that EDS and EELS yield comparable results for relative precision and accuracy. The acquisition time used for the EDS data is however five times larger than for the EELS data.

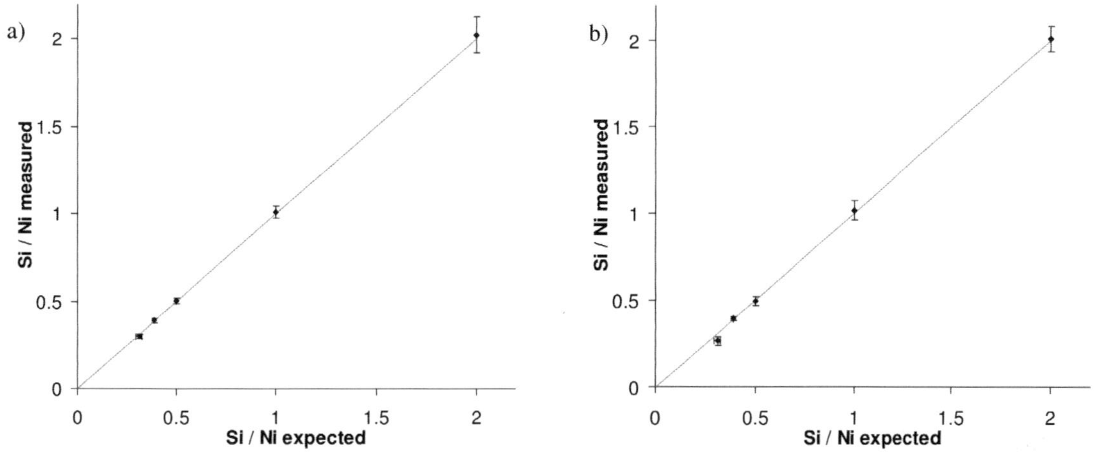

Figure 1. Comparison of a) EDS and b) EELS results. The straight line represents the expected concentration ratios. The y-error bars are given by $(t \cdot St.dev)/\sqrt{N}$ where t defines the 95% confidence interval.

3.3. ELNES analysis

The EELS analysis is combined with a study of the Ni-$L_{2,3}$ edge ELNES of each phase. The ELNES of the phases can be distinguished visually by looking at the anomalous intensity ratio of the Ni-L_2 and Ni-L_3 white lines (figure 2) [4]. This is most clearly visible in the spectra recorded with monochromator, since a far better energy resolution is obtained for these spectra. It is clear that the L_3 peak becomes smaller relative to the L_2 peak if the Si concentration increases. Also, changes in the tail of the L_3 peak are visible. For the Ni_3Si, $Ni_{31}Si_{12}$ and Ni_2Si phases, a shoulder is clearly visible. The NiSi and $NiSi_2$ phases do not show this shoulder. The branching ratio, which is defined as the ratio of the area of L_3 to the sum of the areas of L_2 and L_3, is extracted from the spectra by using the model based approach. The same model is used as for the quantitative EELS analysis, but the spectra are recorded with a smaller dispersion. The results are presented in table 3 and figure 3. It is clear that the Ni-silicide phases can be distinguished based on the difference in branching ratio. For the $NiSi_2$ and NiSi phases the error bars overlap.

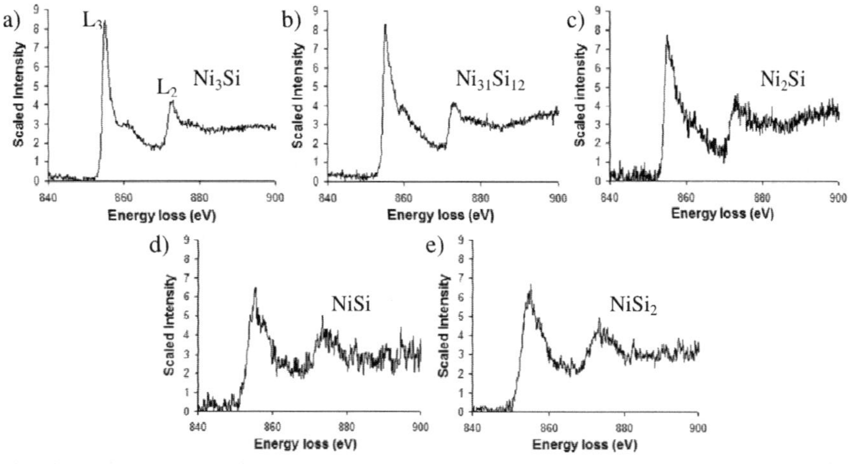

Figure 2. Visual comparison of the ELNES, after background subtraction and deconvolution with the low loss spectrum. a), b) and c) are measured with a monochromator.

Table 3. Results of the branching ratio calculations.

Phase	N	Ratio Si/Ni	Branching ratio	St. dev.
Ni_3Si	30	0.295-0.325	0.854	0.012
$Ni_{31}Si_{12}$	30	0.387	0.816	0.010
Ni_2Si	30	0.500	0.796	0.012
NiSi	30	1.000	0.774	0.007
$NiSi_2$	30	2.000	0.769	0.012

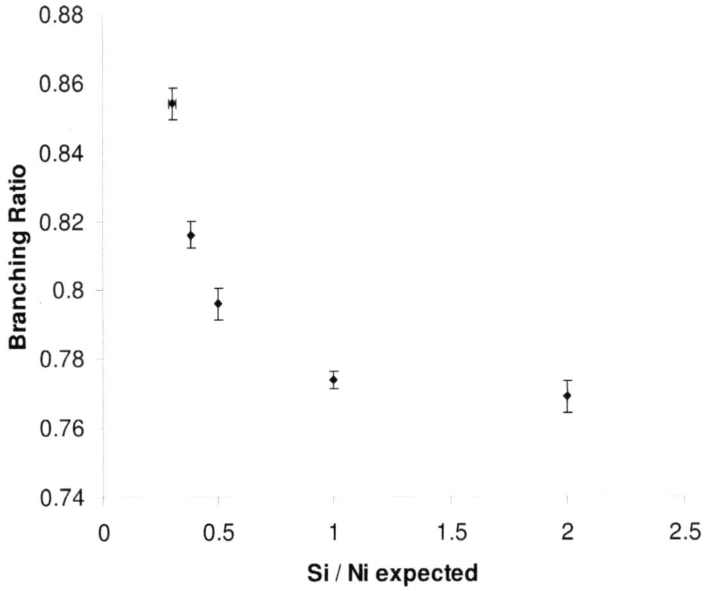

Figure 3. Results of the branching ratio calculations.

4. Conclusions

The results of the characterization of a wide range of Ni-silicide phases by EDS, EELS and ELNES are presented in this paper. It is shown that both the EDS and EELS methodologies yield accurate and precise quantification results. The EELS technique is more straightforward since no standard is needed, and it is not necessary to apply an absorption correction to the data. It is also shown that the Ni-L edge ELNES can be used for phase determination, either visually as fingerprint or by calculating the branching ratio from the white line intensities.

In future research, changes in the low-loss EELS spectrum of the different Ni-silicide phases will be studied. Also, the effect of high spatial resolution on the sensitivity of these methodologies will be tested. This procedure will be extended by combining the EDS and EELS methodologies with a quantification procedure based on HAADF-STEM contrast.

Acknowledgments

The authors acknowledge J. Verbeeck (U Antwerpen), F. Tichelaar and Q. Xu (TU Delft), O. Richard and P. Favia (IMEC) for their constructive advice. The authors acknowledge financial support from the European Union within the Framework 6 program under a contract for an Integrated Infrastructure Initiative for the monochromated EELS measurements at TU Delft. Reference 026019 ESTEEM.

References

[1] Nash P and Nash A 1991 *Masalski T B Binary Alloy Phase Diagrams* 2859
[2] Cliff G and Lorimer GW 1975 *J. Microsc.* **103**, 203
[3] Verbeeck J and Van Aert S 2004 *Ultramicroscopy* **101**, 207
[4] Pearson DH, Ahn C and Fultz B 1993 *Phys. Rev.* B **47**, 8471

Journal of Physics: Conference Series **209** (2010) 012058 doi:10.1088/1742-6596/209/1/012058

Evolution of the interface structure of bonded Si wafers after high temperature annealing

ND Zakharov[1], E Pippel[1], P Werner[1], U Gösele[1], V Vdovin[2,3], M Milvidskii[3], M Ries[4], M Seacrist[4] and R Falster[5]

[1] Max Planck Institute of Microstructure Physics, 06120 Halle (Saale), Germany
[2] Institute for Chemical Problems of Microelectronics, 119017 Moscow, Russia
[3] Institute of Rare Metals 'Giredmet', 119017 Moscow, Russia
[4] MEMC Inc, 501 Pearl Drive, St Peters, MO, 63376, USA
[5] MEMC Electronic Materials SpA, Viale Gherzi 31, I-28100 Novara, Italy

ABSTRACT: The evolution of the interfaces of bonded Si wafers and the corresponding low-angle twist boundary have been analysed in dependence on thermal annealing. Two orientation relations were investigated: i) Si(001)/SiO_2/Si(001) and ii) Si(110)/SiO_2/Si(001). The interfaces were analysed by TEM and STEM/EDXEELS. It is found that the decomposition rate of the intermediate oxide layer and the formation of a Si-Si bonded interface depend very much on the lattice mismatch and on the twist angle. A dissolution of the oxide and the formation of Si-Si boundaries occur much faster in the case of Si(110)/Si(001) bonding than in Si(001)/Si(001). The process of interface fusion and the dissolution of the oxide layer are discussed.

1. Introduction

Tensile strain in silicon channels in complementary metal-oxide-semiconductor (CMOS) devices results in enhanced mobility of both electrons and holes [1]. Direct Wafer Bonding (WB) is a promising way to realize this effect in new types of electronic devices [2]. The formation of a perfect crystalline transition at the interface between the wafers is a crucial problem in this technique [3-5]. Thin SiO_2 layers (native oxide of the wafer surfaces) might hamper this fusion process [6]. In the present work the atomic structure and the composition of Si(110)/SiO_2/Si(001) and Si(001)/SiO_2/Si(001) wafer bonded interfaces were studied by transmission electron microscopy (TEM) as well as by scanning transmission electron microscopy (STEM). The behaviour of intermediate native oxide layers during a high temperature annealing and the interfacial atomic structure are the main issues of this work.

2. Experimental

Samples were fabricated by the direct hydrophilic WB of Si wafers still containing a thin native oxide on their surfaces. The as-bonded structure represented in Fig.1 had the following configuration: i) Si substrates of specific orientations, ii) 2 nm thick SiO_2 layer, and iii) 140-nm thick Si layer. Such samples were thermally annealed in the temperature range 1150 to 1200 °C in vacuum. We could observe a decomposition of the SiO_2 interface layer due to out-diffusion of oxygen and the formation of specific Si(110)/Si(001) and Si(100)/Si(001) interfaces. The bonded structures were investigated by TEM, STEM annular dark field (ADF) using a JEOL JEM4010 and an FEI Titan80-300 electron microscope operating at 400 and 300 kV, respectively. The later one was equipped with an imaging C_s corrector, a monochromator and a Gatan energy loss spectrometer allowing the analysis of the composition with spatial and energy resolutions 0.1 nm and 150 meV, respectively.

3. Results and Discussion

3.1 Si(110)/Si(001) bonding

Fig. 2 shows a typical example of bonded Si wafers after thermal annealing. In this particular case the sample was heated just after bonding at 1150 °C for 15 min. The interface is characterised by a 2 nm thin oxide SiO_x layer between the Si substrate and 140 nm thick Si layer. Beside the thin oxide layer, we found regions where the crystal lattices of both wafers had direct contact (see Fig. 3a).

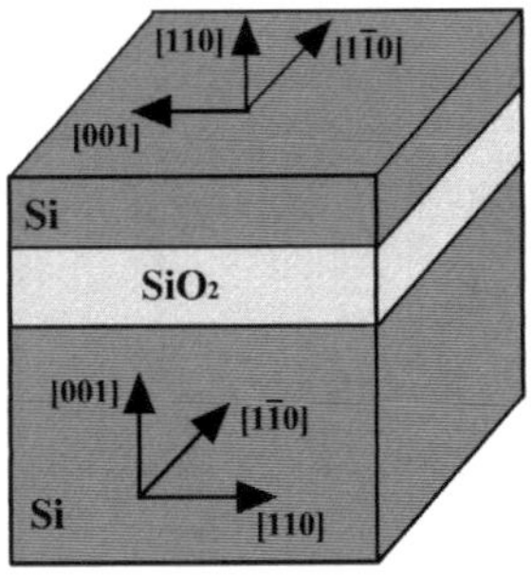

Fig. 1. Schematic view of the bonded structure. Thicknesses of the upper Si layer and the SiO_2 layer amount to 140 nm and 2 nm, respectively.

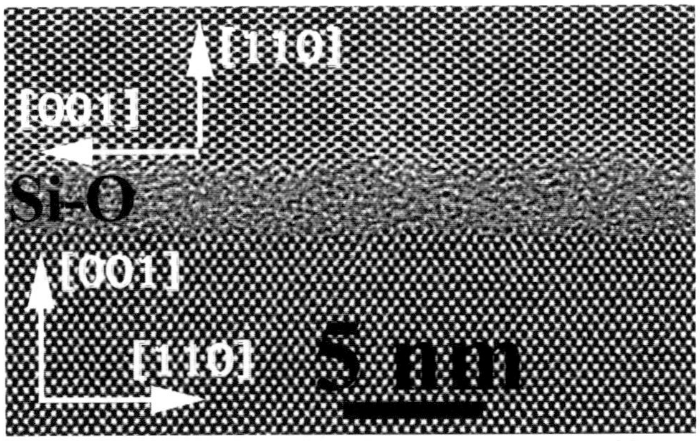

Fig. 2. Cross-sectional image of a Si(110)/Si(001) bonded structure after 15 min annealing at 1150 °C.

This indicates that the oxide layer was partly dissolved ($SiO_x \rightarrow Si + \frac{1}{2}\, x\, O_2$), where the dissociated oxygen could partly diffuse through the 140 nm Si layer and/or to other regions of the interface. In the case of an annealing at 1200 °C for one hour we observed i) extended regions of Si (110)/Si (100), with a directly bonded interface and ii) a rather high number of SiO_x inclusions (see Fig. 3b). From plan-view images it can be shown that these inclusions have a rod-like shape elongated along <110>

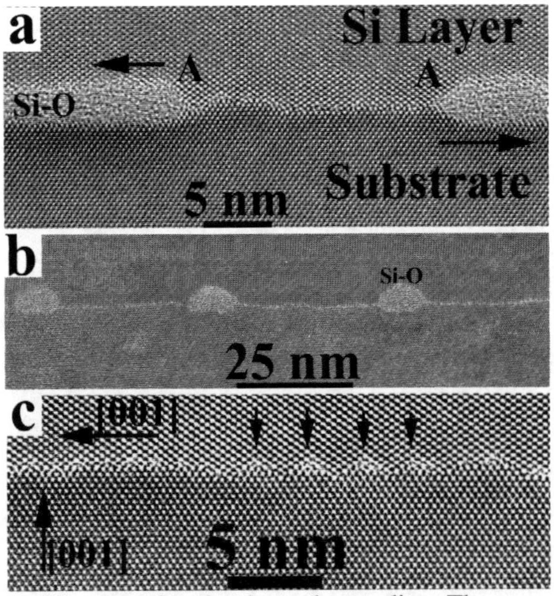

Fig. 3. Evaluation of the bonding interface by thermal annealing. The cross-section images show the Si(110) | | Si(001) interface after 15 min (a), 1 hour (b) and 2 hours (c) annealing at 1200°C. The SiO_x inclusions decrease (a+b) and triangular lattice features are generated (arrows in c).

directions in this particular case. The reason for this behaviour might be that in this direction both crystal lattices have the best fit. Annealing of the bonded structure during 2 hours at 1200°C results in the formation of a Si(110)/Si(001) interface almost free of inclusions (see Fig. 3c).

Analysis by electron energy-loss spectroscopy (EELS) performed with sub-nm spatial resolution reveals a high oxygen concentration in the rod-like inclusions. However, in the directly bonded interfaces (see Fig. 4) no oxygen could be detected within the resolution limit of the EELS measurement (≈ 1 %). Figures 4a and b represent STEM images taken from a relatively thick specimen region (> 15 nm) under high-angle annular dark field (HAADF, a) and low-angle annular dark field (LAADF, b) conditions, respectively. The occurrence of a bright line area in the LAADF image (b) reveals the existence of highly strained regions. According to [7] this contrast feature is caused by a random strain field, i.e. a random shift of the atoms away from the equilibrium position within the related column in beam direction. The resulting de-channelling causes a scattering of the electron waves in small angles only. Thereby bright image details are generated in the LAADF image. In the present case the thickness of the distorted area at the interface amounts to about 1 nm.

Two further findings should be mentioned. First, from the crystallography of the present Si(001)/SiO_x/Si(001) interface follows a huge lattice mismatch parameter of the atoms in the respective surface planes of 34%, which theoretically would demand misfit dislocations with a distance of about 1 nm. However, in the STEM high-resolution micrographs of the crystal lattice structure annealed for 2 hrs we observed triangular structures with a periodicity of about 2.5 nm as marked by arrows in Fig. 3c. The analysis revealed rods along <110> with side facets of low index {111} planes. These structures probably enable a reduction of the interface energy. In this particular case the bonding configuration is characterized by an arrangement of the dislocation cores in the side facets.

Second, we observed a dissolution of the thin SiO_x layer during annealing and the formation of a direct contact between the two Si lattices. This is shown in Fig. 3a, where the direct lattice contact area A-A was increasing. For longer annealing times the number and sizes of SiO_x precipitates decreased (Fig. 3b) and, at the end, vanished (Fig. 3c). Concerning this process we assume that the SiO_x is decomposed as mentioned above: oxygen is diffusing out through the 140 nm thick cap layer and self-interstitials diffuse from the surface to the interface filling up the volume of the precipitates.

3.2 Low twist angle Si{001}/Si(001) bonding

Contrary to the previous example, where the oxide layer disappeared after a nearly one hour annealing at 1200 °C, in the case of low-twisted boundaries the oxide layer does not remarkably change. A plan-view image of a bonded sample after 1.5 hrs annealing at 1200 °C is shown in Fig. 5. Moiré fringes are formed due to the double diffraction of electrons on {220} atomic planes. The outcome of this is a twist angle of $\varphi = 0.18°$, which serves in our case as a misfit parameter. This value is very small; all previous investigations were performed at much larger twist angles [2, 3]. Surprisingly, the oxide layer was retained and its thickness was reduced only by about 0.2 nm. This might correlate to a reduction of SiO_2 decomposition rate.

At the same time a large number of small round and square shaped regions appear – a direct bonding occurs (see A in Fig. 5a). Inside these regions the crystal lattice of the Si cap layer is rotated at an angle $\gamma = -0.18°$ to fit the substrate (compensation of φ). The maximal radius r of these coherently coupled regions never exceeded 120 nm. This gives us the maximal value of tangential elastic displacement on contour of this area $\tau = r \tan\varphi = 0.38$ nm. This value does not exceed the Burgers vector of a screw dislocation, $b = 0.38$ nm. This fact explains the absence of dislocations inside such areas which can be nucleated only if $\tau > b$. The formation of such coherently bonded regions results in the appearance of three-dimensional elastic strain fields in the multilayer system. They are associated with an additional elastic energy proportional to the square of the misfit parameter ~ $n\varphi^2$, where n relates to the number of such features per square area.

4. Conclusions

The dissolution of thin oxide layers at the interface of hydrophilicly bonded Si wafers strongly depends on their orientation or their azimuth: it is faster in the case of Si(110)/Si(001) bonding than in the case of Si(001)/Si(001). The dissolution of SiO_2 can proceed in two ways: (i) perpendicular to the interface – normal growth, and (ii) in the interface plane – tangential dissolution. In the case of Si(110)/Si(001) bonding the process (ii) seems to be much faster than the normal one. The Si layer is growing from SiO_2 supersaturated by Si atoms. The supersaturation is supplied by the decomposition $SiO_2 \Rightarrow Si + 2O$ at 1200 °C. The growing surface is clearly faceted by close packed {111} planes (see

Journal of Physics: Conference Series **209** (2010) 012058 doi:10.1088/1742-6596/209/1/012058

A in Fig. 3a). The extremely high density of misfit dislocation is due to the huge lattice mismatch of ε= 34%. This can also stimulate a fast growth due to dislocations interacting with point defects.

In the case of low-angle twist bonding of Si(001)/Si(001) the dissolution of SiO_2 proceeds very slowly and the formation of the crystalline bonded interface corresponds to the traditional process. The small regions of coherent bonding have a diameter which is always below the critical one. These regions cannot overcome the energy barrier to nucleate screw dislocation.

In summary, oxygen layers of bonded interfaces can disappear much faster when the misfit between both crystal lattices is large. The decomposition rate of oxide layers is slowed down in cases where the interface region contains a high lattice distortion (stored elastic energy). In the present work one has to consider that the second wafer always was quite thin (thickness of 140 nm) in all experiments allowing an interaction/diffusion with the adjacent specimen surface.

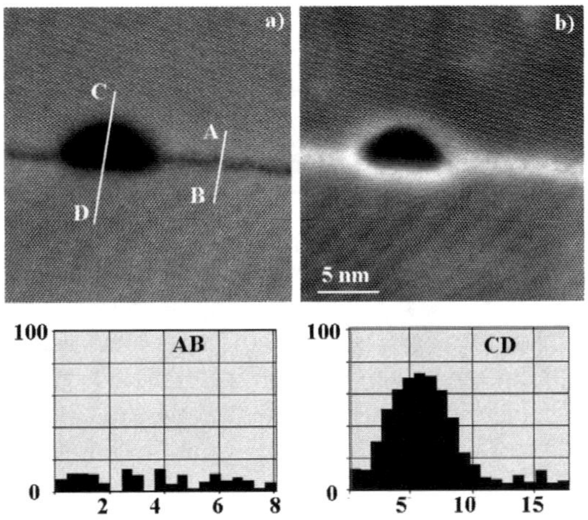

Fig. 4. Images (a) (b) were taken in STEM mode under HAADF and LAADF imaging conditions, respectively. The bright line in (b) indicates the distorted area.
The curves in the lower part represent EELS-profiles of the oxygen concentration (in %) relative to silicon along the marked lines (length in nm).

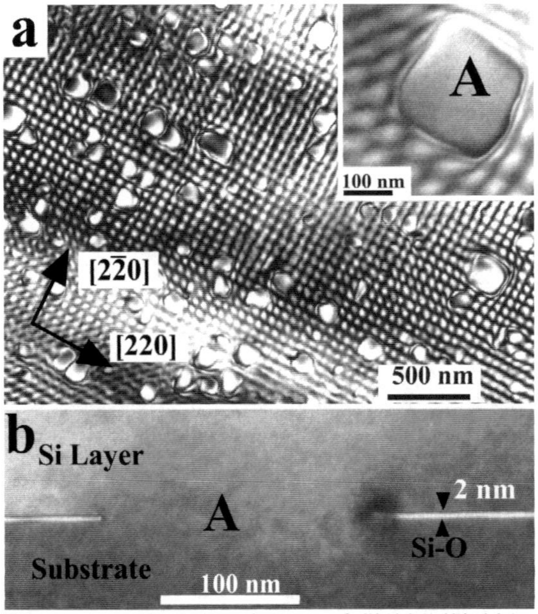

Fig. 5. (a) Plan-view image of Si(001)/Si(001) bonded sample after 2 hrs annealing. Moiré fringes form a square lattice. The regions of coherent coupling of crystal lattices are marked by A. An enlarged image of a coherent region is shown in the upper right corner. (b) Cross-sectional image of coherent regions ‘A’. There are no structural defects at the place of bonding.

References

[1] Sato T, Takeishi Y, Hara H and Okamoto Y 1969 *Phys. Rev. B* **4**, 1950
[2] Toyoda E, Sakai A, Nakatsuka O, Isogai H, Senda T, Izunome K, Omote K and Zaima S 2008 The 5th International Symposium on Advanced Science and Technology of Silicon Materials, Nov. 10-14, Kona, Hawaii, USA
[3] Akatsu T, Scholz R and Gösele U 2004 *Journal of Materials Science* **39**, 3031
[4] Schroer E, Hopfe S, Werner P and Gösele U 1997 *Appl. Phys. Lett.* **70**, 327
[5] Reiche M 2008 *Materials Science Forum* **590**, 57, also online at *http://www.scientific.net*
[6] Ahn K-Y, Stengl R, Tan TY, Gösele U and Smith P 1989 *J. Appl. Phys.* **65**, 561
[7] Zhieng Yu, Muller DA and Silcox J 2004 *Journal of Applied Physics* **95**, 3362

Journal of Physics: Conference Series **209** (2010) 012059 doi:10.1088/1742-6596/209/1/012059

Evolution of porous silicon crystal structure during storage in ambient air

LM Sorokin, VV Ratnikov, AE Kalmykov and VI Sokolov

Ioffe Physical-Technical Institute of the RAS, 26 Polytechnicheskaya St. Saint-Petersburg 194021 Russia

Lev.Sorokin@mail.ioffe.ru

Abstract. Both double- and triple- crystal X-ray diffraction techniques (X-ray DCD and TCD techniques) together with transmission electron microscopy (TEM) were employed for the investigation of structural changes in porous silicon (PS) layers during prolonged periods (up to 6800 hours) of their storage in ambient air. Apart from the Bragg reflection from the Si substrate, the diffraction pattern contains an additional maximum caused by the presence of the PS layer with an increased lattice parameter. The position of this peak shifts to smaller Bragg angles and its intensity decreases as the time of storage in air increases. In addition, the profiles of such peaks become clearly asymmetric. In this case, Gaussian curves were used to reach a fit to the experimental X-ray rocking curves. All samples were biaxially bent due to compressive stresses that arise as soon as 10 min after electrochemical process. The values of lattice strains along the surface normal $(\Delta d/d)_{\perp}$ and lateral deformation $(\Delta d/d)_{\parallel}$ were estimated to be ~ $+10^{-3}$, ~ -10^{-5} respectively. The analysis of diffraction curve evolution shows a gradual destruction of the crystal lattice caused by the air oxidation process.

1. Introduction

It is well known that as-prepared porous silicon (PS) reacts strongly with oxygen in ambient air and both the structural and optoelectronic properties can continuously evolve with the storage time [1]. Therefore, it is important to know what happens with the crystal structure of PS layers during so called "ageing". In [2] it was concluded that the lattice parameter of a porous layer along the normal to the surface is larger than that in bulk Si, i.e. PS experiences some lattice expansion. That gives rise to sample bending caused by biaxial compressive stresses. The goal of this work is to present new data that could concern the changes in the structure of PS layers stored for prolonged periods of time. In this communication, the changes in structural characteristics and the level of PS layer deformation were investigated by double and triple- crystal X-ray diffraction (DCD, TCD) techniques and TEM to elucidate their dependence on the anodization regimes and time of ageing in air.

2. Experimental details

PS layers were formed by the electrochemical etching of boron-doped (p-type) silicon wafers (orientation <100>, resistivity: 10 Ω·cm) under natural lighting. The area exposed to the electrochemical etching solution was about 10 mm in diameter. The electrolyte consisted of 1:3=HF(48%):H_2O mixture. Several micrometer thick porous layers numbered as 1, 2, 3 (see Table 1) were prepared. These samples are characterized by equal total electric charge passing through the electrolytic cell. Another series of samples studied (3, 4, 5) had equal etch times (see Table 1). After the formation of PS layers the samples were rinsed in de-ionized water and dried at 60^0 C for 5 min.

For the X-ray DCD and TCD techniques symmetric (004) and asymmetric (224) Bragg geometries of X-ray intensity measurements were taken using Cu$K_{\alpha 1}$ radiation. All X-ray measurements were carried out on the central exposed area (±4mm) which had a homogeneous porous layer. The total storage time was more than 8 months, and each specimen had been measured at different times, starting from 10 min immediately after anodic etching. The integral specimen porosity was determined by the measurement of photoelectrical absorption coefficient of the X-ray beam passing through the PS layer on Si. The microstructure of PS layers was examined by plan view and cross-sectional transmission electron microscopy (XTEM).

3. Results

3.1. Parameters of porous layers: thickness and porosity

The calculated values of porosity and the thicknesses of PS layers measured by SEM on a cleaved sample surface are listed in Table 1.

sample #	current density mA/cm^2	etching time min	thickness PS μm	porosity %
1	100	1	3.7	65
2	50	2	4.3	60
3	10	10	7.9	56
4	50	10	23	64
5	100	10	47	68

Table 1. Regimes of anodization and parameters of PS layers

In Table 1 it can be seen that the porosity of samples numbered 1-3 decreases as the current density reduces. At the same time, the PS layer thickness increases with longer etching time. The increase in PS layer thickness results from the higher value of charge crossing the electrochemical cell. These dependences are in agreement with previous work [3].

3.2. Stresses and strain in PS layers

Immediately after preparation, all PS samples were bent with positive curvature. The macro curvature of samples was measured by the angular shift of (004) Bragg reflection from the Si substrate while scanning the specimen in the X-ray beam. It was observed that for all stages of ageing in ambient air the bending, caused by biaxial compressive stresses $\sigma_{\parallel}$ in PS leading to lattice expansion along [100], was homogeneous (spherical) and positive. The dependence of stresses calculated according to Stoney's formula during the storage times is seen in Fig. 1. For all samples, their curvature increases with time. Sample 1 shows the largest rate of change in biaxial compressive stress along specimen surface $\sigma_{\parallel}$ which occurred during the first week of storage. This sample has a high value of porosity and is the thinnest PS layer; the rate of change decreases almost to zero after 1.5 months. This observation indicates a gradual distortion of the PS skeleton and a change in the elastic constant of PS.

3.3. Structural changes in PS during its "ageing" in ambient air

In order to understand the behaviour of compressive stresses in PS, its crystal structure was investigated by X-ray DCD and TCD techniques. For all initial PS samples, the first DCD curves both in symmetric (004) and in asymmetric Bragg geometry (224) with grazing incidence and exit angles show a well defined PS peak that lies at smaller Bragg angles than that of the Si substrate ($\Delta\theta \sim -400''$).

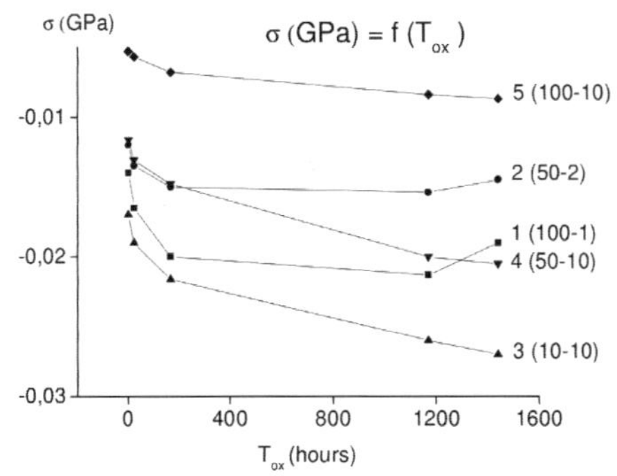

Figure 1. Dependence of compressive stresses $\sigma_{\|}$ in PS on the air oxidation time

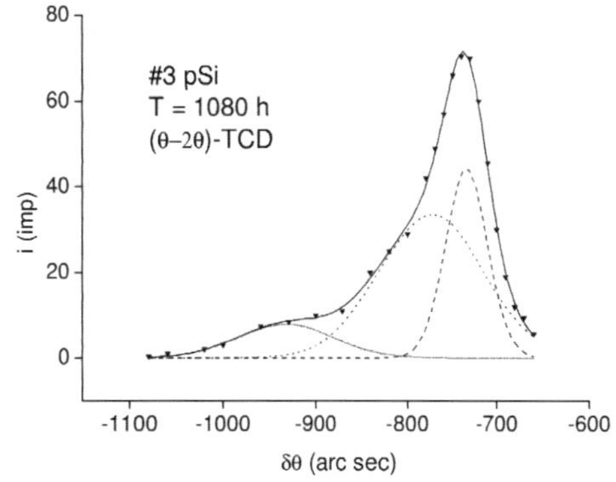

Figure 2. TCD rocking curve (θ-2θ scan) for sample 3

This indicate a lattice expansion in the direction perpendicular to sample surface. In the TCD measurements, an anisotropy of scattering from the PS layer was discovered: FWHM rocking curve for the θ–scan ($w_{PSi} \approx$10-15", due to micro disorientation) was 3-4 times less than that for the θ-2θ scan ($w_{(\theta\text{-}2\theta)} \approx$ up to 50", due to dilatation).This result shows that strain (variations in lattice plane spacing along [100]) is the main reason for broadening of the rocking curve in the initial PS layer. It should be noted that the $\Delta\theta$ value (peak shifting) increases with ageing time. Using Bragg's formula, one can calculate the positive deformation $(\Delta d/d)_{\perp}$ in the PS layer along [100]. For all samples, the rocking curve parameter w_{PSi} increases with duration of air storage, and the DCD rocking curve becomes significantly asymmetric. This asymmetry is related to the appearance of additional PS reflection peaks at smaller Bragg angles. Similar rocking curve behaviour is observed for the TCD (θ-2θ) rocking curves for all PS samples. For example, the changes in $\Delta\theta$, $w_{(\theta\text{-}2\theta)}$, $\sigma_{\|}$ and $(\Delta d/d)_{\perp}$ values for sample 4 are listed in Table 2. As mentioned, the DCD rocking curves as well as TCD ones are gradually transformed to asymmetric profiles, and by using a Gaussian fit the experimental asymmetric rocking curve can be decomposed into several sub-peaks. The TCD (θ-2θ) (004) rocking curve for sample 3 is shown in Fig. 2 as an example. The Gaussian fit makes it possible to reveal three sub-peaks which correspond to different PS layers. Similar rocking curves were observed in [4] for PS layer prepared by the stain-etch technique. The different parts of the PS layer after its air storage for 2660 hours can be seen in the TEM image of Fig. 3. The electron diffraction pattern (inset) shows that the Si skeleton retains its crystal structure with <110> orientation. The weak diffuse rings in Fig. 3 confirm the presence of an amorphous phase which is a result of the PS layer oxidation.

A typical PS structure after prolonged storage in air is presented in Fig. 4. This sample shows a high concentration of small single pores and their agglomeration. The size range of single pores was estimated to be 3 - 8 nm and the Si crystallites were measured to be 3 - 10 nm. As seen in Fig. 4 (inset), the area occupied by the Si crystallites (dark contrast) is significantly less than that of the silicon oxide (grey contrast). This is a result of the PS degradation in air, but nevertheless the Si crystallites keep their initial crystal orientation (Fig. 3, inset).

Measurement of the lateral deformation in PS $(\Delta d/d)_{\|}$, was performed using asymmetrical (224) DCD rocking curves, recorded at both grazing incidence and exit geometries This allows us to calculate the difference between the inclination angles of (112) lattice planes in PS and Si [5]. Thus, the lateral deformation value $(\Delta d/d)_{\|}$ was estimated to be ~ -10^{-5}.

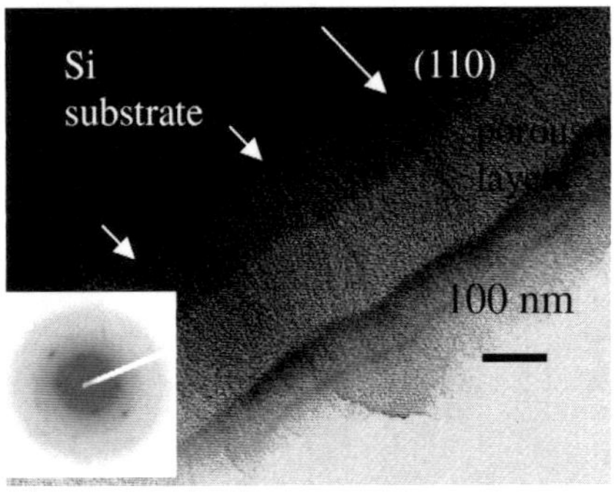

Figure 3. XTEM image of PS (sample 1) Arrows indicate the boundary Si-PS.

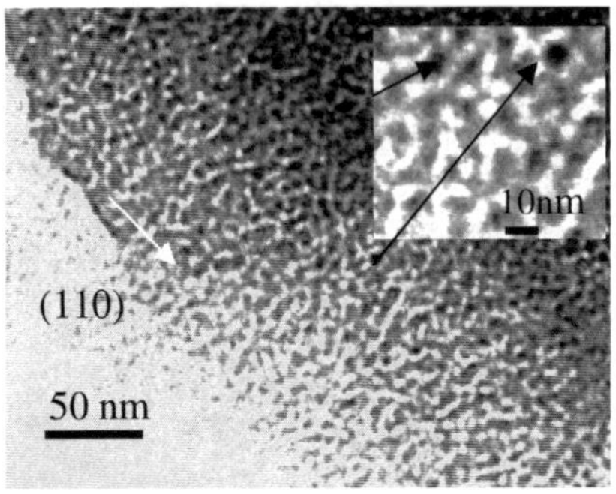

Figure 4. XTEM image of PS (sample 1) Inset: nanocrystallites – dark contrast

t_{ox} (h)	1	24	168	2040	6100
$\sigma_{\|\|}$ (GPa)	-0.012	-0.013	-0.015	-0.021	-0.015
w (arc sec)	34	47	77/55/34	179/48	195/70
$\Delta\theta$ (arc sec)	-400	-420	-611/-544/-500	-972/-904	-1364/-1268
$(\Delta d/d)_{\perp} * 10^{-3}$	+2.81	+2.95	+4.3/+3.8/+3.52	+6.84/+6.36	+9.6/+8.9

Table 2. Biaxial stresses $\sigma_{\|\|}$, (004) FWHM $w_{(\theta-2\theta)}$, $\Delta\theta$ (PS peak shifting) and $(\Delta d/d)_{\perp}$

4. Conclusion

Gradual structure changes in PS during prolonged period of ageing in ambient air (up to 6800 h) have been studied by X-ray techniques. All specimens suffered biaxial compression stresses 10 min after their fabrication. The lattice strains along the surface normal are positive and equal to ~ $+10^{-3}$, two orders of magnitude larger than the lateral strains of ~10^{-5}, i.e. the relaxation of the PS skeleton is insignificant. The analysis of changes in X-ray rocking curves shows a step-by-step distortion of the crystalline PS skeleton by oxidation with an increase in strain and disorientation of nanocrystallites in it. In thick PS layers, aged up to 8 months, two types of crystallite sizes were revealed (~5 nm and ~ 0.15 μm) on the basis of the DCD rocking curve analysis. The existence of a PS multilayer structure exhibited by TEM agrees with the analysis of the X-ray TCD (θ-2θ) asymmetric rocking curves.

Acknowledgement

This work was supported in part by Program OFN RAS "Physics of new materials and structures"

References

[1] Cullis AG, Canham LT and Calcott PDJ 1997 *J.Appl.Phys.* **82**, 909
[2] Barla K, Herino R, Bomchil G, Pfister JC and Freund AJ 1984 *Cryst.Growth* **68**, 727
[3] Beale MIJ, Chew NG, Uren MJ, Cullis AG and Benjamin JD 1985 *Appl.Phys.Lett.***46,** 86
[4] Abramof PJ, Beloto AF, Ueta AY and Ferreira NG 2006 *J. Appl. Phys.* **99**, 024304
[5] Bartels WJ amd Nijman W 1978 *J.Cryst.Growth* **44**, 518

Ge nanocrystals in alumina matrix: a structural study

RJ Kashtiban[1], SRC Pinto[2], U Bangert[1], AG Rolo[2], A Chahboun[2, 3], MJM Gomes[3] and AJ Harvey[1]

[1] Nanostructured Materials Research Group, School of Materials, The University of Manchester, P.O. Box 88, Manchester, M1 7HS, UK
[2] Physics Department, University of Minho, 4710 – 057 Braga, Portugal
[3] Physics Department, Dhar Mehraz Sciences Faculty, BP 1796, Fès, Morocco

E-mail: reza.jalilikashtiban@postgrad.manchester.ac.uk

Abstract. Germanium (Ge) nanocrystals (NCs) embedded in alumina thin films were produced by deposition on silicon (111) substrates using radio-frequency (RF) magnetron sputtering. By changing growth condition and annealing parameters samples with large and small Ge-NCs were produced. The average size of NCs in the sample with larger NCs was estimated to be 7.2, 30.0 and 7.6 nm, and 5.0, 7.0 and 4.8 nm for the sample with smaller NCs, according to X-ray diffraction (XRD), Raman and high resolution transmission electron microscopy (HRTEM) results, respectively. Both, XRD and Raman peak positions of the larger NCs are shifted to higher angles and larger wave numbers in relation to the Ge bulk values, whereas the Raman peak was red-shifted for the smaller NCs indicating phonon confinement. HRTEM shows twinned structures, which is an indication of relaxation. Strain evaluation of the bigger NCs gave values < 0.5 % which is within the estimated error of the evaluation technique.

1. Introduction

Undesirable charge leakage in modern flash memories based on polycrystalline silicon prohibits the use of thin tunnel oxide layers for the reduction of the operating voltage. On the other hand a thick tunnel oxide requires a higher operating voltage, which is a major disadvantage. Si and Ge-NCs embedded in a SiO_2 matrix have attracted great attention because of their potential applications in high performance non-volatile memory devices and light emitting diodes [1-4]. Using isolated NCs in dielectric layers improves the retention time, and thinner oxide layers can be employed resulting in lower operating voltage and faster device speed [3]. Ge has a smaller band-gap than Si which makes it compatible with currently used metal oxide semiconductor (MOS) technology and also improves its write/erase speed [5, 6]. Promising non-volatile properties of Ge-NCs were reported by several groups [7-10].

Since Al_2O_3 has a high dielectric constant, twice that of SiO_2, the thickness of the oxide layer can be reduced to 2-3 nm. Moreover, Al_2O_3 exhibits good mechanical properties and can endure high temperatures; this makes it a good candidate for replacing silica in flash memory systems and thereby improve their performance. Most of the research on Ge-NCs has been done on Ge-NCs embedded in SiO_2 system [3, 5, 7, 9-12]. However, a few studies have been reported on Ge-NCs embedded in Al_2O_3 matrix [13, 14]. In this work we will study Ge-NCs embedded in alumina grown by the RF-magnetron co-sputtering technique.

2. Experimental

Conventional co-sputtering methods were used to grow Ge-NCs embedded in Al_2O_3 in a commercial RF magnetron Alcatel SCM650 machine. Al_2O_3 (99.99%) and polycrystalline Ge (99.99%) were used as targets to produce the doped films simultaneously on 2 inch n-type Si(111) wafers as substrates, kept at 500 °C . Prior to sputtering, a pressure of at least 1×10^{-6} mbar was reached inside the chamber and in order to achieve a clean surface free of any impurities, in situ argon plasma treatment of target and substrate was performed. In order to improve the crystal structure of the Ge phase and achieve structures with large (sample A) and small (sample B) Ge-NC sizes, as-grown samples were annealed at 900 °C and 800 °C, for one hour, under air pressures of 8.0×10^{-3} and 4.6×10^{-6} mbar, respectively. More details are presented elsewhere [13]. Room temperature Raman spectroscopy was carried out using the 514.5 nm and 488.0 nm lines of an argon laser at a power of 50 μW focus on an area of the sample ~1 μm^2 in order to study crystallinity and size distribution of the Ge-NCs. In order to study the crystallographic structure of the materials X-ray diffraction in conventional θ-2θ geometry was performed with a Philips PW1710 apparatus using Cu Kα radiation for sample A and sample B by grazing incidence X-ray diffraction (GIXD) in the European Synchrotron Radiation Facility (ID01 beam line, Grenoble, France) using 11 keV radiation at 0.15 degree of incidence to the sample surface. TEM and HRTEM images were acquired with a Tecnai F30 FEGTEM microscope operating at 300 kV.

3. Results and discussion

Figure 1 represents XRD and Raman spectra of samples A and B. The XRD curve confirms the diamond crystalline structure of Ge. Both, XRD and Raman peak positions of sample A are shifted to

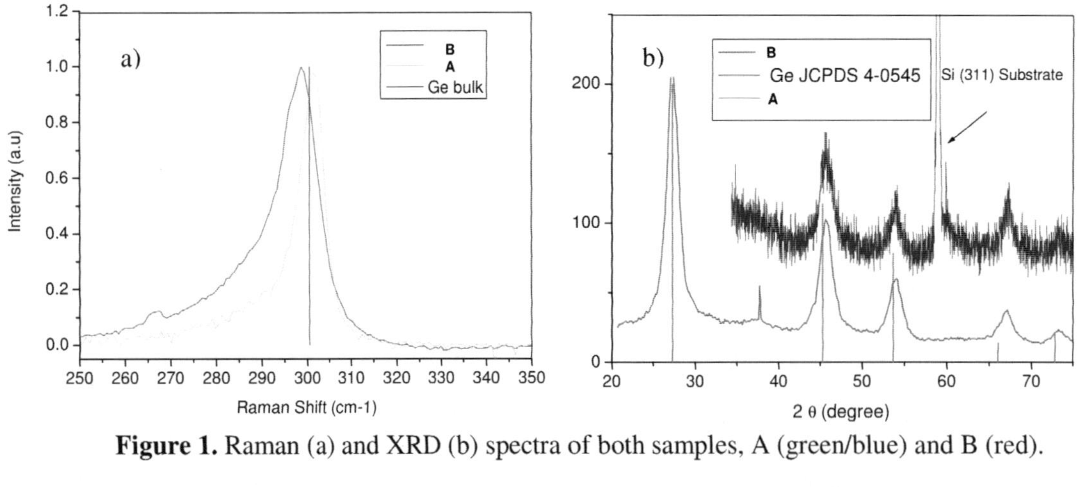

Figure 1. Raman (a) and XRD (b) spectra of both samples, A (green/blue) and B (red).

higher angles and larger wave numbers, respectively, in relation to the Ge bulk. This would indicate that the sample is under compressive stress exerted by the Al_2O_3 matrix. Raman spectra of sample B show a red shift, which is predicted by phonon confinement theory for small NCs.

The average size of NCs in sample A was estimated to be 7.2, 30.0 and 7.6 nm, and 5.0, 7.0 and 4.8 nm for sample B according to X-ray diffraction (XRD), Raman and high resolution transmission electron microscopy (HRTEM) results, respectively. Whereas TEM and XRD results are in agreement, the discrepancy with the Raman results is not clear. It might be partially caused by the elongated shape of the sample-A NCs (>10 nm along the long axis; the value above is a geometrical average) and also by the near-epitaxial relationship between Ge- and neighbouring Al_2O_3-crystallites, giving the appearance of larger crystalline regions.

Figure 2 shows a bright field (BF) and dark field (DF) TEM image of sample B. Formation of crystalline Ge-NCs within the alumina matrix can be seen in the DF image. The silicon substrate, the 600 nm thick alumina matrix and the glue can be seen clearly in the BF image. Diffraction contrast cannot reveal atomic-scale structure of the NCs and the alumina matrix and also not the exact NC sizes. In order to obtain this information, HRTEM images were acquired. A detailed size analysis was carried out on a statistical ensemble of Ge-NCs, which showed the majority of the NCs to be ellipsoidal with the major axis ~11 nm and the minor axis ~5 nm.

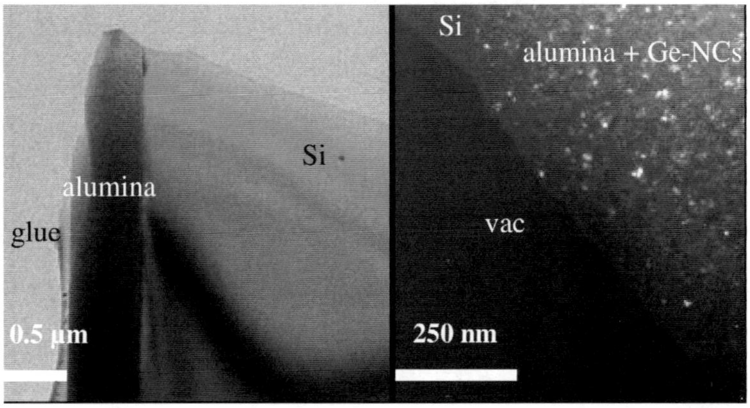

Figure 2. Conventional TEM BF (left) and DF (right) image of sample B.

Figure 3(a) depicts a Ge-NC in sample B. Extensive twinning, observed in most of the NCs, indicates that stress relaxation has taken place and we therefore expect the NCs to be nearly unstrained. Figure (b) and (c) represent images acquired at different foci revealing the polycrystalline nature of the matrix. The circled Ge-NC is the same in both images. The existence of crystalline alumina phases complicates the interpretation of XRD measurements, because of the occurrence of additional peaks close to the position of Ge peaks which make the identification of the latter difficult.

Figure 3 (d) shows an image of big Ge-NC in sample A together with its Fourier transform (FT) pattern representing the (211) zone axis pattern of the diamond crystal structure. Different diffraction spots of this FT pattern were used to measure lattice plane distances in this NC to evaluate the strain which may be exerted by the alumina matrix.

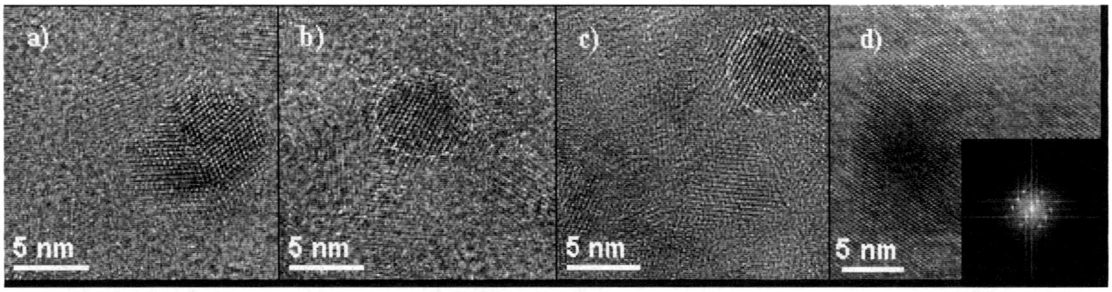

Figure 3. HRTEM images of (a) a relaxed faulty NC in sample B, (b) and (c) the polycrystalline alumina matrix with a Ge-NC (circled) at different focus condition, and (d) a big NC in sample A together with its FT as inset.

To assess the existence of such strains in the Ge-NCs, HRTEM images of the Si substrate were taken as reference in order to estimate the error in the determination of lattice plane distances from FTs of HRTEM images. The lattice spacing of the Si substrate was determined using FTs of HRTEM images with the sample aligned at the (110) zone axis for various reflections and in various locations and compared to calculated values of silicon (a = 5.43 Å). Lattice spacings for Ge-NCs were then obtained from HREM images (FTs) taken at the same magnification as the Si images. Table 1 illustrates lattice planes distances and error values for both, substrate and NCs. As can be seen changes in the lattice distances of the NCs were within the error limit, and hence the strains would be below 1%. The shifts in the XRD peaks in fig. 1b would arise from strains around the ½ % mark; this is the error limit of the HREM technique, and hence we would not be able to confirm these from lattice images.

Planes	Silicon substrate		Error	Ge-NCs		
	Exp.	Calc.		Exp.	Calc.	Displacement
111	3.110	3.135	+0.7%	3.240	3.262	+0.6%
220	1.890	1.919	+1.5%	1.970	1.990	+1.0%
113	-	-	-	1.690	1.700	+0.5%

Table 1. Error values of technique based on Fourier method applied to HRTEM images, and percentage of evaluated strain in the sample A.

4. Conclusion

Ge-NCs embedded in Al_2O_3 films were successfully produced by RF-magnetron sputtering. Samples with two different NC sizes (on average ~5 and ~8 nm) were produced by changing the annealing temperature and gas flow pressure. Raman and XRD results indicate strain in the bigger NCs, however, peaks from the polycrystalline alumina matrix obscure the exact position of the Ge-peaks in the XRD results. Furthermore, the shifts, if genuine, are concomitant with strains < ½ %, which is below the accuracy of HRTEM-based lattice constant determination. HRTEM suggests unstrained structures for the small NCs, because plastic relaxation and extensive nano-twinning have taken place. This explains red shift in the Raman peak position of the sample and is a sign of relaxed NCs.

References

[1] Tiwari S, Rana F, Chan K, Shi L and Hanafi 1996 H *Appl. Phys.Lett.* **69**, 1232
[2] Lu TZ, Alexe M, Scholz R, Talalaev V, Zhang RJ and Zacharias M 2006 *J. Appl. Phys.* **100**, 14310
[3] Kanjilal A., Hansen JL, Gaiduk P, Larsen AN, Cherkashin N, Claverie A, Normand P, Kapelanakis E, Skarlatos D and Tsoukalas D 2003 *Appl. Phys. Lett.* **82**, 1212
[4] Tiwari S, Rana F, Hanafi H, Hartstein A, Crabbe EF and Chan 1996 *Appl. Phys. Lett.* **68**, 1377
[5] Batra Y, Kabiraj D and Kanjilal D 2007 *European Physical J.- Appl. Phys.* **38**, 27
[6] Lee PF, Lu XB, Dai JY, Chan HLW, Jelenkovic E and Tong KY 2006 *Nanotechnology* **17**, 1202
[7] Baron T, Pelissier B, Perniola L, Mazen F, Hartmann JM and Rolland G 2003 *Appl. Phys. Lett.* **83**, 1444
[8] Kanoun M, Souifi A, Baron T and Mazen F 2004 *Appl. Phys. Lett.* **84**, 5079
[9] Rolo AG, Chahboun A, Conde O, Vasilevskiy MI and Gomes MJM 2008 *Physica E* **40**, 674
[10] Hong SH, Kim MC, Jeong PS, Choi SH and Kim KJ 2008 *Nanotechnology* **19**, 305203
[11] Choi WK, Chew HG, Ho V, Ng V, Chim WK, Ho YW and Ng SP 2006 *J. Cryst. Growth* **288**, 79
[12] Park CJ, Cho KH, Yang WC, Cho HY, Choi SH, Elliman RG, Han JH and Kim C 2006 *Appl. Phys. Lett.* **88**, 71916
[13] Caldelas P, Rolo AG, Gomes MJM, Alves E, Ramos AR, Conde O, Yerci S and Turan R 2008 *Vacuum* **82**, 1466
[14] Yerci S, Kulakci M, Serincan U, Turan R, Shandalov M and Golan Y 2008 *Journal of Nanoscience and Nanotechnology* **8**, 759

TEM analysis of Ge-on-Si MOSFET structures with HfO_2 dielectric for high performance PMOS device technology

DJ Norris[1], T Walther[1], AG Cullis[1], M Myronov[2], A Dobbie[2], T Whall[2], EHC Parker[2], DR Leadley[2], B De Jaeger[3], W Lee[3], M Meuris[3], J Watling[4] and A Asenov[4]

[1] Dept of Electronic and Electrical Engineering, University of Sheffield, Sir Frederick Mappin Building, Mappin Street, Sheffield, S1 3JD, UK.

[2] Dept of Physics, University of Warwick, Coventry, CV4 7AL, UK.

[3] IMEC, Kapeldreef 75, B-3001 Leuven, Belgium.

[4] Dept of Electronics and Electrical Engineering, Rankine Building, Oakfield Avenue, Glasgow, G12 8LT, UK.

E-mail: d.j.norris@sheffield.ac.uk

Abstract. In this paper, we present a (scanning) transmission electron microscopy analysis of novel Ge-on-Si MOSFETs which incorporate a high-k HfO_2 dielectric and TaN/TiN metal gate electrodes. A key feature of these devices is the incorporation of a very thin (~1nm) Si passivation layer on top of the Ge virtual substrate, which is partially oxided to form SiO_2 (~0.5nm), before depositing the HfO_2 dielectric and TaN and TiN metal gate electrodes. Our results confirm the architecture of the device structures and the existence of Si passivation.

1. Introduction

In recent years, there has been an increased interest in using Ge as a higher charge carrier mobility alternative to Si, for advanced complementary metal-oxide-semiconductor (C-MOS) device applications, based on hole transport [1].

One consideration is the choice and type of substrate used. Rather than using a pure Ge substrate, it has been found that it is possible to grow a relaxed layer of pure Ge onto a (001) Si wafer thereby producing a Ge-on-Si virtual substrate (VS) onto which MOS field-effect-transistor (MOSFET) devices can be processed [2]. One of the issues associated with fabricating such Ge VS-based devices is the need to form, on its surface, a stable dielectric prior to fabricating the gate electrode. One of the drawbacks of Ge-based devices, however, relates to the poorer thermal oxide quality compared with that of Si, which has been found to give rise to a higher interface trap density [3]. Consequently, methods have been developed to reduce this through the use of Ge passivation [4]. One possible route for such passivation involves the deposition of an ultra-thin epitaxial-Si layer on the surface of the Ge substrate which is then partially oxidised, prior to the growth of a high-k gate dielectric [4]. There have been issues as to whether Ge segregation during growth of the epi-layer will affect the interface trap density (N_{it}); however, charge pumping measurements of fast traps in samples passivated at different temperatures have shown a significant reduction of N_{it} from ~$1.5\times10^{11}cm^{-2}$ (at 500°C) to ~$5\times10^{10}cm^{-2}$ (at 350°C) [5].

In this paper, we present a TEM/STEM analysis of MOSFETs which are processed on a Ge-on-Si VS. Such devices incorporate a partially oxidised Si passivation layer onto which are deposited high-k HfO_2 and metal gate electrodes. The architecture and details of the dielectric region and various other components of the device are measured and verified.

2. Experimental Details

Ge p-MOSFETs were fabricated using a silicon-compatible process flow on (001) Si wafers with a thick (~2 μm) relaxed Ge layer grown on the surface. These were then chemo-mechanically polished (CMP) to remove any cross-hatch surface undulations. After the P well and channel implants, the Ge-on-Si wafer was passivated by growing a thin epi-Si layer (~1nm) which was then partially oxidised at 350°C; see [4]. To avoid further oxidation, a thin layer of HfO_2 (~4 nm) was then immediately deposited by atomic layer deposition on the SiO_2 before growing the TaN (~10 nm) and next the TiN (~70 nm) metal layers. The remaining processes, i.e. the dry metal gate etch, BF_2 extension implants, spacer definition, B implant, NiGe formation in source and drain regions and metal contact fabrication are all detailed elsewhere [6].

The processed wafers include a narrow region of device lines (26 μm wide – 5 mm long) of MOSFETs of various gate lengths which could be processed into TEM specimens. [110] cross-sectional specimens, for the TEM, were prepared in the usual manner with 3 mm disks thinned mechanically to ~30-40 μm followed by Ar^+ ion thinning to electron transparency. Finished specimens were then analysed using a JEOL 2010F field-emission gun instrument, operating at 197kV, and equipped with a GATAN imaging filter, Oxford ISIS X-ray detection system and STEM analysis using bright-field (BF) and high-angle annular dark-field (HAADF) detectors.

3. Results and Discussion

A low magnification image of the MOSFET device structures on top of the Ge-on-Si VS is shown in figure 1. One can see that the Ge VS surface (1.63 μm) is uniformly flat as a result of the CMP. These particular devices have a 325 nm gate length and are spaced ~160 nm apart. A high magnification BF STEM image is shown in figure 2, and shows the various components which make up the device structure. The dark layer at the base of the gate electrode is the heavy TaN layer (~14 nm) under which are the HfO_2 (3.6 nm) and SiO_2 (0.5 nm) gate dielectrics. On top of this is a TiN (~70 nm) metal

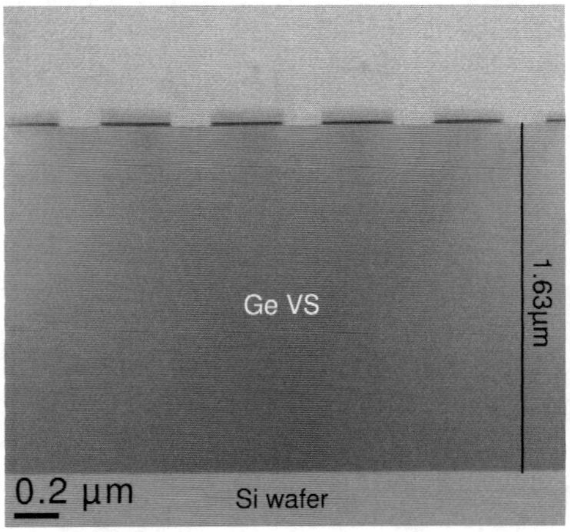

Figure 1. Low mag BF STEM Image of MOSFETs on Ge-on-Si virtual substrate

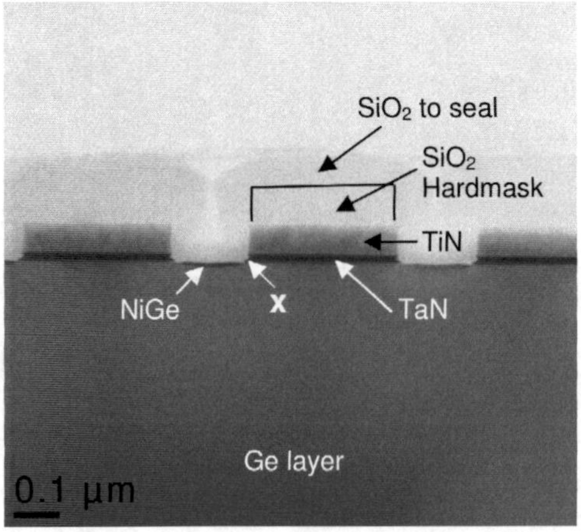

Figure 2. Higher mag BF STEM image of 325nm gate length MOSFET.

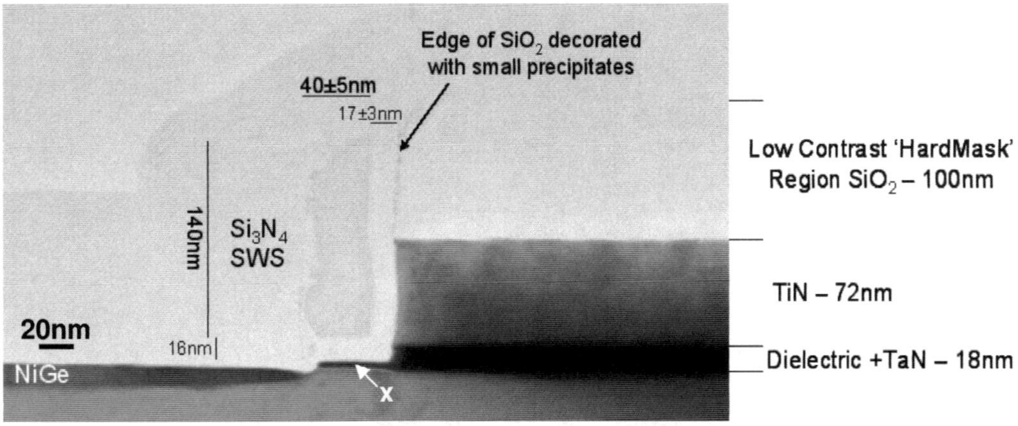

Figure 3. BF STEM image of the edge of larger gate-length (~15µm) MOSFET device structure showing the architecture of the metal gate, side wall spacer (SWS) and Ni germanide.

electrode (as indicated in figure 2) above which is a SiO_2 'hardmask'. In the source and in the drain regions, the NiGe (~12 nm) is clearly visible.

A higher magnification BF STEM image of the edge of a large device is provided in figure 3. Here, the Si_3N_4 side-wall-spacer (SWS) is visible (~40 nm wide and 140 nm high) and is attached to the side of the gate-electrode via 15-20 nm of SiO_2 which acts as a SWS adhesion layer. The purpose of the SWS is to prevent the Ni (for the germanidation of the source and drain regions), deposited after SWS formation, from reacting with the sides of the gate terminal. Upon annealing the Ni reacts with the Ge and not with Si_3N_4. Any residual unreacted Ni is etched away; however, there do appear to be some residual Ni globules decorating the side of the SWS. Under the SWS there is a dark extended HfO_2 layer (labelled X) which indicates that the dry etch for gate-electrode definition terminated earlier than intended; however, this is unlikely to impact greatly on device performance.

A further feature of the present device structures is the occasional presence of stacking fault defects at the edge of the gate into the Ge VS, and a lattice image of this is shown in figure 4. Such features appear to be more prevalent in larger gate-length devices (325 nm and greater). They suggest that the regions at the edge of the gate have a particularly high stress concentration giving rise to the nucleation and expansion of dislocation half-loops. The stacking fault is in the glide plane of such loops.

The dielectric region, of a shorter gate-length device, was imaged using planar illumination under high resolution TEM (HREM) conditions and is shown in figure 5(A). One feature of the present device structures is the incorporation of a Si passivation layer grown on top of the Ge VS. This is partially oxidised to form a very thin SiO_2 layer, which (from figure 5(A)) was measured to be 0.5±0.1nm. However, it is difficult to confirm the presence of the residual unoxided Si from such phase contrast images.

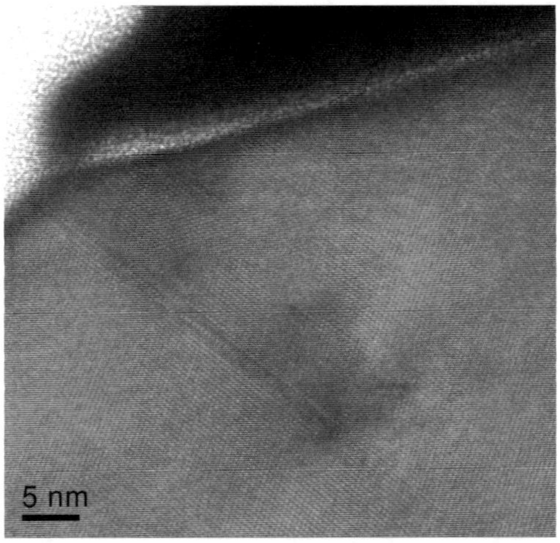

Figure 4. Stacking fault at edge of gate.

An attempt was made, therefore, to determine the presence of a residual Si layer by HAADF (Z-contrast) imaging, and such an image, in a similar region of the device structure, is shown in figure 5(B). One can see clearly the different brightnesses of the various layers of different mass, and the width of the dark region indicated corresponds to the sum of the

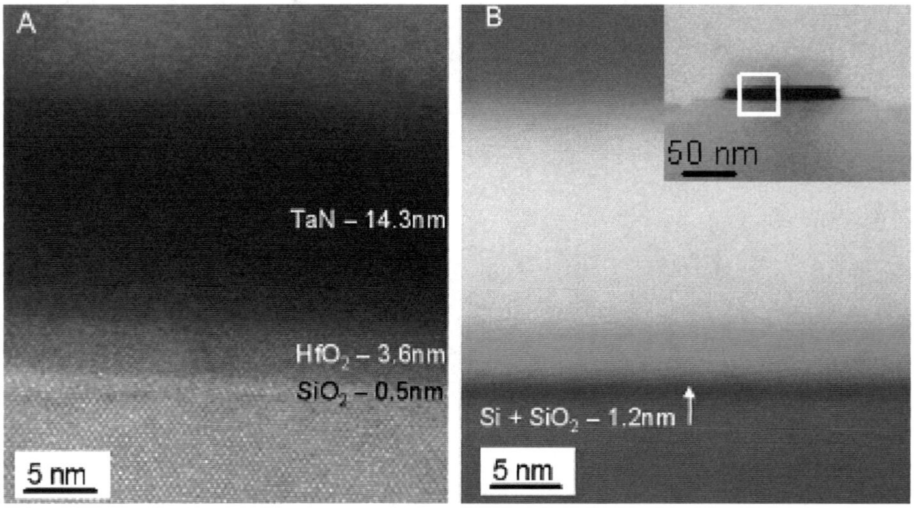

Figure 5. (A) HREM image and (B) HAADF Image of dielectric region under the gate (inset: 145 nm gate device showing region of interest).

thicknesses of the residual Si and SiO_2 layers (1.2±0.2 nm). Hence, a comparison of these images shows that the thickness of the residual Si passivation layer must be 0.7±0.3 nm.

A particularly important issue is whether there is any Ge present within the residual Si and SiO_2 layers. It is known that during growth, Ge may segregate from the surface of the Ge substrate through the epitaxially grown Si layer [7] and may be incorporated into the processed SiO_2 layer, or accumulate at the lower Si/SiO_2 boundary by 'snowploughing'. Verifying this is quite a demanding challenge since the layers concerned are extremely narrow. Our images indicate the presence of a residual Si passivation layer, thus it is confirmed that the Ge distribution does diminish at the Ge/Si boundary before reaching the Si/SiO_2 interface. However, we cannot rule out small amounts of Ge in the ultra-thin Si passivation and SiO_2 layers. We have attempted EELS spectrum imaging to examine the Ge distribution; however, it is extremely difficult to obtain the necessary spatial resolution.

4. Conclusions

Ge-on-Si MOSFET structures have been analysed using TEM-based techniques. A feature of the present devices was the use of Si passivation to reduce interface trap density. This required a very thin epi-layer of Si (1nm) to be grown onto the surface of the Ge-on-Si wafer, which was then partially oxidized. After this a HfO_2 high-k dielectric was deposited, followed by TaN and TiN gate electrodes. Our results verified the architecture of the processed devices structures, and confirmed Si passivation. The residual Si passivation layer was found to be 0.7±0.3 nm, and the SiO_2 layer was 0.5±0.1 nm.

Acknowledgements

The authors would like to thank the Engineering and Physical Sciences Research Council for their financial support of this work under grant number EP/F033893/1 "Renaissance Germanium".

References

[1] On Chui C et al. 2002 *IEEE Electron Dev. Lett.* **23**, 473
[2] Kaczer B et al. 2007 *Microelectronic Engineering* **84**, 2067
[3] Houssa M et al. 2007 *Microelectronic Engineering* **84**, 2267
[4] De Jaeger B et al. 2005 *Microelectronic Engineering* **80**, 26
[5] Leys FE et al. 2006 *Materials Science in Semiconductor Processing* **9**, 679
[6] Zimmerman P et al. 2006 *IEDM Tech. Dig.*, 655
[7] Norris DJ et al. 1999 *J. Appl. Phys.* **86**, 7183

Off-axis electron holography of Si semiconductors prepared using FIB milling

D Cooper, J-P Barnes, J-M Hartmann and F Bertin

CEA-LETI, Minatec, 17 rue des Martyrs, F38054 Grenoble, France.

E-mail: david.cooper@cea.fr

Abstract. Off-axis electron holography is a powerful technique that can be used to provide 2D dopant maps of semiconductor materials. Focused ion beam milling is the only technique that can be used to prepare the parallel-sided specimens with the required site specificity for dopant profiling of state-of-the-art semiconductor devices. However, focused ion beam milling introduces artefacts into the thin TEM specimens, such as a surface amorphous layer, Ga implantation and the creation of defects. Here the effects of these artefacts in the semiconductor specimens are examined leading to a better understanding of the reconstructed phase images.

1. Off-axis electron holography and the FIB.

Off-axis electron holography is a transmission electron microscopy (TEM) based technique that uses an electron biprism to interfere an object wave that has passed through a specimen with a reference wave that has passed through only vacuum. From the resulting interference pattern, or hologram, the phase of an electron can be recovered. As the phase of an electron is very sensitive to changes in the electrostatic potential in a specimen such as from the presence of dopants, off-axis electron holography can in principle fulfill the requirements of the semiconductor industry for a 2D dopant mapping technique with nm-scale resolution. In the absence of magnetic fields and diffraction contrast, the phase an electron that has passed through a semiconductor specimen of thickness, t is given by,

$$\Delta\varphi = C_E \int_0^t V(x, y, z)dz$$

where C_E is a constant dependent on the energy of the electron wave, V is the electrostatic potential and dz the direction of the electron beam [1]. The carrier concentration can then be determined from the changes in potential across a homogeneous semiconductor specimen of constant thickness.
Focused ion beam (FIB) milling is the only specimen preparation technique that can currently be used to prepare specimens for examination by off-axis electron holography due to the ease with which a parallel-sided lamella can be extracted with the required site-specificity. However, the artefacts that are introduced into the specimens during FIB-preparation include Ga implantation and the creation of defects deep in the crystalline regions of the specimen that can trap the dopants. Figure 1(a) shows a montage of phase images that have been acquired of series of 45-nm-gate CMOS devices. A profile has been extracted from across the doped regions averaged over 5 nm, although the differently doped regions can be clearly observed and the electrical gate width can be measured, it is difficult to know how to convert the phase profiles into dopant concentrations. In this paper we have examined

a series of simple Si *p-n* junctions in order to better understand the effects of FIB specimen preparation on the step in phase measured across junctions with different dopant concentrations.

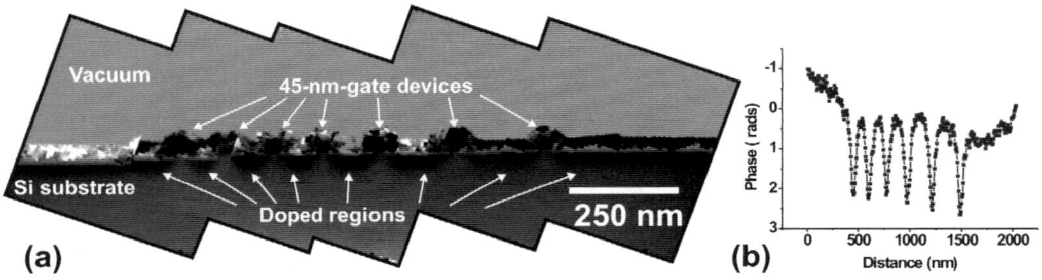

Figure 1. (a) Montage of phase images, revealing the doped regions in a 45-nm-gate CMOS device. (b) Profile averaged across 5 nm taken from across the doped regions.

2. The specimens

Silicon specimens containing differently doped *p-n* junctions were grown for examination using reduced-pressure chemical vapour deposition. Figures 2(a), (b) and (c) show secondary ion mass spectrometry (SIMS) profiles for symmetrically doped *p-n* junctions with concentrations of 2×10^{17} cm^{-3}, 2×10^{18} cm^{-3} and 1×10^{19} cm^{-3} respectively. Parallel-sided specimens were prepared for examination by off-axis electron holography by using an FEI Strata 400 FIB operated at 30 kV. The top surface of the specimen was protected by using a sputtered gold film, before a 1.0 µm thick tungsten layer was deposited directly over the region of interest using Ion Beam Assisted Deposition to protect it from both normal incidence Ga implantation as well as from the edges of the ion beam. The specimens were prepared using conventional FIB milling with lamellas of thickness in the range 200 to 700 nm from each of the differently doped wafers.

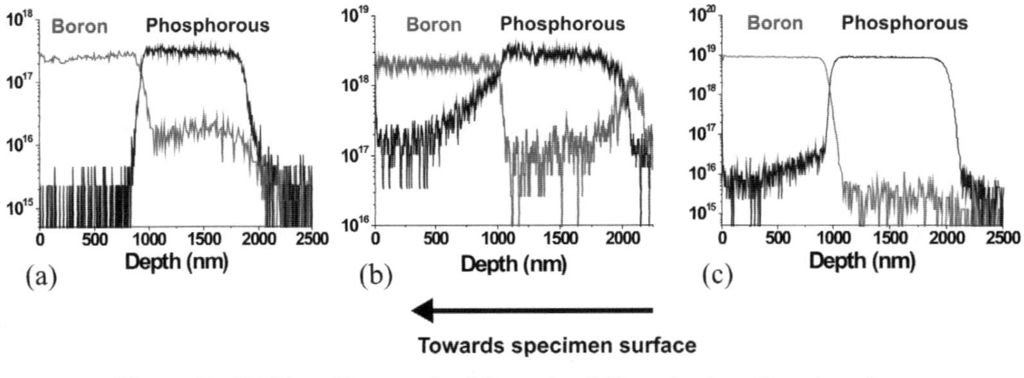

Figure 2. SIMS profiles acquired from the differently doped *p-n* junctions.

3. Results

Electron holograms were acquired using an FEI Titan TEM operated at 200 kV. To extend the field of view a Lorentz lens was used for imaging with the conventional objective lens switched off. Figures 3(a), (b) and (c) show reconstructed phase images of 400 nm thick specimens with different dopant concentrations. What is evident is that for the specimen with the lowest dopant concentration the junction only begins to appear at around 150 nm into bulk, whereas for the higher doped specimens the junction extends almost to the amorphous surface layers. The amorphous layer on these surfaces is approximately 25 nm. This result shown here and elsewhere suggests that the crystalline regions of the specimen have been damaged during FIB-milling and that the dopants in this region are not active

[2]. This is known as the electrically inactive layer and is one of the principle reasons why the measured step in phase across the *p-n* junctions is always much less than expected from theory. This damaged layer will exist on each face of the specimen when examined in transmission. Figures 3(d), (e) and (f) show experimental phase profiles extracted from the 400 nm thick specimens for each of the *p-n* junctions. Also shown are the theoretical phase profiles that would be expected for a perfect specimen with bulk-like properties. The results show that for the lightly-doped specimens, the phase profile measured experimentally is much less than theory, whereas for specimens with higher dopant concentrations the measured approaches the theoretical value. At this stage it appears that the electrically inactive thickness is strongly dependent on the dopant concentration in the specimens.

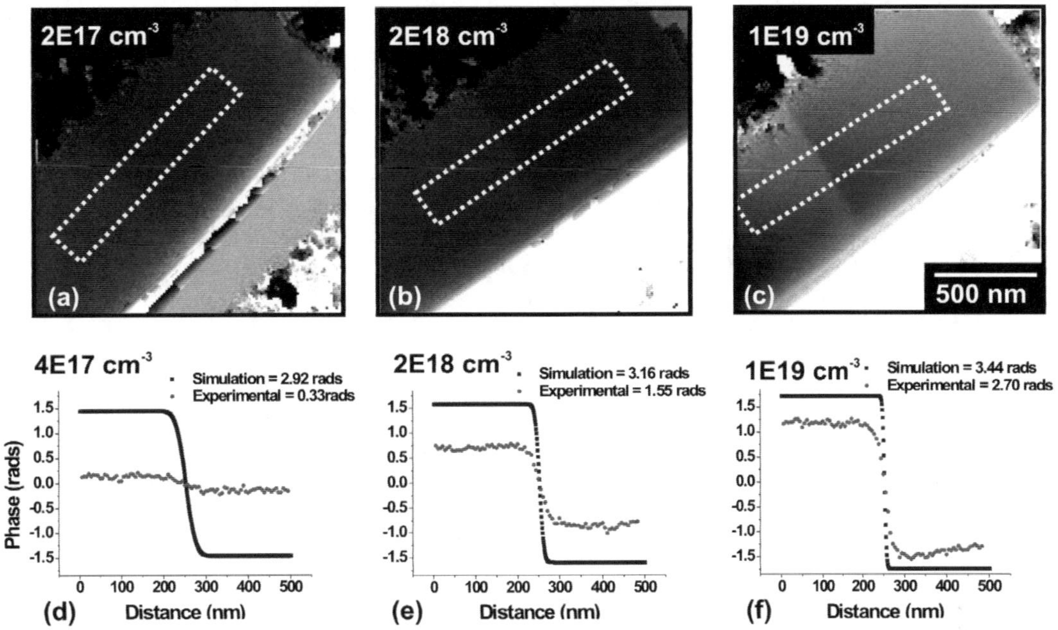

Figure 3. (a), (b) and (c) show phase images of 400 nm thick *p-n* junctions prepared from each of the differently doped wafers. Figure (d), (e) and (f) shows the step in phase measured experimentally from each of the phase images. The profiles were extracted from the regions indicated in (a), (b) and (c). The experimental profiles are compared to the theoretical step in phase for perfect bulk-like specimens.

Figure 4(a) shows the step in phase measured across a series of junctions as a function of the crystalline specimen thickness measured using convergent beam electron diffraction (CBED) for all of the differently doped *p-n* junctions. From the *x*-intercept the electrically inactive thickness for each of the junctions can be measured. It is clear that the electrically inactive thickness here is more than 250 nm (corresponding to 125 nm each face of the specimen) for the junctions with the dopant concentration of 2×10^{17} cm^{-3}, whereas it is around 50 nm (corresponding to 25 nm on each face) for the junctions with a dopant concentration of 1×10^{19} cm^{-3}. From this result, the lower-limit for dopant detection can be determined for FIB-prepared specimens. Figure 4(b) shows the experimentally determined electrically inactive thickness plotted as a function of the dopant concentration in the specimens. Assuming that the effects of surface band bending are only minor contributions to the inactive thickness, the minimum thickness that is required to detect a particular dopant concentration can be determined. For example, it can be seen that to detect a step in phase across a *p-n* junction with a dopant concentration of 1×10^{16} cm^{-3}, a specimen thickness of more than 500 nm would be required. However, in reality it is expected that that the effects of surface band bending will also contribute significantly to the inactive thickness for low dopant concentrations and these effects must also be

Journal of Physics: Conference Series **209** (2010) 012062 doi:10.1088/1742-6596/209/1/012062

accounted for [5]. Nonetheless, these results provide a starting point for the thicknesses of the specimens if low dopant concentrations are present.

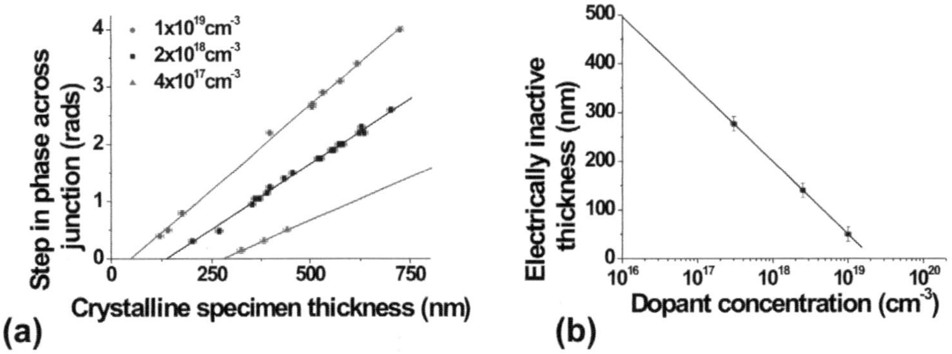

Figure 4. (a) Step in phase measured across the different *p-n* junctions as a function of crystalline specimen thickness measured using CBED. (b) Electrically inactive thickness measured from the *x*-intercept in (a) plotted as a function of the dopant concentration in the specimens.

4. Discussion

It has been shown that FIB milling introduces an electrically inactive thickness into specimens containing *p-n* junctions which is strongly dependent on the dopant concentration. The presence of a variable inactive thickness makes the direct interpretation of the 2D dopant maps complicated. It is known that by annealing the specimens *in situ* at a temperature of only 300 °C, the electrically inactive thickness arising from defects introduced deep in the crystalline regions of the specimens during FIB milling can be removed [3]. However, artefacts such as gallium implantation and from band-bending due to the presence of charged surfaces [4] will remain and must be accounted for if quantitative dopant maps are to be obtained from phase images of FIB prepared specimens.

References

[1] Rau WD, Schwander P, Baumann FH, Hoppner W and Ourmazd A 1999 *Phys. Rev. Lett.* **82**, 2614

[2] Cooper D, Twitchett AC, Somodi PK, Midgley PA, Dunin-Borkowski RE, Farrer I and Ritchie DA 2006 *Appl. Phys. Lett.* **88**, 063510

[3] Cooper D, Hartmann J-M, Rivallin P, Chabli A and Dunin-Borkowski R 2009 submitted to *J. Appl. Phys.*

[4] Cooper D, Ailliot C, Truche R, Barnes J-P, Hartmann J-M and Bertin F 2008 *J. Appl. Phys.* **104**, 064513

[5] Beleggia M, Fazzini P, Merli P and Pozzi G 2003 *Phys. Rev. B* **67**, 045328

Improved accuracy in nano beam electron diffraction

A Béché [1], L Clément [2] and J-L Rouvière [1]

[1] CEA, INAC, SP2M, LEMMA, 17 rue des Martyrs, F-38054 Grenoble Cedex 9
[2] ST Microelectronics, 850 rue Jean Monnet, F-38920 Crolles

E-mail: armand.beche@cea.fr, jean-luc.rouviere@cea.fr

Abstract. Nano beam electron diffraction (NBD or NBED) is applied on a well controlled sample in order to evaluate the limit of the technique to measure strain. Measurements are realised on a 27nm thick $Si_{0.7}Ge_{0.3}$ layer embedded in a silicon matrix, with a TITAN microscope working at 300kV. Using a standard condenser aperture of 50μm, a probe size diameter of 2.7 nm is obtained and a strain accuracy of 6×10^{-4} (mean root square, rms) is achieved. NBED patterns are acquired along a [110] direction and the bidimensionnal strain in the (110) plane is measured. Finite element simulations are carried out to check experimental results and reveal that strain relaxation and probe averaging in a 170nm thick TEM lamella reduces strain by 15%.

1. Introduction

Supported by the microelectronic industry to increase performance of devices [1], perfect control of strain fields at very high spatial resolution is required to characterize the latest generation of transistors [2]. At the present day only transmission electron microscopy (TEM) can get quantitative strain measurement with sufficient spatial resolution and sensitivity. Several TEM techniques are now available, each with their own resolution, sensitivity and experimental drawbacks.

High resolution imaging has a very good spatial resolution (1-2 nm) and a sensitivity of 1×10^{-3}, but its field of view is very narrow (100 × 100 nm^2) and specimens need to be very thin and homogeneous [3, 4]. Convergent Beam Electron Diffraction (CBED) is probably the most accurate TEM technique (sensitivity of 1×10^{-4}) due to information from high indices diffracting planes and has a very fine spatial resolution (2-3 nm) However, the specimen must be tilted slightly off axis (9 to 11°) which leads to structure shadowing and must be rather thick (more than 200nm). Moreover, strains in TEM lamellae relax leading to splitting effects and heavy numerical simulations are then needed to get quantitative strain information [5,6]. A new technique, dark field holography, seems very promising as its spatial resolution is quite good (around 5 nm) combined to a high sensitivity (around 1×10^{-3}) and a large field of view of 250×1000 nm^2 [7]. However, this method will be limited to specimens having a large defect free unstrained region in epitaxy (i.e. same orientation) with the deformed structure. Compared to these techniques, nano beam electron diffraction (NBD or NBED) investigated here, and quite recently applied to strain measurement, presents some advantages [8, 9, 10, 11].

NBED consists of illuminating the sample with a small almost parallel probe to obtain diffraction patterns with quite sharp spots. The spot positions of diffraction patterns from a perfectly known unstrained region are compared to the ones from regions of interest. As diffraction patterns are recorded point by point, a reference picture can be taken anywhere in the sample, or the reference can even be taken on a different sample. Moreover, the complete 2D distortion matrix can be retrieved using main diffraction axis although no particular orientation is required to get strain information.

2. Experimental considerations

Measurements were made using a probe corrected FEI Titan microscope working at 300 kV. The microscope was setup to *μProbe Scan Mode*, probe corrector on with a 50 μm condenser aperture. In these conditions a probe of 2.7 nm in diameter (full width at half maximum) was formed in the image plane. The probe size was evaluated by imaging the probe after its propagation through the [110] Si crystal (fig. 1a and 1b). [110] high resolution lattice fringes can be seen in this probe image. The convergence angle of 0.5 mrad was measured directly in the diffraction pattern (fig. 1c). High angle annular dark field (HAADF) images allowed quickly visualizing the structure and accurately positioning the probe on the sample (fig. 1d). Line scans were realized with the TIA software of the microscope Diffraction patterns were recorded on 2k × 2k Gatan UltraScan cameras using either the Tridiem Gatan image filter (GIF) or not.

A sample composed of a 27 nm silicon-germanium (30% of germanium) layer epitaxially grown on silicon and covered with a 325 nm silicon cap was realized by Reduced Pressure-Chemical Vapour Deposition (RPCVD) [12]. This type of sample, represented in fig. 1d, is perfect for evaluating the NBED technique as the amount of strain, in biaxial condition, is perfectly known from epitaxy and has been measured by different techniques. Moreover, finite element modelling can be easily realized in order to obtain the relaxed structure after thinning. [110] cross-section samples were prepared using an FEI Strata 400 Focus Ion Beam.

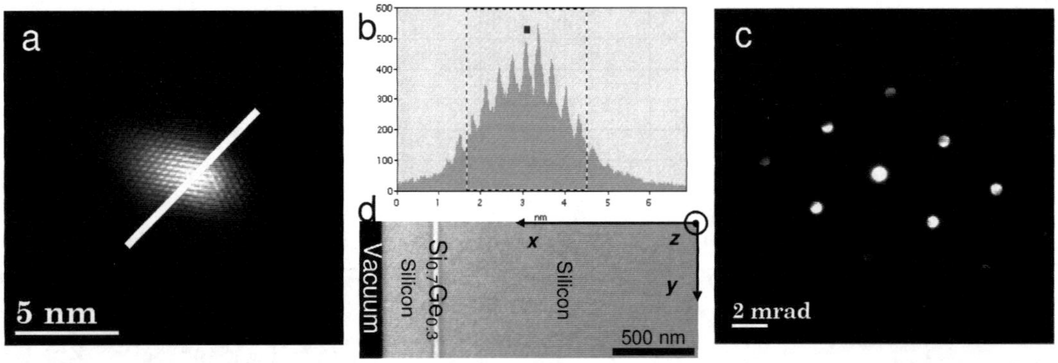

Figure 1. (a). Image of the electron probe in a silicon crystal on [110] zone axis. {111} high resolution fringes are visible within the probe. (b). Profile of the probe as represented by the white line in a. (c). Diffraction pattern in silicon obtained along the [110] axis at 300kV with the same probe as in (a). Semi convergence angle is 0.5 mrad. (d) HAADF image of the sample and definition of geometrical axis.

3. Strain measurement

Diffraction patterns recorded during the experiments were treated by a home written program. The algorithm consists in first searching the maxima positions in the image and improves their positions using a 2D Gaussian fit. This last step allows sub-pixel precision and increases the precision. Two spots are then chosen in the fitted diffraction pattern (presently (004) and (220)) which define two vectors (x and y respectively, in figure 1d). These two vectors define a matrix in reciprocal space which gives access to the complete 2D distortion matrix in the diffraction plane [13, 14].

$$\mathcal{D} = \left(G^t\right)^{-1} G_0^t - Id \qquad \mathcal{E} = \frac{1}{2}(\mathcal{D} + \mathcal{D}^t) \qquad \Omega = \frac{1}{2}(\mathcal{D} - \mathcal{D}^t)$$

Here $\boldsymbol{G_O}$ and $\boldsymbol{G}$ are the matrices composed by the two reciprocal vectors (004) and (220) respectively obtained in the reference and deformed Si crystal, $\mathcal{D}$ is the 2D distortion matrix, $\boldsymbol{\varepsilon}$ and $\boldsymbol{\Omega}$ are respectively the strain and rotation matrices and Id is the identity matrix With this type of algorithm, the strain state is given regarding to the material used as reference.

4. Experimental and numerical results

NBED patterns were acquired along the [110] axis at different points located along a line in the x=[0,0,1] direction. From these patterns strain profiles, such as the one displayed in figures 2 and 3, were extracted. Strain is measured with respect to the unstrained Si crystal. So strain in the SiGe layer is not evaluated with respect to the unstrained equivalent SiGe crystal. With this definition of strain, the strain is concentrated in the silicon germanium layer in the x direction, as expected from epitaxy. Presence of strain in the y direction is attributed to strain relaxation phenomena, typical for the sample thickness considered here (170 nm). In the x direction, sensitivity of the method was investigated by measuring the standard deviation (rms = root mean square) of the measure in the Si crystal far from the SiGe/Si interface. An rms value of 6×10^{-4} is typically found (this is the value calculated for the points taken in the bold part of fig. 2). This value should be compared to the 10^{-3} value generally reported in recent publications [9, 11].

In order to validate nano beam diffraction as a quantitative method for strain measurement, finite element calculations were performed using the Comsol® program [15]. Strain curves obtained numerically were convoluted with a Gaussian probe using the experimental probe FWHM and compared with a biaxial strain state and experimental results as shown in figure 3. There is a very good agreement between the simulated and experimental curves within the SiGe layer. The presence of a reverse strain state in the silicon close to the SiGe layer is also reproduced in the simulation but is less strong than in the experimental measurement. This effect can be due to 3D relaxation effects caused by buckling of the lamella. Further investigation is necessary to validate this hypothesis.

Finite Element simulation allows a comparison of the strain state in the thin lamella and in the bulk wafer. For instance, ε_{xx} is equal to 2.3% in bulk but equal to only 2% in the thin lamella (thickness taken equal to 170nm). Neglecting strain relaxation in the thin lamella would then lead to a relative error of about 15% in strain measurement. Depending on the application, such an error can be admissible.

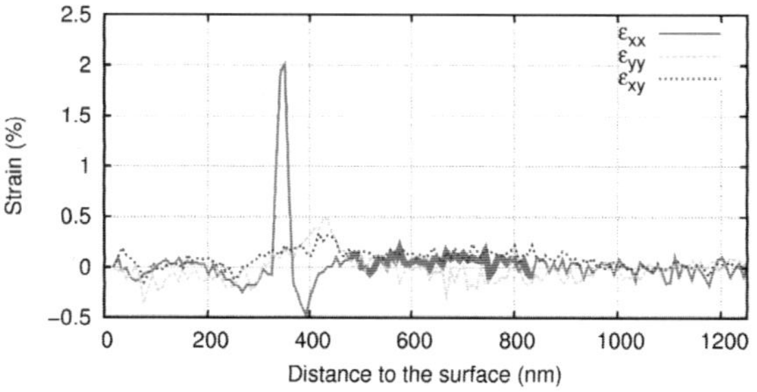

Figure 2. Strain profiles taken across the SiGe layer. The different two dimensional strain components (ε_{xx}, ε_{xy}, ε_{yy}) are drawn.

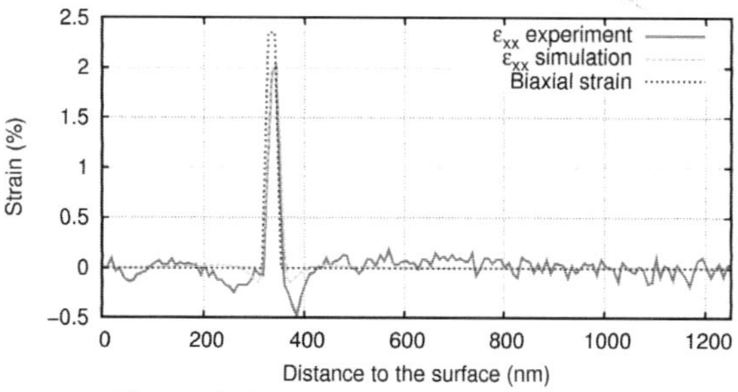

Figure 3. Comparison between experimental, numerical and bulk strain states in x direction.

5. Experimental considerations

Comparison between strain curves extracted from different spots in the diffraction patterns reveals different behaviours as visible in figure 4. If the general shape of strain curves is very similar, small discrepancies appear in the final strain state in the SiGe layer. These differences can be attributed in the present work to double diffraction phenomena which give rise to the (002) and (006) spots.

Energy filtered diffraction patterns exhibit a very high signal to noise ratio and an increased contrast in recorded the image, as effects due to absorption phenomena are removed. However, figure 5 reveals that energy filtered diffraction patterns using the GIF display much noisier strain data than the one recorded without any filtering. This effect is due to image distortions introduced by the imaging system of the GIF which, even though very small, influences the spot positions [16].

6. Conclusion

Nano beam electron diffraction is a very powerful method, easy to use and fast to process. Its quite high sensitivity ($\Delta\varepsilon = 6\times10^{-4}$) and fine resolution (2.7 nm) make it really valuable for quantitative strain measurement at the nanoscale. Simulations have well reproduced the experimental strain profile. In the near future, better lateral resolution and better sensitivity can still be expected by using a smaller condenser aperture and by accurately tuning the probe Cs-corrector of the microscope.

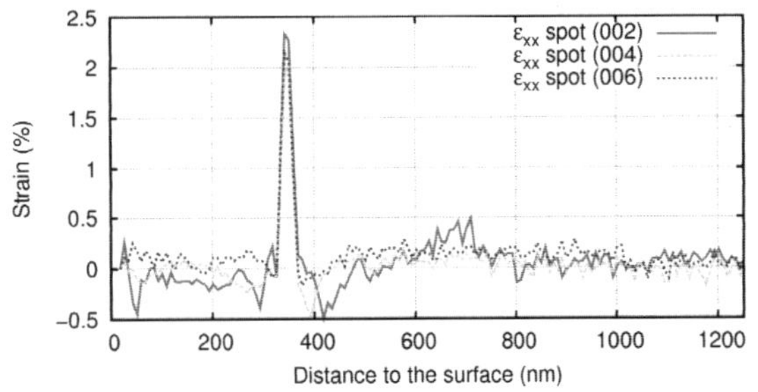

Figure 4. ε_{xx} strain profiles obtained by using different spots: (002), (004) and (006)

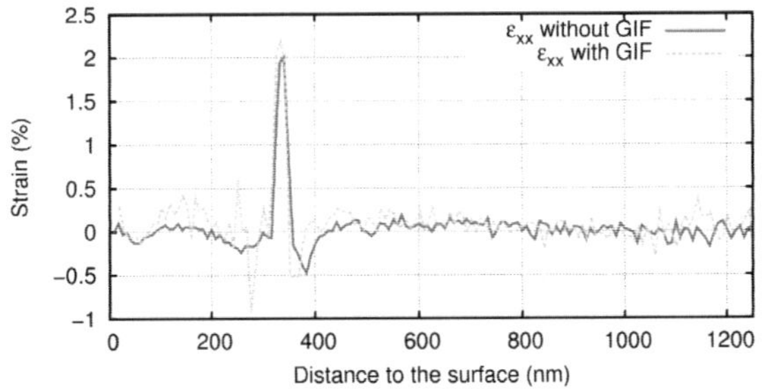

Figure 5. ε_{xx} strain profiles obtained with and without the Gatan image filtering (GIF).

References

[1] Smeys P, Griffin P, Rek Z, De Wolf I and Saraswat K 1999 *IEEE Trans. on Electron Devices* **46**, 1245
[2] Auth C et al. 2008 *Intel Technology Journal* **12**, 77
[3] Bierwolf R, Hohenstein M, Phillipp F, Brandt O, Crook G and Ploog K 1993 *Ultramicroscopy* **49**, 273
[4] Hue F, Hytch M, Bender H, Houdellier F and Claverie A 2008 *Phys. Rev. Lett.* **100**, 156602
[5] Clement L, Pantel R, Kwakman L and Rouviere J 2004 *Appl. Phys. Lett.* **85**, 651
[6] Houdellier F, Roucau C, Clement L, Rouviere J and Casanove M 2006 *Ultramicroscopy* **106**, 951
[7] Hytch M, Houdellier F, Hue F and Snoeck E 2008 *Nature*, **453**, 1086
[8] Usuda K, Mizuno T, Tezuka T, Sugiyama N, Moriyama Y, Nakaharai S and Takagi S 2004 *Applied Surface Science* **224**, 113
[9] Usuda K, Numata T and Takagi S 2005 *Mat. Sci. in Semiconductor Processing* **8**, 155
[10] Armigliato A, Frabboni S and Gazzadi G 2008 *Appl. Phys. Lett.* **93**, 161906
[11] Liu J, Li K, Pandey S, Benistant F, See A, Zhou M, Hsia L, Schampers R and Klenov D 2008 *Appl. Phys. Lett.* **93**, 221912
[12] Hartmann J et al. 2008 *Materials Science and Engineering: B* **154-155**, 76
[13] Rouviere J and Sarigiannidou E 2005 *Ultramicroscopy* **106**, 1
[14] Hÿtch MJ, Snoeck E and Kilaas R 1998 *Ultramicroscopy* **74**, 131
[15] www.comsol.com
[16] Krivanek O, Gubbens A and Dellby N 1991 *Microscopy Microanalysis Microstructures* **2**, 315

Journal of Physics: Conference Series **209** (2010) 012064 doi:10.1088/1742-6596/209/1/012064

Interpretation of electron beam induced charging of oxide layers in a transistor studied using electron holography

F Ubaldi[1], G Pozzi[1], T Kasama[2], MR McCartney[3], SB Newcomb[4] and RE Dunin-Borkowski[2]

[1] Department of Physics, University of Bologna, 40127 Bologna, Italy
[2] Center for Electron Nanoscopy, Technical Univ. of Denmark, Lyngby, Denmark
[3] Department of Physics, Arizona State University, Tempe, AZ 85287-1504, USA
[4] Sonsam Limited, Glebe Laboratories, Newport, Co. Tipperary, Ireland

E-mail: ubaldi@bo.imm.cnr.it

Abstract Off-axis electron holography has been used to characterize a linear array of transistors, which was prepared for examination in cross-sectional geometry in the transmission electron microscope using focused ion beam milling. In reconstructed phase images, regions of silicon oxide that are located between metal contacts show unexpected elliptical phase contours centered several hundreds of nm from the specimen edge. The experimental images are compared with simulations performed using three-dimensional calculations of the electrostatic potential inside and outside the specimen, which take into account the mean inner potential of the specimen and the perturbed vacuum reference wave. The simulations suggest that the oxide layers contain a uniform volume density of positive charge and that the elliptical contours result from the combined effect of the electrostatic potential in the specimen and the external electrostatic fringing field.

1. Introduction

Off-axis electron holography is a specialized electron microscopy technique that relies on the use of an electron biprism to overlap a high-energy electron wave that has passed through an electron-transparent specimen in the transmission electron microscope (TEM) with another part of the same electron wave that passed only through vacuum. The resulting interference fringe pattern can be used to provide a quantitative map of the electrostatic potential within and around the specimen, projected in the electron beam direction. The technique offers the prospect of mapping dopant potentials in semiconductors quantitatively. However, such measurements are often complicated by the effect on the electrostatic potential in the specimen of sample preparation for electron microscopy and both charging and current flow resulting from electron beam irradiation in the TEM. Here we attempt to understand the magnitude, location and distribution of electron beam induced charge in Si oxide layers in a transistor structure by comparing phase images acquired using electron holography with computer simulations of the three-dimensional electrostatic potential within and around the TEM specimen. It is important to study and to avoid such charging effects as they can complicate or preclude measurements of dopant contrast in adjacent regions of the device.

2. Experimental details and results

A Si transistor structure was prepared for TEM examination using focused ion beam (FIB) milling in 'trench' geometry and examined using off-axis electron holography at 200 kV in a Philips CM200-ST field emission gun (FEG) TEM equipped with a Lorentz lens and an electrostatic biprism [1]. Bright-field TEM images of the transistor structure, which contains Si oxide layers located between metal contacts, are shown in Fig. 1 for specimens prepared with nominal thicknesses of 400 and 150 nm. Fig. 2 shows eight-times-amplified phase contours recorded using electron holography from the regions marked '1' in Fig. 1. Unexpectedly, elliptical contours are visible in each oxide region and a fringing field is present outside the specimen edge. Both the elliptical contours and the fringing field are thought to be associated with charging of the oxide as a result of secondary electron emission from the specimen during electron irradiation in the TEM. Fig. 2 also shows similar phase images obtained after coating each specimen on one side with ~20 nm of carbon. The effects of charging now appear to be absent, there is no fringing field outside the specimen edge and the contours in the specimen follow the change in specimen thickness in the oxide. Fig. 3 shows line profiles obtained from the phase images used to form Figs. 2 and 3 along the lines marked '2' in Fig. 1. The dashed and solid lines correspond to results obtained before and after coating the specimen with carbon, respectively, while the dotted lines show differences between these lines. If the charge is assumed to be distributed throughout the thickness of the specimen, then the electric field in the oxide is $\sim 2\times10^{7}$ V m^{-1}, just below the breakdown field for thermal SiO_2 of $\sim 10^{8}$ V m^{-1}. The dopant potential in the source and drain regions of the transistors is always undetectable before carbon coating, whether or not a phase ramp is subtracted from the recorded images.

3. Simulation details and results

The experimental results shown in Fig. 2 were compared with simulated images that incorporated the effects of specimen charging on the three-dimensional electrostatic potential both within and around the specimen. Fig. 4 shows the specimen geometry used for the simulations, which were performed using the commercial software ISE-tCad tools Mesh and DESSIS. The software allows the specimen geometry, dopant concentration and permittivity to be incorporated, to provide numerical solutions for the electrostatic potential in the specimen and the surrounding vacuum by solving Poisson's equation and continuity equations for electrons and holes. The calculated potential was integrated in the electron beam direction to obtain simulated phase images. Periodic boundary conditions were used. A good fit to the experimental results was obtained by modeling the oxide as an insulator with a fixed uniform charge density. The simulations reproduced the elliptical phase contours. However, the true specimen thickness profile had to be incorporated in the simulations to fit the positions of the ellipses accurately, especially for the thicker specimen. Fig. 5 shows best-fitting simulated phase contours, taking into account the known specimen thickness profiles and assuming a mean inner potential of 10 V for Si oxide and a distance of 2 μm to the vacuum reference wave. The simulations suggest that the oxide is charged positively with a uniform volume charge density of $\sim 5\times10^{15}$ cm^{-3}. Fig. 6 shows the separate effects of including the mean inner potential and the perturbed reference wave [2] in the simulations for the 'trench' specimen of nominal thickness 400 nm. The figure shows that both contributions should be taken into account to interpret the experimental images accurately.

4. Conclusions

Elliptical phase contours were observed experimentally at the positions of oxide layers in a transistor structure using electron holography. Simulations of the internal and external electrostatic potential of the TEM specimen, taking into account specimen thickness variations, the influence of the perturbed reference wave and the mean inner potential of the specimen, were used to show that the contours are likely to result from the presence of a uniform positive charge density in the oxide of $\sim 5\times10^{15}$ cm^{-3}. The charging effects appear to be absent when the specimens are coated with carbon. However, further studies are required to fully understand the influence on recorded electron holographic measurements of both charging and current flow in thin TEM specimens.

This work was partially supported by MIUR, FIRB funding RBAU01M97L.

References

[1] Dunin-Borkowski RE, Newcomb SB, Kasama T, McCartney MR, Weyland M and Midgley PA 2005 *Ultramicroscopy* **103**, 67
[2] Matteucci G, Missiroli GF and Pozzi G 1997 *Advances in Imaging and Electron Physics* **99**, 171

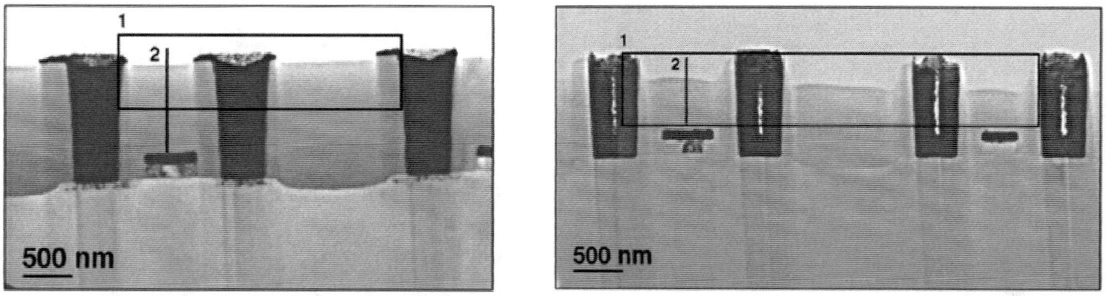

Fig. 1. Bright-field images of PMOS transistors in 'trench-type' FIB-prepared TEM specimens of nominal thickness 400 nm (left) and 150 nm (right).

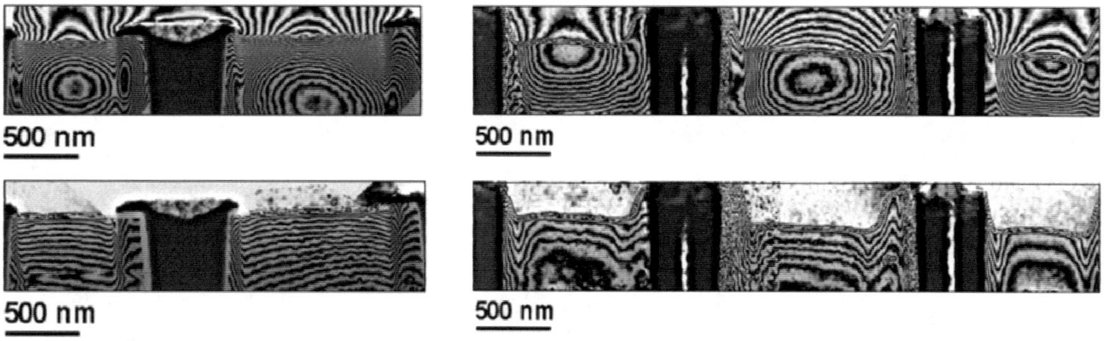

Fig. 2. Eight-times-amplified phase contours measured using electron holography from the regions marked '1' in Fig. 1 for the FIB-prepared TEM specimens of nominal thickness 400 nm (left) and 150 nm (right). Top: before carbon coating. Bottom: after carbon coating.

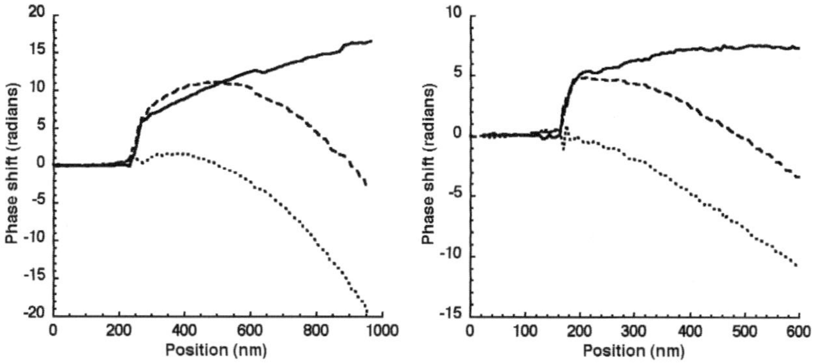

Fig. 3. Phase profiles measured along the lines marked '2' in Fig. 1. The dashed and solid lines were obtained before and after coating with carbon, respectively. The dotted lines show the differences between the solid and dashed lines.

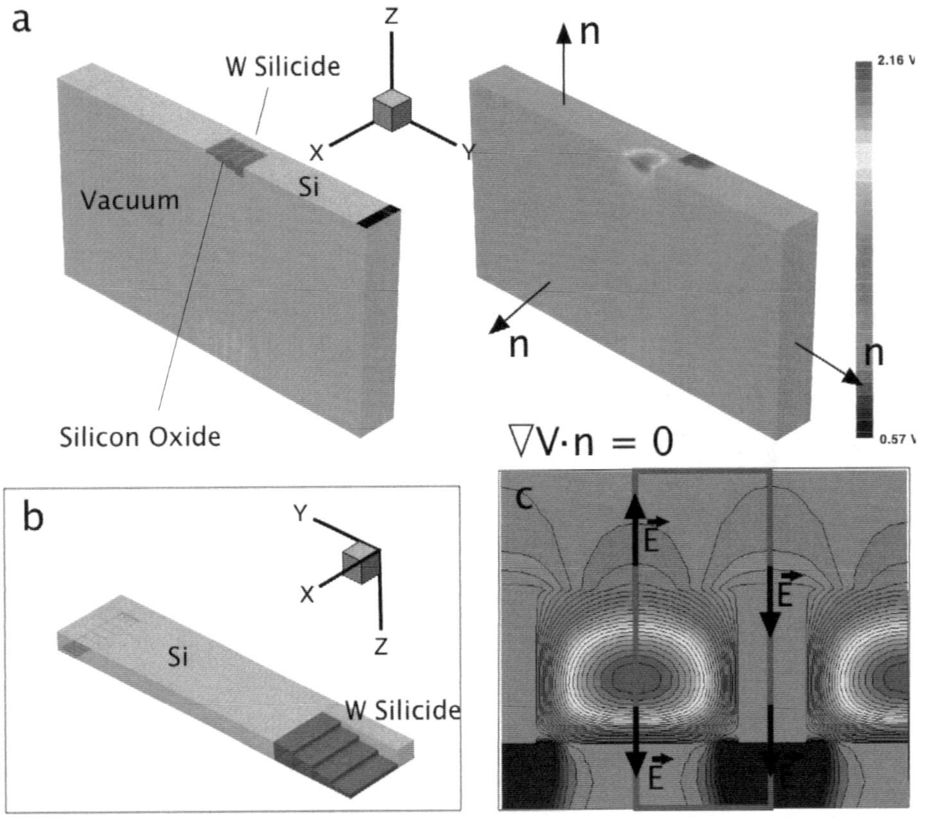

Fig. 4. Simulation geometry used for simulations of electrostatic potentials in TEM specimens containing transistors using the commercial software ISE-tCad tools Mesh and DESSIS, with periodic boundary conditions.

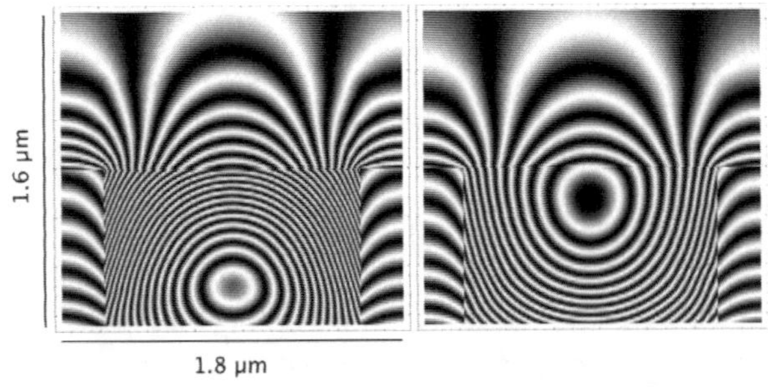

Fig. 5. Simulated phase contours for an oxide thickness that decreases from 400 to ~100 nm at the specimen edge (left) and a constant thickness of 150 nm (right). Best-fitting simulations to the results are for an oxide charge density of 5×10^{15} cm^{-3}. The simulations assume a mean inner potential of 10 V for Si oxide and 2 μm to the vacuum reference wave.

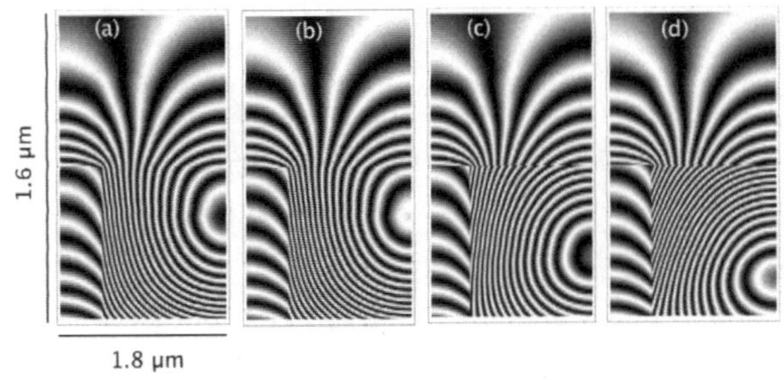

Fig. 6. Simulations for an oxide thickness that decreases from 400 to ~100 nm at the specimen edge: (b) and (d) include the effect of the vacuum reference wave. (c) and (d) include the effect of the mean inner potential. Both contributions to the phase need to be considered for interpretation of the experimental results.

Reliability study on green InGaN/GaN light emitting diodes

ZL Li, PT Lai and HW Choi

Semiconductor Lighting and Display Laboratory, Department of Electrical and Electronic Engineering, University of Hong Kong

Email: hwchoi@hku.hk

Abstract. Although InGaN/GaN green light-emitting diodes (LEDs) are widely available, it is still challenging to grow green LED structures for emission at longer wavelengths due to the difficulty associated with the incorporation of In. The higher concentration of In may also affect the performance and reliability of the device. The reliability of green GaN LEDs with three different indium doping concentrations, with centre-wavelength of 520 nm, 540 nm and 550 nm, is studied in this paper. The electrical properties, including *I-V* characteristics, leakage current and $1/f$ noise were measured. The optical performance of the devices was also evaluated. The devices were subsequently subjected to a 1000 hours continuous stress test. The defect densities of the LED structures were also determined. Our results show that the 520 nm LED, which contains the lowest indium concentration in its quantum wells, produces highest optical output power at 20 mA. It also degrades slower than 540 nm and 550 nm LEDs.

1. Introduction

The development of phosphor-free white light emitting diodes (LEDs) relies on high quality green LEDs at longer wavelengths. Green LEDs are preferred for their high photonic response. LEDs with wavelengths centred at around 555 nm are particularly important for high-brightness, red-green-blue (RGB) based white-lighting systems that coincide with the peak photonic response of the human eye. However, it is challenging to grow InGaN/GaN LED structures for emission at longer wavelengths due to the difficulty associated with the incorporation of indium (In). This difficulty originates from the strain condition in the quantum wells (QWs) which can control the extent of indium incorporation. A higher In content in a QW will lead to a higher compressive strain in the layer, resulting in the difficulty of effective In incorporation [1].

On the other hand, higher indium composition is essential for realizing longer wavelength emission; it also becomes a threat to the device performance and reliability. Firstly, the presence of higher In concentration in multiple QWs (MQWs) can introduce large piezoelectric fields induced by the large strain. This piezoelectric field reduces the overlap of the electron-hole wave functions, and so results in reduced internal quantum efficiency in green LEDs [2]. Secondly, the large lattice mismatch can give rise to high defect concentrations in the structure in the form of dislocations, which results in a degraded material quality. As a result, the higher dislocation densities of green LED introduced by the indium incorporation can often decrease the performance and device reliability.

In this study, we focus on studying the reliability of green LEDs by correlating defect concentrations in devices of different wavelengths ranging from 520 nm to 550 nm with their low frequency noise spectrum and 1000 hour lifetime data. Atomic force microscopy (AFM) surface scans are used to detect the exposed defects after Reactive Ion Etching (RIE) has been conducted to compare the defect densities in different wafers. The electrical properties of the different devices, including *I-V* characteristics, leakage current, 1/*f* noise spectrum were measured. The optical performance of the devices was also evaluated. The devices were subsequently subjected to a 1000 hours continuous stress test. A physical model has been constructed to explain the observed phenomenon.

2. Experimental details

The structure of wafers for LED devices include a GaN buffer layer grown on c-plane sapphire substrate, an n-type GaN layer, the active layers with InGaN/GaN as the quantum well structures, capped by p-type GaN as top contact. The wavelengths of the wafers are 520 nm, 540 nm and 550nm obtained by varying the In composition. Details of the growth will be reported elsewhere. The LEDs were fabricated using a standard micro-fabrication process beginning with the deposition of a semi-transparent current-spreading layer composed of Ni/Au (10nm/10nm) deposited by e-beam evaporation and annealed in an oxygen environment at a temperature of 550 °C for 5 minutes. The 500 × 500 μm^2 emission active area was defined by photolithography and dry etched by RIE using a chemistry comprising CHF_3 and Ar to expose the n-contact regions. Photolithography was carried out to define the *p*-pad and *n*-pad areas. A bi-layer of Ti/Al (50nm/350nm) was deposited as *p*-pads and *n*-pads and annealed in N_2 at 350 °C for 1 min. The chips were diced by laser micro-machining and packaged into TO (transistor outline)-cans. Following wire-bonding, a silicone encapsulant was applied to protect the device.

The electrical properties, including *I-V* characteristics, leakage current and 1/*f* noise spectra were measured. The *I-V* characteristics of the packaged devices were measured with an HP 4156A precision semiconductor parameter analyzer. Optical performance of the devices was also evaluated by an Ocean Optics HR2000 fibre-coupled spectrometer. The 1/*f* noise spectrum was measured by a BTA 9603 FET Noise Analyzer. The devices were subsequently subjected to a 1000 hours continuous stress test. The defect densities of the LED structures were also determined by AFM using a Seiko Instruments Nanopics system, after plasma exposure.

3. Results and discussion

The left part of Fig. 1 shows the measured *I-V* characteristic from the three LEDs. The leakage currents of the devices have been extracted at a reverse bias voltage of –4 V. As the wavelength of the LEDs increases, the leakage current also increases from 0.99 μA in the 520 nm LED to 2.29 μA in the 540 nm LED and 4.45 μA in the 550 nm LED. The slope of the linear region of the *I-V* curve (in the forward bias region) has been extracted to determine the device static resistance. As evident from the *I-V* plots, the slopes increase as the wavelength increases. The extracted static resistances are 23.5 Ω, 33.3 Ω and 52.6 Ω for the 520 nm, 540 nm and 550 nm LEDs, respectively. This consistent trend in leakage current and static resistance can be linked to the increasingly higher defect concentration as the In incorporation increases. The dislocations act as vertical current leakage pathways, while the defects lead to carrier scattering and thus higher resistance. To verify our postulations, additional experiments are conducted to study the carrier properties.

Since low frequency noise is closely related to the presence of impurities and defects in the semiconductors, it can more accurately represent the internal crystal quality than any other electrical parameters. Noise spectra of the three LEDs were measured and are plotted in the right part of Fig. 1. The system floor calibration was conducted for DC bias calculation and 1/*f* noise correction. The range of frequencies was between 3.15 Hz and 105 Hz. The slope can be extracted by the supplied software automatically. From this figure, both the magnitude and slope for the 520 nm LED are lower than for the 540 nm LED and 550 nm LED.

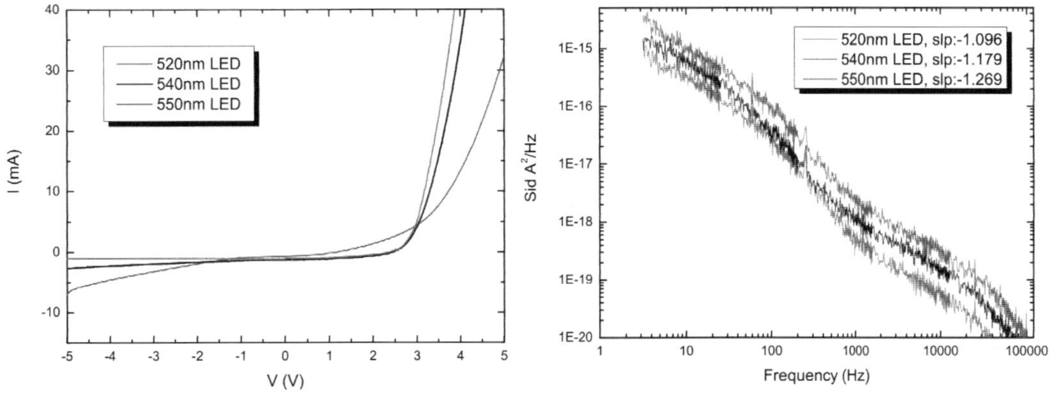

Figure 1. Left: Current-voltage (*I-V*) characteristics and leakage currents of LEDs with different wavelengths. Right: Noise spectrum of LEDs with different wavelengths

In an LED structure, the current spectral density can be expressed by the following equation [3]:

$$S_I(f) = \frac{\alpha q I_D}{f\tau} + \left(\frac{q n_1 A}{I_0}\right)^2 \cdot B \cdot \frac{kTN_t}{\ln(\tau_2/\tau_1)} \cdot \frac{1}{f} \cdot I_R^2 \tag{1}$$

$$\text{where } n_1 = n_i \exp\left(\frac{E_t - E_i}{kT}\right) \tag{2}$$

The first term of equation (1) represents the near-equilibrium steady state conditions whereby there is no net accumulation of charge and the generation rate is equal to the combination rate. α is the Hooge parameter [4], which depends on the quality of the crystal [5], and on the scattering mechanisms. A larger α value generally implies poorer crystal quality, and increases the magnitude of the noise spectrum. I_D is the diode current through the device, τ is the lifetime of the minority carrier lifetime, N_t is the trap density, which can increase the magnitude of the noise spectrum when the trap density increases, while the other terms have their usual meanings. The measured $1/f$ noise spectrum thus indicates that the crystal quality deteriorates as the emission wavelength increases from 520 nm to 550 nm, supporting the theory of enhanced defect densities with increasing In incorporation. A further observation from the slopes of the noise spectra also supports this theory. Moreover, according to the Number Fluctuation Model [6], there is a variation in the generation-recombination rate. The occupation of traps must be taken into account which also affects the total noise spectral density. When the defect density in the InGaN/GaN QW increases, the transitions between traps become more frequent, leading to a drop in the electron mean free time and an increase in the electron number fluctuations. In the noise spectrum, it corresponds to an increase in the slope, consistent with our measured data.

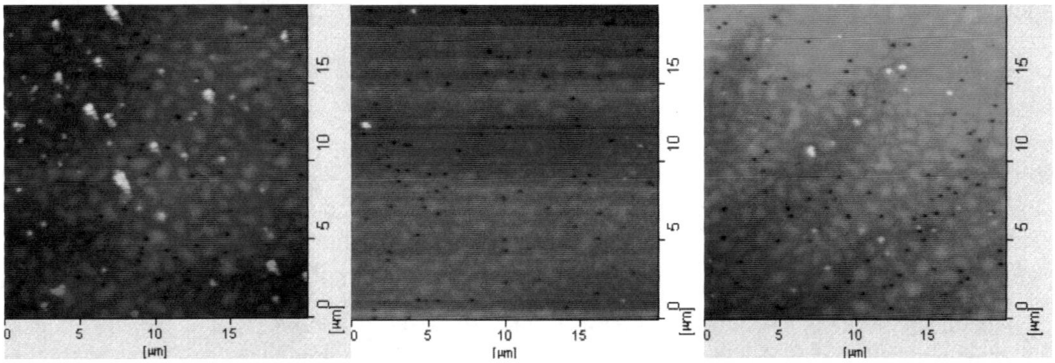

Figure 2. AFM surface scans of the 520 nm, 540 nm and 550 nm wafers (left to right)

The dislocation densities of the wafers are estimated by obtaining AFM scans of plasma-exposed samples. The LED wafers were exposed to plasma in an RIE system to remove about 50 nm *p* type GaN, thus exposing the defects. This is a valid and convenient method of defect estimation since defects propagate through the structure. The densities of etch pits at the surface accurately represent the densities of defects generated underneath [7]. In the AFM surface scans shown in Figure 2, the dark dots are the etch pits propagating from the defects generated at the GaN/sapphire interface due to lattice mismatch, and also from defects generated within the MQWs due to strain. By counting the number of the dark dots within regions of identical areas, a comparative defect density can be obtained. It is estimated that the defect density of the 520 nm, 540 nm and 550 nm wafers are about 1.4×10^7/ cm^2, 1.75×10^7 / cm^2 and 2.22×10^7 /cm^2 respectively. These numbers tie in well with the prediction that increased In incorporation leads to increased defect densities, correlating well with the electrical characteristics presented earlier.

The results can also be correlated to the lifetime of the devices. Figure 3 shows the results of a 1000-hours reliability test, whereby a constant bias current of 20 mA was applied to the various devices. While the 520 nm LED maintains 95% of its initial optical power after 1000 hrs, the number drops to 91% and 82% for the 540 nm and 550 nm LEDs, respectively. Our results show that green LEDs with increasing wavelengths are expected to have increasingly short lifetimes. At the same time, the lifetime of LEDs can be predicted from electrical parameters, particularly from the noise spectrum.

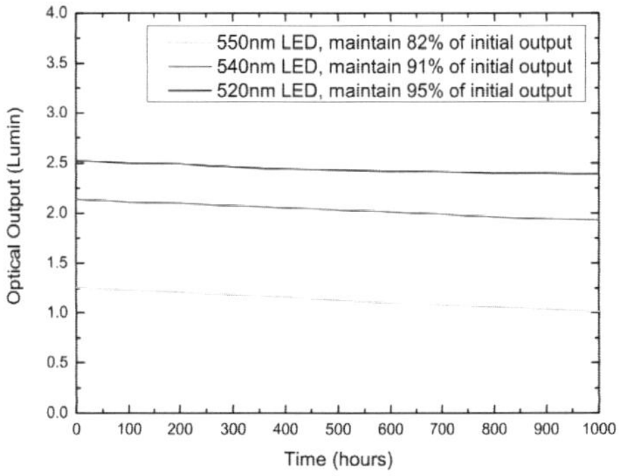

Figure 3. 1000-hours continuous lifetime test

4. Conclusions

In summary, the role of indium incorporation in green InGaN/GaN LED has been studied. While higher indium concentrations in the quantum wells produce LEDs with longer wavelengths (particularly wavelengths around 550nm with high photonic response phosphor-free white LEDs), defects are introduced into the devices. These defects will increase the leakage current, the static resistance and the low frequency noise level, degrading the lifetime of the devices.

This work was jointly supported by a CERG grant of the Research Grant Council of Hong Kong (project HKU 7121/06E) and the University Development Fund (Nanotechnology Research Institute, 00600009) of the University of Hong Kong.

References

[1] Huang CF, Tang TY, Huang JJ et al. 2006 *Appl. Phys. Lett.* **89**, 051913
[2] Dupuis RD, Limb JB, Liu JP et al. 2008 *Proc. of SPIE* **6894**, D8941
[3] Hu J, Du L, Zhang YQ et al 2006 *Acta Physica Sinica* **55**, 1384
[4] Hooge FN 1994 *IEEE Trans. Electron Devices* **41**, 1926
[5] Balandin A, Wang KL, Svizhenko A et al. 1999 *IEEE Trans. Electron Devices* **46**, 1240
[6] McWhorter AL 1957 *Semiconductor interface physics* (Philadelphia: Univ. of Philadelphia Press)
[7] Choi HW, Liu C, Cheong MG, Zhang J and Chua SJ 2005 *Applied Physics A* **80**, 405

Temperature dependence of Ga-assisted oxide desorption on GaAs(001)

F Bastiman[1], R Hogg[1], M Skolnick[2], AG Cullis[1] and M Hopkinson[1]

[1] The University of Sheffield, Department of Electronic and Electrical Engineering, Mappin Street, S1 3JD, UK
[2] The University of Sheffield, The Department of Physics and Astronomy, Hicks Building, Hounsfield Road, Sheffield, S3 7RH, UK

E-mail: f.bastiman@shef.ac.uk

Abstract. Removal of native oxide by conventional thermal desorption causes large scale degradation of the initial surface layer. Ga assisted oxide removal at 440-500°C has been shown to reduce damage and so-called μpit formation. We examine surfaces prepared by this technique over the 400-580°C range with dynamic reflection high-energy electron diffraction (RHEED) and in situ scanning tunnelling microscopy (STM) analysis. Surfaces with oxide clusters, pure GaAs and Ga droplets are found to exist sequentially for sub-monolayer increases in Ga deposition. The STM imaging shows that two distinct, temperature dependent cleaned surfaces exist within the prescribed range. Ga mobility is believed to underpin the mechanisms that produce each surface whilst step-edge density defines the resulting pit density.

1. Introduction

The surface of a GaAs wafer is covered in a native oxide layer. The removal of this oxide layer is an important cleaning step in ensuring the quality of the subsequent buffer layer and abruptness of epitaxial interfaces. Typically, conventional oxide desorption is performed under an As flux to stabilise the surface. Figure 1a shows the extent of the damage with dense arrays of coalescing μpits up to 75nm in diameter and 5-10 nm deep. Figure 1b shows the same surface after 10nm of growth. Whilst the areas around the pits display good (2×4) dimer rows, there are regions with either high step density or deep pits. The high step density is attributed to shallow pits, where material has been removed up to 4 monolayers (ML) deep at step edges, which have subsequently grown out. The μpits are areas where more material was removed and hence are difficult to over grow. It is these μpits that perpetuate throughout buffer layer growth and eventually are thought to lead to large-scale mound formation [1, 2].

Pits formed due to mass transport out of the bulk GaAs [3]. In order to sublime, non-volatile Ga_2O_3 requires Ga in order to generate volatile Ga_2O. The reaction with bulk GaAs is:

$$Ga_2O_3 + 4GaAs_{bulk} \rightarrow 3Ga_2O\uparrow + 2As\uparrow \quad (1)$$

The mass transport of bulk Ga out of the crystal to accommodate oxide desorption is substantial and creates surface-wide damage up to 40 ML (~11nm) deep. However it has been shown recently that externally supplying Ga from a standard effusion cell can reduce the requirement for this mass transport [3-5]. The reaction is characterised as:

$$Ga_2O_3 + 4Ga_{flux} \rightarrow 3Ga_2O \uparrow \quad (2)$$

Whilst the authors reported a reduction of oxide in all cases, ex situ AFM was employed to characterise the surface, inevitably exposing the wafer to substantial oxide regeneration. Furthermore, all work was carried out between 440 and 500°C, which represents a narrow temperature band in the viable range from 400-580°C. In this work in situ STM was performed after Ga-assisted oxide desorption over a range of temperatures. RHEED analysis during clean up indicates there is a strong temperature dependence of the clean-up process.

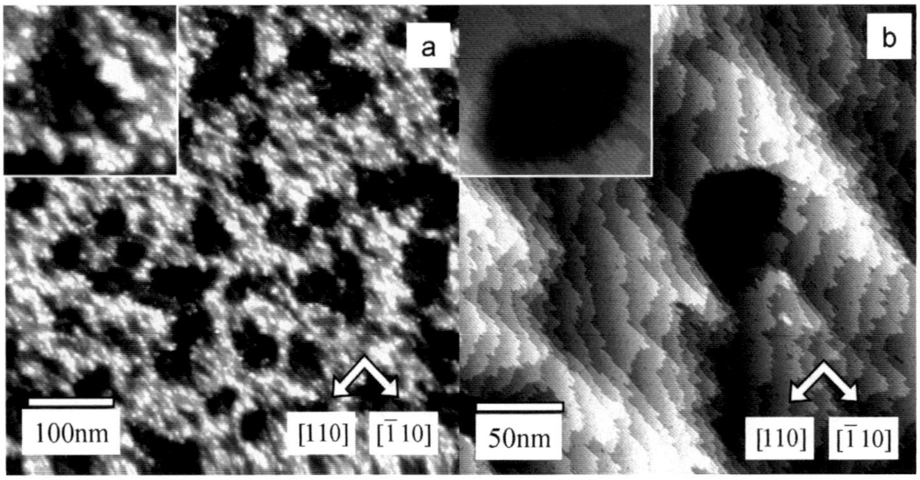

Figure 1: STM images of GaAs(001) surface, a) at clean up with dense pitting formation across the sample. Inset shows high magnification image of a pit, b) after 10nm of growth and a 2 hours anneal at 580°C. Inset shows high magnification of central feature: a 50×60 nm^2 μpit with dimer rows across its floor.

2. Experimental

This work was performed on a commercially available Omicron molecular beam scanning tunnelling microscope (MBSTM) system. The system comprises an interconnected conventional molecular beam epitaxy (MBE) chamber and a variable temperature scanning tunnelling microscope (VT-STM). The former chamber is fitted with standard Ga and As_4 effusion sources and RHEED for dynamic surface analysis. The latter chamber maintains an ultra high vacuum (UHV) pressure of 2.0×10^{-11} mBar to ensure low contamination during post growth imaging. More details of the equipment are published elsewhere [6].

An undoped 2° off-cut GaAs(001) epi-ready wafer was cleaved into 10 × 3.9 mm^2 chips in order to accommodate sample mounting. Heating was achieved through a heating plate, whereby current was passed through a piece of pyrolytic boron nitride (PBN) in close proximity to the sample. The plate has been shown to provide uniform heating when observed by a thermography camera.

With no further ex situ processing, the sample was out-gassed at 300°C for 30 minutes and 400°C for a further 30 minutes. At the end of the outgas stage the chamber background pressure returned to a system typical 1.0×10^{-10} mBar. Sample temperatures between 390 and 520°C were then maintained whilst the sample was irradiated with a 0.05ML/s Ga flux for 0-180s. The surface was monitored by RHEED throughout.

On cessation of the Ga flux, samples were annealed for 10 minutes at the preparation temperature before termination of the heating current and a rapid transfer to the STM chamber. Working in the absence of As_4 allows a background pressure of 1.0×10^{-10} mBar throughout and negates the necessity for complicated quenching procedures [7].

STM tips were prepared with a 2 step process. Firstly, high purity <111> 0.4mm W wire was electro-chemically etched in a 3M NaOH solution. Tips that showed short shanks and sharp apex under light optical microscopy were then placed in a custom built heater plate and inserted into the vacuum system. Six 10s heating cycles were then applied to the apex at 900°C, subliming the oxide.

Scanning into small amounts of residual oxide induces noise distortion into the standard filled state (negative bias) STM images. More stable imaging into the oxide was achieved with empty state imaging. A selection of scan areas from 1500×1500 nm^2 to 250×250 nm^2 was utilised to obtain a broad spectrum of detailed images.

3. Results and Discussion

Four temperature dependent regimes for Ga-assisted oxide desorption were identified in this work. The first (<400°C) and the last (>580°C) regimes proved detrimental to the process. Below 400°C irradiation by Ga flux produces Ga droplets on the sample surface, and hence this range is unsuitable for oxide removal [3]. STM imaging of the sample is hindered by the interaction of the droplets with the tip, resulting in interference in the form of a saw-toothed waveform. Above 580°C oxide thermally desorbs by the conventional mechanism, and hence attempts to utilise Ga assisted oxide desorption are circumvented. The remaining two ranges both prove suitable for the technique, however changes in Ga mobility seem to underpin the surface dynamics in each case.

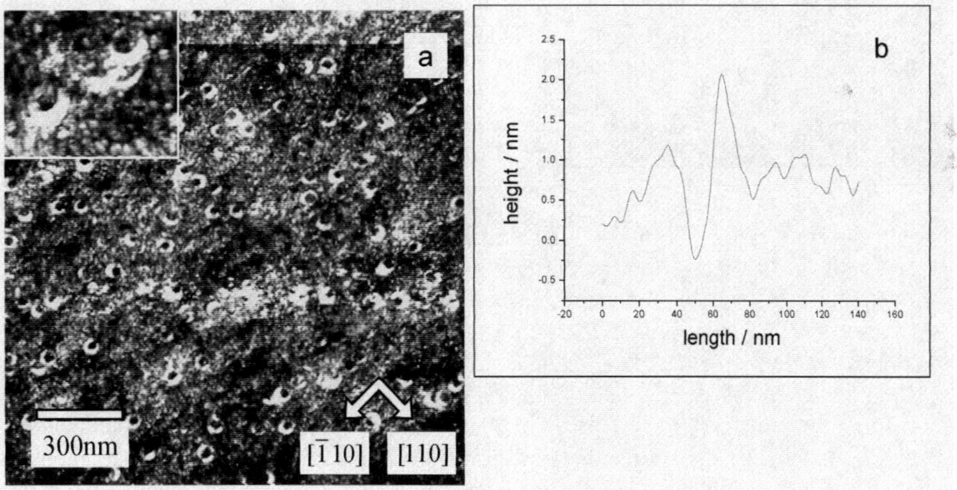

Figure 2. a) STM image of GaAs(001) surface after 8ML of Ga at 460°C. Small pits have formed across the surface up to 2nm deep and 30nm diameter; most have a Ga droplet ring, giving them a volcano shape. The inset shows a high magnification of a volcano-pit, b) height profile across the [110] direction of a pit showing the asymmetry in the volcano lip and a height of 1.5nm.

For the temperature range 400-500°C, the RHEED pattern evolves from a broken 1× to a 2× along [110] after 7.75ML of Ga deposition. STM imaging after quenching at this point reveals a small amount of surface oxide through empty state imaging along with small, circular shaped pits (not shown). An additional 0.25ML of Ga leads to a partial removal of the oxide and the formation of small rings around the circular pits (Figure 2a). The surface is covered in small pits up to 2nm deep and 30nm diameter, however most have a Ga droplet ring around their perimeter, giving them a volcano like structure. The inset of the figure shows a high magnification image of the volcano-pit, showing it is asymmetrical. A cross-section of a pit in the [110] direction is given in Figure 2b; the lip increases the apparent depth, as it stands some 1.5nm above the surface normal. The volcano-pits can be planarized by high temperature annealing under As, however this leads to an increase in their area and depth profile.

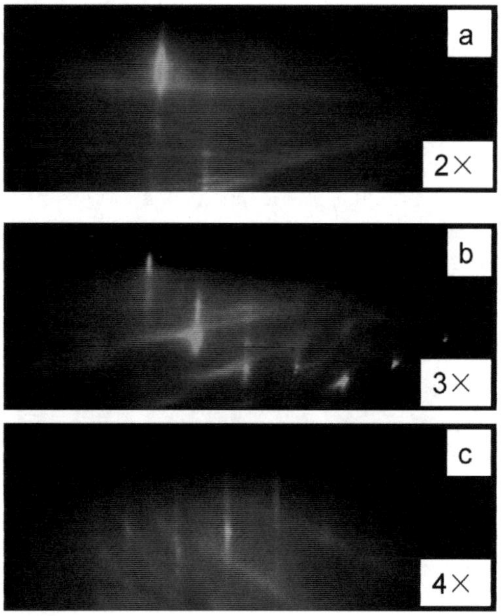

Figure 3. RHEED pattern evolution along [110] azimuth for Ga deposition at 520°C,
a) 7.5ML 2×, b) 7.75ML 3×, c) 8.0ML 4×

For the temperature range 500-580°C, the RHEED pattern is much more sensitive at the clean up point. After 7.5ML of Ga deposition the [110] azimuth reveals a strong 2× pattern (Figure 3a) which corresponds to an As-rich (2×4) reconstruction on the sample surface. The presence of As means that all the deposited Ga has contributed to the oxide removal, and hence none remains for Ga-rich reconstruction formation. After 7.75ML of deposition the RHEED pattern reveals a 3× reconstruction on the [110] azimuth, corresponding to a Ga-rich (3×*n*) reconstruction, where *n* is either 1, 4 or 6 but is unclear from the azimuth. STM imaging of the sample at this stage revealed there is still residual oxide, meaning that whilst Ga is involved in desorbing oxide, the distance between the patches is greater than the average diffusion length and hence not all Ga deposited reaches the oxide. After 8.0ML of deposition the [110] azimuth exhibits a strong 4× pattern, corresponding to Ga-rich (4×2), as expected from depositing 0.25ML of Ga on (3×1) [8]. Very low deposition rates are therefore necessary to prevent over supply of Ga. Occasionally the (4×2) reverts to (3×1) after a short anneal, meaning that excess Ga stored in the reconstruction can still contribute to oxide desorption, and hence deplete the Ga-rich surface. Presence of (4×2) or reversion to (3×1) corresponds to optimal oxide removal with this technique. Further increase in Ga deposition leads to droplet formation on the surface.

Figure 4a shows a surface after 8.0ML of Ga deposition at 520°C. The surface displays anisotropic pits aligned along the [110] direction. The alignment is liable to result from preferred Ga binding along step edges [9]. The pits are 50-60nm long and 20-30nm across with 1-1.5nm depth profiles.

None of the pits imaged showed the volcano structure corresponding to Ga droplet formation. The depth profile in Figure 4b along the [110] direction shows the pit floor has sub-ML height differences possibly due to localised reconstruction formation.

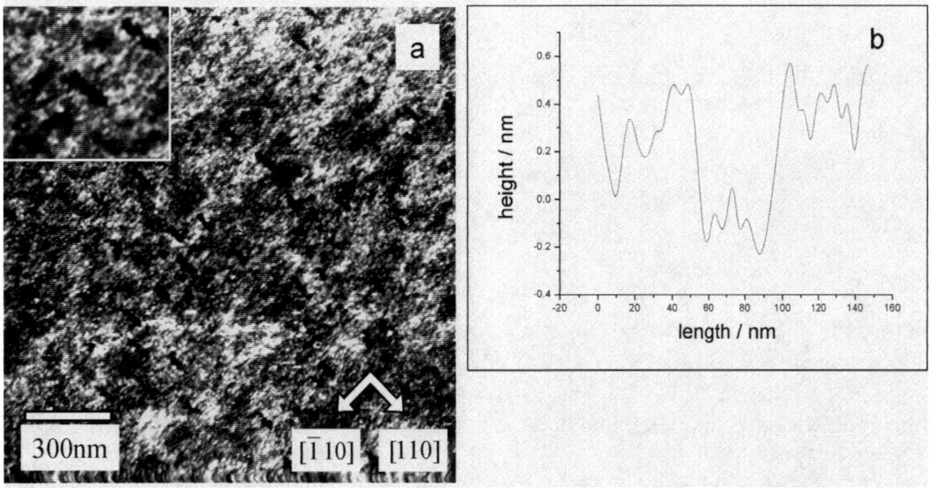

Figure 4. a) STM image of GaAs(001) surface after 8ML of Ga at 520°C. Small anisotropic pits have formed across the surface from 1.0-1.5nm deep and with area 50-60 × 20-30nm^2, b) height profile in the [110] direction showing reconstruction on the pit floor.

All the pits formed after Ga deposition are of a similar density and appear to be linked to the step-edge distribution, which has ~8.5nm spacing for a 2° off-cut sample. Principally, the small pits could form under the exact same mechanism as the larger μpits. Above 400°C Ga_2O thermally desorbs from the surface, and hence there is a small fraction of the bulk GaAs that is consumed to contribute to this thermally desorbed oxide. However since most of the Ga_2O_3 breaks down from externally supplied Ga the pits are smaller and less frequent. It is also possible that these small pits are a consequence of ex situ sample ageing, resulting in a small amount of Ga_2O_3 growth on the surface over time [10].

The difference in the pit shape at the two temperatures can be described in terms of Ga mobility and surface reconstruction formation. The pits represent areas where the oxide has been removed rapidly due to thermal desorption, hence they are areas where surface GaAs is exposed amidst residual oxide. At low temperatures <500°C the mobility of Ga is sufficiently small that deposition on these pitted areas leads to droplet formation, as the excess Ga neither contributes of reconstruction change (simply not enough GaAs is exposed to accommodate the deposited Ga) nor oxide removal (the oxide is further away than the mean diffusion length). Under this mechanism not all the oxide can be removed before droplet formation. However above 500°C, the Ga deposited on these regions can migrate to nearby oxide, reducing the percentage coverage. Excess Ga is temporarily stored in Ga-rich (4×2) before migrating to nearby oxide. Under this mechanism all the oxide is removed before droplet formation.

4. Conclusion

The temperature dependence of Ga assisted oxide removal on GaAs(001) epi-ready wafers has been investigated by RHEED and in situ STM. Two key temperatures have been identified, leading to two distinct shallow pit shapes. Variation of Ga mobility has been invoked to explain the differences between samples prepared in the two temperatures ranges. It has been shown that performing Ga assisted oxide removal above 500°C eliminates nano-sized Ga droplet formation and significantly reduces surface damage induced by conventional oxide removal under an As flux.

Acknowledgements
The authors would like to thank R. Frith for his support and encouragement through the duration of the project and J. Lin for initial co-operation and guidance in this area.

References

[1] Ballestad A, Ruck BJ, Adamcyk M, Pinnington T and Tiedje T 2001 *Phys. Rev. Lett.* **86**, 2377
[2] Coluci VR, Cotta MA, Mendonca CAC,. Landers KMI and de Carvalho MMG 1998 *Phys. Rev. B* **58**, 1947
[3] Asaoka Y 2003 *J. Cryst. Growth* **251**, 40
[4] Lee JH, Wang ZM and Salamo GJ 2006 *Appl. Phys. Lett.* **88**, 252108
[5] Wasilewski ZR, Baribeau JM, Beaulieu M, Wu X and Sproule GI 2004 *J. Vac. Sci. Technol. B* **22**, 1534
[6] Bastiman F, Cullis AG, Hopkinson M and Green M 2008 *Springer Proc. in Physics* (Microscopy of Semiconducting Materials 2007, eds. AG Cullis and PA Midgley) **120**, 471
[7] Yang H, Labella VP, Bullock DW and Thibado PM 1999 *J. Vac. Sci.Techol. B* **17**, 1778
[8] Silveira JP and Briones F 1999 *J. Cryst. Grow*th **201-202**, 113
[9] Atkinson P, Kiravittaya S, Benyoucef M, Rastelli A and Schmidt OG 2008 *Appl. Phys. Lett.* **93**, 101908
[10] Allwood DA, Cox S, Mason NJ, Palmer R, Young R and Walker PJ 2002 *Thin Solid Films* **412**, 76

The impact of ScO_xN_y interlayers on unintentional doping and threading dislocations in GaN

T Zhu, MA Moram, DV Sridhara Rao, H Li, MJ Kappers and RA Oliver

Department of Materials Science and Metallurgy, Pembroke Street, Cambridge, CB2 3QZ, UK

E-mail: tz234@cam.ac.uk

Abstract. To reduce the threading dislocation density in (0001) GaN grown on c-plane sapphire, a series of samples have been grown using scandium oxynitride (ScO_xN_y) interlayers (ILs) on AlN-on-sapphire templates. Scanning capacitance microscopy (SCM) has been employed to investigate the unintentional doping in GaN with varying ScO_xN_y IL thicknesses. The use of ScO_xN_y ILs decreases the threading dislocation density. An unintentionally n-doped layer has been identified by SCM close to the GaN/ScO_xN_y interface. The average width of this conductive layer has been quantified and found to increase as the ScO_xN_y IL thickness increases up to 13 nm.

1. Introduction

Gallium nitride epilayers grown on c-plane sapphire often exhibit unintentional n-type conductivity with carrier densities in excess of 10^{16} cm^{-3}. Depth profiling Hall measurements suggested that this conductivity arises from a region adjacent to the GaN/sapphire interface [1]. Oxygen impurities have been found to be responsible for this conductivity [2], which may originate either from incorporation of impurities from precursor gases during growth [3] or diffusion from the sapphire substrate [4]. For GaN-based high power, high frequency electronic devices, and for optoelectronic devices whose functionality depends on the location and concentration of carriers, we need to both reduce the threading dislocation density and control or eliminate unintentional doping.

Identification and quantification of unintentionally doped regions in GaN epilayers has been recognized as an effective application of scanning capacitance microscopy (SCM) [3], as it measures carrier concentration and type separately. In this report we present an investigation employing SCM to examine unintentional doping in GaN samples in which scandium oxynitride (ScO_xN_y) interlayers (ILs) have been used to reduce the threading dislocation density (TDD). Previous studies have demonstrated the effectiveness of ScN_z ILs in TDD reduction. TDDs as low as 3×10^7 cm^{-2} have been achieved [5]. In that case, faceted voids are generated in the GaN/sapphire template by GaN decomposition during the Sc IL nitridation and the TDD reduction mechanism may have involved dislocation bending during regrowth of the voided region, as well as dislocation blocking at the IL [6]. Here, by using AlN/sapphire templates and a partially oxidised Sc IL, we aimed to prevent void formation and investigate whether TDD reduction is still as effective. The effect of ScO_xN_y ILs on both TDD and unintentional doping are described in this paper.

2. Experimental

Thin layers (5-25 nm) of Sc_2O_3 were deposited by magnetron sputtering of a Sc target in an oxygen/argon atmosphere on 1 µm AlN-on-sapphire templates grown by metalorganic vapour phase epitaxy (MOVPE). (The nominal Sc_2O_3 thickness was determined through interpolation of the deposition rate [7]). Then these layers were annealed in a 6×2 in. Thomas Swan close-coupled showerhead MOVPE reactor using 3 slm NH_3 and 17 slm H_2 at 200 Torr and 1070 °C for 30 minutes to form stable metal oxynitride. Detailed experimental procedures are given elsewhere [8]. Subsequent GaN was grown on the annealed ScO_xN_y covered AlN-on-sapphire template. Three-dimensional island growth was promoted by using a low V/III ratio, and then these islands were coalesced using an increased V/III ratio and temperature. In-situ SiH_4 treatment at 860 °C was then carried out in order to highlight threading dislocation pits on the surface [9]; TDDs were then measured using an atomic force microscopy (AFM) in intermittent contact mode.

SCM measurements were carried out in cross section using a Veeco Dimension 3100 AFM mounted with a commercial SCM module. Cross-sectional sample preparation was achieved by cleaving, following the approach described in reference [10]. SCM was operated at a bias frequency of 90 kHz in the open loop mode and the AC voltages applied were in the range of 0-20 V peak to peak. PtIr-coated etched silicon tips with a nominal tip radius of 20 nm and a metallic layer thickness of 20 nm were used throughout this investigation. Topographic and SCM data have been obtained simultaneously during scanning across the sample's cross-section. The fast scan direction was perpendicular to the AlN/sapphire interface. An automated routine was then carried out to extract the width of the unintentionally doped region in SCM dC/dV phase images as described in reference [2]. For each sample, at least 10 images with 256×256 pixels were taken, and the average width of the conductive layer was obtained. Image processing and data analysis were carried out using WSxM freeware [11].

3. Results and discussion

The threading dislocation density data for samples with initial Sc_2O_3 IL thicknesses from 0-20 nm are shown in Fig. 1. By using Sc_2O_3 ILs, TDD has been reduced from 9×10^8 cm^{-2} to 1.8×10^8 cm^{-2}, showing the effectiveness of the Sc_2O_3 ILs in TDD reduction. The inset graph in Fig. 1 shows that TDD reduction becomes less significant as IL thickness is increased from 5 to 20 nm.

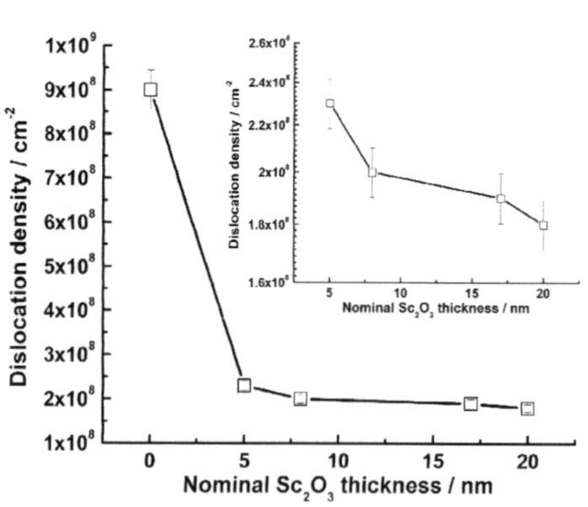

Fig. 1. TDDs measured by AFM as a function of initial Sc_2O_3 layer thickness.

Figure 2 shows montages of SCM phase images of samples with nominal Sc_2O_3 IL thickness from 5-20 nm. All the images are displayed at the same scale. A dark layer with variable thickness is observed for all samples, which suggests that the GaN immediately above the ScO_xN_y interlayer is n-type. (The position of the interlayer was identified from cleavage steps in topographic images, which are not shown here). The AlN and sapphire below and the GaN above the dark region appear grey and noisy, indicating that the carrier density in these regions is below the detection limit of the system.

In Fig. 3, the average width of the conductive layer is plotted against the initial Sc_2O_3 IL thickness. In the analysis, the regions affected by tip changes due to the cleavage steps perpendicular to the growth direction were excluded. However, other factors will also influence the accuracy of this width measurement, including the broadening due to the finite tip size used in SCM, carrier spillage and statistical errors. Figure 3 shows that the average width of the conductive interface layer increases with increasing IL thickness up to 13 nm. Assuming the lower surface of the conductive interface layer to be flat, the root mean square (RMS) roughness of the top surface of the conductive layer was calculated. This RMS roughness is also plotted with Sc_2O_3 IL thickness, showing a trend similar to the conductive layer thickness. It is noticeable that the width of the conductive interface layer first increases and then levels off with increasing Sc_2O_3 thickness, mirroring the behaviour of the TDD.

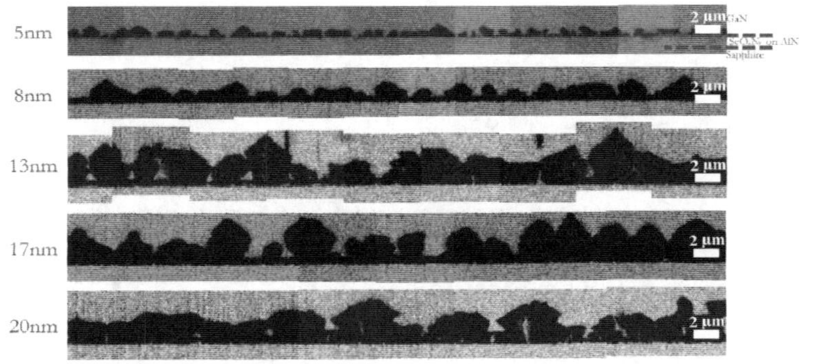

Fig. 2. Montages of SCM d*C*/d*V* phase data for samples with Sc_2O_3 layer thickness from 5-20 nm.

Electron microscopy studies were carried out to check the ScO_xN_y IL thicknesses for two of the samples and clarify the TDD reduction mechanism. Figure 4 shows cross-sectional transmission electron microscopy (TEM) images of the samples with nominal Sc_2O_3 IL thicknesses of 13 and 25 nm. The measured IL thicknesses from TEM are 17±0.8 and 19±0.5 nm, respectively. This discrepancy might partially explain the observed saturation in both TDD and conductive layer thickness. There is no evidence for the formation of voids in the AlN template for either sample.

It appears that the use of $GaN/ScO_xN_y/AlN$ structure results in a less effective dislocation reduction in the overgrown GaN than for $GaN/ScN_z/GaN$ samples. The TDD measured by AFM saturates at around 2×10^8 cm^{-2}, and similarly the SCM data suggest that the 3D-2D coalescence thickness does not vary for the interlayers with a nominal thickness of more than 13 nm. (For other samples, the width of the unintentionally n-doped layer has been found to increase as the coalescence time increases [2].) The absence of voids suggests that the mechanism of TDD reduction is different to that observed for GaN templates. The different mechanism may be due to the oxidation of the Sc or due to the thermal stability of the AlN seedlayer during the nitridation step which may result in the lack of pinholes formed in the interlayer. However, the observed saturation in both the TDD and the conductive layer thickness may not be an outcome of the growth mechanism but may have resulted from instabilities in the sputtering process resulting in inaccuracies in the thickness of the deposited Sc_2O_3.

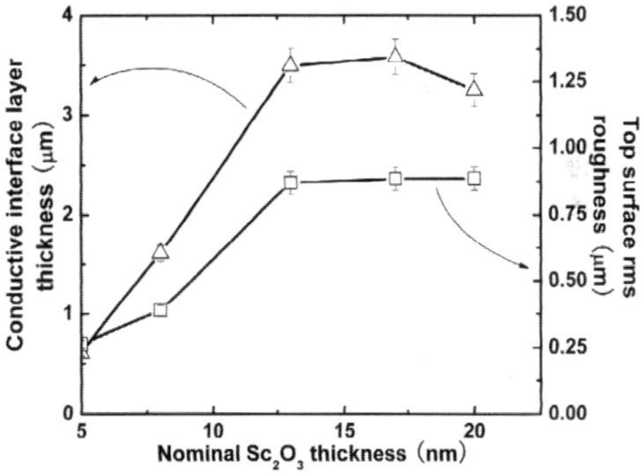

Fig. 3. Variation in conductive interface layer thickness and rms roughness of the top surface of the conductive layer with nominal Sc_2O_3 thickness.

The origin and mechanism of the unintentional doping in GaN are not fully understood. However, it has been suggested that most unintentional doping may be attributed to oxygen impurities [2], which in GaN growth on sapphire could come from contaminated precursors or carrier gases or from diffusion from the sapphire substrate. It has been found that oxygen incorporation is facet dependent and hence more oxygen incorporation may occur in the initial phases of GaN/sapphire growth where inclined facets are present [3]. Hence, the fact that an oxygen-doped region is often seen at GaN/sapphire interfaces does not necessarily imply that the oxygen comes from the sapphire. Inclined facets can influence oxygen incorporation either from gaseous contaminants or diffusion. Despite the fact that the GaN films studied here were grown on AlN-on-sapphire templates with ScO_xN_y ILs rather than directly on sapphire, the ScO_xN_y interlayers might act as a source of oxygen, so we cannot state that in this case the unintentional doping definitely arise from gaseous contaminants. However, further studies of samples in which the oxygen content of the ScO_xN_y ILs is lower and gaseous contamination is different might provide new insights into this question.

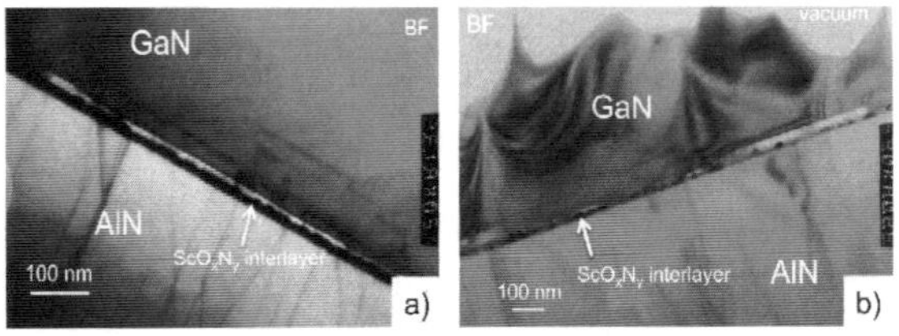

Fig. 4. Bright-field transmission electron micrographs of samples with nominal thickness of (a) 13 nm and (b) 25 nm Sc_2O_3 interlayers.

4. Summary

SCM has been used to identify and quantify the unintentional doping in GaN grown on AlN-on-sapphire templates with ScO_xN_y ILs. The use of ScO_xN_y ILs decreases the TDD. An unintentionally n-doped layer has been identified close to the GaN/ScO_xN_y interface. The average width of this unintentionally doped layer has been quantified. TDD reduction using Sc_2O_3 ILs and AlN templates appears to have a different mechanism to that seen with ScN_z ILs and GaN templates. Most TDD reduction methods – including that studied here – appear to lead to increased unintentional doping.

Acknowledgements

This work has been funded by the Engineering and Physical Sciences Research Council (EPSRC) under Grant no. (EP/E035167/1). RAO would like to acknowledge funding from the Royal Society.

References

[1] Mavroidis C, Harris JJ, Kappers MJ, Sharma N, Humphreys JC and Thrush EJ 2001 *Appl. Phys. Lett.* **79**, 1121
[2] Das Bakshi S, Sumner J, Kappers MJ and Oliver RA 2009 *J. Cryst. Growth* **311**, 232
[3] Sumner J, Das Bakshi S, Oliver RA, Kappers MJ and Humphreys CJ 2008 *Phys. Status Solidi (b)* **245**, 896
[4] Popovici G, Kim W, Botchkarev A, Tang H, Morkoc H and Solomon J 1997 *Appl. Phys. Lett.* **71**, 3385
[5] Moram MA,. Kappers MJ, Zhang Y, Barber ZH and Humphreys CJ 2008 *Phys. Status Solidi (a)* **205**, 1064
[6] Moram MA, Zhang Y, Kappers MJ, Barber ZH and Humphreys CJ 2007 *Appl. Phys. Lett.* **91**, 152101
[7] Moram MA, Barber ZH and Humphreys CJ 2008 *Thin Solid Films* **516**, 8569
[8] Moram MA, Kappers MJ, Barber ZH and. Humphreys CJ 2007 *J. Cryst. Growth* **298**, 268
[9] Oliver RA, Kappers MJ, Sumner J, Datta R and Humphreys CJ 2006 *J. Cryst. Growth* **289**, 506
[10] Sumner J, Oliver RA, Kappers MJ and Humphreys CJ 2007 *Phys. Status Solidi (c)* **4**, 2576
[11] Horcas I, Fernandez R, Gomez-Rodriguez JM, Colchero J, Gomez-Herrero J and Baro AM 2007 *Rev. Sci. Instrum.* **78**, 013705

Progress towards site-specific dopant profiling in the scanning electron microscope

MAE Jepson[1], BJ Inkson[1], R Beanland[2], AKW Chee[3], CJ Humphreys[3] and C Rodenburg[1]

1) Department of Engineering Materials, The University of Sheffield, Sheffield, S1 3JD. UK
2) Department of Physics, University of Warwick, Coventry, CV4 7AL. UK
3) Department of Materials Science and Metallurgy, University of Cambridge, Pembroke Street, Cambridge, CB2 3QZ. UK

Email: m.jepson@sheffield.ac.uk

Abstract. 2-dimensional dopant mapping in the scanning electron microscope is based on the variation of contrast as a function of differences in the number of active dopant atoms. It was put forward as a potential technique for dopant mapping on the nano-scale in response to the challenge of ever-shrinking device dimensions. Its impact, so far, has been limited as the majority of the work has concentrated on sample preparation by cleaving, thus not achieving the site-specificity which is required for nano-scale electronic devices. It was previously shown that cross-sections prepared using a 30 kV gallium focused ion beam (FIB), results in such strongly reduced dopant contrast (DC) that quantification is impossible. We demonstrate that by the use of a final in-situ argon broad ion beam (BIB) milling operation, contrast from a FIB prepared cross-section can be greatly improved. As a result, SEM dopant mapping could find a much more wide-spread application.

1. Introduction

As electronic devices and components shrink in size, the challenge of characterization of their dopant distributions becomes ever more intense. Scanning electron microscopy (SEM) has the potential to solve this challenge by the use of so-called dopant contrast (DC). DC has been known for many years [1] but received increased interest after the publication of results obtained with a field emission gun SEM which is required for successful imaging using DC [2].

The ultimate aim of a dopant mapping technique is to be able to map the location of the dopants within a device in 3 dimensions. Previous work into SEM DC has focused on the characterization of cleaved surfaces. This method of sample preparation has a major drawback; it does not allow site specific investigation of nano-scale devices. One technique which has the ability to create site-specific cross-sections is the focused ion beam (FIB). It was previously shown that FIB preparation of a silicon surface reduced the attainable contrast to levels insufficient for reliable analysis and incomparable to that obtained from a cleaved surface [3]. It was suggested that the reason for this was that the escape depth of secondary electrons (SEs) was only marginally greater than the depth of the amorphous damage caused by the FIB milling. Other dopant mapping techniques which rely on FIB preparation have also reported such effects. For example, users of electron holography have reported an electrically dead zone in a silicon p-n junction to a depth of ~25 nm [4] as well as an amorphous layer.

Many studies into the effect of final FIB energy on damage produced have been conducted [5, 6, 7] and it has been shown that a reduction of FIB energy can reduce damage depth to as little as 1.5 nm [8] after the use of a 2 kV gallium ion beam. To date, a reduction of damage layer thickness has not yet been reported to improve SEM DC.

This paper presents results of the use of a low energy (1 kV) argon ion miller in-situ inside the FIB as a final polishing step to improve the dopant contrast from conventionally prepared FIB cross-sections of doped silicon test structures. We show that the argon ion miller can improve contrast provided that the size of the trench is taken into consideration.

2. Experimental

Two silicon test structures have been produced: Sample 1 is a p-n junction structure produced by chemical vapour deposition (CVD) and Sample 2 is a doping level test structure produced using molecular beam epitaxy (MBE). Sample 1 consists of a 2.5 μm thick boron doped layer (5×10^{18} atoms cm^{-3}) upon an arsenic doped (1×10^{19} atoms cm^{-3}) substrate wafer. Sample 2 consists of 7 boron doped silicon layers with a constant thickness (200 nm) with doping levels reducing from 5×10^{19} cm^{-3} to 4×10^{15} cm^{-3} upon an arsenic doped (1×10^{19} cm^{-3}) substrate. Each layer was separated by a spacer layer and the highest doped layer was situated directly adjacent to the substrate wafer.

Cross-sections were prepared using an FEI Helios with a 30 kV gallium ion beam using successively lower beam currents. The beam energy was then reduced to 10, 5 and 2 kV for polishing; images were collected after each milling step. Imaging was carried out using immersion mode, a 1 kV primary electron beam energy, through-lens detector, a mirror bias of -15 V, a tube bias of 70 V and a working distance of 5 mm.

For the in-situ argon milling experiments, standard FIB cross-sections were polished using an Oxford Applied Research LIon 50 in-situ argon broad ion beam (BIB) miller attached to a Carl Zeiss XB1540 FIB. The BIB energy was 1 kV with an incidence angle of 18.5° to the FIB milled surface. The resulting cross-sections were imaged using a primary electron beam energy of 2 kV, an in-lens detector and a working distance of 5 mm.

To compare the DC to the extent of ion beam damage, expected amorphous layer thicknesses on silicon were calculated by generating stopping / range tables using SRIM. The lateral straggle length was then used to calculate the expected amorphous layer thickness using the expression $T = 3\ \Delta R_t$ where T = amorphous layer thickness and ΔR_t = lateral straggle length as proposed by Wang et al [9].

3. Results and Discussion

Figure 1(a) is a plot of contrast obtained from within a trench cut perpendicular to the p-n junction using varying gallium ion accelerating voltages. The general trend is that as the ion beam energy decreases, the contrast increases. In comparison to this, a plot of the expected amorphous layer thickness is presented which shows the opposite trend. These results demonstrate that the amorphous layer thickness is of utmost importance when considering dopant contrast from FIB prepared surfaces. This is also true for studies using electron holography techniques. Here it has been noticed that, as well as the amorphous layer, an electrically "dead zone" is present on silicon p-n junctions to a depth of 25 nm which causes a dramatic reduction in the measured potential at the junction within this zone [4]. As SEM DC is produced by active dopants [10], dopants in this "dead zone" will make little contribution to the DC obtained. As the damage depth is expected to be reduced when using a lower energy ion beam, a greater proportion of detected SEs will originate from the undamaged silicon and contribute to the dopant contrast. It can be seen from Figure 1(b) that contrast does not substantially increase until a layer thickness of ~10 nm is achieved. This suggests that the majority of SEs which contribute to DC originate from a depth of < 10 nm.

In order to further reduce the amorphous layer and electrically damaged zone thickness produced by the preparation of the cross-section, a low energy, in-situ argon BIB was used. It has been shown by a number of authors that the application of BIB milling after FIB preparation of TEM thin foils can have a dramatic effect on the damage layer thickness [11, 12].

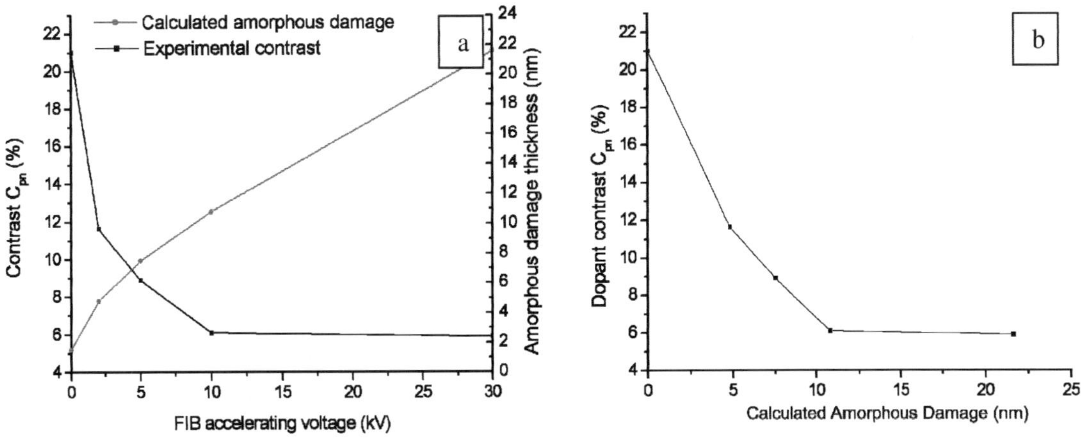

Figure 1. a) Comparison between the dopant contrast achieved and the calculated amorphous layer thickness produced by a gallium FIB at varying accelerating voltages (0 kV accelerating voltage represents contrast from a freshly cleaved specimen) and b) the dopant contrast as a function of calculated damage layer thickness.

Barna et al. [11] showed that by the use of a BIB of 0.25 kV, the amorphous damage depth in silicon can be reduced to as little as 1 nm. Langford et al. [12] compared TEM lattice images of silicon before and after BIB milling and found the image sharpness and contrast to be greatly increased after BIB milling; this is indicative of a reduction in sample damage when compared to conventional FIB preparation. As well as the advantages of reduced damage layer thickness, the use of argon ions also reduces any possible effect of implanting gallium (a dopant atom for silicon) into the silicon target.

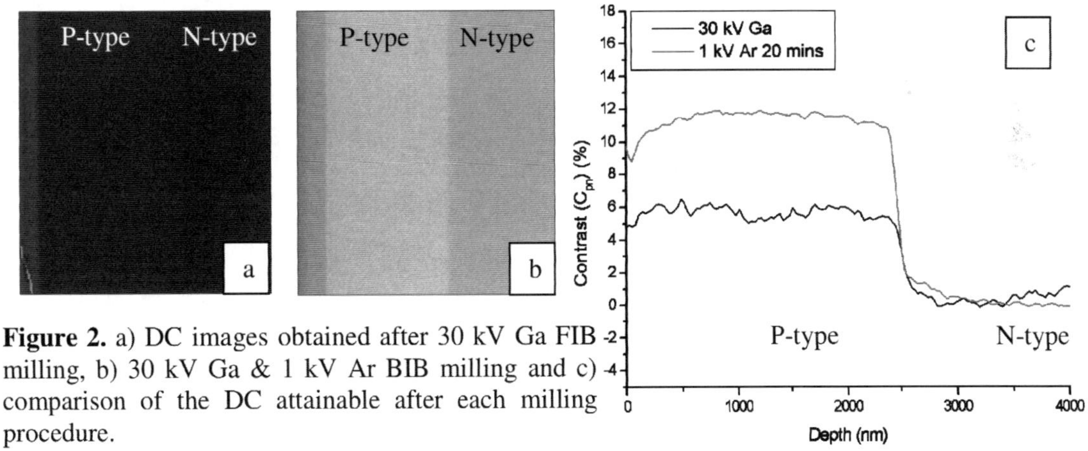

Figure 2. a) DC images obtained after 30 kV Ga FIB milling, b) 30 kV Ga & 1 kV Ar BIB milling and c) comparison of the DC attainable after each milling procedure.

Figure 2 shows images of the p-n cross-section produced after 30 kV gallium milling and 20 minutes of in-situ BIB milling together with the contrast profiles from these cross-sections. The contrast present after gallium ion milling only is approximately 5.5 %. When the same cross-section is exposed to 20 minutes of argon BIB milling, the contrast (for the same imaging conditions) is increased to approximately 11.5 %.

This procedure has also been applied to multilayer Sample 2 (Figure 3). The observed contrast values are hugely reduced for FIB specimens when compared to a freshly cleaved specimen, and the contrast from the layers does not seem to be dependent on their doping level. Although the application of BIB milling does slightly improve the contrast obtained, the contrast of the lower doped layers still does not correspond to their doping level. The trench size may be affect the detection efficiency and hence lead to uneven dopant contrast from the prepared cross-section. In order to assess the effect of the FIB trench size, a larger trench was produced. The contrast profiles from the larger trench show a more expected trend; the highest doped layers have the greatest contrast which decreases with reducing doping level (Figure 3(b)). The large trench also produces an increased contrast. The application of argon milling improves the max. contrast level of the cross-section from ~4.5 % to ~7 %.

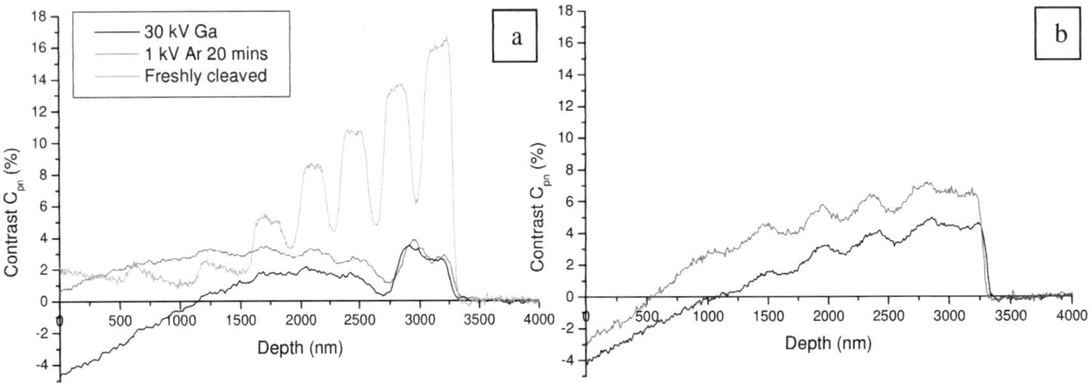

Figure 3. Contrast profiles taken from FIB and FIB+BIB milled surfaces in a) a small trench (20×20×7μm^3) and b) a large trench (50×70×7μm^3).

4. Conclusion

It has been demonstrated that together with careful consideration of the trench geometry, the application of in-situ argon broad ion beam milling can dramatically improve the dopant contrast attainable from a FIB prepared cross-section. These results represent a major step towards achieving site-specific dopant profiling of real nano-scale semiconductor devices.

Acknowledgements

The authors would like to thank the EPSRC (Grant no: EP/E030602/1) and the Royal Society for financial support. Special thanks also to Paul Warburton and Kevin Lee from University College London for providing access to the in-situ argon BIB miller (EPSRC: EP/F019564/1).

References

[1] Chang THP and Nixon WC 1967 *Solid State Electronics* **10**, 701
[2] Perovic DD, Castell MR, Howie A, Lavoie C, Tiedje T and Cole JSW 1995 *Ultramicroscopy* **58**, 104
[3] Kazemian P, Twitchett AC, Humphreys CJ and Rodenburg C 2006 *Appl. Phys Lett.* **88**1 212110
[4] Twitchett-Harrison AC, Yates TJV, Newcomb SB, Dunin-Borkowski RE and Midgley PA 2009 *Nano Letters* **7**, 2020
[5] Rubanov S and Munroe PR 2001 *Journal of Materials Science Letters* **20**, 1181
[6] Kato NI, Kohno Y and Saka H 1999 *J. Vac. Sci. Technol. A* **17**, 1201
[7] Kato NI 2004 *Journal of Electron Miroscopy* **53**, 451
[8] Giannuzzi LA, Geurts R and Ringnalda J 2005 *Microsc. Microanal.* **11**, 828
[9] Wang Z, Kato T, Hirayama T, Kato N, Sasaki K and Saka H 2005 *Appl. Surf. Sci.* **241**, 80
[10] Castell MR, Muller DA and Voyles PM 2003 *Nature Materials* **2**, 129
[11] Barna A, Pécz B and Menyhard M 1998 *Ultramicroscopy* **70**, 161
[12] Langford RM and Petford-Long AK 2001 *J. Vac. Sci. Technol. A* **19**, 982

Journal of Physics: Conference Series **209** (2010) 012069 doi:10.1088/1742-6596/209/1/012069

Site-selective dopant profiling of p-n junction specimens in the dual-beam FIB/SEM system

KWA Chee[1,2], R Beanland[1,3], PA Midgley[1] and CJ Humphreys[1]

[1] University of Cambridge, Department of Materials Science and Metallurgy, Pembroke Street, Cambridge, CB2 3QZ, UK
[2] University of Cambridge, Department of Engineering, Electrical Engineering Division, Centre for Advanced Photonics and Electronics, 9 J J Thompson Avenue, Cambridge, CB3 0FA, UK
[3] Department of Physics, University of Warwick, Coventry CV4 7AL, UK

E-mail: kwac2@eng.cam.ac.uk

Abstract. Results from site-specific dopant profiling in a dual-beam FIB/SEM system are reported. Si specimens containing p-n junctions were milled using Ga^+ ion beam energies ranging from 30 keV to 2 keV, and analysed *in situ* in the vacuum chamber. We compare the dopant contrast observed when milling a cleaved surface to that obtained from a side-wall of a trench cut using 30 kV Ga^+ ions, and using successively lower ion beam energies. The latter technique is suitable for site-specific dopant profiling. We find that lower energy ion beam milling significantly improves contrast, but only achieves 50 % of that observed on a freshly-cleaved surface. Furthermore, the contrast on a side-wall previously milled using high energy Ga^+ ions is less than that of a cleaved surface subjected to the same ion beam energy.

1. Introduction

Secondary electron (SE) imaging of semiconductor samples in the scanning electron microscope (SEM) is developing as a quantitative procedure for the rapid profiling of electrically active dopants over a wide range of 10^{16} – 10^{20} cm^{-3} at sub-10 nm spatial resolution [1]. This technique offers promise in satisfying the requirements of the International Roadmap for Semiconductors for quantitative doping profiling of next generation devices, but only if a site-selective sample preparation procedure is used. At present, site-specificity for doping profiling is achievable only by FIB milling. However, it has been shown that FIB-prepared Si p-n junction cross-sections using 30 keV Ga^+ ions yield far inferior contrast compared to as-cleaved ones, and the contrast is too low for doping profiling [2]. The use of lower ion beam energies is known to reduce surface amorphisation damage on Si [3], and combined SEM and Ga^+ ion source FIB systems with lower energy capabilities are newly available. It is therefore of interest to quantify the effect of low energy ion milling on dopant contrast. Nevertheless, site-specific doping profiling in a dual-beam FIB/SEM system has hitherto not been reported. FIB/SEM dual-beam instruments offer the versatility of milling with the ion beam and imaging with either beam on the same platform. Specimens can be prepared and analysed *in situ*, providing rapid two- (or three-) dimensional dopant profiling with high site-selectivity and without exposure to ambient air. The high resolution imaging provided using the electron beam allows nanometre-scale features to be located in order for low energy ion milling to be performed and monitored precisely. In principle, three-dimensional dopant information may be obtained by the interpolation of a series of two-dimensional data acquired using sequential ion beam polishing and electron beam imaging of cross-sections.

2. Experimental

Experiments were performed using an FEI Helios Nanolab 600 dual-beam microscope. The Si sample examined was a multilayer chemical vapour deposition (CVD) grown epitaxial structure, with dopants incorporated during growth. The substrate was [001] Si with ~ 1×10^{19} cm^{-3} boron doping (p-type), and the epitaxial layer was arsenic-doped (n-type) to ~ 1×10^{18} cm^{-3}. A series of four thin (~ 30 nm) boron-doped layers was grown inside the epitaxial layer with concentrations ranging from 2×10^{16} to 8×10^{18} cm^{-3}, although here we focus on the deepest p-n junction between the epitaxial layer and the substrate. Specimens were cleaved in air to produce a smooth and flat {110} facet and mounted on a tilted stub. This allowed ion beam milling at a glancing angle (0° sample tilt), and imaging with the electron probe incident normally on the cleaved cross-section (38° sample tilt). Samples for site-specific investigation were examined at a fixed tilt of 52°, with the Ga^+ ions incident normally on the surface and the side-wall viewed at 38° by the electron beam. SE images were obtained using a 1 keV, 43 pA electron probe, and the through-the-lens detector (TLD) was operating in the immersion mode, with an extractor voltage of 70 V and a mirror voltage of –15 V, at a working distance of ~ 4.1 mm. Ga^+ ion milling was performed at 30, 20, 10, 5 and 2 keV; the same beam-limiting aperture was used for all ion beam energies, providing ion currents of 93, 59, 33, 16 and 10 pA respectively. This inevitably increased the time required for ion milling at lower energies, which could be as long as ~ 10 minutes. Both electron and ion beam focusing were performed away from the areas of interest to minimise any beam-induced modification of the surface structure. Regions of interest were scanned with the electron probe only once during imaging, and a liquid nitrogen cold trap ('Cryo-can', Fischione Instruments) was used to minimise any beam-induced contamination effects.

3. Results

SE images of cleaved surfaces that were ion milled at a glancing angle of incidence (~1°) are as shown in Fig. 1. In each case, the surface was subjected to milling only at a single beam energy, *ie.* sequential milling of the same surface was not employed here. Milling was performed using the 'cleaning cross-section' (CCS) routine with a nominal depth of 1 μm; due to the glancing angle of incidence this gave a cut ~ 10 μm deep. More than 100 nm of material was removed in order to avoid transient effects which may occur during the initial stages of ion milling. This approach should in principle provide a stable measure of the change in contrast produced by Ga^+ milling of a cleaved surface at different energies. The doped regions are clearly delineated as areas of light (p-type) and dark (n-type) contrast. The absolute contrast value was calculated using:

$$C = \frac{I_p - I_n}{I_p - I_0}$$

where the SE intensities from the p- and n-type regions are denoted as I_p and I_n respectively, and I_0 is the background intensity obtained from the beam-blanked image. The relative contrast data obtained across the p-n junction of the milled cross-sections are as shown in Fig. 3, together with that from a region which was not subjected to any ion milling. It can be observed that the contrast present after ion milling is inversely proportional to the ion beam energy. The 5 kV data point is not shown since the contrast profile was anomalous, probably due to re-deposition of sputtered material on the surface caused by sample drift during the relatively long milling time. It was also found that milling at 2 keV produced a significant slope on the milled surface (~ 10°) due to poor focusing of the ion beam. While a cleaved specimen provides the ideal cross-section for dopant contrast, it is difficult to cleave a device precisely and this is not suitable for measurements of real devices at present. A better approach is to cut a cross-section through a device using the Ga^+ ion beam. The same material was used here to

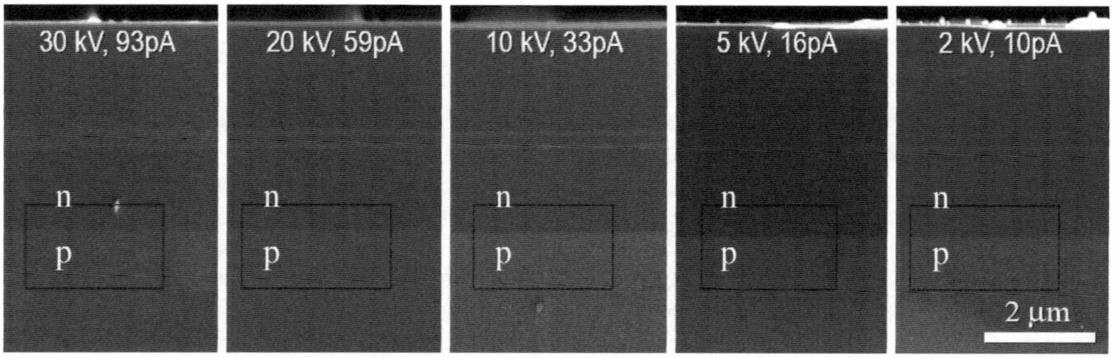

Figure 1. SE images of cleaved Si surfaces after Ga^+ ion milling at various beam energies. The box in each image shows the region used for contrast measurements.

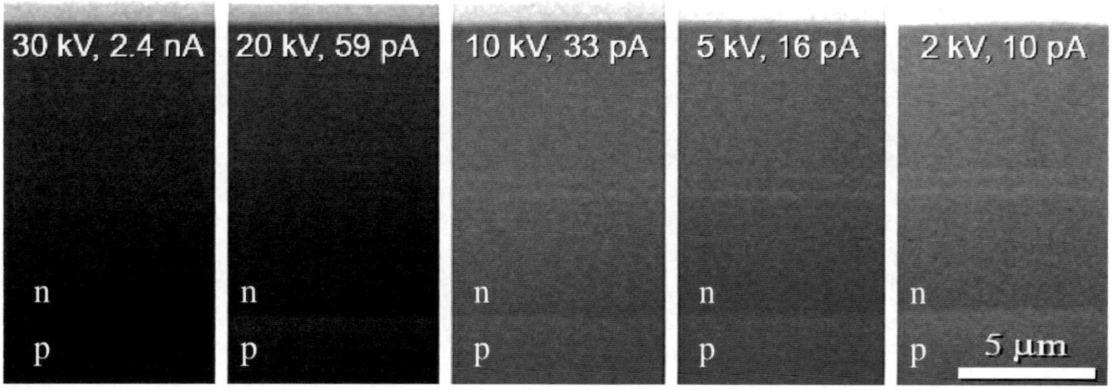

Figure 2. SE images of a trench side-wall after successive Ga^+ ion milling at various energies.

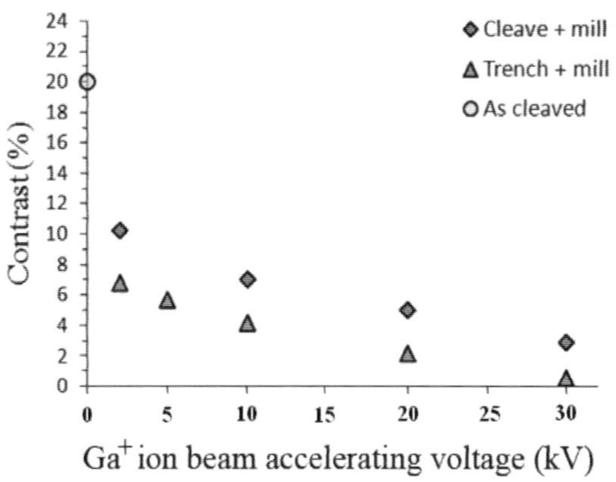

Figure 3. Summary of relative contrast measurements (in %) from cleaved surfaces and a trench side-wall.

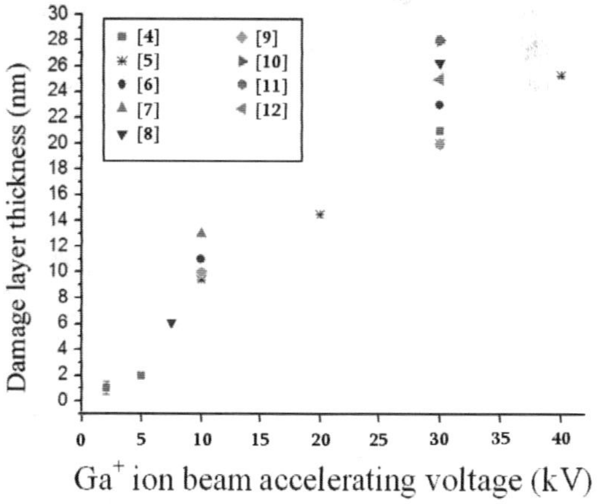

Figure 4. Side-wall damage layer thickness on Si as a function of Ga^+ energy.

provide comparison with the data obtained from the cleaved surface. Before milling, a 15×10 μm^2 protective Pt layer, 1 μm thick, was grown to prevent rounding of the top of the trench wall and "curtaining" effects. The initial 100 nm of this Pt layer was deposited using a 1 keV, 2.8 nA electron probe to avoid Ga^+ implantation into the upper part of the structure, followed by deposition at 30 keV Ga^+ using a probe current of 0.46 nA. A 15×8 μm^2 rectangular trench was then cut 5 μm deep into the wafer, adjacent to the protective Pt layer, using a 30 kV Ga^+ ion beam (6.5 nA). A CCS cut, nominally 1 μm deep, was then made into material covered by the Pt strap using 30 kV Ga^+ and a current of 0.28 nA. SE images of the trench side-wall were captured using identical conditions to that of the foregoing experiments, apart from the 38° angle of incidence of the electron probe (Fig. 2). As before, a series of progressively lower ion beam accelerating voltages was used, although in this case, milling was performed on the same facet, *ie.* previously subjected to milling at a higher Ga^+ energy. It was found that the SE intensity was reduced significantly deeper into the trench, giving a 'background' slope to the data. This obviously causes problems for quantitative measurements; however, we found that the change in contrast was linear with position in the image away from the top surface of the trench and could be subtracted by a linear fit to the data on either side of the p-n junction. It is clear that the contrast obtained from the trench-milling approach is lower than that obtained from a cleaved and ion-milled surface using these conditions (Fig. 3).

4. Discussion and Summary

Experimental measurements by TEM of the amorphous layer on Ga^+ ion-milled Si surfaces show a linear dependence of its thickness on accelerating voltage (Fig. 4). We found that the contrast of the p-type material did not change significantly after ion milling, whereas that of n-type material decreased (see Figs. 1 and 2). This may be explained by the observation that Ga acts as a p-type dopant in Si. It may be expected that this effect has the same dependence on the incident ion energy, and this is consistent with the observation here that contrast is inversely proportional to Ga^+ ion energy. The difference observed between milling a freshly-cleaved surface and that in the trench cut which had been previously milled by a high energy Ga^+ beam is probably caused by a combination of two factors: the dissimilarity in geometries involved in the two measurement methods used, and incomplete removal of the implanted region from the previous high energy milling stages. The side-wall surface was imaged at a tilt of 38° in the site-selective technique, and we also found a dependence of the contrast on the size of the adjacent trench; but no attempt was made to ensure that all material affected by a higher energy milling stage was removed by the lower energy ones. Nevertheless, at best, the contrast found after milling only recovers to 50 % of that seen on a freshly cleaved surface. This result is encouraging and it may be expected that optimisation of the geometry of milling and imaging, as well as the amount of material removed in each step, will significantly improve site-specific dopant contrast measurements.

Acknowledgements

KWA Chee acknowledges the Overseas Research Studentship Award Scheme, the Cambridge Commonwealth Trust, Trinity Hall Cambridge Brockhouse Engineering Studentship and the FEI company industrial bursary for doctoral sponsorship funding.

References

[1] Elliott SL, Broom RF and Humphreys CJ 2002 *J Appl Phys* **91**, 9116
[2] Kazemian P, Twitchett AC, Rodenburg C and Humphreys CJ 2006 *Appl. Phys. Lett.* **88**, 212110
[3] Kato NI 2004 *J. Elec Microsc.* **53**, 451 and references cited therein
[4] Giannuzzi LA, Geurts R and Ringnalda J 2005 *Microsc Microanal* **11** (Suppl 2), 828
[5] Wang Z, Kato T, Hirayama T, Kato N, Sasaki K and Saka H 2005 *Appl. Surf. Sci* **241**, 80
[6] Gao Q, Zhang M, Niou C, Li M and Chien K 2004 *IEEE 42nd Rel. Phys. Symp. Proc.* 613
[7] Rubanov S and Munroe PR 2004 *J. Microsc.* **214**, 213
[8] Kato NI, Kohno Y and Saka H 1999 *J. Vac. Sci. Technol. A* **17**, 1201
[9] Mardinly J and Susnitzky DW 1988 *Mat. Res. Soc. Symp. Proc.* **523**, 3
[10] Venables D, Susnitzky DW and Mardingly AJ 1998 *AIP Conf. Proc.* **CP449**, 667
[11] Pantel R, Auvert G and Mascarin G 1997 *Microelec. Eng.* **37/38**, 49

Monte Carlo modelling of the low-loss electron signal in scanning electron microscopy and comparison with the BSE signal

C Bonet[1], MM El-Gomati[2], JAD Matthew[1] and SP Tear[1]

[1]Department of Physics, University of York, York YO10 5DD, UK
[2]Department of Electronics, University of York, York YO10 5DD, UK

E-mail: spt1@york.ac.uk

Abstract. Nanotechnology places increasing demands on techniques for sample characterisation on the sub-100 nm length scale, and the low-loss electron (LLE) signal may provide one possible way of addressing this need. Simulations of the LLE signal from a line-scan across a semiconductor superlattice structure have been performed using two different Monte Carlo models in order to assess their effectiveness in predicting spatial resolution for compositional imaging. Additionally, experimental measurements of LLE data using a detector added to a scanning electron microscope were made to investigate compositional contrast.

1. Introduction

The growth of nanotechnology in recent years has placed increasing demands on the techniques for the characterization of nanostructures in both a scientific and industrial context. The latter, in particular, often imposes the additional requirement that the technique should be both non-destructive and have a reasonable turn-around time. An area which still poses challenges is the compositional analysis of nanostructures, i.e. providing compositional information from features on a sub-100 nm length scale.

Scanning electron microscopy (SEM) provides fast turn-around and a sub-100 nm probe is easily formed. Imaging using the secondary-electron (SE) signal provides high-resolution topographic imaging but whilst the SE signal is affected by sample composition, the SE contrast mechanism is complex and may be influenced by factors which are not easily controlled. Traditionally, X-ray microanalysis has been used for compositional analysis in the SEM, however the large interaction volume for X-ray generation makes its application to nanostructured samples at best challenging.

The backscattered electron (BSE) signal carries *Z*-contrast information, however this signal originates from a volume of similar size to that associated with the X-ray signal; thus the spatial resolution of the compositional information available from the BSE signal must be questioned. The low-loss electron (LLE) signal is that part of the BSE signal in which the electrons have lost no more than a certain energy, E_{loss}, during their passage through the sample, i.e. they leave the sample with energies from the primary beam energy E_P to $E_P - E_{loss}$, as shown in Fig. 1. The volume from which the LLE signal originates is restricted by virtue of this limit placed on the maximum energy loss. The original experimental work on the LLE signal by Wells [1] concentrated on improving the imaging resolution of the BSE signal; subsequent work has examined the dependence of the LLE yield on composition of bulk specimens under ultra-high vacuum [2,3] and high-vacuum [4] conditions. To provide useful information on nanoscale sample composition, it is important to be able to reliably model the LLE signal. In this paper we discuss the effectiveness of Monte Carlo methods for modelling the spatial resolution attainable using the LLE signal for compositional imaging. We further examine compositional contrast in the LLE signal using experimental data from a high-resolution field-emission gun SEM (FEGSEM).

2. Experiment & simulation

Experimental measurement of LLE images were made in an FEI Sirion XL30 FEGSEM (operated at $E_P = 5$ keV) which had been modified by the installation of a detector for recording the LLE signal. This detector was developed based on an existing design [2] and its energy-filtering action was based on the retarding-field principle. When the detector was operated with the loss energy set to E_{loss}, the recorded signal corresponded to those electrons within the shaded region of the spectrum shown in Fig. 1. The specimen used in this study was a $Si/Si_{0.85}Ge_{0.15}$ semiconductor superlattice structure which had been cleaved in order to permit its observation in cross-section in the SEM. A schematic indicating the form of this sample is shown in Fig. 2; the sample consisted of 12 alloy layers.

Monte Carlo simulations of the LLE and BSE yield were carried out using two different publicly-available codes. The NISTMonte2 code [5] used Mott cross-sections [6] and treated inelastic losses via the continuous slowing-down approximation (CSDA) using the Joy & Luo expression [7] for the stopping power. This approach neglected the effect of energy straggling. The other code used was PENELOPE2006 [8] which similarly used partial-wave cross-sections [9] for σ_{el} but in contrast implemented a discrete-loss model for inelastic scattering. To reduce computational demands the simulated signal was collected from a 2π st solid angle. The aim of using these two different codes was to establish the significance of energy straggling on the LLE signal.

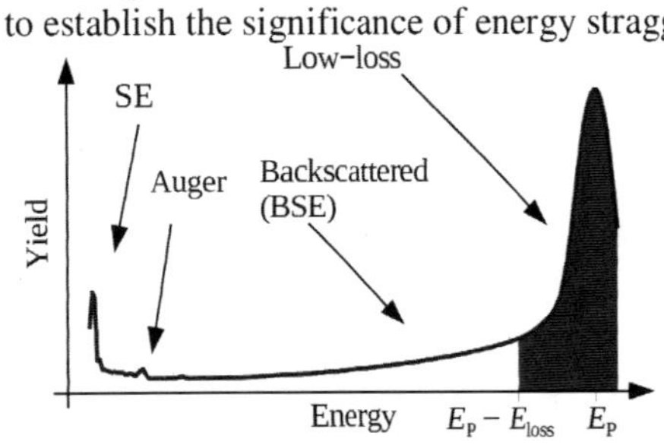

Figure 1. Schematic showing backscattered electron energy spectrum. The signal recorded by the LLE detector is in the shaded region.

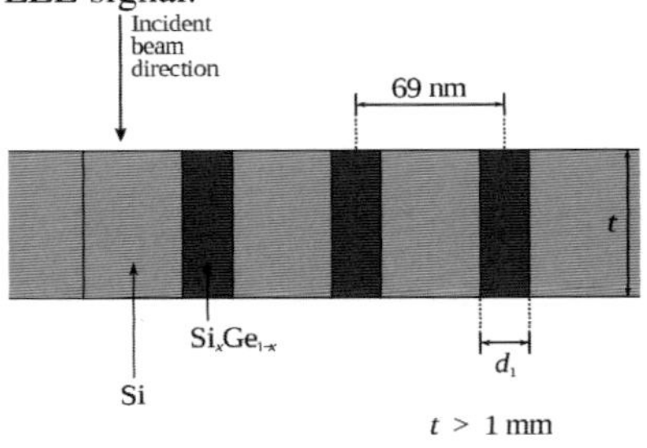

Figure 2. Sketch showing the structure of the $Si/Si_{0.85}Ge_{0.15}$ superlattice sample, viewed in cross-section.

3. Results and discussion

3.1. Monte Carlo simulation of spatial resolution

In order to establish reliable figures for the periodicity and thickness, d_1, of the $Si_{0.15}Ge_{0.85}$ alloy layers in the sample, a thinned section cut from the superlattice wafer was examined in a transmission electron microscope (TEM). Analysis of the resulting TEM image resulted in a figure of 69.2 ± 0.2 nm for the periodicity and $d_1 = 11.5 \pm 0.4$ nm for the alloy layer thickness. These data were used to construct the geometrical models used in the Monte Carlo simulations. After checking for the influence of edge effects, simulations were run for line-scans across a single period of the superlattice structure, for reasons of computational efficiency.

Line-scans generated using the CSDA-based NISTMonte code are shown in Fig. 3. Qualitatively, improved spatial resolution can be seen in the LLE signal compared to the BSE signal and resolution improves further as E_{loss} is decreased. At $E_{loss} \leq 250$ eV the LLE signal reproduces the composition profile quite closely. Line-scans from the PENELOPE code which uses a discrete-loss inelastic model are shown in Fig. 4. The resolution follows a similar trend to the CSDA case, however using this model the composition profile would appear to be not faithfully reproduced until $E_{loss} \leq 100$ eV.

To make a quantitative comparison of the resolution predicted by the two approaches, measures of resolution d_P were used which measured the distance in x over which the signal rose from P % to $(100 - P)$ % of ΔI above I_{min} where $\Delta I = I_{max} - I_{min}$. Here I_{min} and I_{max} are the minimum and maximum signal respectively. Two specific cases of this resolution measure were used, namely d_{25} and d_{10}.

The variation of spatial resolution with E_{loss} using the two different metrics is shown in Fig. 5. A decrease in resolution with increasing E_{loss} is observed, with the discrete-loss model showing a generally poorer resolution than the CSDA results, indicating that in reality, energy resolution is degraded relative to that predicted by a CSDA-based simulation. The influence of E_{loss} on the spatial resolution is also much weaker in the case of d_{25} than in the case of d_{10}. For compositional imaging, it is important that the resolution is such that the measured signal reflects the composition profile of the sample, rather than simply being sufficient that two closely-spaced features are resolved. For this reason, d_{10} is a more relevant measure of resolution for the purpose of compositional imaging and the strong variation of d_{10} with E_{loss} indicates the significance of the latter in this imaging mode.

It can also be seen from Fig. 5 that d_{10} shows the greatest differences between the CSDA-based and discrete-loss simulations of the resolution, suggesting that energy-straggling is an important effect in compositional imaging. The greatest differences in the values of d_{10} are actually found at around 250 eV loss, indicating that the effect is most significant in the low-loss regime. Presumably the influence of energy straggling has a lesser effect in the BSE signal since the electrons contributing to this signal have undergone a significant number of discrete loss processes.

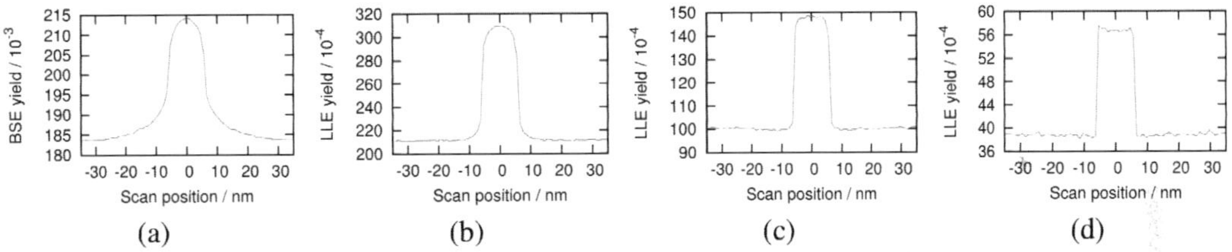

Figure 3. Simulated line-scans across the cross-sectioned superlattice sample generated by the NISTMonte code. (a) BSE signal; (b-d) LLE signal with losses of 500, 250 and 100 eV respectively.

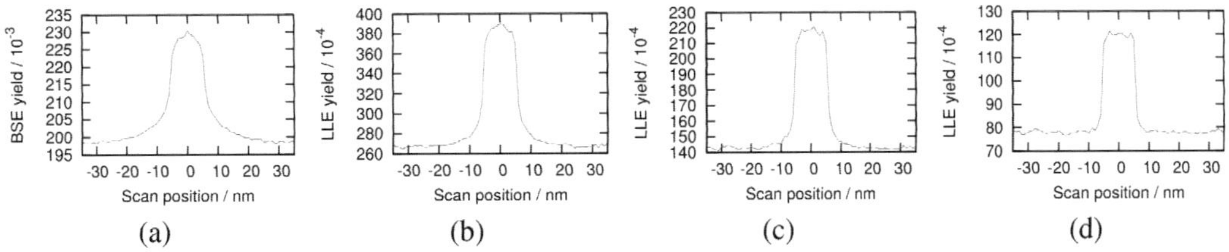

Figure 4. Simulated line-scans across the cross-sectioned superlattice sample generated by the PENELOPE code. (a) BSE signal; (b-d) LLE signal with losses of 500, 250 and 100 eV respectively.

3.2. Contrast in LLE and BSE imaging

The SEM equipped with the LLE detector described earlier was used to collect a series of images of the superlattice sample at a variety of loss energies E_{loss}. In order to permit their analysis with good signal-to-noise ratio, each of these LLE images was projected onto the y-axis to form a line-scan across the superlattice structure. These line-scans were analysed to establish how the contrast between the Si and $Si_{0.85}Ge_{0.15}$ layers varied with E_{loss}. The results of this analysis are illustrated in Fig. 6 which shows a consistent increase in the contrast as the loss energy is decreased, corresponding to a reduction of the volume from which the detected signal electrons emerge. This demonstrates that, in addition to any benefit associated with improved spatial resolution, the LLE signal can be used as a way of increasing compositional contrast compared with the BSE signal when performing compositional imaging on nanostructured materials. Although the data in Fig. 6 suggest that the gains in contrast begin to slow as E_{loss} drops below 100 eV, the influence of sample contamination on this experimental finding must be taken into account. It is suspected that the extreme surface sensitivity of the LLE signal at $E_{loss} \leq 100$ eV could reduce the measured contrast in the poor vacuum environment of a conventional SEM to below the maximum theoretically possible.

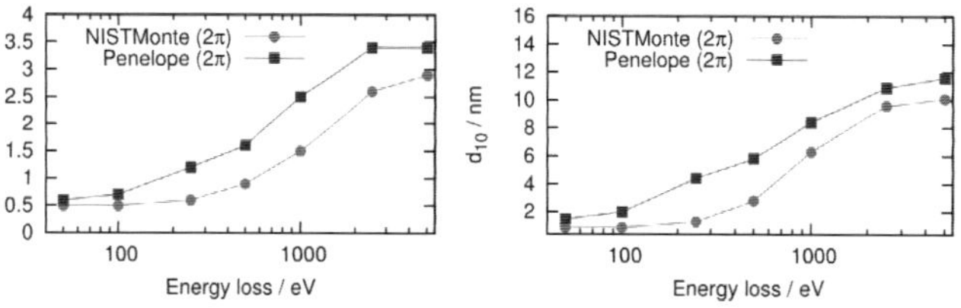

(a) (b)

Figure 5. Variation in spatial resolution with the loss energy E_{loss}, as measured from the line-scans of Figure 3 & 4 using (a) d_{25} and (b) d_{10} (see text for definition).

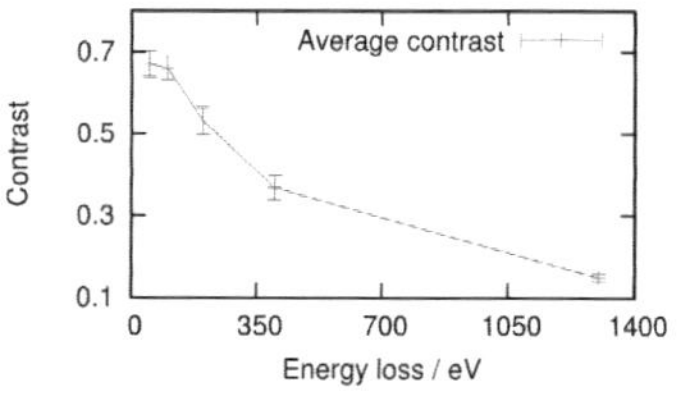

Figure 6. Variation with E_{loss} of experimentally measured contrast in the LLE signal.

4. Conclusions

Monte Carlo simulation showed that the LLE image offers greater spatial resolution for compositional imaging than the BSE image, however, energy straggling means that the achievable resolution is always less than that predicted by simulations based on the CSDA. The d_{10} measure of resolution is more suitable than d_{25} for compositional imaging and the former shows greater differences between the resolutions predicted by CSDA-based simulations and those based on a discrete-loss model. Consideration of the d_{10} parameter has shown that the influence of energy-straggling on spatial resolution is particularly significant in the low-loss regime. Although a SEM with a very small probe size is needed to appreciate the resolution improvement offered by the LLE signal, the LLE image also shows increased contrast in comparison to the BSE image when imaging nanostructures.

Acknowledgements

The authors are grateful for the use of the computational facilities of the White Rose Grid. Financial support of the Engineering and Physical Sciences Research Council (EPSRC) is acknowledged.

References

[1] Wells OC 1971 *Appl. Phys. Lett.* **19**, 232
[2] Barkshire IR, Roberts RH and Prutton M 1997 *Appl. Surf. Sci.* **120**,129
[3] Pratt A, Matthew JAD, El-Gomati M and Tear SP 2007 *Surf. Sci.* **601**, 1804
[4] Bonet C, Pratt A, El-Gomati MM, Matthew JAD and Tear SP 2008 *Microsc. Microanal.* **14**, 439
[5] Ritchie NW 2005 *Surf. Interf. Anal.* **37**, 1006
[6] Jablonski A, Salvat F and Powell CJ 2002 *NIST Electron Elastic-Scattering Cross-Section Database - Version 3.1*
[7] Joy DC and Luo S 1989 *Scanning* **11**, 176
[8] Salvat F, Fernández-Varea JM and Sempau J 2006 *PENELOPE-2006: a code system for Monte Carlo simulation of electron and photon transport.* Nuclear Energy Agency (OECD) Issy-les-Moulineaux
[9] Salvat F, Jablonski A and Powell CJ 2005 *Comput. Phys. Commun.* **165**, 157

Journal of Physics: Conference Series **209** (2010) 011004 doi:10.1088/1742-6596/209/1/011004

Subject Index

Journal of Physics: Conference Series **209** (2010) 011004 doi:10.1088/1742-6596/209/1/011004

Journal of Physics: Conference Series **209** (2010) 011004 doi:10.1088/1742-6596/209/1/011004

AUTHOR INDEX

AUTHOR INDEX

AUTHOR INDEX